# 22nd Australasian Weeds Conference (22AWC)

## A Weed Odyssey: Innovation for the Future

North Adelaide, Australia
25 – 29 September 2022

ISBN: 978-1-7138-6247-5

**Printed from e-media with permission by:**

Curran Associates, Inc.
57 Morehouse Lane
Red Hook, NY 12571

**Some format issues inherent in the e-media version may also appear in this print version.**



Printed with permission by Curran Associates, Inc. (2023)

For permission requests, please contact Council of Australasian Weed Societies Inc. (CAWS) at the address below.

Council of Australasian Weed Societies Inc. (CAWS)
Massey University, Private Bag
11-222 Palmerston North 4442
New Zealand

Phone: +61 6 350 4926
Fax: +61 8 9821 1028

www.caws.org.au

**Additional copies of this publication are available from:**

Curran Associates, Inc.
57 Morehouse Lane
Red Hook, NY 12571 USA
Phone: 845-758-0400
Fax: 845-758-2633
Email: curran@proceedings.com
Web: www.proceedings.com

22nd AUSTRALASIAN WEEDS CONFERENCE
25-29 Sep 2022 - Adelaide Oval
#22AWC | wmssa.org.au/22awc

22AWC | A WEED ODYSSEY: INNOVATION FOR THE FUTURE
WEED MANAGEMENT, RESEARCH, BIOSECURITY, POLICY.

AWC 2022
A WEED ODYSSEY: INNOVATION FOR THE FUTURE

# 2022 AUSTRALASIAN WEEDS CONFERENCE

25 – 29 September 2022

Adelaide Oval, Adelaide, South Australia

wmssa.org.au/22awc

Council of Australasian Weed Societies Inc.

WMSSA
Weed Management Society of South Australia

22AWC Proceedings Sponsor:

Sponsored by
Government of South Australia

Adelaide Oval, Adelaide, South Australia
25 – 29 September 2022

Peer-reviewed proceedings edited by Rachel Melland, Chair, Chris Brodie, Jason Emms, Leah Feuerherdt, Susan Ivory and Shauna Potter. Published by Weed Management Society of South Australia Inc.

Proceedings of the 22nd Australasian Weeds Conference Adelaide

# Table of Contents

## Plenary 9



## 1A Environmental Biosecurity Related Session: Environmental Weed Control Successes and Lessons



## 1B Global, New Zealand and Australian Herbicide Resistance Perspectives & Farming System Behaviour Change Over Time

## 1C Social Research Towards Weed Control — Speakers and Panel Session



## 2A Biological Control — Australia & NZ

## 2B Weed Control in Cotton Farming Systems



## 2C Community Coordination for Weed Management in Arid Landscapes — Speakers and Q&A Session



## 3A Biocontrol Successes, Methods & Modelling

## 3B Cropping Weed Threats, Weed Biology and Seed Dormancy — Speakers and Q&A Session



## 3C CAWS Workshop Series: Weed Remote Sensing Forum



## 4A Environmental Biosecurity Related Session: Weed Invasion, Prevention and Eradication

## 5B GRDC Symposium Stream: Understanding Weeds in Grains



## 5C Ornamental & Nursery Plant Weed Impacts & Strategies



## 5D Weeds of Pasture Systems — Speakers and Q&A Session

## 6A Environmental Biosecurity Related Session: Managing Weeds in Freshwater Ecosystems



## 6B GRDC Symposium Stream: Control Tactics for Weeds in Grain Production

## 6C Advances in Weed Detection Technologies and Eradication Programs

## 7A Rangeland Ecosystem Weed Management



## 7B GRDC Symposium Stream: A Grains Farming Systems Approach to Managing Weeds

## 7C Weed Surveillance Technology Uses & Social Values



## 7D Novel Chemistries, Synthetic Herbicides & Weed Enzymes



## 8A Fire and Weeds: Learnings from the 2019/20 Bushfires & Planning for Next Time — Speakers and Q&A Session

## 8B Area Wide Management of Weeds



## 8C Weeds Biosecurity Risk Assessment Tools

## 8D Weeds in Sugarcane Production Systems



## 9A Biocontrol Host-Specificity Testing & Host Ranges

## 9B Herbicide Resistance Extent & Mechanisms



## 9D Crop Competition & Strategic Tillage for Weed Control

## 10A Biocontrol Establishment & Monitoring



## 10B Weed Control in Cropping Systems

## 10C Looking Forward: Weed Threat Prevention & Abatement



## Poster Presentations

# Developing the National Established Weed Priorities Framework

Shalan Scholfield[2], Shauna Potter[1], Matthew Sheehan[1], John Virtue[1]
[1]Wild Matters Pty. Ltd., Castlemaine, Australia,
[2]Environmental Biosecurity Office, Dept of Agriculture, Water and the Environment, Canberra, Australia
(matt@wildmatters.com.au)

**Summary** The National Established Weed Priorities (NEWP) Framework is an initiative to determine and address shared weed priorities through strategic, nationally coordinated actions. Initiated by the Weeds Working Group of the Environment and Invasives Committee, the Framework seeks enhanced collaboration in established weed management between industry, community and government, from local to national levels.

The NEWP Framework is built around three key delivery streams:

1. The proven Weeds of National Significance (WoNS) model, that is expected to continue through selecting new WoNS and maintaining knowledge and tools for best practice management of the existing 32 WoNS.
2. A new innovative concept of Weed Issues of National Significance (WINS), that takes a multi-species, landscape-scale approach, tackling common challenges in established weed management across Australia.
3. A National Established Weed Action List (NEWAL), provides a mechanism to consolidate and work through short-term management actions of high national benefit that are not otherwise addressed through WoNS or WINS.

The NEWP Framework has been developed to align with the National Framework for the Management of Established Pests and Diseases of National Significance, which sets three overarching criteria in determining national priorities: national impact; feasibility of management intervention; and benefits from taking a nationally coordinated approach.

Fundamentally, the Framework fosters collaboration and co-leadership between community, industry and government stakeholders, supported through a balanced and independent NEWP steering group. This approach extends to nomination and assessment processes to determine priorities, and to the membership of taskforces charged with implementing strategic plans for WoNS, WINS and the NEWAL. These plans will include activities relating to research, development and extension (RD&E), best-practice information and training, prevention, monitoring and supporting networks and partnerships for coordinated weed control programs.

The Framework has been developed with funding support from the Australian Government Department of Agriculture, Water and the Environment.

**Keywords** Established weeds, priorities, assessment, WoNS, WINS

# Using a Multi-pronged Strategy for Wheel Cactus control in Central Victoria

Lee Mead[1]
[1]Tarrangower Cactus Control Group Inc., Australia
(info@cactuswarriors.org)

**Summary** Wheel Cactus (*Opuntia robusta*) has invaded thousands of hectares of natural environment and private properties in Central Victoria. Despite prolonged and relentless community action involving volunteers, landowners and managers, all investing much labour and cash, this noxious weed continues to invade our landscapes. Wheel Cactus is highly invasive, reproduces by both seed and vegetatively, is drought tolerant and has a lack of predators. Combined, these factors make it extremely difficult to control.
One key focus of our group, the Tarrangower Cactus Control Group (TCCG), is to prevent Wheel Cactus reinfesting the Maldon Historic Reserve. We've been using direct injection of herbicide and manual digging techniques for the past 15 years, but still struggling to contain the core infestations on private properties adjacent to the reserve. We've introduced an additional tool, the biological control insect 'Cochineal' (*Dactylopius opuntiae*) hoping that a more integrated strategy will help control this weed. With funding from Landcare Victoria Inc., TCCG completed a project in 2019 where we introduced the biocontrol agent to our local community and land managers. TCCG volunteers have set up a Cochineal nursery and established Cochineal insect populations at six properties located within the core Wheel Cactus infestations. We also designed an individualized, integrated management plan with each of these landowners. We hope this multi-pronged approach will result in less expensive, less labour intense and less chemically focused management of Wheel Cactus, which will also present less risks to humans and our natural environment. A management strategy incorporating a biological control agent should also prove to be a more sustainable and long-term strategy. We'll review the reproduction and spread of the Cochineal insects and the impact of the integrated management plans on these properties, and the impact on the reproduction and spread of Wheel Cactus in Central Victoria.

**Keywords** Wheel Cactus, Biological Control, Cochineal, Integrated management

# Cottonbush invasion in Western Australia: ecology and social perspectives

Mariana Campos[1], Kathryn L. Batchelor[1], Karen L. Bell[1,3], Michael Davy[1] Mike Christensen[3], Raphael K. Didham[1,3], and Bruce L. Webber[1,3]
[1] CSIRO Health & Biosecurity, 147 Underwood Ave, Floreat, WA 6014, Australia
[2] South West Catchments Council, PO Box 5066, Bunbury, WA 6230, Australia
[3] School of Biological Sciences, UWA, 35 Stirling Highway, Crawley, WA 6009, Australia
(Email: mariana.campos@csiro.au)

**Summary** *Gomphocarpus fruticosus* (L.) W.T.Aiton (narrowleaf cottonbush, hereafter referred to as cottonbush) is a weed of concern in Western Australia (WA) threatening agricultural and natural ecosystems. It is a declared weed (C3 Management) in the state, widespread and established, for which eradication is unfeasible. Management currently attempts to reduce abundance or range and sometimes contain spread. However, the extent of the threat and most appropriate control solution(s) remain largely unknown, and there are strongly contrasting views on optimal management. Here we address priority knowledge gaps on biology and ecology of cottonbush invasions in WA and clarify social perspectives on the weed through citizen scientist elicitation and stakeholder survey.

Natural enemies were documented via 13 field surveys across five sites. Two new natural enemies were observed in WA: *Arocatus rusticus* (Stål, 1866), a seed-eating Lygaeidae, and a phytoplasma causing phyllody. Estimated population seed production came to ~1.7 million filled seeds.ha-1.year-1, of which 1.36 million are viable and not dormant. Seeds readily germinate in optimal conditions 1-30 days after forming, indicating no after-ripening period. The stakeholder survey assessing cottonbush impacts with 101 respondents indicated that perceived environmental and economic impacts are not of significant concern. The issue of poorly managed neighbouring properties (private and government) was repeatedly raised.

The outcomes of cottonbush management in WA are ineffective at present and we expect that threats from this weed could worsen. We recommend further work to reassess the risk profile of cottonbush at regular intervals and prioritise improvements in the efficiency, effectiveness, and equitability of management programs.

**Keywords** narrowleaf cottonbush, weed, social science, *Gomphocarpus fruticosus,* impact.

## INTRODUCTION

*Gomphocarpus fruticosus* (L.) W.T.Aiton (narrowleaf cottonbush, hereafter referred to as cottonbush) is an upright perennial plant native to southern and eastern Africa and the Arabian Peninsula with a slender shrubby habit that grows to 2 m, opposite leaves that are narrow or elongated (4-12.5 cm long and 5-15 mm wide), and white or cream-coloured flowers in clusters of 3-10 (ALA, 2021). The primary dispersal mechanism is wind. When released from 2 m high, more than 75% of seeds are dispersed within 10 m from the source, but 6% of seeds are carried beyond 40 m, reaching 100 m (DPIRD, unpublished data). Seeds usually germinate in spring or autumn, but can germinate any time in warm, moist conditions (Lloyd and Rayner, 2012).

In its natural range, it is frequently described as a plant of disturbed areas (Goyder and Nicholas, 2001). Within *Gomphocarpus fruticosus* sensu lato there are five recognised subspecies (Goyder and Nicholas, 2001). Cottonbush is known to hybridise with the closely related *G. physocarpus* E.Mey. (Hussey *et al*., 2007) which is also present in Australia, and hybrid seedlings had viable seeds (Ward *et al*., 2012).

Cottonbush is present in all Australian states and territories, except the Northern Territory (ALA, 2021). Cottonbush has been categorised as a threatening invasive species; and recorded as a weed of the natural environment, agriculture, and an escapee from cultivation (Randall, 2007).

The first records of *G. fruticosus* in Australia date to the late 19th century, with specimens from South Australia, New South Wales and Victoria. It was first declared a noxious weed in WA in 1923, and targeted for control in the Shire of Dardanup by the Agriculture Protection Board from 1966. Its current distribution in WA is from Yanchep to Esperance, mostly from Perth to Busselton (ALA, 2021).

A detailed map of occurrences between 2000 and 2014 showed cottonbush infested an estimated 5,000 hectares or more in WA (Reeves and Dodd, 2014).

Eight natural pests and pathogens of cottonbush have been documented worldwide from the non-native range, two documented in WA: the non-native wanderer butterfly *Danaus plexippus* (Linnaeus, 1758), native to North America, whose larvae feed on cottonbush foliage (Lloyd and Rayner, 2012) and the native lesser wanderer *D. petilia* (Stoll, 1790), which visit the flowers.

Through this work we aimed to widen the ecological understanding on cottonbush and explore social perspectives on it to help inform management.

## MATERIALS AND METHODS

**Demography, phenology, natural enemies** Six sites were selected to assess population density and phenology, and a total of 13 visits were made: Lake Cooloongup (-32.31, 115.78), 13/10/20; Preston Beach (-32.88, 115.66), 07/12/20; Serpentine National Park (-32.36, 116.02), 13/01/21; Wungong (-32.13, 116.06), 03/03, 03/06, 22/07/21; Yalgorup (-32.79, 115.65), 28/05, 21/07, 13/09/21; Glen Mervyn (-33.56, 116.07), 08/12/20, 29/04, 22/07, 13/09/21.

In each site visit, three transects (2 m x 50 m) were established to count number of seedlings, juvenile and mature plants. Ten random adult plants were selected to estimate number of flowers (categories: <100, 100-1,000, >1,000) and fruits (categories: <50, 50-100, 100-200 and >200). Counts of filled and unfilled seeds per fruit were taken (30 fruits from Serpentine, 30 from Wungong).

Natural enemies were recorded at each site (except Lake Cooloongup) and identified by the authors or forwarded to DPIRD for specialist identification.

**Soil cores and germination** Ten soil cores (10 cm diameter, 10 cm depth) were randomly collected beneath each of three cottonbush infestations (Glen Mervyn, Serpentine and Preston Beach) in December 2020 and January 2021. Samples were air dried then wet sieved (1 mm mesh, tap water), and cottonbush seeds separated from debris, visually sorted into intact/damaged via forceps pressure test and counted.

Intact seeds were surface-sterilised (10-minute immersion in 70% ethanol, then 20-minute in 20% bleach), then plated on sterile 90 mm Petri dishes with filter paper soaked with 0.1% Plant Preservation Mixture (Plant Cell Technology®) and placed in an incubator at 30°C, 14D:10N light cycle.

Germination was scored daily for 30 days, seeds considered germinated when the radicle protruded 2 mm from the seed coat. On day 31, ungerminated seeds were dissected to determine if alive or dead by visual inspection of the embryo and endosperm.

Data analyses were undertaken with R software v4.1.0 (R Core Team, 2021); counts of number of seeds per soil core were analysed with a Poisson generalised linear model (GLM), and proportion of seed germination analysed with a binomial GLM.

**Perceived impacts** An online survey (human ethics approval CSIRO 003/21) including questions on the land being managed, impact of cottonbush and other weeds assessed perceived impacts of cottonbush.

The survey link was circulated through South West Catchments Council (SWCC) newsletters, Facebook and Twitter, and promoted through NRM workshops held by collaborators. To reach a wider audience, the survey was also run by SWCC over the phone in an abbreviated format, targeting farmers in their database. The survey ran from 5 October to 14 December 2021. The data were analysed quantitatively when applicable, or for trends and significant responses. Because all questions were optional, sample sizes varied for each question.

## RESULTS

**Demography, phenology, natural enemies** All six cottonbush populations studied were of mixed ages, with seedlings, juvenile, mature (i.e., reproductive adults), and dead plants. This indicates that these stands had been present for multiple years or that recruitment was staggered across multiple time points. When all transects were combined, a total of 2 825 plants were counted in 1.3 hectare; a mean of 2 173 plants per hectare, of which 522 were dead, and 858 were reproductive adults.

No individuals in flower were observed during field visits conducted from April to October, whereas most plants were in flower in December and January, and a single outlier plant at Wungong bore flowers in March. Of 135 plants measured, 27 were flowering: four estimated to have more than 1 000 flowers; ten to have 100-1 000 flowers; and 13 to have fewer than 100 flowers each. Fruits were observed in every field trip, with 50-100% of plants bearing fruits depending on month of visit. Of 135 plants, 71 were fruiting: 51 (40.8%) estimated to have fewer than 50 fruits; 11 (8.8%) to have 50-100 fruits; seven (5.6%) to have 100-200 fruits; and two (1.6%) to have more than 200 fruits. Seed counts showed an average of 55 filled seeds per fruit at Wungong, and 107 at Serpentine.

The natural enemies *Danaus plexippus, D. petilia,* and *Aphis nerii (*Boyer de Fonscolombe, 1841) were observed at four, one and three out of the five sites studied. Two new natural enemies of cottonbush for WA were recorded: *Arocatus rusticus* (Stål, 1866) (four sites) and a phytoplasma causing phyllody (two sites). All were qualitatively observed to cause negative impacts on plant above-ground biomass, but these were not quantified in our study.

**Soil cores and germination** Sites varied significantly in number of seeds per soil core: from an average total seed of 1.1 (Preston Beach, a site that gets inundated periodically) to 9.3 (Glen Mervyn) to 28.7 (Serpentine) per core per site, equating 140 to 3 654 seeds.$m^{-2}$. The average proportion of intact seeds was low (28.4%) and did not vary significantly between sites.

Germination of seeds collected from fresh, nearly erupting fruit from Serpentine was over 54.8% in two weeks, with an additional 27% considered viable under dissection (total viability of 81%).

Germination of intact seeds from soil cores was 33.9% on average, and the proportion of intact seeds germinating within 30 days did not vary significantly between sites. An additional 4% of non-germinated intact seeds were considered viable upon dissection.

**Perceived impacts** We received 57 responses to the online survey and 44 responses from the abbreviated phone survey. Respondents represented 34 Local Government Areas, with a predominance of land holders or managers in 'commercial agriculture' (58 of 105) followed by 'lifestyle property' (30 of 105).

When asked if cottonbush had ever occurred on their land, to the best of their knowledge, 30 respondents answered yes, currently present; 27 said present in the past, and 35 said it was never present. Within the group with cottonbush currently or previously present, the most common control methods were 'hand-pulling only' (35%) and 'combination of herbicide and hand-pulling' (30%). Respondents who answered that cottonbush had never occurred on their property were disproportionately those in large commercial agricultural lands (21 of 35 have >1,000 ha of cropping and livestock).

Estimates of yearly cost and time to control cottonbush ranged from zero to AUD$3,000 per property (average $435; n=37). The estimated number of days controlling cottonbush ranged from zero to 52 days per year. Both cost and time estimates showed no direct relationship with size of property.

Perceived economic impact answers (n=44) were grouped in themes: toxicity was mentioned 12 times, restricted land access and reduced productivity were mentioned 9 times. Eight people believed that cottonbush has little or no economic impact on them. As for perceived environmental impacts, 27 people mentioned negative environmental impacts, 11 respondents believed this species is of little or no ecological concern. The two impacts most commonly mentioned were competition with native flora (7 times) and changes in habitat (4). Finally, the survey asked participants to describe social impacts, where issues of people disagreeing on the need for control (4), lack of awareness (2) or that people don't care (1) were mentioned (n=66).

The answers and comments had a dominant trend of mentions that cottonbush comes from adjacent properties, be that their neighbours (31 mentions), government lands (10), Department of Biodiversity, Conservation and Attractions (DBCA) managed land (4) or Forest Products Commission lands (3).

Cottonbush was ranked as the most impactful weed on the environment by four respondents and was mentioned another 11 times in the second to fifth most impactful categories (n=58). In relation to economic impacts, 8 respondents ranked it as the top weed, and 11 respondents placed it between second and fifth place (n=75). On both fronts, the weed named most impactful overall across the study area was blackberry (*Rubus anglocandicans* A.Newton).

## DISCUSSION

Our study found cottonbush densities an order of magnitude lower than a Botswana study, where it is native but opportunistic on disturbed sites reaching 23 333 plants.hectare$^{-1}$ (Teketay *et al.*, 2021).

The phenology observed for the six sites and times of year visited was in accordance with existing knowledge for WA, although the flowering period range in our study was more restricted than that previously described (Florabase 2021), likely because our study was limited to a time span of one year. Although our study used estimate categories rather than counts, flowers per plant aligned with the 437 mean flowers per plant in the native range (Teketay *et al.* 2021). Teketay at al. (2021) recorded a range of 0-75 (mean 29) fruits per plant in their study, indicating some plants in WA were producing comparatively large amounts of fruits. Seed production at Wungong was half of that previously described, but seed production at Serpentine was close to mean reference values of ovules per ovary (107.5, Wyatt *et al.*, 2000) and seeds per fruit at two sites in Botswana (93 and 105, Teketay *et al.* 2021).

Using our average of reproductive plants and the fruiting body outputs from our results, we can extrapolate that in one hectare at least 20 988 fruits are produced. With an average of 81 filled seeds per fruit, we conservatively estimate ~1.7 million cottonbush seeds produced per hectare per year in a mixed stand. This estimate is far lower than those recorded in a 'weedy' block in the native range of the species, where ~62.3 million seeds per hectare were estimated (Teketay *et al.* 2021); largely due to the higher density of plants recorded in that study.

Our soil results indicate there is significant predation and other damage to the seeds over time, and that because of this a minority of seeds in the soil are viable. When considering soil cores, we were unable to find out the history of the Preston Beach infestation and whether the site has been chemically treated in the past, so the results for that site should be interpreted with caution.

We extrapolated that the number of viable seeds in the soil under a cottonbush infestation was zero (Preston Beach); 891 200 (Glen Mervyn); or 3 819 700 (Serpentine National Park) per hectare. To understand propagule pressure, the seed bank values should be added to the fresh output of seeds, of which approximately 81% were found to be viable in our results, resulting in an added ~1.36 million viable seeds per hectare per year). It will be important to understand seed viability change over time to more comprehensively assess seed bank risk. This work is already underway, with around 50% decline observed in the first 12 months of burial (unpublished data). Our germination trials used the optimum treatment (30°C with no water stress), based on our controlled germination trials on 36 temperature-water potential treatments (unpublished data).

On the social aspects of cottonbush management in WA, we hypothesise that fewer records of cottonbush on large agricultural properties arises from a combination of their relatively lower rainfall and regular weed control. The current control methods available are considered effective for moderately sized infestations (Reeves and Dodd 2014, Petersen 2014). Cost and time to control cottonbush were not proportionate to land size, as these are likely related to infestation size rather than property size. We did not elicit cost of surveillance or monitoring (related to property size) nor account for synergies controlling multiple weeds.

The insight gained from the social survey was in broad agreement with qualitative observations made during our field work and jointly indicate that the environmental and economic impacts from cottonbush are not of the highest order relative to other weeds in the region.

There was agreement between respondents on social impacts, where frustrations were recurrently expressed when adjacent privately- or government-managed properties have no control and act as a reservoir of seeds. We infer that cottonbush lends itself to such social conflicts because it is a large and easily identifiable shrub; it is a prolific producer of seeds; and the wind-dispersed pappi can be seen from afar (whether or not the pappus is carrying a seed or has already detached), which may generate a level of helplessness and discord. Coordinated control of poorly or unmanaged lands appears to be essential to resolve the social issues.

The recommendations arising from our research were to: (1) consider further studies to monitor species distribution with structured annual surveys of presence/density, absence, and control efforts to inform whether the species range is expanding and how active management is influencing distribution; (2) quantify impacts of the natural enemies already present in WA; and (3) investigate approaches and investment to improve the coordination and effectiveness of control.

## ACKNOWLEDGMENTS

We thank DPIRD and CSIRO for funding this research. Simon Merewether, Paul Tholen, Davin Gibellini, Jonelle Cleland, Teele Hooper-Worrell, Kate Duzevich, Shania Willison, and Wolfgang Lewandrowski are thanked for contributions to the project. John Scott, John Moore and Michael Robertson for reviews of the manuscript.

## REFERENCES

ALA (2021). Atlas of Living Australia. Weeds Australia Profiles: *Gomphocarpus fruticosus* (L.)W.T.Aiton. https://profiles.ala.org.au/opus/weeds-australia/profile/Gomphocarpus%20fruticosus. (accessed 9 November 2021).

Florabase (2021). *Gomphocarpus fruticosus* (L.) W.T.Aiton. (The Western Australian Herbarium) https://florabase.dpaw.wa.gov.au/browse/profile/6587 (accessed 9 November 2021).

Goyder DJ and Nicholas A (2001). A revision of *Gomphocarpus* R. Br. (Apocynaceae: Asclepiadeae). Kew Bulletin 56(4):769-836.

Hussey BM, Keighery GJ, Dodd J, Lloyd SG, Cousens RD (2007). Western weeds. A guide to the weeds of Western Australia, Second Edition (The Weeds Society of Western Australia, Victoria Park).

Lloyd S and Rayner B (2012). Narrow-leaf cotton bush and its control. Farm Note 498 (Department of Agriculture and Food, Western Australia).

Petersen E (2014). Industry economic performance impact assessment on narrow leaf cotton bush (*Gomphocarpus fruticosus*). Community consultation report. Prepared for Department of Agriculture and Food, Western Australia by Advanced Choice Economics Pty Ltd.

R Core Team (2021). R: A language and environment for statistical computing. https://www.R-project.org/ (R Foundation for Statistical Computing, Vienna, Austria).

Randall RP (2007). The introduced flora of Australia and its weed status. CRC for Australian Weed Management (Department of Agriculture and Food, Western Australia).

Reeves A and Dodd J (2014). Situation statement: cotton bush in Western Australia (Department of Agriculture and Food, Western Australia).

Teketay D, Mafote T, Puodipe KK, Mompati PB, Nkoborwane N, Mathowa T, Mojeremane W (2021). Reproductive traits that foster the invasion of four alien plant species in Botswana. Chapter 2 in Invasive Alien Species: Observations and Issues from Around the World, Volume 1: Issues and Invasions in Africa, First Edition, eds T. Pullaiah and Michael R. Ielmini (John Wiley & Sons Ltd).

Ward M, Johnson SD, Zalucki MP (2012). Modes of reproduction in three invasive milkweeds are consistent with Baker's Rule. Biological Invasions 14(6):1237-50.

# Boneseed and bitou bush in Western Australia: a tale of two *Chrysanthemoides*

Kathryn L. Batchelor[1], John K. Scott[1] and Bruce L. Webber[1,2,]
[1] CSIRO Health & Biosecurity, 147 Underwood Ave, Floreat, Western Australia 6014, Australia
[2] School of Biological Sciences, University of Western Australia, 35 Stirling Highway, Crawley, Western Australia 6009, Australia
(kathryn.batchelor@csiro.au)

**Summary** Boneseed and bitou bush are present in Western Australia (WA) and subject to eradication, however prospects for success are quite divergent. Boneseed (*Chrysanthemoides monilifera monilifera* (L.) T.Norl.) has been present since at least 1950. Initial control efforts commenced in the late 1980s after the weed was discovered at several wheatbelt towns and in the Perth hills. Targeted control commenced in 2006 with the appointment of a national *Chrysanthemoides* co–ordinator, which led to the identification of 42 infestation locations. There is no doubt subsequent control efforts have helped reduce abundance and spread. However, on numerous occasions these efforts have regressed due to insufficient and short–term funding, breaks in management that allowed the seedbank to be refreshed, or not completing adequate delimitation. Boneseed is now known at 47 locations, and our review of past management efforts has established a current baseline and management plan that will deliver localised extirpation with an eventual goal for state–wide eradication.

The situation with bitou bush (*Chrysanthemoides monilifera rotundata* (DC.) T.Norl.) is a complete contrast. When bitou bush was first discovered at the industrial port of Kwinana in 2012, a systematic delimitation survey was undertaken. This initial survey found over 1,200 plants and seedlings over ca. 2.5 km$^2$, but a greater surveillance buffer area was added to the invasion footprint to achieve local delimitation. Subsequently the delimitation and buffer areas have been surveyed annually with new plants found yearly since. As of 2022 the seedbank is likely depleted, suggesting local eradication is feasible. Prospects for successful eradication of both species are at a point where management needs are critical, but the continuity of sufficient resources to deliver this outcome is uncertain. We discuss the remaining challenges for eradication of these WoNS species, the strategy to find the last plants and the data–driven approach that will enable future survey effort to deliver greater efficiency of resources without compromising effectiveness.

**Keywords** *Chrysanthemoides*, containment, delimitation, eradication, surveys, Western Australia.

## INTRODUCTION

*Chrysanthemoides monilifera* is a South African plant with at least six subspecies (Norlindh 1943) of which two are established in Australia: boneseed and bitou bush. The taxonomic groupings by subspecies have been supported by molecular genetics studies (Barker *et al.* 2009). Consequently, we treat the two subspecies as separate taxonomic entities.

Both boneseed and bitou bush are Weeds of National Significance (WoNS) in Australia (Thorp and Lynch 2000). Currently, under WA legislation (the *Biosecurity and Agriculture Management Act 2007*; Government of Western Australia 2007), *C. m. rotundata* (bitou bush) has been declared as category C1 (plants which should be excluded from part or all of WA), whereas *C. m. monilifera* (boneseed) is in category C2 (plants which should be eradicated from part or all of WA).

Despite this legislation, both taxa have naturalised in WA. In this paper we compare and contrast the outcomes of past management for *Chrysanthemoides* in WA and assess the feasibility of eradication as an end goal for both species.

## BONESEED MANAGEMENT

Boneseed has been present in WA for at least 75 years, 50 years longer than bitou bush (1 site, inferred to have started 1995). Consequently, it is far more widespread, covering 47 locations as of 2022, which are best described as comprising 89 sites (*cf* populations) because the management approach is different. The weed is mostly found around regional WA townships and residential gardens in the Perth Hills (Figure 1), with populations relatively small in extent (<50 m$^2$ up to 10 ha). Targeted control commenced in 2006 with the appointment of a National Boneseed Coordinator who profiled the distribution, extents and produced the WA Boneseed Eradication Strategy (Cherry 2010). Since then, boneseed management has been mostly resourced with short–term (not necessarily consecutive) NRM grants in urban areas and in an *ad hoc* way by Department of Primary Industries and Development (DPRID) biosecurity staff in regional areas. Sites were visited most years, but missed years have occurred due to breaks in funding and staff changes, resulting in refreshed seedbanks and extents not being fully translated.

**Figure 1.** Distribution in 2022 of boneseed and bitou bush in Western Australia.

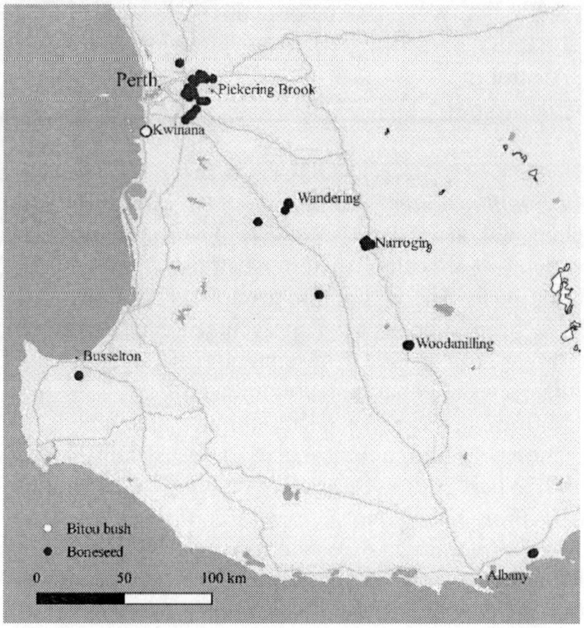


In 2020, the authors commenced a data aggregation project to generate a complete understanding of the extent of boneseed in WA, along with management and other influential events (e.g., fire) that have occurred at each site. We extracted data from diverse sources; including individual GPSs of people who have made surveys, government databases across multiple agencies and by interviewing landholders and retired staff. This process resulted in a database of over 2,000 records of weed removal, with detailed information on the demography of 700 records. From this synthesis, a historical timeline was developed for each site together with a future projection of extirpation likelihood, based on the last known fruiting event and assuming effective annual control is deployed (Figure 2). We also produced a risk heat map for each site, using locations where seedlings are recurrent suggesting a live seedbank.

Based on this aggregation and attending all the populations as part of this work, including 7 past extirpated sites, we estimate that it will take a minimum of 2,100 hours of on–site surveillance over the next 16 years (starting from at least 212 hours in 2022, decreasing to 64 hours in 2036) to achieve full extirpation for the 47 locations across the state. Some sites require delimitation before extirpation can declared. This estimate is based on a cautiously predicted viable seedbank of up to 15 years (anecdotal evidence suggests 10 years, L. McMillan pers. comm.), the effectiveness of existing control methods and a sustained management program. This duration may be reduced if methods to enhance the depletion of the soil seedbank, or drone–based detection of isolated plants are successfully developed; work that is currently underway.

Our baseline assessment reinforces that biocontrol is not a logical solution to pursue for boneseed in Western Australia while eradication remains feasible. However, we recommend the genetics of boneseed across Australia be examined in case other states were to restart their biocontrol development programs. With a single aggregated spatiotemporal database now available for all boneseed populations in Western Australia, we are now able to deploy a robust and realistic eradication program, but one that must remain informed by adaptive management.

## BITOU BUSH MANAGEMENT

The situation with bitou bush is a complete contrast to that of boneseed in WA. Bitou bush was discovered in the state in 2012 as an established population of some 1,700 plants in the coastal industrial area of Kwinana, south of Perth (Scott and Batchelor 2014). CSIRO, recognizing a unique opportunity to study a species at an early invasion phase, proposed to undertake a delimitation survey to realise the extent of the population and removed plants along the way. This initial survey found over 1,200 plants and seedlings over ca. 250 ha, but a greater surveillance buffer area was added to the invasion footprint to achieve local delimitation. The delimitation and buffer areas have been surveyed annually with new plants found every year since (Figure 3). Very few seedlings have been found since 2017 and none since 2020. Three large plants were found in the 2022 annual survey carried out between April–June, all within the delimitation area and well hidden amongst dense vegetation, one of which was only discovered by drone, demonstrating the bitou threat is not over.

The annual decline in bitou bush numbers with each annual survey was reported in Scott *et al.* (2019b). Between 2012 and 2018 we surveyed over 253 ha of land and removed 1,766 bitou bush plants. The seed bank was measured using soil cores and by 2018 the seed bank was undetectable with the standard sampling methods used. By 2022, 1,792 plants have been removed and we expect the bitou bush seedbank to be depleted, based on seed viability of 5–7 years (K. French, pers. comm.), suggesting successful extirpation is a near term possibility.

This targeted and systematic approach to annual surveying and confirming delimitation has been undertaken in a consistent and evidence–based way.

However, despite familiarity with the species and known hot–spot locations, large plants are still missed apparently obscured by vegetation and infrastructure in previous years. Extirpation is not expected until 2026, as the last possible fruiting event occurred in 2019.

Could extirpation have been achieved earlier for bitou bush in WA? Unfortunately, fire as a management tool (for stimulation of the seedbank and reduction of above ground biomass and seed) was unfeasible due to the close proximity of petrochemical plants and other industry. The program has been assisted over the survey period by a reduction in the risk footprint (and therefore survey area) due to an increasing cover of concrete and industrial development.

**Figure 2.** A past and planned future management timeline for boneseed control at sites near the town of Narrogin. Red line is current year (2022).

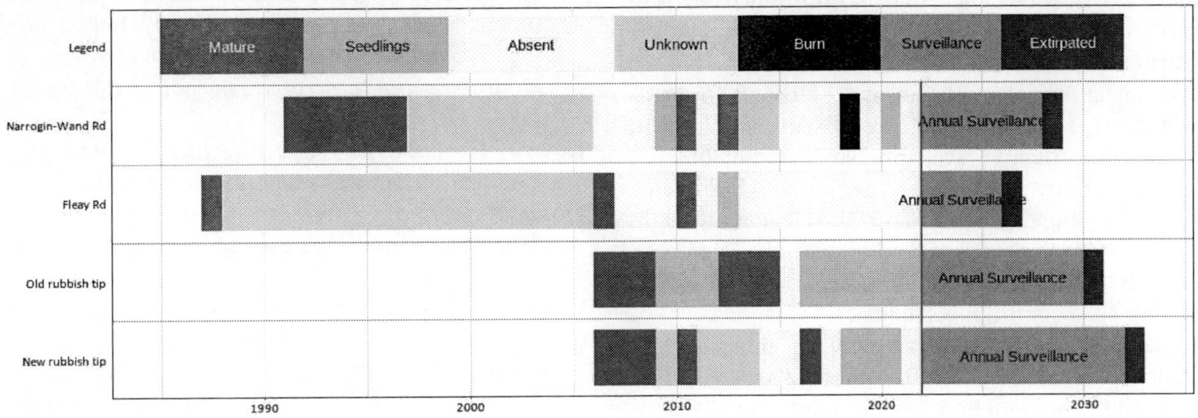


**Figure 3**. Log comparison of bitou bush plants and seedlings found at Kwinana between 2012 and 2022. Plant size categories refer to maximum crown diameter.

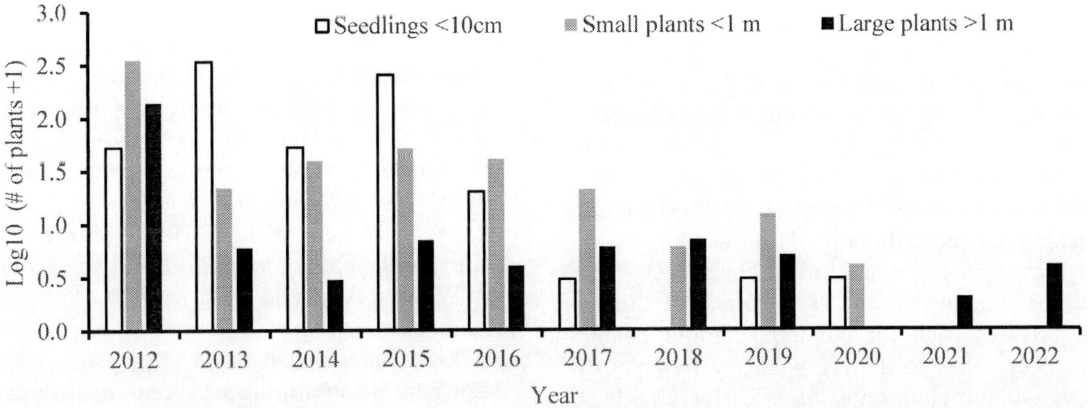


## DISCUSSION

Bitou bush can be considered effectively contained in WA while being annually surveyed. However, bitou bush is at the stage where the survey effort to find the last plants is at its greatest, while boneseed remains at risk of becoming unfeasible to eradicate if current surveys are suspended or infestations are not adequately delimited.

One of the main challenges for finding the last plants is their ability to meld in with other vegetation and not be seen over multiple years. In Scott *et al.* (2019a) we give the example of a pair of bitou bush plants growing under a clump of *Acacia* and not spotted until they were 3 m tall and in flower. Boneseed is equally challenging to detect during surveys but has the additional challenge of being distributed across residential properties, which have access challenges.

Positively, bitou bush is an obligate outcrossing taxon (Gross *et al.* 2017; Scott *et al.* 2019a). This means that isolated plants do not produce seeds until another individual germinates nearby and flowers (i.e., subject to allee effects due to pollination limitations). It is not known if boneseed is likewise outcrossing. This fact needs to be determined as it has significant implications for boneseed management when population numbers become very low.

An important element of delimitation is understanding how re–invasion could occur. This means understanding the original invasion progress. Currently we are assessing three lines of evidence on the re – invasion issue; history, based on documented records of bitou bush, nuclear DNA genome variation, and chloroplast DNA genome variation.

Both boneseed and bitou bush are bird and rodent dispersed, but given the length of time they have been in WA, it is surprising they aren't more widely distributed. One possibility is the lack of suitable long–distance volant dispersers. Starlings (*Sturnus vulgaris* L.) for example, are absent in WA and parrots are effective seed predators. Bitou bush is also in a highly industrialised area, and while plants were often found under bird perches (fences and light posts) evidence of rodent feeding was observed with gnawed endocarps found in the soil cores, providing another seed control mechanism.

Research on seed bank longevity is underway (K. French pers comm) and its outcome is critical to predicting the length of both control programs. Schoeman *et al.* (2010) showed bitou seeds have reduced resilience compared to boneseed, and the two sub-species should be considered separately when designing effective control measures.

## CONCLUSION

We have adopted a data driven approach for improving control outcomes for both boneseed and bitou bush. Taking this adaptive management approach to delivering successful extirpation also provides useful motivation and feedback through the program via evidence of progress, even if small. Establishing the contribution of the seed bank to ongoing invasion risk was critical for the strategy adopted for bitou bush (Scott *et al.* 2019a). A similar data–driven approach for boneseed should enable future survey effort to deliver greater efficiency of resources without compromising effectiveness. Past management efforts have been effective at containing boneseed, as since Cherry (2010) only four addition sites have been found. Moreover, with effective future management 20 sites are likely to be extirpated within three years. Significant resources have been invested in removing boneseed across WA since 2006. Without a long–term management commitment, we predict that infestations will revert to their pre–2006 state within a decade. The value proposition of ongoing control thus appears a most attractive proposition, positioning WA to achieve eradication for two Weeds of National Significance.

## ACKNOWLEDGMENTS

This research was supported by CSIRO. We thank Fremantle Port Authority and DPRID for access to sites and funding, Perth NRM, DBCA and SERCUL for their assistance with boneseed management.

## REFERENCES

Barker N.P., Howis S., Nordenstam B., Källersjö M., Eldenas P., Griffioen C. and Linder H.P. (2009). Nuclear and chloroplast DNA based phylogenies of *Chrysanthemoides*Tourn. ex Medik. (Calenduleae; Asteraceae) reveal extensive incongruence and generic paraphyly, but support the recognition of infraspecific taxa in *C. monilifera*. *South African Journal of Botany* 75, 560–572.

Cherry H. (2010). Western Australia Boneseed Eradication Strategy. Unpublished Report.

Government of Western Australia (2007) 'Biosecurity and Agriculture Management Act 2007.'

Gross C.L., Whitehead J.D., de Souza C.S. and Mackay D. (2017). Unsuccessful introduced biocontrol agents can act as pollinators of invasive weeds: Bitou Bush (*Chrysanthemoides monilifera* ssp. *rotundata*) as an example. *Ecology and Evolution* 7, 8643–8656.

Norlindh T. (1943). 'Studies in the Calenduleae. 1. Monograph of the genera *Dimorphotheca*, *Castalis*, *Osteospermum*, *Gibbaria* and *Chrysanthemoides*.' (CWK Gleerup: Lund).

Schoeman J., Buckley Y.M., Cherry H., Long R.L. and Steadman, K.J. (2010). Inter–population variation in seed longevity for two invasive weeds: *Chrysanthemoides monilifera* ssp. *monilifera* (boneseed) and ssp. *rotundata* (bitou bush). *Weed Research* 50. 67–75.

Scott J.K., Batchelor K.L. (2014). Management of *Chrysanthemoides monilifera* subsp. *rotundata* in Western Australia. *Invasive Plant Science and Management* 7, 190–196.

Scott J.K., Batchelor K.L., Webber B.L. (2019a). Long term monitoring of recruitment dynamics determines eradication feasibility for an introduced coastal weed. *Neobiota* 50, 31–53.

Scott J.K., Batchelor K.L., Jucker T., Webber B.L. (2019b). Aerial photography and dendrochronology as tools for recreating invasion histories: do they work for bitou bush (*Chrysanthemoides monilifera* subsp. *rotundata*)? *Biological Invasions* 21, 2983–2996.

Thorp J.R., Lynch R. (2000). 'The determination of weeds of national significance.' (National Weeds Strategy Executive Committee: Launceston, Tasmania).

# Accelerating hawkweed eradication: Innovation, collaboration and persistence

Hillary Cherry, Mark Hamilton, Paul Skeen[1], Jo Caldwell[1], Janelle Jenkins[1]
[1]NSW National Parks and Wildlife Service, Australia
(mark.hamilton@environment.nsw.gov.au)

**Summary** The hawkweed (*Pilosella aurantiaca* and *P. officinarum*) eradication program is the largest weed eradication in New South Wales (NSW) and, together with the Victorian hawkweed eradication program, aims to eradicate hawkweeds from mainland Australia. Eradication began in NSW in 2009, and the program recently gained significant long-term support to accelerate eradication efforts. Fast tracking eradication through greater innovation, research to fill knowledge gaps and refining best practice are key to ensuring commitment is maintained and the objective reached. Weed eradication as a management objective is challenging but attractive as it has an end point, a low cost to benefit ratio and avoids future long-term management costs. Key factors affecting eradication feasibility and success include, 1) the ability to prevent reinvasion; 2) availability of effective controls; 3) suitable biological characteristics of the target species; 4) manageable infestation size; 5) target species detectability; and 6) socioeconomic factors, including political and community support. Hawkweeds in NSW are measured against these criteria, and a 2017 program review concluded eradication is feasible. However, key challenges remain to ensure eradication occurs rapidly and cost-effectively, including to ensure: i) the entire infestation is delimited; ii) seed set is prevented; and iii) infestations are rapidly progressed from active to extirpated status. Earlier innovations such as detection of hawkweed by colour drone imagery and scent detector dogs are assisting to delimit the infestation, but more is required. This presentation outlines how enhanced drone surveillance and technology; dispersal and habitat suitability modelling; time to eradication modelling; hawkweed biology and ecology research; volunteer surveillance and improved weed hygiene contribute to accelerating eradication of hawkweeds from NSW.

# A decade of new weed records in South Australia: an overview from the State Herbarium of South Australia

Chris Brodie[1], Peter J. Lang[1]
[1]State Herbarium of South Australia, Kent Town, Australia
(chris.brodie@sa.gov.au)

**Summary** Since 2009 there has been a major effort made by the State Herbarium of South Australia in collecting weed specimens to document the occurrence, establishment and spread of new weed species, including plants that are showing early signs of self-establishing. Scientific verification and official recognition of weed taxa is achieved through the lodgement of voucher specimens for identification by taxonomic botanists in the State Herbarium. For taxa not previously recorded for South Australia, an entry is then made in the Census of South Australian Vascular Plants, Algae and Fungi (http://flora.sa.gov.au/census.shtml). These new weeds are typically found at only one or few locations, and in low numbers. Early recognition of new incursions offer land managers the best opportunity for containment or eradication of potential new weed threats. We reviewed new weed herbarium records spanning 10 years from July 2009 to June 2019. Over this period, 217 new weed taxa were added to the Census. These come from 69 plant families, with 37 represented by a single taxon. However, 11 plant families each contain five or more taxa and together account for 126 taxa, over 55 % of the total. For each taxon, initial invasion pathways into South Australia were assigned, based on collection details and context, to the following categories: 'garden/planted', 'agricultural', 'both' or 'unknown'. The 'garden/planted' category has the overwhelming majority of taxa (184 or 84.79%): gardens and other non-agricultural plantings are now the most common pathway for plants becoming established as weeds in South Australia. The 'agricultural' pathway only has three taxa (1.38 %). Six taxa (2.76 %) were assigned to the 'both' category and 24 (11.06 %) to the 'unknown' category. These results are illustrated with case studies and we discuss the role of surveillance to help to prevent the establishment of further invasive weeds in this State.

**Keywords** New weeds, identification, taxonomic, herbarium, South Australia, voucher specimens

# From tactical to strategic herbicide use: On the systematic challenges of herbicide resistance management in New Zealand's arable sector

Martin Espig[1], Robyn Dynes[1], Roxanne Henwood[1], Trevor James[1]
[1]AgResearch, Lincoln, New Zealand
(martin.espig@agresearch.co.nz)

**Summary** Synthetic herbicides are crucial weed management tools in farm systems worldwide that have enabled significant productivity gains and new practices like minimum tillage. However, since their first commercial release in the 1940s there has been an accelerating increase in herbicide resistant weeds, with now 266 confirmed resistant species in 71 countries. Weed scientists and extension professionals have for decades promoted integrative weed management strategies to reduce the over-reliance on specific herbicides, including more diverse crop and herbicide rotations. Yet, herbicide resistance continues to increase, threatening the sustainability of food and fibre production. Much of New Zealand's agricultural sector has been regarded as less susceptible to resistance due to relatively diverse crop and livestock rotations. However, recent surveys identified unexpectedly high incidences of resistance across cropping farms in the Canterbury region, which cautions against resistance as a growing challenge and prompts critical reflections on current herbicide use practices. This paper investigates the complexities of weed management and herbicide resistance prevention within New Zealand's arable sector, particularly the challenges associated with more strategic herbicide use. We draw on qualitative social research with arable farmers, rural professionals, and weed scientists to identify factors that determine current weed management practices and contextualise the problem of herbicide resistance. We outline the drivers of herbicide use through a multi-level perspective that systematically considers i) the individual psycho-social level, ii) farm systems level, and iii) the national agricultural systems level. These interconnected drivers highlight that farmers' herbicide applications are influenced by diverse biophysical, technological, and sociocultural factors, with some hindering best practice herbicide use. Our findings demonstrate that integrated systems-based approaches are needed to holistically understand herbicide resistance as a critical first step in collaborative efforts to shift from tactical to more strategic herbicide use.

**Keywords** Herbicide resistance, systems thinking, integrated weed management, practice change, sustainability

# Differences Between New Zealand and Australia in Development of Herbicide Resistance

Kerry Harrington[1], H Ghanizadeh[1], I Heap[2]
[1]Massey University, Palmerston North, New Zealand,
[2]International Survey of Herbicide Resistant Weeds, Corvallis, United States
(K.Harrington@massey.ac.nz)

**Summary** Cases of herbicide resistance have been developing for many years in Australia, with multiple issues in *Lolium r*igidum being reported since the early 1980s from cereal crops. The first cases of herbicide resistance in New Zealand were also reported at about this time, but these were initially triazine resistant weeds, mainly *Chenopodium album*, from maize crops. Herbicide resistance has been reported on a regular basis since then for both countries, but there have been a number of differences between the countries. There have been far more cases reported from Australia than New Zealand, which is to be expected due to the larger size of Australia, differences in farming systems and possibly because of more weed scientists in Australia working on the problem than in New Zealand. *Lolium rigidum* has been a species that appears particularly susceptible to evolving resistance to herbicides, but this species is almost non-existent in New Zealand. However problems have been developing in recent years within New Zealand with the closely related species of *Lolium perenne* and *Lolium multiflorum*, which are both grown extensively for pasture production. The main concerns in New Zealand for several decades were resistance in maize and pastures, though in the past decade, herbicide resistance within vineyards and cereal crops have also becoming an issue. But over this time, Australia has had extensive issues in many situations, including cereal crops, lucerne, lupins, canola and vineyards. This paper will discuss some of these differences that have occurred over the past 40 years.

**Keywords** Herbicide resistance, New Zealand, Australia, research, crop, pasture

# A macroecological perspective of herbicide resistance in weeds

Philip Hulme[1]
[1]Bio-Protection Research Centre, Lincoln University, Canterbury, New Zealand
(philip.hulme@lincoln.ac.nz)

**Summary** Current knowledge of herbicide resistant weeds has been built over many decades through painstaking case studies examining the mechanisms of herbicide resistance of individual weeds. Considerable data now exist on the frequency of herbicide resistance to different modes of actions in different weeds species worldwide comprehensively captured in the Herbicide Resistant Weed Database. These data now permit a macroecological perspective addressing broader generalisations and synthesis of key global patterns in herbicide resistance. Three recent examples illustrate the value of a macroecological approach: a) the role of weed species traits in the likelihood of herbicide resistance evolution; b) patterns in the occurrence of multiple resistance across different herbicide modes of action; and c) global economic drivers of the number of herbicide resistant weed worldwide. Boosted regression tree analysis has highlighted that species traits associated with rapid evolution (chromosome number, self-incompatibility) are more closely linked to herbicide resistance than traits normally associated with weediness such as annual life-history and self-compatibility. However, the most important trait was the prevalence of the species, the more prevalent the more likely to be herbicide resistant. Hierarchical cluster analysis distinguished three primary clusters of modes of action within which multiple resistance is most likely. A clear message when mixing herbicides is to select between rather than within clusters. Finally, an information theoretic analysis of the numbers of herbicide resistant weeds in different countries pointed out that in many cases numbers are likely to be underestimated due to low research capability in many countries and in general also simply reflects the time since herbicide resistance was first detected. The intensity of herbicide use was also important, but the sampling effects suggest the level of herbicide resistance is underestimated worldwide. The foregoing illustrates that macroecological approaches and machine learning techniques have considerable potential to provide insights into herbicide resistance.

**Keywords** Glyphosate, machine learning, maize, species traits, wheat

# Are you sure? A test of how herbicide resistance testing can inform weed management decisions

Rick Llewellyn[1,2], Fiona Dempster[2], Masood Azeem[1], Roberto Busi[2]
[1]CSIRO, Urrbrae, Australia,
[2]University of Western Australia, Perth, Australia
(rick.llewellyn@csiro.au)

**Summary** Herbicide resistance testing services that confirm the resistance status of weed populations using submitted seed samples have been available in Australia for over 25 years. Despite growers in all cropping regions facing increasingly diverse forms of resistance across a widening range of weeds there is ongoing concern that testing services are underutilized. In this study we examine the use and information value of resistance testing from the grower and agronomist perspective with the aim of identifying opportunities for increased use and value. The study included 51 farms and 15 agronomists involved with the collection of 230 weed samples (annual ryegrass, wild radish, brome grass, barley grass and capeweed) in 2020 from 128 Western Australian cropping paddocks. These were tested at University of Western Australia and classified as Resistant, Developing or Susceptible to a total of 50 herbicide treatments. To evaluate how paddock-specific herbicide resistance testing information can inform existing knowledge of resistance status, the perceived resistance status of tested populations was elicited from twenty-five growers and 15 agronomists prior to testing. Sixty percent of the growers had undertaken some herbicide resistance testing in the past 10 years, although usually not regularly, and most relied on visual observation to determine resistance status. Resistant populations (based on test results) were very rarely perceived to be susceptible (and vice versa). There was a greater tendency for growers to overestimate the developing resistance status of susceptible populations. Perceptions and test results were less well aligned for forms of resistance that can be more difficult to observe in the field e.g. pre-emergence herbicides and broadleaf herbicides commonly used in mixes. Opportunities are identified where test results can offer the most potential value and barriers to testing can be reduced.

**Keywords** Herbicide, resistance, testing, adoption, economics, behaviour

# Herbicide resistance in perennial pasture systems – The horse has bolted

Jo Powells[1]
1 Local Land Services, PO Box 26, Cooma, NSW 2630, Australia
(jo.powells@lls.nsw.gov.au)

**Summary** Whilst herbicide resistance is well documented problem in cropping systems, far less is known about its presence and extent in perennial pastures. Perennial grass weeds such as serrated tussock (*Nassella trichotoma*) and African lovegrass (*Eragrostis curvula*) are particularly problematic in the Southern Tablelands of NSW, with minimal herbicide options available for control.

Following the identification of widespread resistance to flupropanate in serrated tussock on the Monaro in 2017, seed samples from African lovegrass were collected in Autumn 2020 and 2021 from locations across the Monaro region and subsequently tested for resistance to flupropanate.

Testing identified plants at multiple locations with low-high resistance levels as well as sites with plant still susceptible to flupropanate at the higher label rate (3 L $ha^{-1}$). Testing using lower rates of flupropanate (1.25–2 L $ha^{-1}$) resulted in reduced control of plants from some locations sampled. These results present further challenges for land managers struggling to control perennial grass weeds within perennial pasture systems.

**Keywords** Herbicide resistance, serrated tussock, African lovegrass, flupropanate

## INTRODUCTION

There has been an exponential increase in the occurrence of herbicide resistance in weeds since 1975 (Heap 2022) and whilst the cropping sector has been managing these issues since the 1970s, managers of pasture-based systems have only had to contend with the issues since the early 2000s (Noble 2002).

Since their introduction into the country over 100 years ago, the perennial grasses serrated tussock (*Nassella trichotoma*) and African lovegrass (*Eragrostis curvula*) have posed ongoing challenges to graziers in their pasture systems in the tablelands of NSW. The group 0 (previously group J) herbicide flupropanate, has been the most successful and remains to be the most widely used chemical for control of these species in pastures. However, it is this over-reliance on a single herbicide that has driven the development of herbicide resistance. McLaren et al. (2008) noted that the continual use of group 0 herbicides (e.g. flupropanate) for more than 15-20 years has driven the development of herbicide resistance serrated tussock.

A survey investigating the extent of herbicide resistance in serrated tussock in 2004 identified 3 sites (2 in Victoria and 1 in New South Wales) where flupropanate resistance had developed (McLaren et al. 2008). This was the first finding of herbicide resistance in a perennial weed found predominantly in perennial pasture systems and prompted further investigations into the plant's biology and other herbicide control options.

Following the identification of this herbicide resistant serrated tussock, McLaren et al. (2010) surveyed a 5 km radius around the 3 known serrated tussock resistance sites and found that the resistance had become widespread. They concluded that land managers would have to increasingly deal with the issue of herbicide resistance and posed the question, "was the genie was out of the bottle?" when it came to resistance in perennial pasture weeds.

**The Monaro** Local archived reports indicate that NSW Agriculture commenced serrated tussock control trials using flupropanate in the late 1970s to early 1980s near Dalgety in the Monaro region. The herbicide remains in common use today across Australia however in recent times, limited availability of the chemical is supressing it's use.

Monaro land managers raised concerns about the efficacy of their flupropanate use in 2016, resulting in Local Land Services undertaking a multi-year project to test resistance to herbicides in serrated tussock grass. Herbicide resistance in African lovegrass was then investigated in 2020 following increasing concerns about management of the weed that was rapidly spreading across the Monaro.

## MATERIALS AND METHODS

**Serrated Tussock** In 2016, an initial thirteen samples of serrated tussock seed were collected from sites across the Monaro region. This seed was sent to Plant Science Consulting (South Australia) and germinated with the seedlings sprayed with the herbicide flupropanate (745 g a.i. $L^{-1}$) at the equivalent rates of 1.25, 2 and 3 L $ha^{-1}$. Wetting agents were not used. Assessment of the effectiveness of herbicide control was made nine weeks after treatment based upon the percentage of plants surviving in herbicide treatments compared to the control. Plants with slight or no biomass reduction were classified as "resistant", biomass reductions of between 40-80% with plant recovery were classified as "developing resistance" and plants with full biomass suppression and plant death were classified as "susceptible". Under controlled conditions, the control seedlings grew strongly, so any reduction in growth was attributed to the impact of herbicide.

The following year, in an attempt to identify just how widespread the herbicide resistance issue may be across the region, a further forty-one seed samples were collected from additional locations across the Monaro and another 5 collected from outlying sites further north of the region that had been experiencing serrated tussock control challenges. The same methodology used in 2016 was repeated for this year's sampling and testing.

**African lovegrass** As part of a pilot study, seeds from African lovegrass plants were collected from twelve sites in autumn 2020 from the northern and central areas of the Monaro where the oldest and most dense populations of African lovegrass are located. Plant Science Consulting again germinated these seeds and then sprayed the seedlings at the four-leaf stage with flupropanate (745 g a.i. $L^{-1}$) at the equivalent rates of 1.25, 2 and 3 L $ha^{-1}$ (no wetter was added). Assessment of the effectiveness of herbicide control was made using the same methodology as described for serrated tussock above.

This work was then expanded in 2021 with a more extensive collection of African lovegrass seeds from across the Monaro region. The same methodology used in 2020 was repeated for this second year of sampling and testing.

## RESULTS

**Serrated Tussock** In the first year, five of the thirteen samples were found to have a high level of resistance to flupropanate with these samples originating from north and western areas of the Monaro. Increasing the rate of application of flupropanate from 1.25 to 2 and 3 L $ha^{-1}$ did not significantly improve the control of these resistant biotypes which confirmed the high-level of resistance (it should be noted that 2 L $ha^{-1}$is the maximum on-label/ on-permit flupropanate rate for use on serrated tussock). The remaining eight samples were found to be susceptible to flupropanate.

In the second year of testing, twenty-three of the forty-six locations sampled tested had a high level of resistance to flupropanate. Fifteen further locations were identified as developing resistance and eight locations were still susceptible to the herbicide. For the locations where resistance was identified, increasing the rate of flupropanate application from 1.25 to 3 L $ha^{-1}$ only resulted in small increases in control in most instances (e.g. 50 % control at 1.25 L $ha^{-1}$, 70 % control at 2 L $ha^{-1}$ and 80 % control at 3 L $ha^{-1}$). However, plants from one location showed zero control at any of the rates of flupropanate used.

**Table 1.** Combined results for seed testing for flupropanate resistance in serrated tussock (2016-2018) and African lovegrass (2020-2021) populations collected from the Monaro region of NSW at maximum on-label control rates.

| | Serrated tussock | African lovegrass |
|---|---|---|
| Resistant | 28 | 7 |
| Developing resistance | 15 | 22 |
| Susceptible | 16 | 14 |
| Total | 59 | 43 |

The more widespread sampling of locations across the district revealed that localities with historical and higher densities of serrated tussock and a longer historical use of flupropanate, generally had higher instances of herbicide resistance. However, locations with plants still susceptible and with those identified as developing resistance to flupropanate were scattered consistently across the Monaro region.

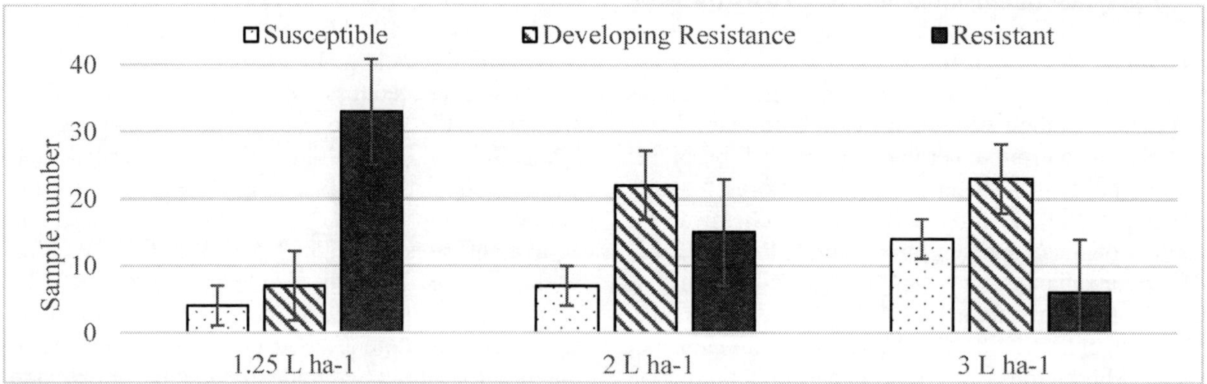


**Figure 1.** Resistance status of African lovegrass samples collected from the Monaro, NSW in 2020-2021 at different rates of flupropanate (745 g a.i. $L^{-1}$) application (standard error bars shown).

**African lovegrass** Of the samples collected in 2020 and 2021, seven of the forty-three locations were found to have strong resistance to flupropanate (Table 1). A further twenty-two locations showed low to mid-level resistance and fourteen locations were identified as still being susceptible to the herbicide (at the 3 L $ha^{-1}$ rate).

At those locations where resistance was noted or was developing, many samples showed distinct responses to varying rates of herbicide application. Many showed no instances of plant death at the lower application rates of 1.25 and 2 L $ha^{-1}$ (note that the maximum on-label rate for flupropanate application on African lovegrass is 3 L $ha^{-1}$) however, plants at 85% of the sites tested were controlled at an application rate of 3 L $ha^{-1}$.

There were only 4 sites in total where control was achieved at all of the flupropanate application rates tested.

## DISCUSSION

The confirmation of flupropanate resistance in serrated tussock confirmed some land managers' concerns about resistance within the weed populations that they were attempting to control, whilst for others it came as a surprise. Many of the land managers involved in the survey believed their poor weed control outcomes were a result of recent higher than average rainfall events, on poor spray application or both. However, by using the herbicide resistance testing process, variables affecting herbicide control outcomes in these weeds - such as rainfall, soil type, water quality, application rate and temperature - were all able to be controlled giving, clear indications about herbicide resistance.

It has been proposed that continual use of group 0 herbicides (such as flupropanate) for a fifteen-to-twenty-year period results in the development of resistance (McLaren et al. 2008) so with both serrated tussock and African lovegrass having a long historical presence and history of attempted control in the Monaro region, these results should not come as a surprise. NSW Agriculture records document serrated tussock herbicide control trials using flupropanate commencing in the late 1970s, so given the over forty years use of this herbicide in the region, the results found are well within McLaren et al. (2008) predictions.

Resistance testing of seed samples from both serrated tussock and African lovegrass sites identified a mixture of resistant, developing resistance and susceptible populations (Figure 1, Table 1). The herbicide susceptible sites were generally located in areas with less history of flupropanate usage, however sites where herbicide resistance was identified had a variety of weed control histories.

Label and permit rates for flupropanate use on serrated tussock in NSW are 1.5-2 L $ha^{-1}$ and for African lovegrass 1.5-3 L $ha^{-1}$ (APVMA PER9792). Lower rates of the herbicide are recommended for lighter soil types (granite and sedimentary type soils) and in situations where desirable species or only young weed seedlings are present. It is worth noting that for many of the sites where of resistance was confirmed or found to be developing, the rate of plant survival was much higher at low rates of herbicide application for both serrated tussock and African lovegrass. In most, but not all instances, increases in plant control were achieved when the application rate was increased (up to 3 L $ha^{-1}$). Whilst lower application rates are usually recommended to try and limit damage to desirable vegetation (usually valuable pasture species) surrounding the target weeds, these results are concerning and highlight the risk of more rapid selection for resistance in these weed populations when low rates of herbicide are applied without other management follow-up undertaken to remove surviving plants.

If low rates of flupropanate are continually used over large populations of perennial grass weeds without the follow-up control of surviving plants, we are likely to see a rapid increase of herbicide resistant populations. Ramasamy et al. (2010) found that serrated tussock plants primarily self-fertilised and that serrated tussock plants resistant to flupropanate would go on to produce at least 85–90 % herbicide resistant seeds. This rate of self-replication indicates that once a resistant population of serrated tussock has developed, it is likely to keep increasing in the absence of a change in control tactics. Possible dilution of the resistant gene pool by introducing susceptible plants is unlikely to yield significant changes in populations of herbicide controllable plants.

The rate of self-fertilisation of African lovegrass is unknown so the rate of resistance expansion within the population of this species cannot be discussed. It is also unknown if herbicide resistant populations of either serrated tussock or African lovegrass are less or more competitive in the environment. Early investigations into the fitness of herbicide resistance African lovegrass plants have commenced with no variation in germination rates recorded to date and other characteristics including the rate of photosynthesis, water-use efficiency, above and belowground rates of biomass production and plant reproduction rates still under investigation (J. Brown, personal communication, January 12, 2022).

Despite warnings from researchers who had identified three separate populations of flupropanate resistant serrated tussock back in the early 2000s, little has changed in the management of perennial grass weeds in pasture systems. This "head in the sand" approach to managing herbicide resistance in pasture systems has resulted in the development of resistance to a key herbicide used to control both serrated tussock and African lovegrass.

The confirmation of flupropanate resistance in serrated tussock (Powells 2018) and African lovegrass would not come as a surprise to those working in the field of weed ecology and the development of herbicide resistance. What is now needed is a focus on slowing down further development of herbicide resistance, thus allowing more time for additional chemistries to be identified (Gaines et al. 2021). A focused and more comprehensive approach to the application of integrated weed management principles is also required along with the consideration of any additional, new or novel control options for perennial grass weeds.

In the more challenging parts of the landscape, control of perennial grass weeds may never be cost-effective for an individual land manager and engaging with all types of land managers to prompt action regarding their responsibility for weed management remain a challenge at all levels. However, the true environmental cost of not taking action and allowing continued weed spread landscape degradation to occur cannot be overstated.

The results of these herbicide resistance studies have significant implications for how serrated tussock and African lovegrass are managed not only the Monaro, but also in other tablelands perennial pasture systems into the future. They reinforce the need for integrated approaches to be taken for the control of perennial grass weeds and the importance of herbicide resistance testing in pasture systems.

The confirmation of herbicide resistance in both serrated tussock and African lovegrass in perennial grass pastures indicates that not only is the "genie out of the bottle" as suggested by McLaren et al. (2010) but the widespread scale of the resistance across the Monaro, Southern Tablelands and more recently locations within the Central Tablelands of NSW (Upper Macquarie County Council 2022) suggests that the herbicide resistance ship has sailed, and the horse has truly bolted.

## ACKNOWLEDGMENTS

The cooperation of the many land managers from across the Monaro region, Biosecurity Officers from Snowy Monaro Regional Council and members of Monaro Farming Systems are gratefully acknowledged.

Herbicide resistance testing was undertaken by the team at Plant Science Consulting with many thanks for their assistance and feedback.

This work was funded by Local Land Services, Monaro Farming Systems, and the Boco Rock Wind Farm Community Fund.

## REFERENCES

Gaines, T.A., Busi, R., and Küpper, A. (2021). Can new herbicide discovery allow weed management to outpace resistance evolution? *Pest Management Science*, 77(7), 3036-3041.

Heap, I. (2022). The International Herbicide-Resistant Weed Database. Available www.weedscience.org (accessed 30 January 2022).

McLaren, D.A., Merton, E., Pritchard, G., Grech, C. and Bonilla, J. (2008). Evaluating the extent of serrated tussock (Nassella trichotoma) resistance to the herbicide, flupropanate in Australia. *Proceedings of the 16th Australian Weeds Conference,* eds R.D. van Klinken, V.A. Osten, F.D. Panetta and J.C. Scanlan, pp. 94-6. (Queensland Weeds Society, Brisbane).

McLaren, D.A., Grech, C.J., Bonilla, J., Butler, K. and Ramasamy, S. (2010). "Serrated tussock resistance to flupropanate in Australia – is the genie out of the bottle?", *Proceedings of the 17th Australasian Weeds Conference*. http://caws.org.nz/old-site/awc/2010/awc201012661.pdf

Noble, S. (2002). An investigation into the herbicide resistance of serrated tussock. Honours thesis, Department of Applied Biology and Biotechnology, RMIT University, Victoria, pp. 1-84.

Powells, J. (2018). Flupropanate resistant serrated tussock (*Nassella trichotoma)* identified at multiple locations in the Monaro, New South Wales. *Proceedings of the 21st Australasian Weeds Conference*, Sydney NSW. http://caws.org.nz/old-site/awc/2018/awc201813971.pdf

Ramasamy, S., McLaren, D.A., Preston, C. and Lawrie, A. (2010). Heritability of flupropanate resistance in *Nassella trichotoma*. In *Proceedings of the 17th Australasian Weeds Conference* (pp. 210-273).

Upper Macquarie County Council (2022). Herbicide Resistance in Serrated Tussock. Media Release: 20 January 2022. https://www.umcc.nsw.gov.au/news-public-exhibition/130-herbicide-resistance-in-serrated-tussock

# Scaling up and applying qualitative social research on lay knowledge in invasive plant management

Nicholas Gill[1], Laurie Chisholm[1], Jenny Atchison[1], Gina Hawkes[1], Sonia Graham[1], Shaun McKiernan[1], Lesley Head[2]
[1]University Of Wollongong, Australia,
[2]University of Melbourne, Melbourne, Australia
(ngill@uow.edu.au)

**Summary** Invasive plants are a key feature of global environmental change. Research on the social dimensions of invasive plants is proliferating. Both physical and social scientists argue for the importance of understanding the social dimensions of invasive plants and their consequences for invasive plant governance and for governance of environmental change. Much of the social research on invasive plants uses qualitative methods. This work generates valuable insights into the values and experiences of lay people and landowners, and valuably highlights complexity, context, and contingency through in-depth studies. These features, however, pose challenges to effectively incorporating research insights into policy and practice, and to acceptance in the natural sciences. To 'scale up' qualitative invasive plant research we conduct a meta-ethnography of lay experiences and perceptions of invasive plants. Meta-ethnography is a systematic meta-analysis method for qualitative research developed in education and health sciences and largely used in those fields. It aims to "produce novel interpretations that transcend individual study findings, rather than aggregate findings". Our meta-ethnography of nineteen qualitative invasive plant studies generated six meta-themes that demonstrate an underlying coherence to this research without losing the nuance of the individual studies. These meta-themes represent higher level structuring concepts to the content and findings of qualitative invasive plants research. Our synthesis makes the contribution of qualitative research on invasive plants clearer to policy makers and natural scientists. It potentially provides a platform for dialogue in invasive plants governance and for asking new research questions. Further, the meta-themes speak to lay rationalities of invasive plant assessment and management. They thus also represent capacity – inquiry, reflexivity, experimentation, observation, and engagement – to navigate the issues posed by invasive plants in uncertain environmental futures

**Keywords** Invasive plants, meta-ethnography, qualitative research, social science, lay experience, reframing, environmental management

# Social successes in weed management

Sonia Graham[1], Jenny Atchison[1], Zhao Ma[2], Leonie Miller[1], Tanya Howard[3], Tom Bach[4]
[1]University of Wollongong, Wollongong, Australia,
[2]Purdue University, West Lafayette, United States,
[3]Department of Planning, Industry and Environment, Queanbeyan, Australia,
[4]Environmental Systems Solutions, Melbourne, Australia
(sonia.graham1@gmail.com)

**Summary** Each year untold numbers of land managers, government staff and volunteers work individually and collectively to manage invasive plant species. Yet metrics of success often focus on ecological and economic outcomes, with little consideration of the social processes and outcomes. Where social metrics have been developed, they often measure the number of hours spent on surveillance and management activities or the number of participants who attended trainings and working bees. Such narrow metrics fail to account for the diverse social successes that underpin weed species management, such as establishment of social networks, connection to Country, skills development, etc. This presentation provides a critical reflection of dominant framings that guide weed management, such as risk minimisation and best practice management, and overlook the social processes and outcomes. We propose that weed management needs to be redesigned to incorporate three principles--diversity and justice, learning and failure, and social relationships--into its design. Emerging insights from other fields of conservation practice indicate that re-envisioning weed management in this way can enhance social, environmental and ecological processes and outcomes over the long-term.

**Keywords** Collaboration, participation, area-wide management, biosecurity

# Outdoor recreation and plant biosecurity in the Snowy Mountains region of NSW Australia

Jen Smart[1], Nicholas Gill[1], Leonie Miller[1], Sonia Graham[1], Isabelle Wolf[1] and Hillary Cherry[2]
[1]University of Wollongong, Northfields Ave, Wollongong, NSW, 2522
[2]NSW National Parks and Wildlife Service, 4 Parramatta Sq, 12 Darcy St Parramatta, NSW 2150

**Summary** We report on research conducted to understand how people undertaking outdoor recreation activities might perceive their role in contributing to biosecurity management. The research explored visitors and residents' knowledge, experience, meanings and practices of recreation and biosecurity in the Snowy Mountains region of NSW, Australia. An online survey investigated the routines and activities that occur when packing for, undertaking and unpacking from outdoor recreation trips. The survey looked at where, why and how people undertake some form of action to clean their recreation clothing and equipment, including how these routines might change or stay the same for different types of recreation activities. Biosecurity is the product of not just individual behaviour but is the result of social cultural factors in everyday recreation life. The results have relevance for managing landscapes where the issues of biosecurity and recreation are reflected.

**Keywords** plant biosecurity management, outdoor recreation, snowy mountains, equipment, online survey

## INTRODUCTION

Globally, recreation and nature-based tourism contributes to the risk and threat of unwanted or invasive species moving beyond their natural or normal range. Invasive microscopic algae, pathogens, and plants can move and be moved intentionally and unintentionally through ecological pathways of water, wind, soil, animals as well as via socio-economic pathways of trade and transport of goods and material, human mobility and changes to land use and via tourism and recreation (Robbins 2004, Jussaume and Ervin 2016).

Strong evidence points to the role of recreation-users as unintentional carriers of invasive plant material on their clothing and equipment (see (Whinam *et al.* 2005, Pickering and Mount 2010, Pickering *et al.* 2011, Auffret and Cousins 2013; Ansong and Pickering 2014, Smith and Kraaij 2020).

Biosecurity is a response to manage the negative impact of invasive species on the environment and the tourist economy. Biosecurity management refers to actions to identify, search for, report, clean, brush down, or otherwise support activities that prevent the spread of plant material (NSW Department of Planning and Industry 2012).

The research investigated to what extent biosecurity features in the minds and actions of outdoor recreation users in the Snowy Mountains Region. This research has wide application to all agencies managing biosecurity and recreation activity. The Snowy Mountains region has high environmental values in the face of increasing demands and pressures for recreation-tourism use and development (NSW Government Department of Environment and Conservation 2006). The research has relevance not only to managing the protected area of Kosciuszko National Park (KNP), but also across alpine regions globally where the issues of biosecurity management and recreation use are reflected.

## MATERIALS AND METHODS

The study site comprised the protected area of Kosciuszko National Park (KNP) and the adjoining Local Government Areas (LGA) of Snowy Valleys to the west of KNP and Snowy Monaro on the eastern flank of the park.

Outdoor recreation residents and visitors to the region were surveyed using on online questionnaire starting in January 2022. The survey was run until 31 May 2022. Preliminary data was downloaded 30 April 2022. The questionnaire was administered via Qualtrics. Survey distribution was via a QR code and an anonymous html link. Additionally, a website (https://recreation.study) provided potential respondents with information on the research and a link to the survey.

Printed advertising material was provided to recreation users at various points within the region including trailheads, mountain bike parks, camping areas, boat ramps and at backcountry huts in KNP. Printed material was placed at local shops, cafes and accommodation places throughout the region as well as outside the LGA. Promotion also included the use of social media sites and print media in regional newspapers. Emails were sent to recreation clubs and land/environment groups. This distribution strategy aimed to access a wide cross section of the community who may have undertaken outdoor recreation activities in the region at any time.

**Data analysis** 233 responses were analysed using SPSS v27 for descriptive and cross tabulations.

RESULTS

Descriptively, respondents were male (54 %), female (45 %) or preferred not to say. Respondents ranged in age from 20-75 years. Generally, most respondents reported as visitors (69 %) to the region who on average have been visiting for 24 years. A proportion of respondents were permanent residents whilst others were visitors with property or employment connection within the region as they own land/house, have seasonal employment or have membership within an organisation/group (Table 1).

**Table 1.** Description of respondents who have undertaken outdoor recreation activities within the Snowy Mountains region.

| Description | # | % |
|---|---|---|
| Female | 104 | 44.6 |
| Male | 126 | 54.1 |
| Prefer not to say | 2 | 0.9 |
| Resident Snowy Mtns | 31 | 13.3 |
| Own property and visit | 12 | 5.2 |
| Lodge member and visit | 24 | 10.3 |
| Live seasonally and visit | 9 | 3.9 |
| Visitor | 161 | 69.1 |

Walking/hiking was the activity most undertaken (86 %), followed by camping (66 %), cycling (38 %), nature photography (38 %), picnicking (31 %), swimming in lake/river (30 %) and fishing (26 %). The most recent activity for 60 % of visitors and residents was walking/hiking. Smaller proportions of respondents were cycling/mountain biking (11 %), hunting (9 %), combination of other land-based activities (6 %), camping (5 %), or fishing (5 %) (Table 2).

**Table 2.** Proportion of respondents who have undertaken all types of activities and their most recent activity in the Snowy Mountains region.

| Activities | All | Recent |
|---|---|---|
| Walking/hiking | 85.8 | 59.2 |
| Cycling/Mtn biking | 38.2 | 10.7 |
| Trail running | 9.0 | 1.3 |
| Orienteering | 3.4 | 0 |
| Rock climbing | 6.0 | 0 |
| Horse riding | 4.7 | 0.4 |
| 4WDriving | 12.0 | 0.9 |
| Trail bike riding | 2.1 | 0 |
| Hunting | 13.3 | 8.6 |
| Fishing | 25.8 | 4.7 |
| Boating (motorised) | 8.6 | 0 |
| Canoeing/kayaking | 24.9 | 0.4 |
| Water/wake boarding | 3.4 | 0.4 |
| Sailing | 4.7 | 0 |
| Swimming (lake/river) | 30.0 | 0.9 |
| SUP Boarding | 7.3 | 0.4 |
| Camping | 66.1 | 5.2 |
| Picnicking | 31.3 | 1.3 |
| Bird watching | 21.0 | 0.9 |
| Photography (nature) | 37.8 | 2.1 |
| Art (outside) | 2.6 | 3.4 |
| Food collecting | 6.0 | 0 |
| Other | 13.7 | 3.4 |

Respondents used a variety of roads to get to the locations they visited in the Snowy Mountains region. Most were on four-wheel drive trails in Kosciuszko NP (26 %) and Snowy Monaro LGA (31 %) and single track in Snowy Valleys LGA (22 %). In both LGAs, one in five respondents were on non-formed trails and single tracks (Table 3). Further cross tabulation indicated that a third of those on non-formed trails were walking/hiking (32 %), trail running (33 %), horse riding (100 %), hunting (55 %), fishing (18 %), camping (33 %) or bird watching (50 %).

**Table 3.** Cross-tabulation of types of roads used to get to locations in Snowy Mountains region.

| Road/trail | KNP | SV | SM | Other |
|---|---|---|---|---|
| Paved | 9.6 | 5.9 | 3.8 | 8.7 |
| Gravel | 18.6 | 13.7 | 9.6 | 17.4 |
| 4wd/service | 26.3 | 15.7 | 30.8 | 17.4 |
| Single | 19.8 | 21.6 | 19.2 | 17.4 |
| Unmarked | 10.8 | 19.6 | 13.5 | 8.7 |
| Non formed | 13.8 | 21.6 | 21.2 | 26.1 |
| Other | 1.2 | 3.0 | 1.9 | 4.3 |

With regard to the equipment being used, a majority of respondents used their own equipment when undertaking outdoor recreation activities in the Snowy Mountains (Table 4). More respondents say they wash/clean their equipment after the trip (73 %) than before (52 %). A small proportion do not wash/clean equipment when packing (10 %) or unpacking (6 %) (Table 5).

Three quarters say they remove mud/soil or plant material from their footwear (71 %), socks (57 %), clothing (56 %) whilst they undertake the recreation activity. Only a small proportion (9 %) use a disinfectant on their shoes (Table 6). In terms of using infrastructure for cleaning, less than a third use cleaning stations at the trip start and end but over forty percent say they would use it if there was a cleaning station available (Table 7).

**Table 4.** Source of equipment for the recreation activity in the Snowy Mountains region.

| Source equipment | % |
|---|---|
| Use own equipment | 98.3 |
| Borrow before leave home | 3.4 |
| Borrow/hire from tourism operator | 1.3 |
| Hire from shop in SMR | 2.1 |
| Other | 1.7 |

**Table 5.** Proportion who check, clean/wash some or all of their equipment when packing for and unpacking from an activity.

| Equipment maintenance | Packing | Unpacking |
|---|---|---|
| Check some or all | 59.7 | 49.4 |
| Wash/clean some or all | 51.5 | 73.4 |
| Dry some or all | N/A | 39.1 |
| New equipment (so no) | 6.0 | 2.1 |
| Un/pack without doing | 10.3 | 6.4 |
| Other | 7.7 | 2.6 |

**Table 6.** Proportion who have undertaken biosecurity activity whilst in the Snowy Mountains region.

| Personal equipment | % |
|---|---|
| Remove mud/soil or plant material from footwear | 70.8 |
| Disinfect footwear | 8.6 |
| Remove from socks | 56.7 |
| Remove from clothing | 55.8 |
| Remove from Velcro | 32.2 |
| Brush gaiters[A] | 23.2 |
| Other | 12.4 |
| None of the above | 17.6 |

[A] gaiters asked only of walkers/hikers, trail runners and those orienteering.

**Table 7.** Proportion who have used biosecurity infrastructure whilst in the Snowy Mountains region.

| Infrastructure use | % |
|---|---|
| Cleaning station start of trip | 27.0 |
| Cleaning station at end of trip | 17.6 |
| Would use but not available | 45.1 |
| Carry own brush/equipment | 10.3 |
| Never use a cleaning station | 9.4 |
| Other | 15.5 |

## DISCUSSION

The wide range of multiple land and water-based activities being undertaken by respondents reflects the nature-based touristic value of the Snowy Mountains region. Outdoor activities are based mostly in Kosciuszko National Park where there are high environmental values. Activities also take place across the Snowy Valleys and Snowy Monaro LGAs with trails that cross the borders of KNP and the LGAs. Recreation is therefore important to the region and protecting the environmental and touristic values requires consideration of biosecurity management.

Recreational users undertake biosecurity activities or actions when packing, unpacking and when in the Snowy Mountains region. Around half of the respondents reported they checked, cleaned/washed or dried their equipment when packing and unpacking for activities. Cleaning usually occurred when unpacking rather than at the packing stage. A study undertaken previously in KNP also found respondents more likely to clean after a walk rather than before (Gill *et al.* 2020). The online survey found that cleaning of equipment can be about the upkeep of equipment rather than a response to biosecurity management actions. Some respondents reported that cleaning routines when packing/unpacking are often about checking and maintaining equipment for serviceability whilst being ready for the next recreation outing. As one respondent stated '*It is important not to overstate this. While I do clean and dry equipment as necessary, it is with a view to keeping equipment in good repair, rather than ensuring biological cleanliness.*' Hence there is key link between the cleaning for maintenance and biosecurity purposes in managing land use.

Additionally, checking equipment when unpacking can also reflect the weather conditions, length of trip and the type of trail that recreationalist experienced during their activity. More than half of respondents removed mud/soil or material from shoes, socks and clothing during their recreational activity. This action, removing plant material may reflect the high use of non-formed or unmarked trails where plant material can potentially attach to shoes and clothing. Respondents reported generally that they are using a wide variety of trails including gravel, single and four-wheel drive tracks. These trail types may increase the likelihood of soil or mud attaching to footwear and clothing, depending on the weather. Thus, biosecurity management needs to consider trail conditions and drainage where these might contribute to how biosecurity actions are carried out and routinised.

Further understanding of biosecurity actions in the Snowy Mountains region, places consideration on those who use fixed in-situ cleaning infrastructure whilst undertaking recreation. The survey results indicated that less than a third used a cleaning station at the start of their trip and even less at the end of the trip. Although, a higher proportion of respondents reported that they would use infrastructure if it was available. Anecdotally, during the survey promotion, people indicated that they would use infrastructure as it was a reminder for them of the biosecurity issues in the region. This was similarly noted in the study by Gill *et al.* (2020). However, observations and interactions at some sites indicated that use was low where the infrastructure was positioned to the side of the track. Therefore, whilst cleaning infrastructure can contribute to uptake of cleaning it needs to be located in such a way that people want to use it.

The results presented in this paper contribute to understanding when, why and how people undertake some form of action to clean their recreation clothing and equipment. The doing or not doing of biosecurity actions such as check, clean/wash and/or dry occur for a range of reasons pertinent to equipment needs and locations visited. The results have relevance for the design and management of walking and mountain bike trails within the Snowy Mountains region. Design can impact the extent to which drainage contributes to build-up of mud sections. Design can also contribute to the provision of integrated cleaning systems at trailheads or nearby which may help people to maintain their equipment but also lead to biosecurity action such as removal of potential invasive plant material.

It is likely that the results may be limited by the use of the self-reporting questionnaire as people may disclose information with a response bias. This bias may contribute to reliability issues. Further study within this research project will consider qualitative research methods that dig deeper to understand the validity and reliability of the study and contribute to understanding the sociocultural impacts on biosecurity management within the Snowy Mountains region.

ACKNOWLEDGMENTS

The research was impossible without the generosity of respondents completing the survey. The research is part of the Hawkweed Eradication Program and was supported financially by this program through the NSW National Parks and Wildlife Service.

REFERENCES

Ansong, M., and Pickering, C. (2014). 'Weed seeds on clothing: A global review', *Journal of Environmental Management*, 144, 203-11.

Auffret, A.G., and Cousins, S.A.O. (2013). 'Humans as long-distance dispersers of rural plant communities', *PLoS One*, 8(5), e62763-e62763.

Gill, N., McKiernan, S., Lewis, A. Cherry, H. and Annunciato, D. (2020). 'Biosecurity hygiene in the Australian high country: footwear cleaning practices, motivations and barriers among visitors to Kosciuszko National Park', *Australasian Journal of Environmental Management*, 27(4), 378-95.

Jussaume, R.A. and Ervin, D. (2016). 'Understanding weed resistance as a wicked problem to improve weed management decisions', *Weed Science*, 559-69.

NSW Department of Planning and Industry. (2012). *New South Wales Biosecurity Strategy 2013-2021*, by NSW Department of Planning and Industry, NSW Government.

NSW Government Department of Environment and Conservation. (2006). *Kosciuszko National Park Plan of Management*, Sydney South.

Pickering, C. and Mount, A. (2010). 'Do tourists disperse weed seed? A global review of unintentional human-mediated terrestrial seed dispersal on clothing, vehicles and horses', *Journal of Sustainable Tourism*, 18(2), 239-56.

Pickering, C.M., Mount, A., Wichmann, M.C. and Bullock, J.M. (2011). 'Estimating human-mediated dispersal of seeds within an Australian protected area', *Biological Invasions*, 13(8), 1869-80.

Robbins, P. (2004). 'Comparing Invasive Networks: Cultural and Political Biographies of Invasive Species', *Geographical Review*, 94(2), 139-56.

Smith, K. and Kraaij, T. (2020). 'Research note: Trail runners as agents of alien plant introduction into protected areas', *Journal of Outdoor Recreation and Tourism*, 31, 100315.

Whinam, J., Chilcott, N. and Bergstrom, D.M. (2005). 'Subantarctic hitchhikers: expeditioners as vectors for the introduction of alien organisms', *Biological Conservation*, vol. 121(2), 207-19.

# A regional perspective on NZ's National Collective funding model for Biocontrol of weeds research

Shane Hona[1]
[1]Bay of Plenty Regional Council, Rotorua, New Zealand
(shane.hona@boprc.govt.nz)

**Summary** The National Biocontrol collective (NBC) is a collaborative partnership that has been funding biocontrol of weeds research in New Zealand since 2002. It is comprised of Regional Councils, unitary authorities and the Department of Conservation, and funds a national biocontrol programme. Funding is pooled and collaborative decision making is undertaken annually to determine where best to focus these resources. Since its inception, 25 biocontrol agents have been released, targeting 14 different weed species. Bay of Plenty Regional Council has been a member of this collective since its establishment, and this presentation provides a regional perspective on the history, track record, achievements and challenges of the NBC. In addition, future challenges and opportunities are discussed, focusing on how can the delivery of biocontrol agents be improved, and potential options to increase funding for the research and delivery of biocontrol tools.

**Keywords** Biocontrol of weeds, collective funding model

# Validating the New Zealand Biocontrol Risk Model for Australia: Systematic surveys for non-target host use

Jackie Steel[1], Hasan Rahmani[1], John Ireson[2], Dianne B.J. Taylor[3], Quentin Paynter[4], Caitlin Selleck[1], Francesco Martoni[1], Raelene M. Kwong[1]
[1] Department of Jobs, Precincts and Regions, AgriBio Centre, Bundoora, Victoria, Australia.
[2] TasWeed Biocontrol, Hobart, Tasmania, Australia.
[3] Department of Agriculture and Fisheries, Ecosciences Precinct, Brisbane, Queensland, Australia.
[4] Manaaki Whenua – Landcare Research, Auckland, New Zealand.

**Summary** This research aims to determine whether weed biological control agents attack non-target plants that supported some development in host specificity testing. Field surveys were conducted for agents released to control six major Australian weeds: ragwort (*Jacobaea vulgaris* Gaertn.) English broom (*Cytisus scoparius* (L.) Link), gorse (*Ulex europaeus* L.), alligator weed (*Alternanthera philoxeroides* (Mart.) Griseb.), Montpellier broom (*Genista monspessulana* (L.) L.A.S.Johnson), and St John's wort (*Hypericum perforatum* L.). A total of 10 agents, including mites and insects, previously released for biological control were examined, and host specificity checks were conducted. Overall, 15 potential alternative host plants were examined for agent attack as well as the target plants. All agents are well established in Australia and had been released from 12 to 44 years earlier. There was no evidence that the non-target species included in this survey were able to support full development of the agents, although an incidental finding of non-target host use requires further survey effort.

**Keywords** biological control, non-target, host specificity testing, systematic surveys.

## INTRODUCTION

The biggest risk from introducing a weed biocontrol agent from one country to another is that the agent could damage non-target plants, especially crops or native species. Globally, of the 332 weed biocontrol agents that have established, 60 (18 %) were recorded causing non-target attack (Schwarzländer *et al.* 2018, Hinz *et al.* 2019). In Australia the incidence is lower with 11 agents recorded causing non-target damage of the 142 that have established (8 %) (Hinz *et al.* 2019, Winston *et al.* 2021).

Despite this concern, non-target surveys are rarely carried out and are usually reactive; in response to a recorded attack, opportunistic establishment, or incidental finds (Hinz *et al.* 2014, 2019). The literature generally agrees that major non-target impacts would already have been observed and reported if they had occurred. Systematic surveys for non-target damage in New Zealand did not discover any further agents causing major damage to valued species than those already recorded in the literature (Fowler *et al.* 2000, 2004, Paynter *et al.* 2004, 2015, 2018).

Whilst the risk of non-target attack is real, biocontrol is a useful tool that has a proven track record of reducing the impact of some of the most intractable weed problems (Schwarzländer *et al.* 2018). A better understanding of the circumstances under which biocontrol agents cause non-target damage will improve the risk analysis process that underpins the approval of biocontrol releases (Paynter *et al.* 2015). To this end we used a literature survey to develop a list of non-target plant species on which certain biocontrol agents were able to develop under laboratory conditions. We then used field surveys to determine if they were attacked by the agent after release.

## MATERIALS AND METHODS

Surveys were conducted during 2020-2022 in Tasmania, Victoria, and Queensland in areas where each of the agents were known to have become widely established on the target weed. The target weed was first examined for the presence of the agent and, if present, the nearest non-target plant located was examined. Methods used to detect the presence of agents on both the target and non-target plants included field and laboratory examination of invertebrates collected using a pooter after beating the target plant over collection trays, the presence of webbing, frass, larvae, pupae and associated stem damage, galls, or webbed larval shelters.

Some non-target plants were also returned to the laboratory, cut into sections, and placed in Tullgren funnels (width 30 cm diameter; depth to grid 10 cm; distance from grid to 2.5 cm opening, 50 cm; 60 W incandescent light source 25 cm from grid) for 3-4 days and any extracted invertebrates examined under a compound microscope.

Non-target plants included in the survey list were both native: *Dillwynia glaberrima* Sm., *Hypericum gramineum* G.Forst., *Oxylobium ellipticum* (Labill.) R.Br., *Pultenaea juniperina* Labill., *Senecio quadridentatus* Labill., *S. linearifolius* A.Rich., *S. pinnatifolius* var. *alpinus* (Ali) I.Thomps., *S. pinnatifolius* var. *maritimus* (Ali) I.Thomps.; and naturalised: *Alternanthera sessilis* (L.) R.Br. ex DC., *Cytisus proliferus* L.f., *Genista monspessulana*, *Lupinus angustifolius* L., *L. arboreus* Sims, and *Spartium junceum* L.

## RESULTS

Surveys were conducted for 29 combinations of biocontrol agents and the non-target plants they were able to develop on under laboratory conditions. For 24 agent/non-target combinations no non-target plant use was observed (Table 1). Biocontrol agents were observed on five non-target plant species (Table 2). *Sericothrips staphylinus* was located on *P. juniperina* during spring 2020 but not during follow up surveys in early and late autumn 2021 when the agent was still active on the target weed. *Arytinnis hakani* was located on *C. scoparius* and *S. junceum* in Victoria. Surveys in Tasmania did not find *A. hakani* present on *S. junceum* whilst the agent was active on the target weed within 20 m of the non-target. One *B. villosus* was netted from *C. palmensis*. In Queensland, surveys for *A. hygrophila* on *A. sessilis* resulted in observations of the agent on *A. denticulata* R. Br. at one site.

## DISCUSSION

All the biological control agents included in the survey were well established in each state and are widespread, having been released from 12 to 44 years earlier. The biological control agents released on *S. jacobaea* have had a highly significant impact on their target plants (Ireson *et al.* 2007). For instance, in southern Tasmania, *S. jacobaea* was once a major problem in pastures throughout the Huon Valley and is now confined to a few minor infestations as well as in orchards where insecticides are used. Densities of the biological control agents have declined accordingly, but the agents are still active and usually present where ragwort can be located. Given the decline in *S. jacobaea* densities, alternative hosts would have been open to agent exploitation if the agents were not host specific to ragwort. However, there was no evidence that any of the biocontrol agents targeted in the survey could complete their life cycle on any of the potential alternative hosts examined during the survey.

Although adults of *S. staphylinus* were found on *P. juniperina* in spring, they were not found on samples taken the following autumn *when S. staphylinus* adults were still active on gorse (Ireson *et al.* 2008). This suggests that *S. staphylinus* had crossed over to *P. juniperina* from neighbouring gorse during spring dispersal (Ireson *et al.* 2008). Its absence from the later samples is indicative that it cannot complete its life cycle on this plant.

**Table 1**. Summary of surveys for biocontrol agents where non-target host use was not observed.

| Non-target plant | No. survey sites | Distance from agent (km) |
|---|---|---|
| Agent: *Aceria genistae* Nalepa | | |
| *Cytisus proliferus* | 2 | 0-0.5 |
| *Lupinus angustifolius* | 1 | 0 |
| *Lupinus arboreus* | 4 | 0-2 |
| Agent: *Bruchidius villosus* Fabricius | | |
| *Genista monspessulana* | 1 | 0 |
| Agent: *Arytinnis hakani* Loginova | | |
| *Lupinus arboreus* | 3 | 0.5-2.8 |
| Agent: *Chrysolina quadrigemmina* Suffrian | | |
| *Hypericum gramineum* | 1 | 0.01 |
| Agent: *Cochylis atricapitana* Stephens | | |
| *Senecio quadridentatus* | 3 | 5-5.5 |
| Agent: *Platyptilia isodactyla* (Zeller) | | |
| *S. pinnatifolius* var. *alpinus* | 1 | 22 |
| *S. pinnatifolius* var. *maritimus* | 1 | 24 |
| *S. linearifolius* | 3 | 0.5-1.9 |
| Agent: *Agonopterix umbellana* (Fabricius) | | |
| *Cytisus proliferus* | 2 | 0.3-5 |
| *Genista monspessulana* | 4 | 0-1.7 |
| *Lupinus arboreus* | 1 | 1 |
| *Pultenaea juniperina* | 1 | 0.015 |
| Agent: *Sericothrips staphylinus* Haliday | | |
| *Cytisus proliferus* | 2 | 0.5 |
| *Dillwynnia glaberrima* | 1* | 0.2 |
| *Lupinus angustifolius* | 1 | 2 |
| *Lupinus arboreus* | 5 | 1-2 |
| *Oxylobium ellipticum* | 2 | 1 |
| Agent: *Tetranychus lintearius* Dufour | | |
| *Cytisus proliferus* | 2 | 0.5 |
| *Lupinus angustifolius* | 1 | 1 |
| *Lupinus arboreus* | 5 | 1-2 |
| *Oxylobium ellipticum* | 2 | 1 |
| *Pultenaea juniperina* | 2 | 0-0.1 |

* Site surveyed twice, 9 months apart

**Table 2.** Detail of surveys where non-target host use was observed

| Non-target plant | Survey site | Survey date | Distance from agent (approx.) | Agent present on non-target? |
|---|---|---|---|---|
| ***Agasicles hygrophila* Selman & Vogt targeting *Alternanthera philoxeroides*** | | | | |
| *Alternanthera denticulata* | Coorparoo, Brisbane (Qld) | 28/10/2020 | 8 km | N |
| | Cliveden Ave, Corinda, Brisbane (Qld) | 13/01/2021 | 2 km | N |
| | Cliveden Ave, Corinda, Brisbane (Qld) | 3&10/03/2021 | 2 km | Y |
| | Cliveden Ave, Corinda, Brisbane (Qld) | 12/05/2021 | 2 km | Y |
| | Kendall St, Corinda, Brisbane (Qld) | 10/03/2021 | 3 km | N |
| | Eddystone Rd, Oxley, Brisbane (Qld) | 10/03/2021 | 4 km | N |
| | Bill Moore Park, Brisbane (Qld) | 10/03/2021 | 2 km | N |
| ***Bruchidus villosus* Fabricius targeting *Cytisus scoparius*** | | | | |
| *Cytisus proliferus* | Beechworth (Vic) | 23/11/2021 | 0 km | Y* |
| ***Arytinnis hakani* Loginova targeting *Genista monspessulana*** | | | | |
| *Cytisus scoparius* | Thomas's lookout, Daylesford (Vic) | 10/02/2021 | 0 km | Y |
| *Spartium junceum* | Botanic Gardens, Daylesford (Vic) | 25/12/2020 | 1 km | Y |
| | Botanic Gardens, Daylesford (Vic) | 10/02/2021 | 1 km | N |
| | Reservoir, Melbourne (Vic) | 5/10/2021 | 3 km | Y |
| | Dynnyrne (Tas) | 13/02/2021 | 20 m | N |
| | Dynnyrne (Tas)` | 12/02/2022 | 20 m | N |
| ***Sericothrips staphylinus* targeting *Ulex europaeus*** | | | | |
| *Pultenaea juniperina* | Mount Nelson (Tas) | 20/03/2020 | 100 m | N |
| | Mount Nelson (Tas) | 30/11/2020 | 15 m | Y |
| | Mount Nelson (Tas) | 15/03/2021 | 15 m | N |
| | Mount Nelson (Tas) | 18/05/2021 | 15 m | N |

*One beetle collected from non-target. Follow up survey required to confirm utilisation (or not).

Similarly, *A. hakani* was not found on *S. junceum* in Daylesford in a follow up survey. The non-target had finished flowering by that stage, and it is likely that the agent was dispersing. At the location where *A. hakani* was collected from *C. scoparius* the non-targets were interspersed with the target. Although the agent appeared less abundant on the non-target, follow up surveys at sites where the two species are more distant from each other would clarify whether the agent can fully utilise this non-target as a host.

The single *B. villosus* collected from *Cytisus proliferus* likely indicates that non-target host use is occurring on this species in Australia as it does in New Zealand but follow up surveys are required to confirm this (Syrett 1999).

The discovery of *A. hygrophila* on *A. denticulata* was incidental. This native non-target plant was not included in the survey list as host testing data for this species was not available. Native plants were not routinely included in host testing in the 1970s when this agent was introduced to Australia. This non-target host use was not unexpected, given *A. denticulata* is attacked in NZ, and unpublished post-release testing demonstrated it could be a host. Larvae feed on this species but are unable to pupate on *A. denticulata* as it lacks hollow stems.

Aside from this incidental discovery, none of the agents observed or collected from non-target plants was having a major impact on those species. This confirms the notion that surveying for non-target host use is unlikely to uncover major damage to valued plants. The damage to the native plant species *A. denticulata* has potentially been overlooked as being attack on the invasive look-a-like *A. sessilis* (Heenan and de Lange 2004).

The results of this survey will be included, alongside previously published survey data, in an analysis of the comparison between the host testing results for an agent and its ability to utilise non-target species as hosts.

ACKNOWLEDGMENTS

This project was funded by Agriculture Victoria, the Australian Government Department of

Agriculture, Water and the Environment as part of its Rural R&D for Profit Program and AgriFutures Australia. We would like to thank Michael Day for providing literature and Andrew McConnachie for advice on survey candidates.

REFERENCES

Fowler, S.V., Syrett, P. and Hill, R.L. (2000). Success and safety in the biological control of environmental weeds in New Zealand. *Australian Journal of Ecology* 25, 553–562.

Fowler, S.V., Gourlay, A.H., Hill, R.H. and Withers, T. (2004). Safety in New Zealand weed biocontrol: a retrospective analysis of host-specificity testing and the predictability of impacts on non-target plants. Proceedings of the XI International Symposium on Biological Control of Weeds, eds. J.M. Cullen, D.T. Briese, D.J. Kriticos, W.M. Lonsdale, L. Morin, and J.K. Scott, pp. 265–270. (Canberra (Australia): CSIRO Entomology).

Goeden, R.D. and Ricker, D.W. (1986). Phytophagous insect faunas of the two most common native *Cirsium* thistles, *C. californicum* and *C. proteanum*, in southern California. *Annals of the Entomological Society of America* 79, 953-962.

Heenan, P.B. and de Lange, P.J. (2004). *Alternanthera denticulata* (Amaranthaceae) in New Zealand: a new addition to the indigenous or naturalised flora? *New Zealand Journal of Botany* 42, 739-745.

Hinz, H.L., Schwarzländer, M., Gassmann, A. and Bourchier R.S. (2014). Successes we may not have had: a retrospective analysis of selected weed biological control agents in the United States. *Invasive Plant Science and Management* 7, 565-579.

Hinz, H.L., Winston, R.L. and Schwarzländer, M. (2019). How safe is weed biological control? A global review of direct nontarget attack. *The Quarterly Review of Biology* 94, 1-27.

Ireson, J.E., Davies, J.T., Friend, D.A., Holloway, R.J., Chatterton, W.S., Van Putten, E.I. and McFadyen, R.C. (2007). 'Weeds of pastures and field crops in Tasmania: economic impacts and biological control. Technical Series no. 13' (CRC for Australian Weed Management, Adelaide).

Ireson, J.E., Gourlay, A.H., Holloway, R.J., Chatterton, W.S., Foster, S.D. and Kwong, R.M., (2008). Host specificity, establishment and dispersal of the gorse thrips, *Sericothrips staphylinus* Haliday (Thysanoptera: Thripidae), a biological control agent for gorse, *Ulex europaeus* L. (Fabaceae) in Australia. *Biological Control* 45, 460-471.

Johnson, D.M., and Stiling, P.D. (1998). Distribution and dispersal of *Cactoblastis cactorum* (Lepidoptera: Pyralidae), an exotic Opuntia-feeding moth, in Florida. *Florida Entomologist* 81, 12-22.

Paynter Q., Fowler S.V., Gourlay A.H., Haines M. L., Harman H.M., Hona S.R., Peterson P.G., Smith, L.A., Wilson-Davey J.R.A., Winks C.J., and Withers T.M. (2004). Safety in New Zealand weed biocontrol: a nationwide survey for impacts on non-target plants. *New Zealand Plant Protection* 57, 102–107.

Paynter Q., Fowler S.V., Gourlay A.H., Peterson P.G., Smith L.A. and Winks C.J. (2015). Relative performance on test and target plants in laboratory tests predicts the risk of non-target attack in the field for arthropod weed biocontrol agents. *Biological Control* 80, 133–142.

Paynter Q., Fowler S.V. and Groenteman R. (2018). Making weed biological control predictable, safer and more effective: perspectives from New Zealand. *BioControl* 63, 427–436.

Schwarzländer M., Hinz H.L., Winston R.L. and Day M.D. (2018). Biological control of weeds: an analysis of introductions, rates of establishment and estimates of success, worldwide. *BioControl* 63, 319–331.

Syrett P. (1999) Broom beetle gets taste for tagasaste. *Biocontrol News and Information* 20, 51.

Turner, C.E., Pemberton, R.W. and Rosenthal, S.S. (1987). Host utilization of native *Cirsium* thistles (Asteraceae) by the introduced weevil *Rhinocyllus conicus* (Coleoptera: Curculionidae) in California. *Environmental Entomology* 16, 111-115.

Winston, R.L., M. Schwarzländer, H.L. Hinz, M.D. Day, M.J.W. Cock and M.H. Julien, Eds. (2021). Biological Control of Weeds: A World Catalogue of Agents and Their Target Weeds. Based on FHTET-2014-04, USDA Forest Service, Forest Health Technology Enterprise Team. Available online at https://www.ibiocontrol.org/catalog/ [Accessed 4 January 2021].

# Assessing the risks of biological control to crop and ornamental cultivars

Greg Lefoe[1,2], Cindy E. Hauser[2], Jackie Steel[1], Umar Lubanga[1], Anthony T. Slater[1], Raelene Kwong[1], Libby Rumpff[2]
[1]Agriculture Victoria, AgriBio, Centre for AgriBioscience, Bundoora, Australia,
[2]School of BioSciences, The University of Melbourne, Parkville, Australia
(greg.lefoe@agriculture.vic.gov.au)

**Summary** The introduction of a new biological control agent to Australia is preceded by rigourous risk assessment. For weed biological control, there is a need to estimate the possible consequences of introduction to native plants, crops and ornamentals. This is a particularly challenging task for weeds such as silverleaf nightshade *Solanum elaeagnifolium* that have many closely-related native and crop species that could be at risk. In these cases, it is necessary to prioritise non-target plant species for testing against new biocontrol agents. However, crops such as potato *S. tuberosum* and tomato *S. lycopersicum* have thousands of cultivars that could also vary in their susceptibility to damage. In these cases, cultivars must also be selected and prioritised for testing. Despite this requirement, we could find no detailed or consistently applied guidelines for selecting which cultivars to test. We propose a decision support tool to prioritise cultivars for biocontrol agent host-specificity testing. We demonstrate the application of the decision tool on a large, complex cultivar group, and argue that our approach will result in cultivar test lists that are transparent, defensible, and feasible to study.

**Keywords** Biological control, silverleaf nightshade, risk assessment, cultivars

# Glyphosate resistance in Australian cotton farming systems, what are the surveys telling us? The then and now.

Eric A. Koetz[1], MD. Asaduzzaman[1], Jeff A. Werth[2] and Graham W. Charles[3].
[1] NSW Department of Primary Industries, WWAI, PMB, Pine Gully Road Wagga Wagga NSW, 2650
[2] Leslie Research Centre, Queensland Department of Primary industries, Toowoomba, 4350
[3] NSW Department of Primary industries, ACRI, Narrabri, NSW 2390

**Summary**
During the last 20 years several industry wide surveys have been conducted to determine the frequency and distribution of common weeds in Australian cotton farming systems. A shift in species occurrence has been recorded with problematic broad leaf weeds replaced with hard to control grass weeds. The three most common in-crop weeds from a 1991 grower survey were, Noogoora burr, Cyperus species and Bathurst burr (Charles, 1991). In 2001 Peachvine, Bladder ketmia and Nutgrass were the top three (Charles, 2001) and in 2010 Flaxleaf fleabane, Sowthistle and Peachvine were the most abundant (Walker *et al.* 2010). In 2015 the three most common weeds recorded were feathertop Rhodes grass, Awnless barnyard grass and Windmill grass.

In addition to the shift in species occurrence, weed surveys conducted since 2015 have confirmed high levels of glyphosate resistance in five hard to control weeds in cotton farming systems. In 2015-16, >95% of Flaxleaf fleabane populations were resistant and in 2017-18, >75% of populations were resistant. For grass weeds Awnless barnyard grass populations from 2015 (>70%), 2017 (65%) and 2018 (>66%). Feathertop Rhodes grass populations are of the most concern with an increase in glyphosate resistance measured from 20% in 2015, 35% in 2017 and 40% in 2018. Resistance levels in Windmill grass were 45% from 2017 and 2018 whilst samples from 2015 were 90% resistant. The high levels of glyphosate resistance are concerning and reinforce the importance of the cotton industry's Herbicide Resistance Management Strategy (HRMS) which promotes a dynamic, integrated approach to weed control focusing on preserving the long-term efficacy of glyphosate.

**Keywords** glyphosate resistance, species-shift, grass weeds, Herbicide Resistance Management Strategy.

## INTRODUCTION

The introduction of Roundup Ready® varieties to Australia in the 2000-01 cotton season was quickly adopted by the industry and now encapsulates 99% of all cotton areas sown. The rapid adoption of this technology around the world coincided with a shift toward minimum and no till farming (Givens, *et al.* 2009). In conjunction with changes to tillage, the inclusion of genetically modified (GM) crops around the world has seen a shift in species composition (Culpepper 2006, and Werth et.al.2013).

Many conventional tactics including in-crop tillage, pre-emergent residual herbicides and hand chipping were replaced with glyphosate as the primary form of weed control. The reliance on glyphosate, often as the only active ingredient used, has resulted in weed species shifts and the evolution of glyphosate resistant weeds in GM crops (Johnson *et al.* 2009). The Australian cotton industry has followed a similar pattern with broadleaf weeds replaced by grass weeds as the most problematic and hard to control.

In the 1990s Australian cotton farmers used a range of chemical and mechanical tools to control weeds in their fields. Weed control relied on large inputs of mostly residual herbicides followed by the inevitable hand chipping to control escapes or survivors. Charles reported in 1991 that the cost of weed control from a survey of growers in 1989 averaged $187 per hectare with herbicides accounting for $76 and chipping $67. In 2018 the Australian Cotton Comparative Analysis reported herbicide cost was $134 per hectare and chipping was $4 per hectare. The relative reduction in costs can be attributed to the introduction of Roundup Ready® Cotton in the early 2000's. This system evolved quickly to a relatively simple weed management plan heavily reliant on glyphosate. However, the impressive control provided by glyphosate may also turn out to be its Achilles heel.

In the United States widespread resistance to glyphosate developed within 10 years of the introduction of glyphosate tolerant varieties. The Australian cotton industry is approaching the 20-year mark since the introduction of Roundup Ready® cotton, and as such a need was identified to quantify the resistance status of key weeds in Australian fields.

Industry surveys have been conducted to try and get an understanding of the level of glyphosate resistance in cotton fields. In a 2016 survey growers reported that they suspected resistance (approximately 74%) to glyphosate, however resistance testing is only at relatively low levels (30%). A follow up survey in 2019 has seen an increase in suspected resistance, with 85% of growers reporting issues with weed control. This is supported with 52% of growers confirming glyphosate resistant weeds in the 2020-21 season (CCA Qualitative report, 2020-21).

To validate these grower results funding from the Cotton Research and Development Corporation and NSW Department of Primary Industries allowed researchers to collect samples from cotton fields for herbicide resistance testing to quantify these findings.

## MATERIALS AND METHODS

In the 2015-16 cotton season a random survey was conducted sampling 144 fields on 50 farms across 7 cotton farming regions in Queensland and NSW. Sampling was coordinated to occur after post-emergent application of glyphosate to collect seed from survivors and a record of other weeds present at each site was noted. Targeted weed surveys from 43 fields were conducted in 2017 and 70 samples were collected across cotton growing regions in 2018/19. In addition to the collection of weed seeds a record of the incidence of weed species was also compiled for each field.

Seeds of these populations were sown on soil surface of plastic pots (25 cm in diameter) pre-filled with potting mix with the top 2 cm field soil. Seeds were placed on the substrate surface, watered, and covered with paper towel and maintained in a glasshouse. Plants from each pot were transplanted to trays with same peat substrate (6 alternating spots on the tray) at two to four-leaf stage. Each population had 18 experimental units (6 plants or units per replication). When seedlings were at the rosette stage (8-10 cm diameter for broadleaf weeds and early tillering for grass weeds), they were sprayed with 0.68 kg ae $ha^{-1}$ of glyphosate,http://www.wssajournals.org/doi/full/10.1614/WT-D-09-00080.1 - n3#n3 which is a commonly used rate for general fallow weed control in Australia (Walker *et al*. 2011).

The herbicides were applied using an automated laboratory sized cabinet sprayer with a moving boom applying a water volume of 77 L $ha^{-1}$ equivalent from a flat fan nozzle at 300 kPa pressure. The irrigation was turned off one day prior to spraying and turned back on one day post spray application. Trays were arranged in a completely randomized design with three replications. Seedling survival was assessed at 28 days after treatment (DAT) using a scale ranging from 0% (zero control or no difference from control) to 100% (complete control or plant death). Total number of survived plants for each population was counted and converted to a percentage value at 28 DAT. Populations with plant survival >20% after spraying was considered ''resistant'. Plant survival of 10% to 19% was ''developing resistance" and populations with plant death and necrosis >90% or survived less than 10% considered as "susceptible".

## RESULTS

Results from the 2015 survey for sowthistle recorded over 20% assessed as resistant or developing resistance to glyphosate and remaining at similar levels across all three surveys (2017/18 survey only had 6 viable samples tested). Very high levels of glyphosate resistance (>95%) were recorded in Fleabane in 2015 and greater than 75% in 2016 and it appears to have a naturally high tolerance to glyphosate, and as such has not been included in further surveys. Windmill grass and Barnyard grass are either developing or have high levels of resistance to glyphosate (Table 1).

The biggest concern from the surveys has been the increase in glyphosate resistance in feathertop Rhodes grass samples collected with an increase in three seasons from 20% to 40% of samples resistant (Table 1).

**Table 1.** Percentage of populations of six problem weeds resistant to glyphosate from surveys in 2015 to 2018.

| Weeds | 2015/16 | 2016/17 | 2017/18 |
|---|---|---|---|
| Fleabane | 97 | 75 | Not tested |
| Sowthistle | 22 | 10 | 28 |
| Barnyard grass | 72 | 65 | 57 |
| Windmill grass | 90 | 45 | 44 |
| Feathertop Rhodes grass | 20 | 35 | 40 |
| Annual ryegrass | Not tested | Not tested | 83 |

From the historical to the most recent surveys there has been a significant shift in species prominence from the broadleaf weeds of the pre-Roundup Ready® era to an increase in grass weeds (Table 2). The increase in glyphosate resistance in grass weeds follows a similar pattern to that observed in minimum till broadacre grain systems in northern NSW. In the latest surveys we are now starting to see fleabane and sowthistle hard to control, especially in dryland cotton systems where minimum or zero tillage dominates.

Table 2. The three most common weed species through time in Australian cotton fields

| 1991 | 2001 | 2010 | 2016 |
|---|---|---|---|
| Noogoora burr | Peachvine | Flaxleaf fleabane | feathertop Rhodes grass |
| Cyperus spp | Bladder ketmia | Sowthistle | Awnless barnyard gass |
| Bathurst burr | Nutgrass | Peachvine | Windmill grass |

## DISCUSSION

The surveys have identified weeds that are developing glyphosate resistance. The reliance on glyphosate as the main weed control tactic has placed increased selection pressure from using a single mode of action. The result is a species shift from fields with mostly broadleaf weeds (Werth et.al.2013) to the recent surveys showing an increase in grass weeds. Over 80% of growers are using glyphosate as the only knock down herbicide prior to planting, this coupled with in-crop applications, places a lot of pressure on glyphosate to do all the heavy lifting for weed control. Many weeds are now proving difficult to control in a glyphosate dominant system.

Consequently, we have seen a shift in species occurrence with grass weeds the most common survivors of glyphosate applications. In response we are seeing a move to more diverse weed control tactics including pre-emergent, in-crop and layby herbicides with different modes of action. The addition of targeted tillage, shrouded sprayers and optical spray technology has added extra diversity and robustness to weed control.

Results from the latest CCA surveys report growers are using a diverse range of herbicides across a number of MOA's The CCA qualitative report from 2020-21 reports that almost 70% of growers are applying a pre plant residual and 43% are applying a residual herbicide at planting. Additionally, 64% of growers are incorporating more than two and three other modes of action into their weed control program, however what is concerning is that we still see 33% of growers using glyphosate plus only one other mode of action.

The Australian cotton industry has been very proactive in developing a stewardship program around integrated weed management. It is just as important to target the non-cropping phase of the rotation and implement robust and diverse tactics for weed control, including the use of at least two non-glyphosate herbicides in fallow and in crop (Thornby *et el.* 2013). The addition of in-crop tillage has also proven to be a useful tool for controlling late emerging weeds in crop and controlling survivors of herbicide application. This approach is the cornerstone of the (HRMS) developed by the cotton industry.

As a result of the stewardship adopted by Australian cotton growers, we still have good efficacy with glyphosate. We have detected a species shift toward grass dominant weed communities which have proven difficult to control with glyphosate alone. Of greatest concern is the increase in glyphosate resistance for feathertop Rhodes grass. This hard to control weed is spreading across the whole farming system and is now the weed of greatest concern amongst growers.

The weed survey results are important to guide decision making now and into the future as we grapple with developing glyphosate resistance.

The importance of an integrated approach to weed management. is critical to protect the long - term efficacy of glyphosate. As always, the focus remains on controlling all survivors from glyphosate applications

## ACKNOWLEDGMENTS

The assistance of NSW DPI technical staff in collection and preparation of samples is appreciated. The role of CottonInfo Regional Extension Officers in sourcing fields and facilitating grower access is appreciated. Funding was provided by NSW DPI and CRDC.

## REFERENCES

Charles, G. W., 1991. A grower survey of weeds and herbicide use in the New South Wales cotton industry. *Australian Journal of Experimental Agriculture,* 1991,**31**,387-92

Charles GW, Taylor IN (2004) Herbicide resistance and species shift in cotton: the need for an integrated weed management (IWM) approach. Pages 817-828 in Swanepoel A, ed. Proceedings of the World Cotton Research Conference–3. Cape Town, South Africa: Agricultural Research Council – Institute for Industrial Crops

Crop Consultants Australia (2021) Qualitative report on the 2020-21 cotton season, A survey of consultants. CRDC

Culpepper, A. (2006). Glyphosate-Induced Weed Shifts. *Weed Technology, 20*(2), 277-281.

Givens.W., Shaw.D., Kruger.G., Johnson.W., Weller.S., Young.B., Wilson.R., Owen.M., and Jordan.D. 2009, Survey of Tillage Trends Following the Adoption of Glyphosate-Resistant Crops **Vol 23**, Weed Technology Pg 150-155.

Johnson.W.G., Davis.V.M, Kruger.G.R and Weller.S.C. (2009) Influence of glyphosate-resistant cropping systems on weed species shifts and glyphosate-resistant weed populations. Europ. J. Agronomy **31** 162–172

Thornby David, Werth Jeff, Walker Steven (2013) Managing glyphosate resistance in Australian cotton farming: modelling shows how to delay evolution and maintain long-term population control. *Crop and Pasture Science*.**64**, 780-790.

Walker.S.,[A] Werth.J.,[A] McDonald.C.[B] and Charles.G.[C] (2010) Changes in weeds and practices since the introduction of herbicide tolerant cotton in Australia [A]Agri-Science Queensland, Department of Employment, Economic Development and Innovation, PO Box 2282, Toowoomba 4350 [B] Monsanto, PO Box 433 Narromine NSW 2821 [C]I&I NSW, Australian Cotton Research Institute, Narrabri. Australian Cotton Conference.

Walker.S., Bell.K., Robinson.G., and Widderick.M. (2011) Flaxleaf fleabane (*Conyza bonariensis*) populations have developed glyphosate resistance in north-east Australian cropping fields. Crop Protection. Vol 30, Issue 3, 2011, Pages 311-317,

Werth.J., Boucher.L., Thornby.D, Walker.S, and Charles.G. (2013) Changes in weed species since the introduction of glyphosate-resistant cotton. *Crop & Pasture Science*, 2013, **64**, 791-798

# Effectiveness of glufosinate, dicamba and clethodim on glyphosate-resistant and susceptible populations of five key weeds in Australian cotton systems

Jeff Werth[1], David Thornby[2], Michelle Keenan[1], James Hereward[3], Bhagirath Singh Chauhan[4]
[1]Queensland Department Of Agriculture And Fisheries, Toowoomba, Australia,
[2]Innokas Intellectual Services, Upper Coomera, Australia,
[3]The University of Queensland, St Lucia, Australia,
[4]Queensland Alliance for Agriculture and Food Innovation, Gatton, Australia
(jeff.werth@daf.qld.gov.au)

**Summary** Cotton with herbicide resistance to dicamba and glufosinate in addition to glyphosate (XtendFlexTM) may soon become commercially available in Australia. These traits will allow two additional modes of action to be applied over-the-top of the cotton crop. We proposed that an effective way to include these herbicides as part of an integrated weed management strategy is to adopt the double knock tactic with glufosinate as the follow-up herbicide in place of paraquat which is commonly used in fallow. In a glasshouse experiment that was repeated, treatments containing glyphosate, dicamba and clethodim (for grasses) and glyphosate mixtures with dicamba or clethodim were applied with the follow-up glufosinate applied at 1, 3, 7 and 10 days after initial applications. These combinations were applied to glyphosate-resistant and -susceptible populations of *Conyza bonariensis*, *Sonchus oleraceus*, *Chloris virgata*, *Chloris truncata* and *Echinochloa colona*. Total control of *Conyza bonariensis* was achieved with dicamba and glyphosate+dicamba followed by glufosinate at all timing intervals. Effective control of Sonchus oleraceus was also achieved dicamba and glyphosate+dicamba followed by glufosinate and all timing intervals. Effective control of *Chloris virgata* was achieved with glyphosate, clethodim or glyphosate+clethodim followed by glufosinate 7 and 10 days later. Control of *Chloris truncata* was inconsistent, with the most effective treatment being glyphosate+clethodim followed by glufosinate 10 days later. *Echinochloa colona* was controlled with all treatments apart from glyphosate alone on the glyphosate-resistant population. This experiment showed no consistent evidence of reduced control of glyphosate-resistant populations with dicamba, clethodim or glufosinate when used alone or in combination compared to glyphosate-susceptible populations for each species. As a result, the addition of dicamba and glufosinate in XtendFlexTM cotton should prove beneficial when used in combination in-crop along with existing weed control tactics.

**Keywords** Double knock, weed control, integrated weed management

# Using the critical period for weed control to establish a weed threshold in irrigated cotton

Graham W. Charles[1] and Ian N. Taylor[2]
[1]Weed Biosecurity Research, NSW Department of Primary Industries, Locked Bag 1000, Narrabri 2390, New South Wales, Australia
[2]Cotton Research and Development Corporation, PO Box 282, Narrabri 2390, New South Wales, Australia
(graham.charles@dpi.nsw.gov.au)

**Summary** Pest control thresholds are widely used in various disciplines and have been the aspiration of weed scientists. However, weed control thresholds have been challenging to establish as the results from competition experiments, on which thresholds are based, are almost always specific to species, site and season. This is especially true for crops where seasonal conditions, in particular rainfall, have an overwhelming influence on both weed and crop growth, such that the density of weeds required to cause economic damage to the crop can vary widely over years. This level of seasonal variation is generally not seen in fully-irrigated cotton crops in Australia, where inputs are managed to optimise cotton lint yield. Hence, quantifying a weed control threshold for these crops should be feasible.

The results from our research culminated in a dynamic, multi-species weed competition model, relating relative crop yield to weed height and biomass, together with the times of weed emergence and weed removal. Additional research has modelled the effect of successive germination and weed control events during the cropping year. These models enable cotton growers to determine the yield loss caused by a given weed population at any stage during crop growth and to estimate the cost of delaying weed control. A variable weed control threshold can be applied to the models, triggering control decisions according to crop value and the cost of control.

**Keywords** mimic weeds, weed succession, yield-loss threshold.

## INTRODUCTION

The modern Australian cotton industry began in 1961, with 120 ha of cotton planted near Wee Waa, NSW. Competition from weeds and flooding were major issues for this crop and only 26 ha was harvested. Two years later, the first large-scale planting occurred, with 1700 ha of cotton planted, but only 400 ha harvested, with the remainder lost to unmanageable competition from weeds (Jones and Shaw 2014, Marshall 2015).

Over the next four decades, cotton growers developed an integrated approach to weed management, combining residual and over-the-top herbicides, cultivation, hand-hoeing, crop rotations and fallow weed control to develop a robust weed management system. This approach gave acceptable levels of management for most weeds and the weed seedbank declined over time, but the system was expensive and not infrequently resulted in crop damage, often from residual herbicides (Taylor *et al.* 2003). The weed management system was also largely prescriptive, with heavy reliance on residual herbicides applied prior to crop planting in anticipation of weed issues, and the use of in-crop tools was largely limited to the first half of the crop season due to crop size and the need to avoid trafficking wet soils.

The introduction in the 2001/2 season of cotton varieties including the Roundup Ready® trait, which confers tolerance to glyphosate, and the later introduction of the Roundup Ready® Flex trait in 2007/8, radically changed in-crop weed management for most Australian cotton crops, with glyphosate use replacing most or all other herbicides and many in-crop cultivation passes (Werth *et al.* 2013).

The use of glyphosate-tolerant varieties simplified in-crop weed management for cotton and contributed to the increase in yields seen in Australian crops. However, the ideal timing of in-crop glyphosate applications with Roundup Ready Flex cotton was unclear. Glyphosate is a broad-spectrum herbicide that can control nearly all in-crop weeds and can be applied at up to 1 kg ai. $ha^{-1}$ at any stage of crop growth, with up to four in-crop applications allowed per season. Most weeds can be controlled with this robust rate of glyphosate at any growth stage, provided the weeds are actively growing, which is the normal situation in irrigated cotton. Hence, weed management with glyphosate in cotton is not limited by the typical window constraining herbicide applications to small, actively growing weeds, and the ideal timing for glyphosate applications in irrigated cotton is not well defined.

The need to optimise the timing of in-crop glyphosate applications is important to minimize crop yield losses due to weed competition, thus optimising the level of weed control achieved from a maximum of four spray applications, but also to ensure herbicide is not applied unnecessarily, increasing production costs and the environmental footprint of cotton production.

**Defining a weed control threshold** A series of competition experiments was conducted in cotton crops between 2003 and 2015 at the Australian Cotton Research Institute, Narrabri NSW, using three mimic weeds: Japanese millet (*Echinochloa esculenta*), a 'grass weed'; mungbean (*Vigna radiata*), a medium-sized 'broadleaf weed'; and sunflower (*Helianthus annuus*), a large 'broadleaf weed'. The mimic weeds were planted to achieve densities of 1 to 200 plants $m^{-2}$, planted with the cotton or at predetermined times following crop emergence and removed later in the season. Charles and Taylor (2007a) explored the potential of a series of published competition models to define a weed control threshold for irrigated cotton in Australia using this data. They found that although existing models could be used to define a threshold, the outputs from the models were invariably season and weed species specific, needing to be redefined for each weed species and season. More sophisticated crop growth models are also available and could be used to develop a weed control threshold, but Dean et al. (2003) found that the greatly increased complexity of these models did not improve the accuracy of the model's outputs.

Charles and Taylor (2007a) subsequently develop a multi-species statistical model that related crop yield loss (*Y*) as a function of crop growing degree days (*T*), weed height (*H*), weed leaf area (*A*), and crop height (C), where:

$$Y=0.0297+0.000282T+0.00199H+0.00161A^{0.5}+0.00234C$$

While this statistical model was a valuable step towards developing a weed threshold model for cotton, the inclusion of weed leaf area in the model makes the model difficult for cotton growers to use.

Charles and Taylor (2007b, 2007c) later released a weed control threshold to the Australian cotton industry for the 2007/8 season using the critical period for weed control (CPWC), based on weed density and weed type (large broadleaf weeds, medium broadleaf weeds and grass weeds). These threshold models represented a further big step in developing a weed control threshold for cotton, but were species (weed type) specific and required cotton growers to estimate the average density of a dominant weed type in a paddock. Also, they were not predictive, in that they did not predict when an under-threshold weed population would reach threshold, and they assumed all weeds emerged at the same time, not allowing for differing weed size from successive germination events.

The thresholds were based on a set yield-loss of 1% of crop yield (Charles and Taylor 2007c). The yield-loss threshold (YLT) was determined by a combination of the predicted bale value and cotton yield, together with the applied cost of weed control, and will vary, responding to changes in the value of the crop and the cost of the control tool under consideration. The models of Charles and Taylor (2007c) used a 1% YLT based on 2007 values and did not allow the YLT to be varied.

**The critical period for weed control (CPWC)** The concept of the CPWC was first developed by Nieto et al. (1968) and has since been applied to a wide range of weeds in a range of crops.

The CPWC is developed from crop yield loss data obtained by: (1) allowing weeds to emerge with the crop and then removing these weeds at intervals during crop growth, and; (2) allowing weeds to emerge at intervals during crop growth and compete till harvest, when the subsequent crop yield is recorded. Analysis of this data enables the period of the season during which the crop is most sensitive to weed competition to be described, relating the yield losses from weed competition to a YLT.

The CPWC is delimited by the critical time of weed removal (CTWR) and the critical weed free period (CWFP). The CTWR defines the period during which the crop can tolerate early-season competition before unacceptable crop losses occur, and the CWFP defines the period during which the crop needs to be kept weed free to avoid unacceptable

**Figure 1.** The CPWC for 50 large weed (sunflower) plants per $m^2$ in cotton. The CPWC (shaded area) extended from 49 to 937 growing degree days after crop planting. From Charles and Taylor (2021).

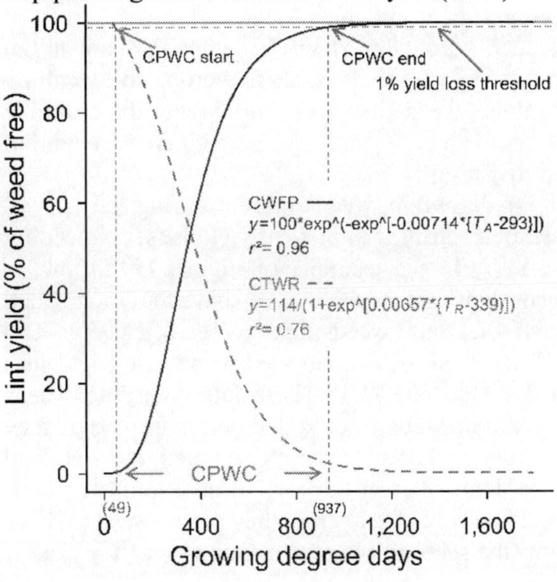

losses (Charles and Taylor 2021). The CPWC is delimited by the intersections of the CTWR and CWFP lines with the YLT. An example CPWC relationship taken from Charles and Taylor (2021) is shown in Figure 1.

The value of the CPWC approach was further enhanced by Charles *et al.* (2021) who developed a multispecies threshold model by including weed height and weed biomass in extended logistic and Gompertz equations. Their relationships allow a cotton grower to determine the CPWC for any species or mix of species of weeds of size from 1 to 200 cm tall, weighing 10 to 2000 g $m^{-2}$, although the models were only tested on three mimic weeds. The value of these relationships could be further enhanced if growth curves were developed for the major weeds of cotton. Including growth curves would allow the relationships to be used predictively, identifying when in the future a weed population which is under the YLT would grow to exceed the YLT.

These models (Charles *et al.* 2021) go a long way to achieving the aim of delivering a weed control threshold to Australian cotton growers that can be applied in the field, although they require the input of data on weed height and weed biomass for each field. However, the relationships do not allow for successive germination or weed control events during the cropping season.

**Successive weed germination and understanding the CPWC relationships** The CPWC concept is based around the competitive effect of weeds that emerge in a single cohort with the crop and are removed at some time post-emergence, or weeds that emerge in a single cohort post-crop emergence and continue to grow throughout the season. However, in a cotton field, in the absence of heavy rates of residual herbicides, there is normally an initial emergence of weeds with the crop, followed by additional weed emergence throughout the cropping season, with emergence flushes triggered by rainfall and irrigation events.

By definition, weeds present during the CPWC must be controlled to prevent yield losses exceeding the YLT. In the example of Figure 1, this means weeds that emerge with the crop have to be controlled by 49 GDD and weeds that emerge after 937 GDD will not need to be controlled to prevent yield loss greater than the YLT. These late emerging weeds may still need to be controlled before harvest as they can host pests and diseases, contaminate lint, and cause difficulties at harvest. In addition, all weeds should be controlled before they can set seed to drive down the seedbank over time and reduce the risk of herbicide resistance developing.

Thus, the in-field understanding of the CPWC has been that in-crop weed management needs to commence by the start of the CPWC and should be maintained until the end of the CPWC, as weeds present during the CPWC cause economic damage exceeding the YLT.

This understanding, however, does not take into account the impact of multiple weed emergence and weed removal events. Taking the relationships of Figure 1 as an example, weeds that emerge at 936 GDD are emerging within the CPWC, and so will need to be controlled to prevent a yield loss that exceeds the YLT. However, these weeds will not compete sufficiently with the crop that they need to be controlled by 937 GDD (the end of the CPWC), as the weeds will still be at cotyledon stage at this time. By definition, the competitive effect of these weeds will not be sufficient to require their control until just before harvest, as conversely, weeds that emerge at 938 GDD don't compete sufficiently to require removal during crop life.

Charles and Taylor (2021) explored the relationship between successive weed germination events and the CPWC using the mimic weed, sunflower, in irrigated cotton. They were able to describe weed succession in their CPWC models by including additional terms in the CTWR equations. These new equations were able to define the CPWC for weeds emerging or being controlled at any time during crop life. An example CPWC relationship

**Figure 2.** The CPWC for 2 large weed plants (sunflower) per $m^2$ in cotton. Example CTWR curves are shown for weeds emerging 0, 200, 400 and 600 GDD after crop planting. The relationship can generate curves for weeds emerging at any time in the season. From Charles and Taylor (2021).

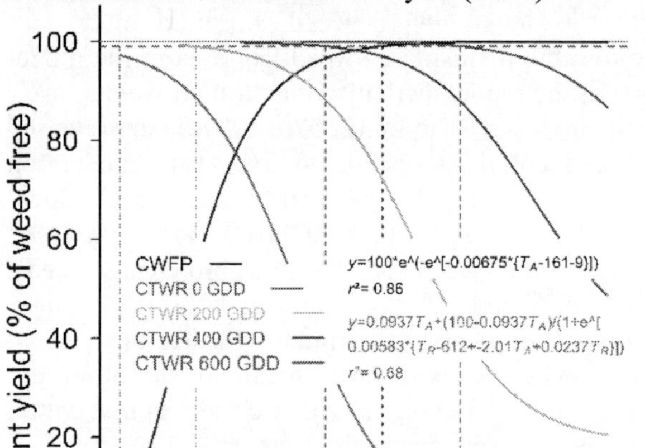

with later weed emergence events taken from Charles and Taylor (2021) is shown in Figure 2.

In this example (Figure 2), the CTWR curve for weeds that emerge at 200 GDD exceeds the YLT at 249 GDD, that is, these weeds do not need to be controlled until 249 GDD, even though the CPWC is 0 – 861 GDD. Weeds that emerge at 600 GDD, do not compete sufficiently to cause yield loss exceeding the YLT until 1118 GDD, some 257 GDD after the end of the CPWC. Hence, these curves for later emerging weeds are not defining the CPWC, they are defining the critical time of weed control.

**Delivering a CPWC relationship for cotton growers** Our aim is to deliver to Australian cotton growers a CPWC relationship that includes models allowing for multiple weed emergence and in-season control events (Charles and Taylor 2021), into a multispecies threshold model (Charles *et al.* 2021), with the addition of growth models for the major weeds found in cotton production. Incorporating these approaches will deliver a weed threshold model that is multispecies, allows for successive weed emergence and control events, and is predictive. A combined model will be of immense value to cotton growers in allowing them to optimise their weed control inputs.

We propose that such a model could be delivered to cotton growers using a spread-sheet interface that would allow them to determine their YLT according to their expected crop yield and value, and the cost of their anticipated weed control input. The model would be limited by the number of weed growth models available but could provide valuable guidance around weed thresholds delivered to the paddock. Differences in the growth rate of crops and weeds over the now geographically widely spread cotton industry in Australia are allowed for in the models by the use of growing degree days as the measure of time, making these models equally valuable throughout the industry.

ACKNOWLEDGMENTS

Our research was made possible through the support of the NSW Dept. Primary Industries and the Cotton Research and Development Corporation. We also acknowledge and thank the many support staff who contributed to this work.

REFERENCES

Charles, G. and Taylor, I. (2007b). Understanding the "critical period for weed control" concept. *The Australian Cottongrower* 28, 48-50.

Charles, G. and Taylor, I. (2007c). Applying the "critical period for weed control" in the field. *The Australian Cottongrower* 28, 51-53.

Charles, G.W. and Taylor, I.N. (2007a). Developing a threshold model for controlling weeds in glyphosate resistant cotton. Proceedings of the 4th World Cotton Research Conference, ed M. Stephens, pp 22. (International Cotton Advisory Committee, Lubbock, Texas, USA).

Charles, G.W. and Taylor, I.N. (2021). Extending the critical period for weed control model to better include weed succession using common sunflower as a mimic weed in high-yielding cotton. *Weed Technology*, 35, 1029-1037.

Charles, G.W., Sindel, B.M., Cowie, A.L. and Knox, O.G.G. (2021). Developing a multispecies weed competition model for high-yielding cotton. *Weed Technology* 35, 202-698.

Deen, W., Cousens, R., Warringa, J., Bastiaans, L., Carberry, P., Rebel, K., Riha, S., Murphy, C., Benjamin, L.R., Cloughley, C., Cussans, J., Forcella, F., Hunt, T., Jamieson, P., Lindquist, J. and Wang, E. (2003). An evaluation of four crop : weed competition models using a common data set. *Weed Research* 43, 116-129.

Jones, P. and Shaw, W.S. (ed) (2014). A life in cotton: the birth of the modern cotton industry. A documentary. Paper Moose Film. https://vimeo.com/102796184. Accessed: May 27, 2020.

Knezevic, S.Z., Evans, S.P., Blankenship, E.E., Van Acker, R.C. and Lindquist, J.L. (2002). Critical period for weed control: the concept and data analysis. *Weed Science* 50, 773-786.

Marshall, A. (2015). Cotton's man of fibre dead at 96. Farmonline National. https://www.farmonline.com.au/story/3380377/cottons-man-of-fibre-dead-at-96/. Accessed: May 27, 2020.

Nieto, J.H., Brondo, M.A. and Gonzales, J.T. (1968). Critical periods of the crop growth cycle for competition from weeds. *Pest Articles and News Summaries* (C) 14, 159-166.

Taylor, I., Charles, G. and Inchbold. B. (2003). Reducing residual pre-emergent or pre-plant herbicide use in cotton through the development of weed thresholds. Proceedings of the 3rd World Cotton Research Conference, ed A. Swanepoel, p. 835-845. (Agricultural Research Council – Institute for Industrial Crops, Pretoria, Cape Town, South Africa).

Werth, J., Boucher, L., Thornby, D., Walker, S. and Charles. G. (2013). Changes in weed species since the introduction of glyphosate-resistant cotton. *Crop and Pasture Science* 64:791-798.

# A community approach to tackling Hudson pear

Jo-Anna Skewes[1], Andrea Fletcher-Dawson, Peter Dawson[3], Wendy Gibney[4], Mat Savage[5], Andrew McConnachie[6], Royce Holtkamp[7]
[1]Northern Slopes Landcare Association, Australia,
[2]Castlereagh Macquarie County Council, Australia,
[3]Tamworth Agricultural Institute, Australia,
[4]Wollongbar Primary Industries Institute, Australia,
[5]Castlereagh Macquarie County Council, Walgett Shire, Australia,
[6]NSW Department of Primary Industries, Australia,
[7]Horizon Ecological Consulting, Australia
(nwcactus@nsla.net.au)

**Summary** Hudson pear is an invasive cactus in Australia and is one of eight species in the genus listed as Weeds of National Significance (WONS). Various agencies are currently working together to implement an integrated management strategy to tackle this menace, including Northern Slopes Landcare Association NW Cacti Control Co-ordinator (CCC), North West Local Land Services (NW LLS), Castlereagh Macquarie County Council (CMCC) and NSW Department of Primary Industries (DPI). The strategy seeks to reduce the distribution and density of Hudson pear using chemical (herbicides) and biological (biocontrol) approaches, which in turn will benefit the environment, agriculture, mining and tourism industries. One of the challenges with implementing the strategy was the diverse community which is affected by Hudson pear. Before formal community consultation began, the biggest barrier faced was the 'blame game' and the feeling of hopelessness considering the scale of the Hudson pear incursion. The perceptions and concerns of the community needed to be explored in order to bring about the behaviour change required to give them ownership of the solutions to the problem. To achieve this, workshops and surveys were deployed to help guide community engagement with biocontrol and chemical control for Hudson pear. A total of 48 people from the affected areas attended the workshops and 38 completed the survey. Respondents included active Landholders, land managers, small business owners, miners and others who manage Hudson pear. Identified behaviours included: mapping, reporting, spraying, releasing biocontrol agents, and monitoring. Surveys included questions about barriers for each individual behaviour. Barriers identified in the surveys were addressed by education and awareness programs (including facilitation around issues or concerns, providing support with control and overall best practice management) to great success. The community as a result has taken ownership and is on board with the plan to tackle Hudson pear.

**Keywords** Hudson pear, community engagement, community, biocontrol

# Buffel grass invasion across arid lands: evidence of impacts on ecological and cultural values

Ellen Ryan-Colton[1], Christine Schlesinger[1], John Read[2], Glenda Wardle[3], Kris French[4]
[1]Charles Darwin University, Alice Springs, Australia,
[2]University of Adelaide, Adelaide, Australia,
[3]University of Sydney, Sydney, Australia,
[4]University of Wollongong, Wollongong, Australia
(ellen.ryan.colton@gmail.com)

**Summary** Across arid ecosystems, where traditionally weeds have been slower to take hold, highly invasive grasses are dominating the landscape and negative impacts are emerging. Exotic grass invasion is changing fire regimes and reducing floristic diversity, however, the more indirect impacts on animals, ecosystem processes and cultural values of indigenous peoples whose lives are closely intertwined with country have not been well documented. A̲nangu people in the A̲nangu Pitjantjatjara Yankunytjatjara (APY) lands in north-west South Australia are concerned about the rapid invasion of buffel grass (*Cenchrus ciliaris*) throughout much of their ancestral lands. Culturally A̲nangu see invaded areas as buffel grass desert where wildflowers and food plants now do not exist. In response, our research aims to investigate the ecological impacts of invasion on animal and plant communities, including aspects which are important to A̲nangu. Birds, reptiles, small mammals, invertebrates including ants, and vegetation were all re-monitored at sites originally surveyed in 1994 and 1995 before buffel grass took hold, with some sites now heavily invaded and some are still free from invasion. We worked collaboratively with A̲nangu to document traditional ecological knowledge of how landscapes have changed with buffel grass invasion and how this relates to the survey results. Drone imagery was also used assess the level of invasion for each site. Preliminary results show whilst plant communities are converging into a similar habitat type with buffel grass invasion, animal communities are varied in their response. A̲nangu knowledge, such as concern for the loss of trees through buffel-fuelled fires, and the change in vegetation structure which affects their ability to hunt bush foods, is combined with the dataset to identify species and values most at risk from buffel invasion. Drone imagery provides strong evidence for how buffel grass is dominating arid ecosystems at the patch and landscape scale.

**Keywords** Impacts, Cultural values, biological survey, grasses, arid ecology, Indigenous

# Alinytjara Wilurara Landscape Board and Indigenous Rangers protecting millions of hectares of pristine desert from Buffel Grass invasion

Oak Valley Rangers[3], Adam Wood[1], Stephen Rapp[2]
[1]Alinytjara Wilurara Natural Resource Management Board, Australia,
[2]Maralinga Tjarutja, Australia,
[3]Oak Valley Ranger team
(adam.wood3@sa.gov.au)

**Summary** The Alinytjara Wilurara (AW) Landscape Board region in north west of South Australia covers more than 280,000 km², stretching from the Northern Territory and West Australian borders south to the Great Australian Bight. The land is mostly dedicated to conservation and traditional Aboriginal use and occupation. This includes Aṉangu Pitjantjatjara Yankunytjatjara (APY) Lands, Maralinga Tjarutja, Yalata, and co managed Parks and Reserves. Buffel grass (*Cenchrus ciliaris* and *C. pennisetiformis*), or mamu tjanpi/tjanpi kura (Pitjantjatjara: devil grass/bad grass) is an introduced perennial tussock grass species that has emerged as a significant threat to the culture and safety of remote communities in the region. This unwanted invader has colonised large areas of the APY Lands. It outcompetes native grasses, shrubs and threatens woodlands, communities and infrastructure with destruction caused by high fuel load fires. Heavy infestations prevent traditional hunting, foraging and cultural activities. It establishes a dense monoculture, unsuitable as habitat and unpalatable to wildlife. It's now recognised as one of the worst environmental threats in Australia's arid rangelands. The Great Victoria Desert (GVD) is the largest desert in Australia and contains significant biodiversity and cultural assets. Early intervention in Buffel grass control has ensured that it remains one of last desert regions in the rangelands to have a limited distribution of buffel grass. This is both a challenge and opportunity for all stakeholders including Indigenous ranger groups. Project coordination has many challenges due the area's vastness, remoteness, weather dependencies combined with minimal resources to undertake the work. Increased regional collaboration and the establishment of indigenous ranger groups has led to AW Landscape Board committing to new Buffel Free GVD project to be funded over the next 3 years. This presents the means to implement a coordinated approach across WA, NT and SA and apply the lessons learnt from the last 10 years to reduce the spread and to control/eradicate Buffel grass in the GVD. This presentation will discuss this new planned approach.

**Keywords** Buffel grass, *Cenchrus ciliaris*, Great Victoria Desert, Oak Valley Rangers, Maralinga Tjarutja, traditional hunting, cultural activities, Alinytjara Wilurara

# New insights from population genomics into the invasive *Lantana camara* L species complex

Patricia Lu-Irving[1], Francisco Encinas-Viso[2], Jason Callander[3], Michael Day[3], Johannes Le Roux[4]
[1]Research Centre for Ecosystem Resilience, Australian Institute of Botanical Science, Royal Botanic Garden Sydney, NSW, Australia
[2]Centre of National Australian Biodiversity Research, CSIRO, Canberra, ACT, Australia
[3]Department of Agriculture and Fisheries, GPO 267, Brisbane, QLD, Australia
[4]School of Natural Sciences, Macquarie University, Sydney, NSW, Australia

**Summary** Extensive morphological diversity in the invasive *L. camara* species complex has resisted clear taxonomic characterisation, yet molecular studies to date have revealed minimal genetic structure. Analysis of thousands of genome-wide markers successfully detects differentiation among sub-lineages within the complex, revealing that two of the morphological variants in Australia comprise multiple, distinct lineages. The common pink-flowering morphotype appears to be monophyletic, whereas the pink-edged red flowering morphotype does not. Pending further investigation and expanded sampling, these findings hold promise for improving weed management through a deeper understanding of the systematics of the *L. camara* complex (e.g., by enabling selection of biological control agents that are best matched with target host populations).

**Keywords** biological control, biosecurity, invasive species, weeds

## INTRODUCTION

*Lantana camara* is weed of national significance in Australia and has been a target for biological control for over 100 years. However, the diversity of forms that make up this species complex has limited the success of biological control programmes, with natural enemies performing better on some varieties than others (Day et al. 2003a). Resolving the evolutionary diversity within the *L. camara* species complex is expected to lead to improved management of these globally significant weeds (e.g., through better biological control agent-host matching). However, natural species boundaries within the complex have been obscured by anthropogenic hybrid introductions, and subsequent spontaneous hybridisation among native and naturalised plants. Studies to date of both morphological and molecular variation in *L. camara* have yet to reveal clear and consistent patterns which can be used to identify invasive populations across the complex's entire geographic distribution.

Within Australia, *L. camara* is thought to consist of multiple taxa (Smith & Smith, 1982) and varietal groups (Day et al. 2003b): suggesting that the complex comprises distinct sub-lineages that are divergent enough to warrant recognition. However, molecular studies to date of up to ~200 genetic markers (Scott et al. 1997, Watts 2009) and Sanger sequencing of 16 nuclear and chloroplast loci (Watts 2009) have found insufficient evidence to support this view.

Here, we report preliminary results from analysis of genome-wide markers in the *L. camara* complex (hereafter referred to by the common name, "lantana"). We used a Genotyping-By-Sequencing (GBS) approach to discover thousands of bi-allelic Single Nucleotide Polymorphisms (SNPs) in 101 Australian and extra-Australian lantana samples. We analysed these data to investigate whether there is detectable genetic structure in Australian lantana, and, if so, the extent to which it aligns with flower colour (the morphological character most used to define subgroups within lantana). Day et al. (2003b) described five broad varietal groups based on flower colour: common pink, pink-edged red, red, orange, and white. Since these concepts have yet to receive formal taxonomic recognition, we use the term "morphotype" to refer to them.

## MATERIALS AND METHODS

**Sampling** Healthy leaf tissue was collected in the field and preserved by desiccation. Most samples analysed here were collected from the invaded range in eastern Australia (87 samples across 23 sites from latitude 37.0°S to 17.5°S ); 6 plants were sampled per morphotype per site at 10 sites (9 sites with 1 morphotype, 1 site with 2); 1 plant was sampled per site at 12 sites. The remainder of the samples originated from northern and western Australia (3), Hawaii (2), South Africa (2), and the Americas where the genus *Lantana* is native (North America: 5; Caribbean: 1; South America: 1).

Sampling focused on the common pink-flowering morphotype (52 samples over 15 sites), and the pink-edged red flowering morphotype (16 samples over 6 sites).

**Genotyping** Tissue samples were submitted to Diversity Arrays Pty Ltd. (DArT) for DNA extraction, sequencing, and SNP calling using the

DArTseq GBS pipeline; this approach is described in detail by Rossetto et al. (2019).

**Population genomic analysis** All data processing and analyses were conducted in the R environment (R Core Team 2020). The SNP loci returned by DArT were filtered by reproducibility, representation among samples, and independence; SNPs passed filter if they were >96% reproducible, missing from =<20% of samples, and not co-located on sequencing tags.

The filtered data were analysed using Splitstree v4.14.6 (Huson & Bryant 2006) to infer a splits graph of evolutionary relationships among samples. This approach enabled representation of a non-treelike (i.e. potentially reticulate) evolutionary history (Huson & Bryant 2006), which is to be expected in lantana, given its history of extensive hybridisation.

For sites/morphotypes where six samples were collected, pairwise population $F_{ST}$ values were calculated using the relative beta estimator implemented in the "SNPrelate" v1.20.1 package (Zheng et al. 2012). T providing a measure of genetic differentiation among populations.

## RESULTS

The raw SNP matrix consisted of 101 samples and 60,634 loci, 42,958 (71%) of which were found in less than 80% of samples (i.e., were missing from 20% of samples or more). After filtering, 10,847 SNPs remained in the final data set.

Multiple, distinct lineages were resolved in a phylogenetic network (Fig. 1). The most clearly defined lineage consisted entirely of eastern Australian common pink-flowering lantana (the morphotype with broadest geographic distribution; "pink " hereafter).

Genetic differentiation (pairwise $F_{ST}$) between populations of pink lantana was close to zero, even over large geographic distances (>1,000 km). In comparison, $F_{ST}$ between populations of the pink morphotype and pink-edged red morphotype ("PER" hereafter) were very high, even in areas where they occurred in close proximity or in sympatry (Table 1).

**Figure 1 (opposite).** Phylogenetic network of 101 lantana samples inferred from 10,847 SNPs. Clusters of eastern Australian pink-flowering and pink-edged red flowering lantana samples are indicated, as are samples from the native range.

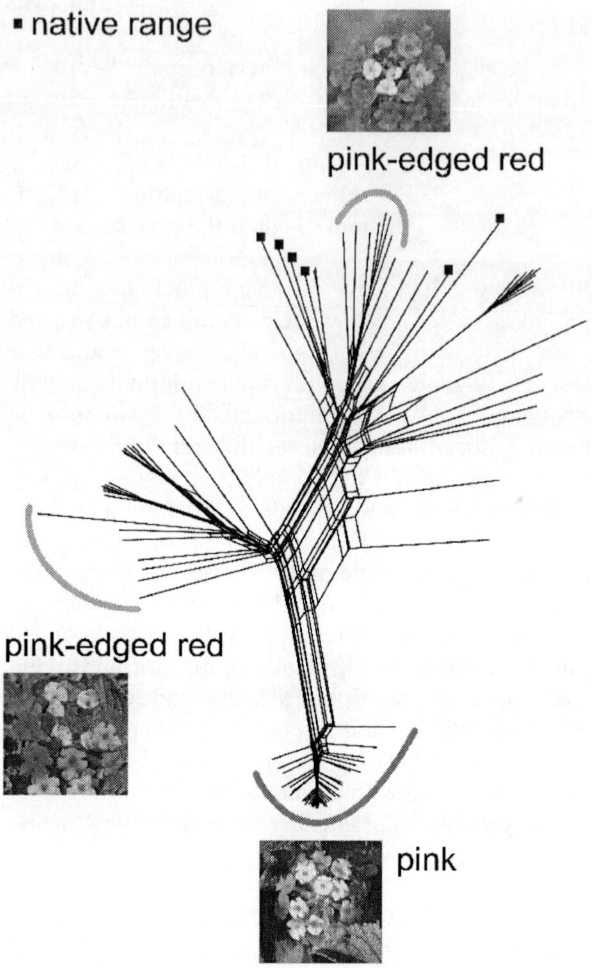


**Table 1.** Pairwise $F_{ST}$ and geographic distance (km) among selected populations of pink-flowering and pink-edged red flowering morphotypes of *Lantana camara*. Boonah was the one site at which two morphotypes were sampled in sympatry; of the other sites with 6 samples per morphotype, Mt. Fisher was most geographically distant from Boonah.

| | MT FISHER (PINK) | BOONAH (PINK) | BOONAH (PER) |
|---|---|---|---|
| MT FISHER (PINK) | 0 KM | 1,369 KM | 1,369 KM |
| | $F_{ST}$= 0 | $F_{ST}$= -0.03 | $F_{ST}$= 0.56 |
| BOONAH (PINK) | - | 0 KM | 0.32 KM |
| | - | $F_{ST}$= 0 | $F_{ST}$= 0.51 |

## DISCUSSION

The unprecedented quantity of data yielded by our GBS approach enabled detection of patterns in genomic variation which was not possible with prior molecular approaches. Our analysis of genome-wide markers provided clear evidence of distinct lineages within Australian lantana (i.e., monophyletic groups which are genetically divergent from other such groups). That 71% of SNPs discovered by DArTseq were not found across all samples is consistent with this interpretation (i.e. a large proportion of loci were unique to one lineage or a subset of lineages).

The lineages revealed show strong correspondence with described lantana morphotypes (*sensu* Day et al. 2003b). This is in contrast with prior molecular studies reporting limited differentiation among morphotypes (Scott et al. 1997; Watts 2009), which we attribute to the methodological advances our study was able to apply. We report here on our findings specifically concerning the two most widespread morphotypes in Australia, for which we had greatest sampling effort: pink (52 samples) and PER (16 samples); further work is ongoing which will report in greater detail on a more comprehensive sample including other morphotypes as well as extra-Australian populations.

Samples of pink lantana were phylogenetically distinct from other samples (Fig. 1), and we found no genetic differentiation between populations of this morphotype (even between populations >1,000 km apart; Table 1). However, populations of pink and PER morphotypes were strongly differentiated, even when in sympatry (Table 1). We conclude that the common pink-flowering morphotype corresponds with a distinct lineage, a meaningful biological and evolutionary unit. On the other hand, the PER morphotype does not meaningfully identify a lineage, with at least two distinct and divergent groups of plants appearing to share this phenotype (Fig. 1).

**Biological control implications** Since the pink lantana morphotype represents a uniform host genetic background, agents which prefer it (e.g. *Prospodium tuberculatum* (Spegazzini) Arthur) would be expected to affect different populations of this morphotype consistently. However, agents preferring the PER morphotype (e.g. *Teleonemia scrupulosa* Stål and *Aceria lantanae* Cook) might not be expected to express this preference consistently among host populations, because this morphotype encompasses multiple and genetically-distinct lineages of lantana. Further study is required to test these predictions and report in depth on the applications of the findings presented here.

## ACKNOWLEDGMENTS

We thank the colleagues and volunteers who contributed samples to this project, in particular the 18 individual members of Landcare groups including Upper Clarence Combined Landcare and Landcare Illawarra. This project was supported by the Department of Agriculture, Water, and the Environment (activity ID 4-FY9KQZ3).

## REFERENCES

Day, M.D., Wiley, C.J., Playford, J., Zalucki, M.P., 2003a. Lantana current management status and future prospects. ACIAR Canberra. Monograph 102.

Day, M.D., Broughton, S., Hannan-Jones, M.A., 2003b. Current distribution and status of Lantana camara and its biological control agents in Australia, with recommendations for further biocontrol introductions into other countries. Biocontrol News and Information 24(3): 63N – 76N.

Huson, D.H., Bryant, D., 2006. Application of phylogenetic networks in evolutionary studies. Molecular Biology and Evolution 23(2): 254 – 267.

R Core Team, 2020. R: A language and environment for statistical computing. R Foundation for Statistical Computing, Vienna, Austria. URL https://www.R-project.org/.

Rossetto, M., Bragg, J., Kilian, A., McPherson, H., van der Merwe, M., Wilson, P.D., 2019. Restore and Renew: a genomics-era framework for species provenance delimitation. Restoration Ecology 27(3): 538 – 548.

Scott, L.J., Graham, G.C., Hannan Jones, M., Yeates, D.K., 1997. DNA Profiling resolves the limited importance of flower colour in defining varieties of Lantana camara. Electrophoresis 18: 1560-1563.

Watts, R., 2009. Testing the hybrid swarm model for weedy lantana. Unpublished report.

Zheng, X., Levine, D., Shen, J., Gogarten, S., Laurie C., Weir, B., 2012. A high-performance computing toolset for relatedness and principal component analysis of SNP data." Bioinformatics 28(24): 3326-3328.

# Biological control of Navua sedge (*Cyperus aromaticus*) in Australia

Kunjithapatham Dhileepan [1], Daisuke Kurose [2], Dianne Taylor [1], Boyang Shi [1], Marion Seier [2], Yu Pei Tan [1,3] and Roger Shivas [3]
[1] Department of Agriculture and Fisheries, Biosecurity Queensland, Ecosciences Precinct, Dutton Park 4102, Queensland, Australia
[2] CABI, Egham, UK
[3] Centre for Crop Health, Institute for Agriculture and the Environment, University of Southern Queensland, Toowoomba 4350, Queensland, Australia.
(k.dhileepan@qld.gov.au)

**Summary** Navua sedge (*Cyperus aromaticus*), a native of equatorial Africa, is an extremely aggressive perennial sedge, affecting beef, dairy, and sugarcane industries in the wet tropical regions of northern Queensland, Australia. Navua sedge is also a major weed of crops and pastures in Fiji and other South Pacific Islands. Navua sedge is unpalatable and forms dense stands that can replace palatable tropical pasture species. Current management options are mechanical and chemical, which are expensive and offer only short-term relief. Biological control is considered the most cost-effective and long-term management option. Navua sedge has been approved as a target for biological control in Australia, where a biological control program was initiated in 2017.

Native range surveys in Kenya, Nigeria and Tanzania found three promising biological control fungi; specifically, a smut fungus (*Cintractia kyllingae*) that infects flower heads and seeds; a rust fungus (*Uredo kyllingae-erectae*) that attacks leaves and stems; and an inflorescence-colonising ascomycete (*Curvularia tanzanica*). Field surveys have only recorded these fungi in association with Navua sedge. For effective biological control of Navua sedge, multiple agents that target different parts of the sedge may be needed to reduce seed production and minimise its impact and spread. These three fungal pathogens have been exported to CABI-UK, where host-specificity testing for *C. kyllingae* is in progress; and testing for *U. kyllingae-erectae* will commence soon. If proven to be host specific, the pathogens will be released in Australia. Current research in Australia has focused on the search for local pathogens on Navua sedge that may have potential as mycoherbicides. Several fungi of interest have been found in Australia, including species of *Curvularia*, *Epicoccum*, *Fusarium*, *Neopestalotiopsis*, *Nigrospora*, and *Phaeosphaeria*, as well as other fungal pathogens yet to be identified.

**Keywords** Navua sedge, *Cyperus aromaticus*, weed, biological control, Australia.

## INTRODUCTION

Navua sedge, *Cyperus aromaticus* (Cyperaceae), is an extremely aggressive and invasive sedge of tropical environments. In the Australian wet tropics, Navua sedge is of concern to the beef, dairy, and sugarcane industries (Shi *et al*. 2021). Navua sedge forms dense strands, and strongly competes with crops and pastures for nutrients, light and moisture (Vitelli *et al.* 2010). Navua sedge provides little nutritional value for cattle and replaces palatable tropical pasture species (Vitelli *et al.* 2010). In Fiji, Navua sedge lowered the carrying capacity of pastures by up to 40 %, which led to reduced milk production (Karan 1975; Kerr *et al.* 1995).

Current management options are mechanical and chemical, which are expensive and offer only short-term solutions. Mechanical control methods require repeated applications and often increase dispersal of the seeds, and thus further contamination. Halosulfuron is the only approved herbicide for Navua sedge control (Vitelli *et al.* 2010, Vogler *et al.* 2015). This chemical kills the aerial part of Navua sedge while established underground rhizomes remain viable (Chadha *et al*. 2022). There is also an issue of herbicide residues and prolonged withholding periods for both beef and dairy cattle.

As biological control may be a cost-effective and long-term management option for Navua sedge, the invasive weed has been approved as a target for biological control in Australia. In 2017, a biological control program commenced that aims to (1) conduct native range surveys and identify prospective biological control agents, (2) conduct host-specificity tests for prioritised agents, and (3) seek the release of approved biological control agents in Australia.

## MATERIALS AND METHODS

**Native range studies** Surveys were conducted in Kenya, Tanzania (May – June 2018, December 2019, and March 2021) and Nigeria (June 2019, March – November 2020, and June 2021), focusing on locations based on herbaria records. Navua sedge was sampled for plant pathogens, phytophagous insects, and mites. Other co-occurring sedges were also sampled to ascertain the field host range of prospective agents associated with Navua sedge in the field. Prospective biological agents were prioritised based on disease/infestation severity and observed field-host range.

**Test plants for host specificity** The centrifugal phylogenetic method (Wapshere 1974) was used to compile a plant list for host-specificity testing (Taylor and Dhileepan 2021). Test-plant species were selected based on phylogenetic relationships, following the familial classification of the Angiosperm Phylogeny Group (mobot.org/MOBOT/research/APweb/). Within the phylogenetic framework, an emphasis was placed on native and economically important species with a biogeographic overlap with Navua sedge in Australia.

**Host-specificity tests** Multiple consignments of Navua sedge plants (as seeds and rhizomes) sourced from Queensland were exported to quarantine facilities at CABI, UK to establish *in planta* cultures of the respective pathogens and determine their life cycles. Plants of other species (as seeds and bare rooted plants) were also sent to CABI, UK for host-specificity testing. Specimens of flower smut and leaf and stem rust from multiple sites in Nigeria and Tanzania were exported to CABI, UK. The inoculations methodology for the flower smut and the leaf and stem rust were developed and spores of the fungal pathogens were stored in liquid nitrogen for future use. Flower smut and leaf rust strains from Nigeria and Tanzania were assessed on Australian Navua sedge populations to select the most virulent ones for host-range testing. Life-cycle studies and host-specificity testing of the smut and life-cycle studies for the rust fungus are currently in progress in quarantine at CABI, UK.

**Australian native pathogens** Surveys for pathogens on Navua sedge in Queensland were conducted in September 2020, April 2021, and March 2022. Samples of leaves and stems showing symptoms of disease (leaf spots, blight, necrosis, discoloration) were examined for potential pathogens. Fungal pathogens from the samples were cultured for identification by morphological and molecular analysis. Pathogenicity and host specificity of promising pathogens as prospective mycoherbicides are being screened in containment glasshouses.

## RESULTS

**Biological control agents** Three prospective biological control agents for Navua sedge were determined from surveys in Africa. These were (i) the smut fungus *Cintractia kyllingae* (Basidiomycota, Anthracoideaceae) that infects the inflorescences and destroys florets and seeds, (ii) the rust fungus *Uredo kyllingae-erectae* (Basidiomycota, Pucciniaceae) that attacks leaves and stems, and (iii) the fungus, *Curvularia tanzanica* (Ascomycota, Pleosporaceae) that colonises inflorescences (Crous *et al.* 2021, Kruse *et al.* 2021, Dhileepan *et al.* 2022).

**Test plants for host specificity** There are about 125 native *Cyperus* spp. in Australia, as well as numerous naturalised species. Based on centrifugal phylogenetic methods, the test list for Navua sedge agents includes 38 species, of which 21 are *Cyperus* spp. and 13 are from other genera of sedges (Cyperaceae). Four representatives from other families have been included as outlier species, including rice and sugar cane (Taylor and Dhileepan 2021).

**Smut fungus *Cintractia kyllingae*** This smut fungus was found in Kenya, Tanzania, and Nigeria. It infects some or all of the florets of the inflorescence on Navua sedge (Figure 1). In the field *C. kyllingae* was not found on other co-occurring sedges, which indicates that the pathogen may be host specific. Field samples of four smut strains from Nigeria and two smut strains from Tanzania were screened for viability and virulence on Navua sedge plants sourced from Australia, Nigeria, and Tanzania. Inoculation studies confirmed that the Australian Navua sedge is susceptible to the smut strains from both Nigeria and Tanzania. Inoculation studies also showed that young flower heads of Navua sedge plants were highly susceptible to infection by *C. kyllingae*, with many seeds being destroyed. A strain of the smut from Tanzania that was found to be particularly virulent towards Australian Navua sedge was selected for host-specificity tests that are currently in progress.

**Rust fungus *Uredo kyllingae-erectae*** This rust fungus infects the leaves and stems of Navua sedge (Figure 2) in Nigeria and Tanzania. Life-cycle studies and its evaluation as a potential biological control agent for Navua sedge in Australia have commenced under quarantine conditions at CABI, UK. To date, preliminary inoculation studies using four Nigerian strains and two Tanzanian strains have produced symptoms of rust infection on Navua sedge plants collected from their respective countries, but not on Australian sourced plants. Other Nigerian stains are being evaluated to identify a rust strain(s) that is (are) pathogenic to the Australian population(s) of Navua sedge.

**Figure 1.** Inflorescences of Navua sedge. Healthy (left) and infected by the smut fungus *Cintractia kyllingae* (right).

**Figure 2.** Leaf and stem rust, *Uredo kyllingae-erectae*, on Navua sedge in Nigeria.

**Flower blight *Curvularia tanzanica*** Flower blight (Figure 3) was found only in Tanzania, sometimes together with the smut *C. kyllingae*. *Curvularia tanzanica* was not found in Kenya and Nigeria. *Curvularia tanzanica* colonised the inflorescences of Navua sedge and was superficially very similar in appearance to the flower smut. The flower blight samples were exported to CABI, UK, for isolation of the pathogen and its culture was stored in liquid nitrogen for future use.

**Figure 3.** *Curvularia tanzanica* on blighted inflorescence of Navua sedge in Tanzania.

**Australian native pathogens** Surveys for endemic or established fungal pathogens on Navua sedge conducted at 12 sites in northern Queensland yielded 41 fungal isolates. Molecular phylogenetic methods were used to identify the isolates as culturable ascomycetes in the genera *Curvularia* (5 isolates), *Epicoccum* (3), *Fusarium* (6), *Neopestalotiopsis* (4), *Nigrospora* (16) and *Phaeosphaeria* (1). Six isolates have yet to be identified to the rank of genus. These isolates were cryopreserved in the Queensland Plant Pathology Herbarium. An isolate of *Curvularia asiatica* and three isolates of unidentified *Curvularia* species, were spray-inoculated onto seedlings of Navua sedge in glasshouse pathogenicity tests. Only minor leaf spotting was observed, and therefore these fungi were not considered further as potential biological control agents. Testing of the remaining pathogens for pathogenicity is in progress.

## DISCUSSION

Native range studies in equatorial Africa have identified three fungi as potential biological control agents, specifically, a flower smut (*C. kyllingae*), a leaf and stem rust (*U. kyllingae-erectae*) and a flower blight (*C. tanzanica*). For effective biological control, multiple agents that target different parts of Navua sedge are needed. We have prioritised two pathogens that target foliar and reproductive parts of Navua sedge for further investigation. The flower smut attacks the inflorescence destroying the seeds and thereby reducing the seed bank in the soil; and the rust targets leaves and stems thereby reducing the plant vigour. Smit and rust fungi are obligate biotrophic pathogens that have previously been exploited successfully as weed biological control agents (Hershenhorn *et al.* 2016). To date, phytophagous insects or mites feeding on Navua sedge have not been observed in the native range of

Navua sedge nor has an agent been found that attacks underground parts of the sedge (roots and rhizomes).

The three fungal pathogens, *C. kyllingae*, *C. tanzanica*, and *U. kyllingae-erectae*, have been exported to CABI, UK for further research. Among them, the inflorescence smut (*C. kyllingae*) and the leaf and stem rust (*Uredo kyllingae-erectae*) have been prioritised for life cycle and host-specificity studies. Host-specificity tests for the flower smut as well as the selection of virulent strains for the leaf and stem rust are currently in progress. If one or more of the pathogens are shown to be specific to Navua sedge, an application seeking approval for their release in Australia will be submitted.

## ACKNOWLEDGMENTS

We thank Paul Mutuku Musili, Fredrick Munyao Mutie (National Museums of Kenya, Nairobi, Kenya), John Elia Ntandu (National Herbarium of Tanzania, Arusha, Tanzania), Emmanuel Chukwuma (Forest Herbarium Ibadan, Forestry Research Institute of Nigeria, Ibadan, Nigeria) and Ocholi Edogbanya (Kogi State University, Nigeria) for native range field surveys. The project was supported by AgriFutures Australia, with funding from the Australian Government Department of Agriculture, Water, and the Environment as part of its Rural Research and Development for Profit Program; the Australian Government Department of Agriculture, Water and the Environment as part of Enhancing National Pest Animal and Weed Management Program, under the Federation Funding Agreement – Environment; and the Queensland Department of Agriculture and Fisheries.

## REFERENCES

Chadha, A., Florentine, S.K., Dhileepan, K., Turville, C. and Dowling, K. (2022) Efficacy of halosulfuron-methyl in the management of Navua sedge (*Cyperus aromaticus*): differential responses of plants with and without established rhizomes. *Weed Technology* 36: doi:10.1017/wet.2022.29.

Crous, P.W., Cowan, D.A., Maggs-Kollings. G., et al. (2021). Fungal Planet Description Sheets:1182-1283. *Persoonia* 46, 313–528.

Dhileepan, K., Musili, P.M., Ntandu, J.E., Chukwuma, E., Kurose, D., Seier, M.K., Ellison, C.A. and Shivas, R.G. (2022). Fungal pathogens of Navua sedge (*Cyperus aromaticus*) in equatorial Africa as prospective weed biological control agents. *Biocontrol Science and Technology* 32, 114–120.

Hershenhorn, J., Casella, F. and Vurro, M. (2016). Weed biocontrol with fungi: Past, present, and future. *Biocontrol Science and Technology*, 26, 1313–1328.

Karan, B. (1975). Studies of Navua Sedge (*Cyperus aromaticus*). 1. Review of the problem and study of morphology, seed output and germination. *Fiji Agricultural Journal* 37, 59–67.

Kerr, D.V., Fell, R.F., Murray, A.J., & Chaseling, J. (1995). An assessment of factors associated with increased productivity of dairy farms in Fiji. *Asian-Australasian Journal of Animal Science,* 8, 481–487.

Kruse, J., McTaggart, A., Dhileepan, K., Musili, P.M., Mutie, F.M., Ntandu, J.E., Edogbanya, O., Chukwuma, E.C., Shivas, R.G. (2021). Broad and narrow host ranges in resolved species of *Cintractia limitata* s. lat. (Anthracoideaceae Ustilaginomycotina) on *Cyperus*. *Mycological Progress* 20, 191–201.

Shi, B., Osunkoya, O.O., Chadha, A., Florentine, S.K., Dhileepan, K. (2021). Biology, Ecology and Management of Invasive Navua sedge (*Cyperus aromaticus*) – A Global Review. *Plants* 10, 1851, doi.org/10.3390/plants10091851.

Taylor, D.B.J. and Dhileepan, K. (2021). Proposed plant host test list for assessing risk of biological control agents for Navua sedge, *Cyperus aromaticus*. Biosecurity Queensland, Department of Agriculture and Fisheries, Australia, January 2021

Vitelli, J.S., Madigan, B.A. and van Haaren, P.E. (2010). Control techniques and management strategies for the problematic Navua sedge (*Cyperus aromaticus*). *Invasive Plant Science and Management* 3, 315–326.

Vogler, W.D., Carlos, E.H., Setter, S.D., Roden, L., and Setter, M.J. (2015). Halosulfuron-methyl: A selective herbicide option for the control of the invasive *Cyperus aromaticus* (Ridley) Mattf. and Kukenth (Navua sedge). *Plant Protection Quarterly* 30, 61–66.

Wapshere, A.J. (1974). A strategy for evaluating the safety of organisms for biological weed control. *Annals of Applied Biology* 77, 20–211.

# Controlling cacti with cochineal in the SA Arid Lands

Paul Hodges[1], C Foulton[2]
[1]South Australian Arid Lands Landscape Board, Australia,
[2]Natural Resources SA Arid Lands, Australia
(paul.hodges@sa.gov.au)

**Summary** Cochineal, as a biocontrol agent, has become a vital tool in tackling Opuntioid cacti infestations in the SA Arid Lands region. Traditionally, foliar spray or stem injection of herbicides was used for control, with varying degrees of success. Chemical control proved time and labour intensive, was expensive, difficult in steep terrain and required ongoing follow-up control due to regrowth. In recent years, host-specific cochineal has proven an effective biocontrol for specific *Opuntia* and *Cylindropuntia* species and is providing a reinvigorated approach to tackling *Opuntia*. Cochineal has been used effectively on *Opuntia robusta*, *O. stricta* and *Cylindropuntia imbricata* in the North Flinders district for over 10 years, supported by dedicated volunteers. Following on from this success, we have been collecting and spreading cochineal on properties in the region for the past five years with promising results. Since 2017, Biosecurity Queensland has supported South Australian Arid Lands Landscape Board by providing *Dactylopius tomentosus* lineages suitable for *Cylindropuntia fulgida* var. *mamillata*, *Cylindropuntia prolifera* and *Cylindropuntia pallida*. These lineages have been released and monitored at a number of properties. A partnership with Port Augusta City Council allowed the development of a cochineal nursery where four species of cochineal are bred. This cochineal has been provided to pastoral properties, mining enterprises, Department of Defence properties and townships within the SA Arid Lands, as well as neighbouring Landscape boards for release in their regions. This presentation looks at the progress of our cochineal breeding and release program and discusses the benefits, partnerships, community engagement and issues encountered as part of an integrated management plan. It will feature case studies where cochineal has given land managers fresh hope in eradicating Opuntioid cacti from their properties.

**Keywords** Biocontrol, Cochineal, Opuntioid cacti, Cactus, SA Arid Lands

# Biology of Globe Chamomile (*Oncosiphon piluliferum* (L.f.) Källersjö)

Alexandra Douglas[1] and David F. Nicholson[2]
[1]Department of Primary Industries and Regional Development, 444 Albany Hwy, Albany, Western Australia 6330, Australia
[2]Department of Primary Industries and Regional Development, 75 York Rd, Northam, Western Australia 6401, Australia
(alex.douglas@dpird.wa.gov.au)

**Summary** A number of experiments were conducted to investigate triggers for germination and seed bank persistence of Globe Chamomile (*Oncosiphon piluliferum*) in Western Australia.

Globe Chamomile germinated between 10°C and 25°C and not at other temperatures tested. It is unlikely that Globe Chamomile will germinate in summer. Globe Chamomile seed survives well when buried at 2-10 cm, and darkness inhibits germination under laboratory conditions.

Seed has been shown to survive in the soil for several years if buried. There may be implications for summer fallow management if cultivation is used and the use of soil inversion at crop establishment as these practices will bury seed to depths where it can remain viable.

**Keywords** *Oncosiphon Spp*, Globe Chamomile, Calomba Daisy, Matricaria, germination, seed bank.

## INTRODUCTION

*Oncosiphon* Källersjö (Asteraceae: Anthemideae) is a genus of aromatic herbs that are endemic to the Cape region of southern Africa (Kolokoto and Magee 2018). Two species, *Oncosiphon suffruticosum* (L.) Källersjö and *Oncosiphon piluliferum* (L.f.) Källersjö, have been reported as invasive in Australia (Western Australia and South Australia) (ALA 2021) and in California and Arizona in the USA (Hedrick and McDonald 2020).

Species of *Oncosiphon* are commonly referred to as stinkruid in Afrikaans (meaning stink-weed) and stinknet in the US. In Australia *O. suffruticosum* is referred to as Calomba Daisy and *O. piluliferum* as Globe Chamomile. In the Western Australian grain belt they are referred to collectively as Matricaria.

The two species of Matricaria in WA can be distinguished by their flower heads. Columba Daisy (more 'club' shaped flower heads) and Globe Chamomile (rounder, globe shaped flower heads). Both species of Matricaria are erect annual herbs with bright yellow flowers and look very similar until the flowers start to form. They both have a strong unpleasant smell, form dense stands and are considered unpalatable to grazing livestock (although not known to be toxic) (Parsons and Cuthbertson 1992). Only one species, Globe Chamomile, has been used in these studies as it is the most common in the region.

Since appearing in the eastern grain belt of Western Australia in the late 1960s, Matricaria has spread widely in that region and is now a serious weed (Dodd 1990). Matricaria is spreading into the neighbouring Northern agricultural region where the farming systems and seasonal conditions are similar and is considered an emerging threat (Michael et al 2011).

This research aimed to identify triggers for germination and potential longevity of the seed bank of Globe Chamomile (extrapolated to Calomba Daisy), as an aid to management plans in cereal cropping systems.

## MATERIALS AND METHODS

Seeds from four Globe Chamomile populations were subjected to various conditions and the effect on germination assessed. All four populations had been collected during the same week in December 2017, with three of the collected populations growing within 25 km of each other north of Merredin (31.44°S, 118.27°E) and the fourth population located in Mukinbudin (30.93°S, 118.34°E). Two populations were collected from roadsides, Knungajin-Merredin Rd (31.46°S, 118.27°E) and Nangeenan Rd (31.49°S, 118.16°E), and the remaining two collected within paddocks adjacent to trial areas.

**Temperature on germination** Five temperatures were assessed: constant 5°C, 15°C, 25°C, 35°C and 45°C, for a two week period. Three replicates with 50 seeds each were placed in petri dishes with filter paper and distilled water. The test was conducted in an incubation cabinet with an alternating 12 hour dark/light cycle. The test at 5°C was conducted in a refrigerator.

**Light or dark conditions and scarification** Further samples of seed from the populations above, were subjected to light and dark conditions (petri dishes wrapped in aluminum foil) and scarification (the seed was rubbed gently between two rubber mats for 10 seconds, before being placed in petri dishes).

Samples were placed in cabinets with a 10/20°C 12 hour dark/light cycle, for two weeks.

**Depth of burial over time** A field trial was established at Northam (31.65°S, 116.69°E) to determine the effect of depth of seed burial on the persistence of Globe Chamomile. Three new Globe Chamomile populations were sampled for this experiment; Nangeenan/Connell (31.45°S, 118.15°E), Nangeenan/Fitzpatrick (31.42°S, 118.14°E) and Nokanning (31.37°S, 118.18°E). Fifty seeds from each of the three populations were placed in nylon bags and buried at different depths (0, 2, and 10 cm) in June 2017. Bags were collected in September and December 2017, June 2018, June 2019 and a final collection, at 36 months, June 2020.

Following collection, seeds were removed from the bags and placed in petri dishes with filter paper, distilled water and gibberellic acid (0.1g/1L solution). Globe Chamomile seeds were placed in an incubator and subjected to a 12 hour, 10/20°C dark/light temperature regime. Dishes were checked 10 days after the test commenced, germinated seedlings were counted and removed and the test continued for a further 10 days, when the final count was made. The percentage germination was calculated from total number of seeds initially placed in the bags at the time of burial.

## RESULTS AND DISCUSSION

**Temperature on germination** Globe Chamomile germinated at 15 and 25 degrees only. No germinations were recorded for the samples kept at 5°C, 35°C and 45°C (Table 1). Matricaria are autumn germinating species and this range of temperatures fits with autumn conditions. It is less likely that Matricaria will germinate in the summer months to become a summer weed.

**Table 1.** Effect of temperature (°C) on germination (%), for seed from four populations of Globe Chamomile.

| | Temperature | | | | |
|---|---|---|---|---|---|
| Population | 5 | 15 | 25 | 35 | 45 |
| Knungajin-Merredin Rd | 0.0 | 0.0 | 36.0 | 23.3 | 0.0 |
| Nangeenan Rd | 0.0 | 0.0 | 12.7 | 18.7 | 0.0 |
| Mukinbudin | 0.0 | 0.0 | 24.7 | 20.7 | 0.0 |
| Merredin | 0.0 | 0.0 | 24.7 | 11.3 | 0.0 |
| LSD (5%) Temperature 3.7, Population 3.3 | | | | | |

**Light or dark conditions and scarification** Seeds kept in darkness showed much lower germination than those seeds exposed to light (up to 3.3% compared to up to 49.3%; Table 2). Scarification did not increase the level of germination under the standard conditions (light). Use of cultivation at crop establishment/fallow maintenance is unlikely to stimulate additional germination. The action of tillage may bury Matricaria seed leading to increased longevity of the seed in the soil seed bank.

**Table 2.** Effect of light or dark conditions or scarification on germination (%) of seed from four populations of Globe Chamomile.

| Population | Light | Dark | Scarification |
|---|---|---|---|
| Knungajin-Merredin Rd | 49.3 | 3.3 | 51.3 |
| Nangeenan Rd | 41.3 | 0.7 | 40.0 |
| Mukinbudin | 37.3 | 0.0 | 28.7 |
| Merredin | 29.3 | 1.3 | 36.7 |
| LSD (5%) Population 6.4, Treatment 6.4 | | | |

**Depth of burial over time** There was no difference in the behaviour of the three populations over burial depth and length of time the seed was buried. Depth of burial did have a real effect on the level of Globe Chamomile germination recorded at each of the three collection times (Figure 1).

A greater number of Globe Chamomile seeds persisted (remained viable) when buried at 2 cm and 10 cm than those on the soil surface. This was consistent for the first three collection times. This may indicate that there is some secondary dormancy being exhibited by these Globe Chamomile populations. However, following 24 months of burial, this changed, with more Globe Chamomile germinations recorded for the surface and 2 cm samples. There may be some breakdown in the dormancy of the Globe Chamomile seeds after an extended period, greater than 24 months depending on population.

Data from work conducted in the late 1980s (Dodd and Lloyd 1988 and 1989) demonstrated that seeds can remain viable in the soil seed bank for at least five years, indicating that the seeds of this weed, although small, are capable of persisting for several years, if buried.

Initially there appeared to be a trend for a decline in the percentage of seeds that will germinate over time when the seeds were buried at 2 and 10 cm. However, testing of the final seed collection, following burial for 36 months, 50% of buried seed germinated from the samples buried at 2 cm. Low levels of seed were still viable following burial at 10 cm or storage on the soil surface after 36 months.

Having seed that retains viability for a number of years while buried, makes Matricaria a more serious weed issue in farming systems. Seed longevity will influence the time required to manage field populations as recruitment from the soil seed bank will extend the time needed to reduce the population. There may also be implications for summer fallow management if cultivation is used and the use of soil inversion at crop establishment as these practices will bury seed to depths where it can remain viable.

**Figure 1.** Germination (%) of Globe Chamomile seed, averaged over three populations, seeds collected following burial at three depths for 3, 6, 12, 24 and 36 months (standard error bars shown).

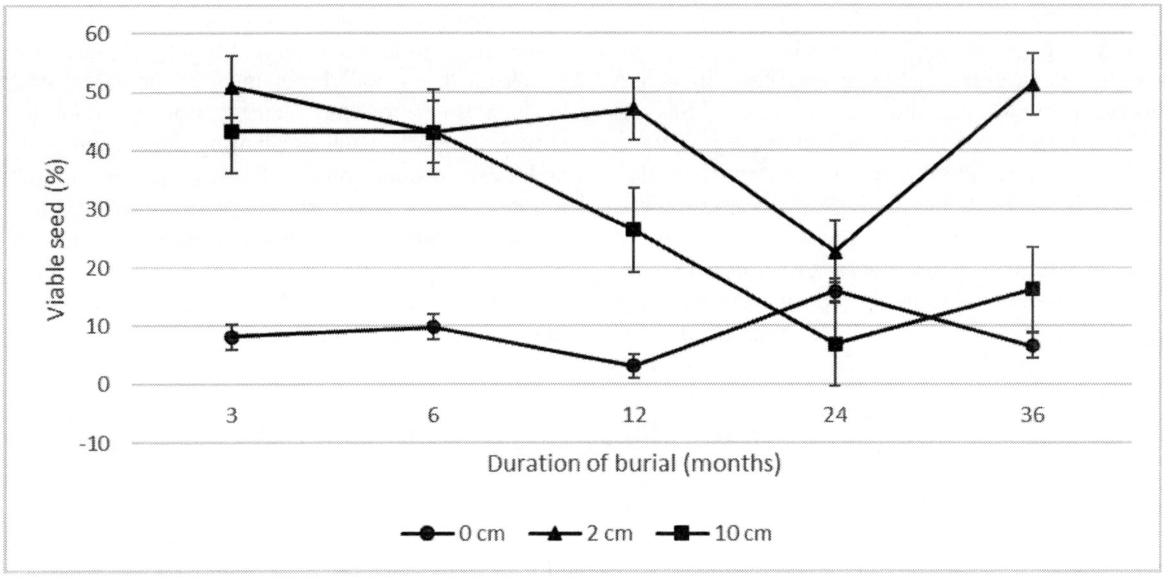



ACKNOWLEDGMENTS

This research formed part of the Locally Important Weeds project, DAW00257, and was made possible by the significant contributions of growers through both trial cooperation and combined DPIRD and GRDC investment, the author would like to thank them for their continued support.


REFERENCES


Atlas of Living Australia (2021). INTERNET http://bie.ala.org.au/species, Accessed 15 November 2021.

Dodd, J. (1990). Pentzia (Oncosiphon) spp. In Western Australia: History, distribution and significance. Proceedings of the 9th Australian Weeds Conference, ed J.W. Heap p. 451. (Crop Science Society of South Australia, Adelaide).

Dodd, J. and Lloyd, S.G. (1988 and 1989). Department of Agriculture WA, Plant Industries Experimental Summaries.

Hedrick, P.W. and McDonald, C.J. (2020). Stinknet, a new invasive, non-native plant in the southwestern United States. *Desert Plants* **36**, 5-16.

Kolokoto, R. and Magee, A.R. (2018). Cape stinkweeds: Taxonomy of Oncosiphon (Anthemideae, Asteraceae). *South African Journal of Botany* **117**, 57-70.

Michael, P.J., Yeoh, P.B., Ota, N. and Scott, J.K. (2011). Climate change impacts on agricultural weeds in Western Australia. RIRDC publication No. 11/059. RIRDC Project No. AWRC 08-85.

Parsons, W.T. and Cuthbertson, E.G. (1992). Noxious weeds of Australia, Calomba Daisy pp. 296-298. Inkata Press.

# Biology of feathertop Rhodes grass (*Chloris virgata*)

Bhagirath Chauhan[1]
[1]The University Of Queensland, Gatton, Australia
(b.chauhan@uq.edu.au)

**Summary** Feathertop Rhodes (FTR) grass is the most difficult to control grass weed species in summer crops and fallows. Its initial infestation was along roadsides, which has now entered cropping fields. This weed costs >AUD 7 million per year to the grain growers in the northern region of Australia. FTR grass has been observed in recent years to grow during the winter season. Several populations have evolved resistance to glyphosate in summer fallows and glyphosate-tolerant cotton systems. If the incidence of FTR grass populations with extended seasons of growth continues to increase, this could become a great concern for winter cropping systems. These observations suggest that there is a need to develop sustainable and effective management programs for FTR grass in different crops and seasons. However, to develop such programs, there is a need to better understand the biology of this weed. A series of trials are in progress at the Gatton farm of the University of Queensland to understand its biology. Results of the following trials will be discussed: the effect of the frequency of mowing, planting time (in relation to temperatures) and wheat planting density on growth and seed production of FTR grass. In addition, information on the morphology and seed production of glyphosate-resistant and glyphosate-susceptible populations of FTR grass will also be discussed.

**Keywords** FTR, mowing, resistance, seasonality expansion, planting time

# Resistance surveys and commercial testing services — similarities and differences for annual ryegrass (*Lolium rigidum*) across south-eastern Australia

Peter Boutsalis[1,2], John Broster[3], Gurjeet Gill[1] and Christopher Preston[1]

[1]School of Agriculture, Food and Wine, University of Adelaide, PMB 1, Glen Osmond SA 5064, Australia
[2]Plant Science Consulting (PSC), Prospect, SA 5082, Australia
[3]Gulbali Institute, Charles Sturt University, Boorooma Street, Wagga Wagga, NSW, 2678, Australia
(peter.boutsalis@adelaide.edu.au)

**Summary** Annual ryegrass (*Lolium rigidum*) is the most problematic weed species across southern Australian states with estimated crop production losses of over 90 million dollars per annum (Llewellyn et al., 2016). In this paper we compare the incidence of herbicide resistance in ryegrass between GRDC funded random weed surveys and commercial testing services between South Australia (SA), Victoria (VIC), New South Wales (NSW) and Tasmania (TAS) from samples collected in the most recent surveys. The information is presented as percent of resistant samples to the Group 1 FOP, DIM and DEN herbicides, Group 2 sulfonylurea herbicides, trifluralin, triallate/prosulfocarb, pyroxasulfone and glyphosate. Several hundred ryegrass samples were tested with each herbicide. The random weed surveys detected similar levels of resistance to the commercial testing services for wheat selective Group 1 and 2 herbicides, trifluralin, triallate/prosulfocarb and pyroxasulfone. However, large discrepancies between the two approaches were detected for resistance to clethodim and glyphosate. Reasons for these differences will be discussed.

Llewellyn RS, Ronning D, Ouzman J, Walker S, Mayfield A and Clarke M (2016) *Impact of Weeds on Australian Grain Production: the cost of weeds to Australian grain growers and the adoption of weed management and tillage practices* Report for GRDC. CSIRO, Australia.

# Weed managers guide to Remote Detection: Understanding opportunities and limitations of technologies for remote detection of weeds

Jane E. Kelly [1,2], Motiur Rahaman[1,2], Juan Sandino[1,3] , Lihong Zheng[1], Hillary Cherry[4], Mark Hamilton[4], Remy Dehaan[1,2], Felipe Gonzalez[3], Wendy Menz[4], Liesl Grant[4]
[1] Charles Sturt University, Boorooma St, Wagga Wagga NSW Australia
[2] Gulbali Institute Land Water and Environment, Boorooma St, Wagga Wagga NSW Australia
[3] QUT Centre for Robotics, Queensland University of Technology, Brisbane City, QLD 4000 Australia
[4] NSW Department of Planning and Environment, NSW Australia
(janekelly@csu.edu.au)

**Summary** Remote detection is an influential tool for weed management, however accessing current technology can be costly, heterogeneous and unattainable for land managers. A new project aims to break down these barriers by investigating the limitations of this technology for remote weed detection in complex landscapes and create a *Community of Practice* for knowledge and information sharing that is accessible to all land managers. This paper presents an overview of the project methods and objectives, together with preliminary results and conclusions drawn from early analyses of recently acquired red, green, blue (RGB) and multispectral hawkweed imagery. Initial results emphasise the promise of RGB and multispectral sensors mounted on Remotely Piloted Aircraft Systems (RPAS), and supervised machine learning (ML) models for detecting hawkweed flowers with high accuracy in a rich set of landscapes.

**Keywords** remote sensing, Hieracium pilosella, Remotely Piloted Aircraft Systems (RPAS), machine learning (ML), object detection, community

## INTRODUCTION

Conventional surveillance methods (e.g. field surveys) for invasive plant species, or weeds are time-consuming, dangerous and expensive, resulting in a lack of quantitative information about weed distribution in Australia (Campbell 1991). Such a lack of updated data hampers effective weed management (Coutts-Smith and Downey 2006). Traditional remote sensing efforts to detect weeds using aerial photography and multispectral imagery have obtained mixed success, with low spatial (from 10 to 30 m/pixel) and spectral (~100 nm) resolutions (e.g., Spot, Landsat) (Lamb and Brown 2001, Thorp and Tian 2004). Even with higher spatial resolution, satellite multispectral sensors (e.g., IKONOS and WorldView) have low instrument signal to noise ratios (SNRs), limiting their use to only large-scale weed infestations.

Hyperspectral imaging is a cutting-edge remote sensing tool that can obtain many spectral measurements (from 50 to 400 bands) in one pass. The resulting images allow separation of weeds from desirable vegetation and provide semi-quantitative abundances in plant and soil mixtures (Boardman 1998), showing considerable promise for identifying and mapping weed abundance (Miao *et al.* 2006, Dehaan *et al.* 2007). However, analysis using airborne and satellite systems can be costly and resolution is not always acceptable. Further, previous trials have demonstrated the deployment of active optical sensors in aerial platforms (Lamb *et al.* 2009), where detection of greenness by multispectral sensors typically worked well in crops when weeds are easily differentiated. In landscapes where weeds are mixed with other vegetation in heterogeneous situations, multispectral systems have shown less detection reliability.

Remotely Piloted Aircraft Systems (RPAS) and sensor technologies are now commercially available, achieving spatial resolutions from 2 to 50 cm/pixel, and with an increased flexibility to collect quantitative data at lower costs than traditional methods. In addition, machine learning (ML) may offer the ability to model relationships between low-resolution satellite imagery and corresponding higher-resolution images. This will allow enhancement of low-resolution satellite imagery, improving ability to detect weeds using this lower cost imagery.

Research in this space has traditionally been widespread and segregated across Australia, with no existing mechanism to bring findings and resources together to improve uptake. This paper describes a new research project that aims to address this gap by investigating the limitations and opportunities of existing remote sensing technologies now available for detecting weeds in heterogeneous landscapes. The paper also presents preliminary results of recent hawkweed imagery analysis in the sections below.

## MATERIALS AND METHODS

**Project description** The project, entitled *"The weed managers guide to Remote Detection: Understanding the opportunities and limitations of multi-resolution and multi-modal technologies for remote detection of weeds in heterogeneous landscapes"* aims to investigate opportunities for cost-effective use of high-resolution red, green, blue (RGB), or colour, multispectral and hyperspectral technologies across various airborne platforms (drone, aircraft, satellite), paired with multi-modal ML analyses to detect weeds in heterogeneous landscapes. Three nationally significant 'model' weeds: 1) (hawkweed, (*Pilosella aurantiaca*); 2) African lovegrass (*Eragrostis curvula*); and 3) bitou bush (*Chrysanthemoides monilifera* subsp *rotundata*) will be used to test each technology, with the objective of determining practicable methods for land managers to use remote sensing for weed detection, aiding different management objectives (i.e. eradication, containment, asset protection). The project aims to grow extensive national partner networks, and to develop a national community of practice and portal to share learnings and advice on remote detection of weeds.

RPAS-mounted RGB, multispectral and hyperspectral imagery will be collected for each weed species. Field sites have been established in complex ecological landscapes where the weeds are present in varying densities. Sites will be sampled over 18 months in accordance with physiological or phenological changes that may allow improved detection of target species.

Analysis with multispectral and hyperspectral imagery will comprise the discrimination of key spectral bands and vegetation indices per weed species against other vegetation, as well as developing a pipeline to autonomously detect and map such weeds for a range of landscape ecosystems applying digital image processing, and supervised ML techniques such as gradient boosting and convolutional neural networks (CNNs). The development and outcomes of these pipelines will be validated with ground-based, or on-site data from experts. The first data capture for this project was at orange hawkweed sites in December 2021.

**Site description** Hawkweed drone imagery was obtained from the Port Phillip hawkweed study site during December 2021 in Kosciuszko National Park (148.5875990°E 35.6923769°S), NSW, Australia. Much of the infestation at this site is enclosed within an approximate 20 m x 20 m area, which encompassed the flight region for data capture.

**Ground truthing** To facilitate the validation of all species captured within imagery, white plastic reference quadrats (1 m x 1 m) were placed across the site in areas representing variable botanical composition and hawkweed density. Ground images (Nikon D600 DSLR camera) of each quadrat were captured as reference images and quadrat features were recorded, including GPS location, plant species composition, plant height, species phenological stage and percentage ground disturbance. Cloud cover, wind speed, humidity, temperature and altitude were also recorded.

**Imagery acquisition** Imagery was captured on the 16th and 17th of December 2021 using DJI M300 and M600 drones. A number of different camera systems were mounted to these drones including high-resolution RGB cameras (Phase One-100MP, DJI P1-45MP and Fuji GFX 100s-100MP), multispectral (Micasense Altum) and hyperspectral (Specim AFX VNIR covering 400-1000nm of the electromagnetic spectrum) cameras. Each payload configuration was flown at different heights (20 m to 120 m) to facilitate various ground sampling distance (GSD) resolutions (0.22 to 5 cm/pixel). Ground calibration panels were placed in the field to help with spectral calibration for the multispectral and hyperspectral data. Various ML models were applied to the captured images, establishing separate processing pipelines for high-resolution RGB, and multi/hyperspectral data.

**RGB imagery analysis**

**1.Training and optimisation** To provide reasonable accuracy, 128 sample images from the RGB dataset were selected for model training, beginning with those of highest resolution (0.22 cm/ pixel). Bounding-box annotations were generated for each hawkweed flower appearing per sampled image. Several parameters such as the initial learning rate, final OneCycle learning rate, momentum, weight decay, obj loss gain, focal loss gamma, batch size, epoch, and confidence were optimised to obtain the maximum accuracy of the selected deep models.

**2. Testing and Prediction** Model performance was evaluated using several metrics, including precision, recall, and mean average precision (mAP) for an intersection over union (IOU) of 0.5 (50%). Precision measures the number of correctly predicted boxes, while recall measures the number of true boxes correctly predicted.

Flower detections of hawkweed within images were validated by weed experts. Detected hawkweeds within the ten ground-truthed reference quadrats were counted and visually cross-checked with ground images. Accuracies of all quadrats were subsequently averaged, producing the overall accuracy of the hawkweed detector model. An illustration of detected flowers is shown in Fig. 1.

**Fig. 1.** Visual outlook of a tested high-resolution RGB image with hawkweed (quadrat dimensions 1m x 1m). Spatial resolution 11648 x 8736 pixels, and GSD 0.22 cm/pixel.

**Multispectral imagery analysis** Spectral signatures of hawkweed flowers and rosettes were identified, and a ML-based supervised model was tuned to map the weed using a data fusion approach between high-resolution RGB and multispectral orthomosaic rasters.

**1. Orthomosaics and raster alignment**
Initial image processing consisted of generating an orthomosaic for the site, georeferencing the resulting raster using ground control points (GCPs) and overlaying the multispectral orthomosaic to achieve pixel-level alignment between both rasters.

**2. Data labelling** Pixel-wise image labelling was performed over the reference quadrats containing hawkweed presence. Given that the spatial resolution of the multispectral raster is considerably lower than the high-resolution RGB imagery, the labelling task was supervised by on-ground weed experts. Challenges in labelling the data were addressed by applying a data-fusion approach. To simplify the spectral analysis, a total of four classes were compiled for hawkweed assessments, namely hawkweed rosettes, flowers, other vegetation, and non-vegetation.

RESULTS

The hawkweed detector model achieved an overall accuracy of 98.67% in the detection of hawkweed flowers within RGB imagery (0.22cm/pixel resolution) acquired at the Port Phillip site. Similarly, preliminary results on labelled data with the multispectral ML model reported a pixel-wise classification accuracy of 98.67% in the detection of hawkweed rosettes, and an overall accuracy of 98% percent for all hawkweed classes labelled. An example of the mapped classes extrapolated to the entire multispectral imagery is shown in Fig. 2.

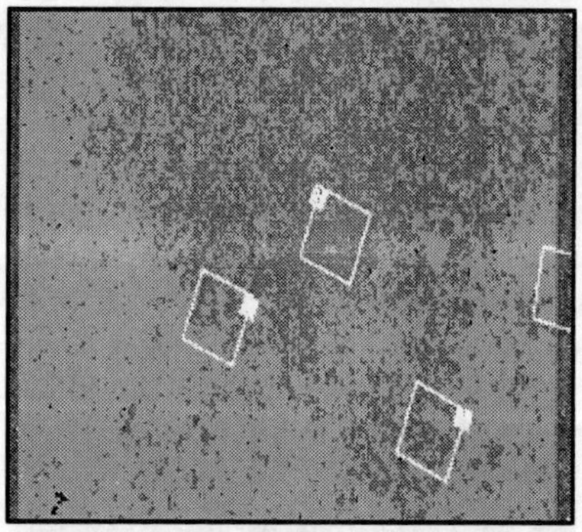

(a)

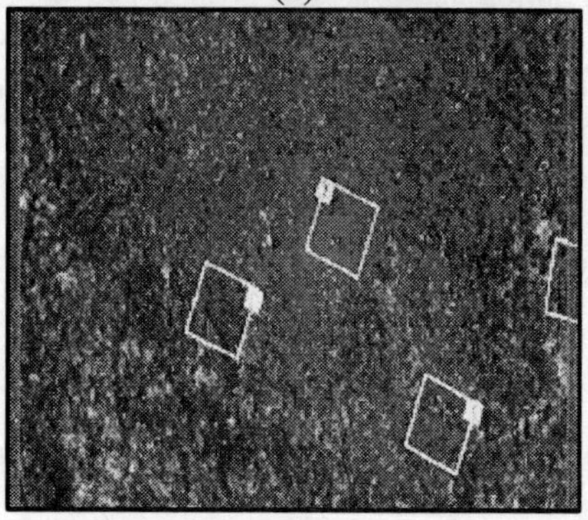

(b)

**Fig. 2.** Image preview of predicted areas of hawkweed rosettes using multispectral imagery. **(a)** Highlighted areas of hawkweed rosettes in red, other vegetation in green, and non-vegetation in white. **(b)** Highlighted areas with the presence of hawkweed rosettes in red. Quadrat dimensions 1m x 1m.

High precision, recall and mAP values for hawkweed detection from RGB imagery were of 93%, 97% and 97.3% respectively. Similarly, high precision and recall values were also achieved from multispectral imagery, at 97% and 99% respectively.

## DISCUSSION

Study results support the value of RPAS-mounted RGB and multispectral sensors for detecting hawkweed plants at flowering stages within an alpine heterogeneous landscape. Findings are consistent with other studies investigating RGB and multispectral sensors for hawkweed detection (Hamilton *et al.*, 2018; Ajamain *et al.*, 2021). The accuracy metrics on the ML model are preliminary and indicate that further validation on prediction of hawkweed using multispectral imagery with a wider range of vegetation is required. Results also highlight the significance of image resolution in relation to image clarity and detection accuracy when applying deep learning models to remotely sensed data. Since high resolution imagery acquisition can be costly, there remains a need to identify models capable of improving clarity and detecting species within lower resolution imagery.

Despite the potential for detection at flowering stages, optimal control of weeds largely relies on detection during the vegetative stage, a goal historically less successful using RGB and multispectral technology (Hamilton *et al.*, 2018; Ajamain *et al.*, 2021). Such challenges point to the potential of RPAS-borne hyperspectral sensors as demonstrated by the capacity for spectroradiometers to distinguish hyperspectral profiles of hawkweed leaves to an accuracy of 80% (Ajamain *et al.*, 2021).

Future project work will continue to investigate remote sensing technologies and their application to each model weed system, including: 1) the development of detection models for low-resolution images; 2) the development of image pre-processing pipelines and models to improve image quality, and 3) the development of an image super-resolution method to upscale low resolution imagery for improved weed detection. Project methodologies, imagery and results will be collated into a set of guidelines, an online portal and community of practice for the sharing of resources associated with the Remote Detection of Weeds.


## ACKNOWLEDGMENTS

Authors gratefully acknowledge the funding support of the Department of Agriculture, Water and Environment and the assistance of Erin Fardy and Andrew Murrell, the Spatial Analysis Network and NSW National Parks and Wildlife Service for their assistance in hawkweed data acquisition.


## REFERENCES


Ajamian, C.; Chang, H.-C.; Tomkins, K.; Farebrother, W.; Heim, R.; Rahman, S. (2021). Identifying Invasive Weed Species in Alpine Vegetation Communities Based on Spectral Profiles. *Geomatics* 1, 177–191.

Boardman, J.W. (1998). Leveraging the high dimensionality of AVIRIS data for improved sub-pixel target unmixing and rejection of false positives: mixture tuned matched filtering. Proceedings of the. 7th Annual JPL Airborne Geoscience Workshop, Pasadena, California, 12–14 January 1998, pp. 55–56.

Campbell, M.H. (1991). Weed control in pastures — are we winning? *Plant Protection Quarterly* 6 (2), 55–63.

Coutts-Smith A. & Downey, P. (2006). Impact of Weeds on Threatened Biodiversity in NSW, Technical Series 11, CRC for Australian Weed Management, Adelaide

Dehaan, R. Louis, J. Wilson, A. Hall, A. & Rumbachs, R. (2007). Discrimination of blackberry (*Rubus fruticosus* sp agg.) using hyperspectral imagery in Kosciuszko National Park, NSW, Australia, ISPRS *Journal of Photogrammetry and Remote Sensing* 62, 13–24.

Hamilton, M., Matthews, R. and Caldwell, J., (2018). Needle in a haystack–detecting hawkweeds using drones. Proceedings of the 21st Australasian Weeds Conference, pp. 9-13 (Weeds Society of NSW, Sydney).

Lamb, W. & Brown, B. (2001) PA -Precision Agriculture: Remote-Sensing and Mapping of Weeds in Crops, *Agricultural Engineering Research*, 78 (2), 117-125.

Lamb, D.W., Trotter, M.G. and Schneider D.A. (2009) "Ultra low-level airborne (ULLA) sensing of crop canopy reflectance: A case study using a CropCircleTM sensor". *Computers and Electronics in Agriculture* 69, 86-91.

Miao, X. Gong, P. Swope, S.M. Pu, R. Carruthers,

R.I. & Anderson, G.L. (2006). Estimation of yellow starthistle abundance through CASI-2 hyperspectral imagery using linear spectral mixture models, *Remote Sensing of Environment* 101 (3), 329–341.
Thorp, K.R. & Tian, L.F. (2004). A review on remote sensing of weeds in agriculture, *Precision Agriculture* 5, 477–508.

# Progress in the eradication of *Miconia calvescens* from Australia

Simon Brooks[1] and Kimberley Erbacher[2]
[1] Tropical Weeds Research Centre, Biosecurity Queensland, Department of Agriculture and Fisheries, PO Box 976 Charters Towers, Queensland 4820, Australia
[2]Biosecurity Queensland, Department of Agriculture and Fisheries, PO 20 Box, South Johnstone Queensland, 4859 Australia.
(simon.books@daf.qld.gov.au)

**Summary** *Miconia calvescens* (Miconia) is a target of the nationally cost-shared National Tropical Weeds Eradication Program which commenced in late 2003. *Miconia calvescens* infestations have been found in wet tropical areas between the Whyanbeel Valley and Tully in north Queensland, as well as in southeast Queensland and northern New South Wales. Updates to the methods for measuring eradication and progress towards removing this rainforest invader from a challenging environment in some of the wettest parts of Australia are presented below.

**Keywords** tropics, declaring eradication, rainforest, Miconia

## INTRODUCTION

*Miconia calvescens* DC. (Miconia) is a small, invasive, shade tolerant, frugivore dispersed, rainforest tree. With attractive bicoloured leaves the plant has been cultivated as a botanical curiosity in tropical and subtropical areas and become invasive outside its native Central and South America (Brooks and Jeffery 2010). *Miconia calvescens* is one of the targets of the National Tropical Weeds Eradication Program (NTWEP) in Australia. This program also targets multiple infestations of *Limnocharis flava* (L.) Buchenau and *Mikania micrantha* Kunth, and single locations of *Miconia racemosa* (Aubl.) DC. and *Miconia nervosa* (SM.) Triana. Single infestations of the Miconia shrubs co-occur within two large *M. calvescens* infestations (Jeffery and Brooks 2016). Survey and control of *M. calvescens* utilizes over 75% of NTWEP resources.

The historical and practical field aspects of *M. calvescens* incursion, as well as the last update on eradication progress, were presented by Brooks and Jeffery (2010) and Jeffery and Brooks (2016). Many of the subsequent changes in the NTWEP are documented in internal reports or presentations, so updates to the NTWEP methodology and *M. calvescens* eradication progress data are presented.

## ERADICATION FIELD METHODS

Field crews survey areas of intact forest types around 1000 m from waypoints known or suspected to have had at least one mature *M. calvescens* recorded. The survey extends outwards to 1500 m in fragmented vegetation types such as riparian areas amongst agricultural land uses. Different portions of larger infestations are surveyed on different days, months, or years. The survey frequency is approximately 24 months but varies between 12 to 36 months, depending on the number of plants controlled. Two-yearly intervals allow two opportunities to detect a small proportion of fast-growing plants before they mature, with average growing plants taking 6 to 8 years to mature (S. Brooks unpublished data). To cover over 3000 ha of ground surveillance a year, surveys are conducted year-round but concentrated in the cooler and drier months between April and October each year with additional seasonal workers. Field surveys are conducted by Biosecurity Queensland staff, casual labour contractors, and Rous County Council (under contract) with in-kind support from local government and Queensland Parks and Wildlife Service.

## ERADICATION REPORTING METHODS

Presence or absence is one of the parameters derived from field records collected at points within infestations. A new unique site identification is issued for any plant found more than 30 m away from all previously recorded sites. For the purposes of program reporting, site records are aggregated into static one-hectare cells (100 m x 100 m), generated as a 'grid layer' across the entire incursion. These cells are used in NTWEP reporting as 'management areas', they are not survey units. From 2010, eradication progress reporting changed from an infestation-based system to a grid-based system of management areas. Field data collected from 2003 to 2010 was re-analysed on the finer reporting scale. The management area system allows NTWEP reporting to be spatially consistent over time. The term 'loci' is still used for broad scale reporting, these are discrete occurrences of *M. calvescens*, more than 2 km apart.

At the end of each financial year, in every management area, point records are summarized to allocate a 'control phase' status where plants are present, or 'monitoring phase' status where plants are absent. Management areas only enter a monitoring phase when all records in the last 12 months show plants were absent, so progression is via evidence of absence. The time that management areas have been in the monitoring phase is categorised as 'years in monitoring phase'. If plants are recorded in a management area which is in the monitoring phase, then it relapses to a control phase, for a minimum of 12 months. The number of years of monitoring before a relapse is tallied to determine 'monitoring relapse' data. As the *M. calvescens* revisit frequency is greater than 12 months, management areas that do not have a visit recorded in the previous 12-months default to the control or monitoring status that they were in the previous reporting period. The status of management areas remains in control (default) until there are absence records, but years in monitoring phase can accrue between visits. Once management areas reach the sixth year in the monitoring phase they are classified as 'provisionally eradicated' for the purpose of eradication reporting.

The NTWEP has also started using the 'time since last reproduction' as a measure of eradication progress (Brooks and Jeffery 2018b). In cases where no seed production has been observed the discovery date is used to calculate time since last seed production. The time since last seed production (or discovery) accrues annually unless there is a seed production event (reproductive escape) causing the management area to suffer a 'reproductive relapse'. The last reproduction data is determined at the end of each financial year from a single (discovery or reproductive relapse) date for each management area. The last detection and last reproduction or discovery data have the same sample size and appear similar (Table 2) but are calculated differently. The data presented are examples from annual reporting to cost-share partners to the end of June 2021.

## BROAD ERADICATION PROGRESS

There are three large loci in north Queensland with more than 100 management areas each (Table 1). The remaining naturalized infestations cover between 1 and 82 management areas and are considered small loci. Non-naturalized occurrences such as potted, garden or nursery specimens are single site identification waypoints that occupy one management area each. There are records of cultivated specimens from public gardens in Sydney, Melbourne, Brisbane, Mt Tamborine and Townsville. Single large garden specimens become a small locus if additional plants are detected.

There have been no new loci or potted specimens in north Queensland since December 2014. The NTWEP has conducted extension and awareness activities during this time. These include TV advertisements, social media posts, landholder mailouts, newspaper articles, public displays of potted specimens, targeted group presentations (also with live specimens) and stakeholder awareness sessions. These activities have generated reports of garden plants and plants near known infestations, so they are important to build regional delimitation confidence. Four processes for detection of new infestations were identified by Brooks and Galway (2008). In the intervening years, the proportions have not changed dramatically with 46.6% of new *M. calvescens* loci detected by workers in a weed related field, 43.8% by information from the public and 9.6% from tracing information, mostly historically such as botanical garden records.

**Table 1.** Number and type of *Miconia calvescens* infestations in Australia to June 2021.

| | State | | |
|---|---|---|---|
| Infestation type | QLD | NSW | Total management areas |
| Large loci | 3 | | 417 |
| Small loci | 30 | 5 | 695 |
| Non naturalised specimens* | 21 | 10 | 33** |

*Includes botanical gardens, nurseries, and potted specimens, plus two locations in Melbourne. **Four historical botanical garden specimens are not in the management area database.

In the last six years, three locations (currently considered garden specimens) have been identified in northern New South Wales, including two in 2020–21. Across both states it will be important to maintain the capacity to detect unmanaged infestations, that originate in cultivated situations. That capacity currently remains with trained professional officers and members of the public. Given the detection effort that has been maintained, low numbers of non-naturalized locations increase confidence that the extent of the incursion is known.

## TRENDS IN MANAGEMENT AREA DATA

Of the 1141 management areas 63.7% had progressed to a monitoring stage, plant absence for more than a year by June 2021 (Table 1). There was a 5-year surge in discovery of new management areas from 2012–13 (Figure 1). This corresponded with an increase in field resources and field crews were instructed to record new site waypoints every 30 m to define infestations as the 'grid layer' was fully implemented. This was also the time increased seedling emergence was recorded after Tropical Cyclone Yasi (Brooks and Jeffery 2018a). Over the last four years, annual totals of newly discovered management areas have continued to taper, with most new areas discovered in the vicinity of known areas. In 2020–21 there were 28 new management areas recorded, with 20 of these sharing a boundary or a corner with a known management area, and 6 were within 1 km of a previously known point. Two new occurrences were in New South Wales, and more than two kilometres apart.

**Figure 1**. Discovery of new *M. calvescens* management areas (1 ha each) and the percentage of management areas with plants (control phase) at 12 monthly intervals. Some areas were being managed when the program commenced.

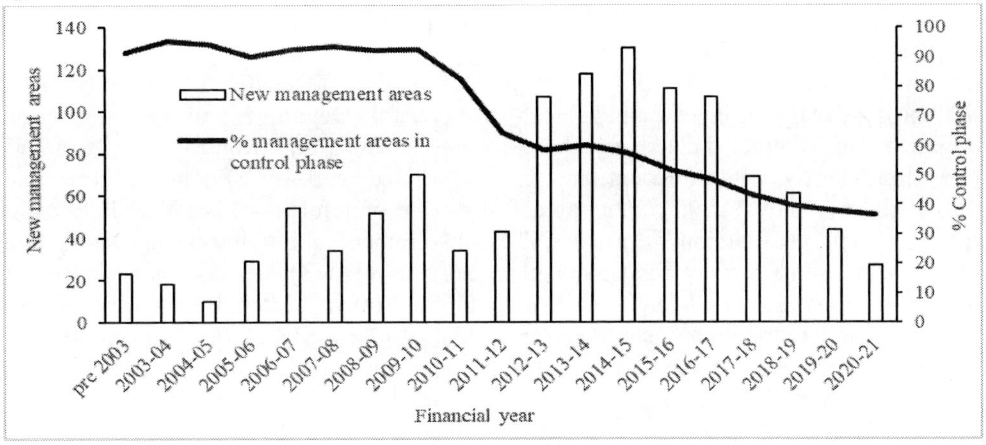


**Table 2.** Summary of years in monitoring phase and years since last mature plant was detected or the management area was discovered for 1141 *M. calvescens* management areas as of June 2021.

| Years | Management areas categorized by years in monitoring phase | Management areas categorized by years since last reproduction or discovery |
|---|---|---|
| 0 | 414 (control phase) | 28 |
| 1 | 109 | 50 |
| 2 | 83 | 64 |
| 3 | 118 | 71 |
| 4 | 93 | 116 |
| 5 | 81 | 115 |
| 6 | 84 | 135 |
| 7 | 37 | 112 |
| 8 | 25 | 108 |
| 9 | 39 | 47 |
| 10 | 27 | 44 |
| 11 | 13 | 95 |
| 12 | 6 | 51 |
| 13 | 6 | 34 |
| 14 | 0 | 27 |
| 15 | 1 | 17 |
| 16 | 2 | 6 |
| 17 | 1 | 11 |
| 18+ | 2 | 10 |

Typically, new *M. calvescens* management areas have small seedling counts which suggest periodic recruitment from past frugivorous dispersal, rather than the presence of locally mature plants. Only 26.6% or 304 of 1141 management areas had mature plants present at discovery. As such, the last reproduction or discovery column in Table 2 largely reflects the discovery patterns evident in Figure 1, with a surge 4 to 8 years ago. Fruiting and potentially mature plants (based on the basal diameter measurement) have been detected at a rate of between 0 and 0.4% of known cells in the last five financial years (data not shown). These reproductive relapses have only had negligible impact on the NTWEP search areas and did not require additional resources.

Many of the new management areas added since 2012 have low plant numbers, a short control phase and have progressed to monitoring from 2015 onwards. This progression to the monitoring phase, shown by the declining % control line in Figure 1, has continued to balance or outpace new and relapsing management areas.

NTWEP annual reporting includes the total proportion of management areas in long term monitoring (6 years +) and deemed provisionally eradicated for the purpose or reporting to the cost-shared partners. This was 243 management areas (14.6%) in June 2021 (Table 2) and has grown consistently since 2014. The number of management areas approaching six years monitoring should continue to see this value grow. The longer management areas spend in monitoring phase the less likely plants are to reoccur (a monitoring relapse event). More than 91% of relapse events (518 cases) have occurred from the first five years of monitoring, and there are two cases over 8 years of monitoring.

These counts may increase as the long-term monitoring sample size grows and with surveys years apart. *Miconia calvescens* seedlings can also grow very slowly in the forest understory and effectively form a difficult to detect 'seedling bank' (S. Brooks unpublished data).

## DISCUSSION

Aspects of the reporting data highlight the biological reality of eradicating this species. Many of the loci have been effectively managed for 10 to 15 years, during which the discovery of new management areas has continued. This protracted discovery phase in areas surveyed previously, with no recent mature plant records and close to known management areas results from the sporadic germination from the soil seedbank. *Miconia calvescens* forms a persistent soil seedbank, which is impossible to sample at low densities and inherently variable over a small scale, let alone the scale and complexity of the current field operational area. As such the NTWEP is developing multiple criteria to consider reductions in visit frequencies and eventually no visits. For example, Table 1 shows subsets of management areas with more than 10 years in the monitoring phase and more than 16 years since the last reproduction or discovery. In addition to the progress data, management areas will have to be considered in spatial groups rather than isolation. Further criteria include NTWEP confidence in the frequency, extent and recency of surveys and associated absence surveillance records. Within loci, adjacent management areas return a range of values for last reproduction and monitoring parameters. Such that any decision on declaring eradication is based on combinations of field observation experience, data and program research.

Further NTWEP refinements include more nuanced planning of ground searches. Once fully implemented this will see field surveys planned with habitat suitability layers and susceptibility information derived from time since last mature data.

To date, no new *M. calvescens* loci have been found through direct survey, reflecting the disparate nature of infestations resulting from the cultivation of plants. Unmanned aerial vehicles continue to be investigated to detect plants around known infestations, particularly on the margins in combination with the development of an AI (Artificial Intelligence) model. The NTWEP has been developing the capacity to collect and automatically screen aerial rainforest imagery for *M. calvescens* leaves as part of an AI system development.

The program is addressing the challenges of detecting all occurrences of this small tree in gardens or rugged terrain, via extension and awareness activities and from the ground or remotely from the air. The longevity of the soil seed bank and a wide potential dispersal buffer means the eradication program remains a long term and intensive proposition. The overall progress made towards the eradication of this serious tropical weed is prompting discussions about the type and duration of resources that are deployed to survey areas with continuous records of plant absence.

## ACKNOWLEDGMENTS

We thank Department of Primary Industries New South Wales and Rous County Council for their continued work on Miconia in New South Wales. We are grateful to all NTWEP field officers, Local Government and Queensland Parks and Wildlife Service (QPWS) officers for all their efforts in the field. Moya Calvert (Biosecurity Queensland) and Jens Froese (CSIRO) for continued development of a 'riskmapr' model for planning *M. calvescens* surveys.

## REFERENCES

Brooks, S.J., and K.E. Galway. (2008). Processes leading to the detection of tropical weed infestations during an eradication program. Proceedings of the 16th Australian Weeds Conference, eds R.D. van Klinken, V.A. Osten, F.D. Panetta and J.C. Scanlan. pp. 424-6. (Queensland Weeds Society, Brisbane).

Brooks, S.J. and Jeffery, M. (2010). Status of *Miconia calvescens* and the eradication program in Australia. Proceedings of the International Miconia Conference. eds L.L Loope, J.-Y. Meyer, B.D. Hardesty and C.W. Smith. (Maui Invasive Species Committee and Pacific Cooperative Studies Unit. Hawaii).

Brooks, S.J and Jeffery, M. (2018a). The effects of cyclones on a tropical weed eradication program. Proceedings of the 21st Australasian Weeds Conference, eds S Johnson, L Weston, H Wu and B Auld. pp. 119-23. (The Weed Society of New South Wales. Sydney).

Brooks, S.J and Jeffery, M. (2018b). Progress in the eradication of *Mikania micrantha* from Australia. Proceedings of the 21st Australasian Weeds Conference, eds S. Johnson, L. Weston, H. Wu and B. Auld. pp 350-3. (The Weed Society of New South Wales).

Jeffery, M. and Brooks, S.J. (2016). Eradication in the tropics: constantly changing and adapting. Proceedings of the 20th Australasian Weeds Conference, eds R. Randall, S. Lloyd and C. Borger. pp. 23-7. (Weeds Society of Western Australia, Perth).

# Horizon scan for incoming weeds into Queensland, Australia

Olusegun O. Osunkoya* [1], Christine Perrett[1], Moya Calvert[1] & Steve Csurhes[1]
[1] Department of Agriculture and Fisheries, Biosecurity Queensland, Ecosciences Precinct, Dutton Park 4102, Queensland, Australia
(*Olusegun.osunkoya@daf.qld.gov.au)

**Summary** Invasive alien species (IAS), of which weeds are a subset, are often threats to natural and managed ecosystem services, including agriculture. The distribution and impacts of IAS are expected to exacerbate in response to increasing human connections (globalisation, commerce) and climate change. Unfortunately, once firmly established across large areas, populations of invasive species tend to become highly resilient and preventative measures become generally unaffordable or unrealistic. Hence, there is a need to identify emerging threats that are still in an early phase of invasion. These emerging threats can be candidates for preventative action; either complete eradication or early containment. Using the grey literature and the Web, including "The weed flora of Australia and its weed status" and "CABI's invasive species compendium", we carried out horizon scanning for ~230 weed species that have been identified as potential IAS in an early stage of invasion in Queensland (QLD), Australia. The majority of these potential IAS are of South and tropical North America in origin, and their present invaded ranges are wide (North/South America, Oceania, Asia, and Africa). Potential impacts are deemed generally negative (especially on environment/ecology, biodiversity, livestock, and economy/livelihood) to neutral, but positive impacts (on livestock, cultural amenity, and economy) were also identified. Introduction mode and pathways of entry are likely to be deliberate via nursery/horticultural trade (40 %), agroforestry for soil stabilization/habitat restoration (26 %) and mail-order/Internet (17.3 %). Once in their invaded ranges, further spread (dispersal) can be expected via mammals (especially by birds and rodents), soil disturbance/waste disposal and aquatic systems. Using the dataset on impact and spread of the focal IAS in invaded ranges around the globe and adjusting for countries/regions whose climates match closely to that of QLD, we derived a state-wide horizon weed priority list of high, medium and low impact scores for policy, research and management.

**Keywords:** Climate-matching, Horizon-scanning, Invasion-pathway, Pest-risk-assessment, Prioritisation, Weeds

## INTRODUCTION

The spread and impact of invasive alien species remain unabated in most habitats around the globe because of increasing human connections and habitat modifications (Beaury *et al.* 2021, Osunkoya *et al.* 2021). Hence for maintenance of the integrity of natural and/or managed ecosystems, biosecurity risks need regular assessments and prioritization for both established and potentially incoming IAS. Using horizon-scanning methodology (Cuhls 2019), we assessed the potential impact of entry and/or spread of new weeds in the State of Queensland (QLD). Horizon or environmental scanning warns us about impending change. The term 'horizon-scanning' evokes images of lookouts on old ships or modern-day radar scanning the horizon. The horizon scanner is to the future what the lookout is to the sea. Most change does not occur suddenly, out of the blue, even if it initially appears that way. Horizon scanning attempts to break the habit of ignoring the early signs of change. It forces people to look at the novelty happening around them and report those signs that could have a significant impact on the enterprise (i.e., on ecosystem services and goods), not just those changes that are sure to have an impact.

For the assessment, we initially relied on a horizon weed list of ~230 species compiled and regularly updated by Biosecurity Queensland (BQ) staff of the Department of Agriculture & Fisheries (DAF) (Csurhes 2021). Our aims were:

1. Explore some of the issues that are deemed current and applicable in IAS management as stated in Neve *et al.* 2020 (climate change, invasiveness (spread), pathways, relative role of species traits (biology), human interaction (sociology/economy), and habitat ranges (geography)).
2. Identify likely threats by the horizon weeds in QLD, and rank them for proactive management, including eradication where feasible.

## MATERIALS AND METHODS

At the inception of the project, a tentative list of tasks were drawn up, as below:

1. Use BQ compiled list of horizon weeds (~230 species) as a starting point;

2. Review the pest management plans of all 72 local government areas in the state for emerging species;
3. Cross-check the list with introduced species in Australia listed in Randall (2007);
4. Present the list to stakeholders (impacted farmers, natural resource managers and biosecurity officers) via either online or physical meetings;
5. Review the grey and scientific literature (online) for the global distributions of listed weeds, noting each weeds: (a) native vs invaded ranges, (b) impact on agriculture, nature conservation, health, social-wellbeing and economy, and (c) pathways, including dispersal modes;
6. Predict the potential distribution of listed species in QLD;
7. Combine data on realised/potential distribution of the weeds worldwide and in QLD, and their (perceived) impacts to generate a prioritized list and actions required;
8. Examine the feasibility of eradication.

Tasks 1, 3, 5, 6 have been completed, to some extent, and will be the focus of this paper. For task one we extracted the species list from Csurhes (2021). For task three we used Randall (2007) to confirm taxonomic status, life forms, and weed status. For task five we used online sources including, The Global Biodiversity Information Service (GBIF), Global Register of Invasive and Introduced species (GRIIS), Plants of the World Online (POWO), and CABI Invasive species compendium (ISC). We found CABI the most comprehensive in terms of data needed, and hence our reports are based mainly on extracts from this database. From the CABI-ISC database, we were able to extract information on native and invaded countries for each species, their documented impact, and invasion pathways. For task six we used species and habitat modelling software of the Centre of Excellence for Biosecurity Analyses (MESS and EX-DET rather than the popular CLIMEX) to project/match the climate of the state of QLD to those of invaded countries around the globe (see https://apps.cebra.unimelb.edu.au/climate_matcher/).

We combined indices of spread (based on native and invaded ranges globally), documented impact, pathways, and QLD habitat/climate suitability (similarity) to native ranges (i.e. country) of the weeds to derive a state-wide horizon weed priority list of high, medium and low impact scores.

RESULTS

Of the ~ 230 species of concern in QLD, 197 of 232 (85 %) also appear on Randall's (2007) list. Of these 197 species, we were able to compile comprehensive information on global spread, impact, habitat ranges (using country as surrogates), and invasion pathways for 132 species (132/197 = 67 %).

As seen in Figure 1, the majority of potential IAS of QLD are from South and tropical North America, and to some extent Indian and Chinese subcontinents. Currently, their invaded ranges are broad (North/South America, South and East Asia, Oceania, and Africa) (Figure 2).

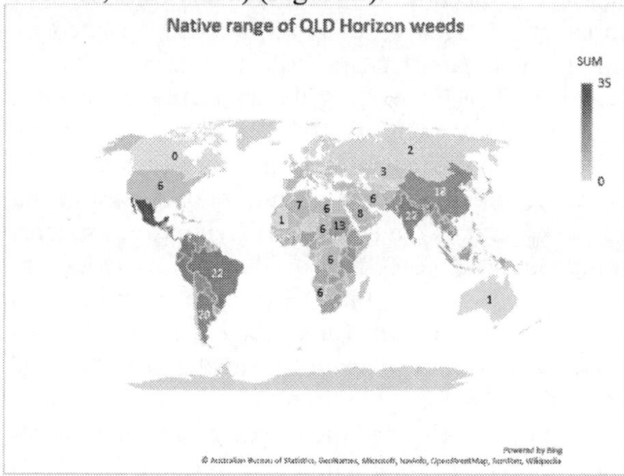


**Figure 1.** Native ranges (countries) of horizon weeds of QLD. Number on maps refers to number of weeds out of 132 originating from that particular country.

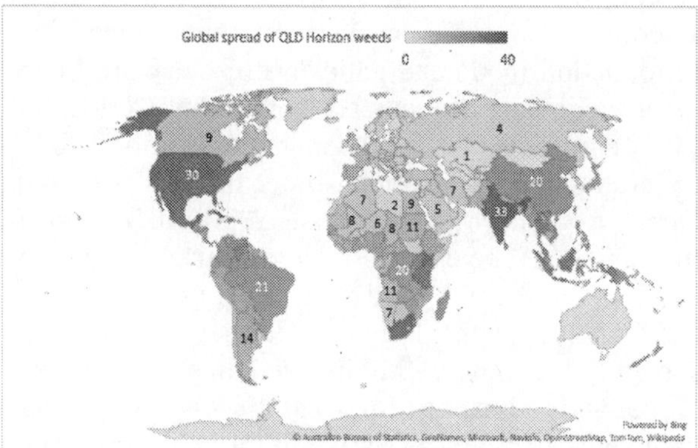


**Figure 2.** invaded ranges (countries) of horizon weeds of QLD. Number on maps refers to number of weeds out of 132 spreading into that country.

Introduction mode and pathways of entry of QLD horizon weeds are likely to be deliberate via nursery/horticultural trade (40 %), agroforestry for soil stabilization/ habitat restoration (26 %) and mail-order/ internet (17.3 %) (Figure 3). Once in their invaded ranges, further spread (dispersal) can be expected via mammals (especially by birds and rodents) (18.5 %), soil disturbance/ waste disposal (20.2 %) and aquatic systems (18.5 %) (Figure 4).

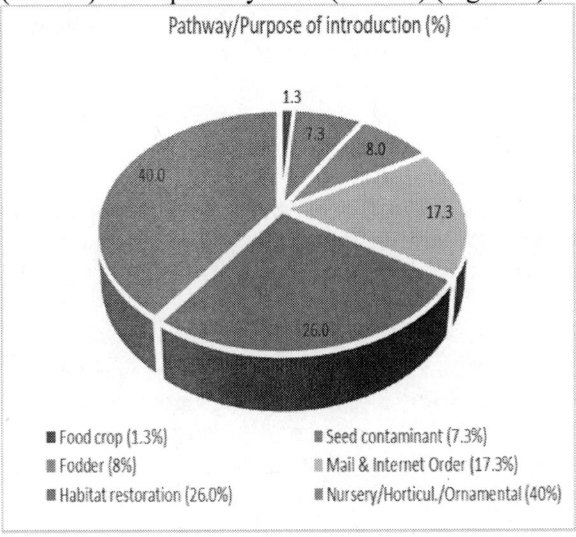


**Figure 3.** Identified pathways of entry for horizon weeds of QLD.

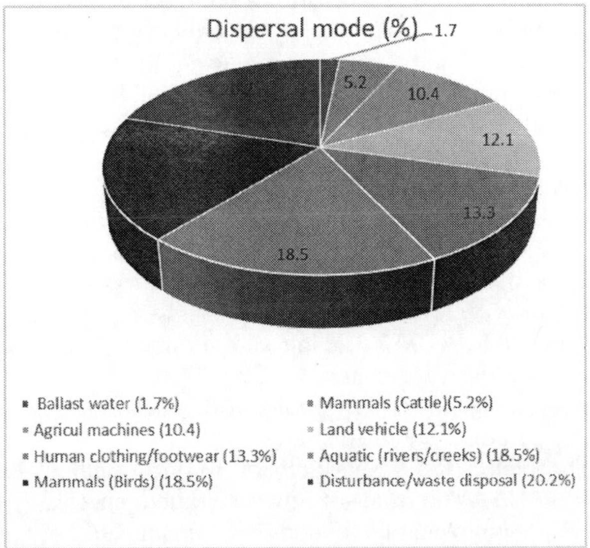


**Figure 4.** Dispersal vectors for horizon weeds in QLD.

Potential impacts are deemed generally negative (especially on environment/ ecology, biodiversity, livestock, and economy/ livelihood) to neutral, but positive impacts (on livestock, cultural amenity, and economy) were also identified (Figure 5).

Using a simple summation on spread, impact and pathways indices, the top horizon weeds worthy of immediate attention span all life forms, although trees and shrubs dominate (Table 1).

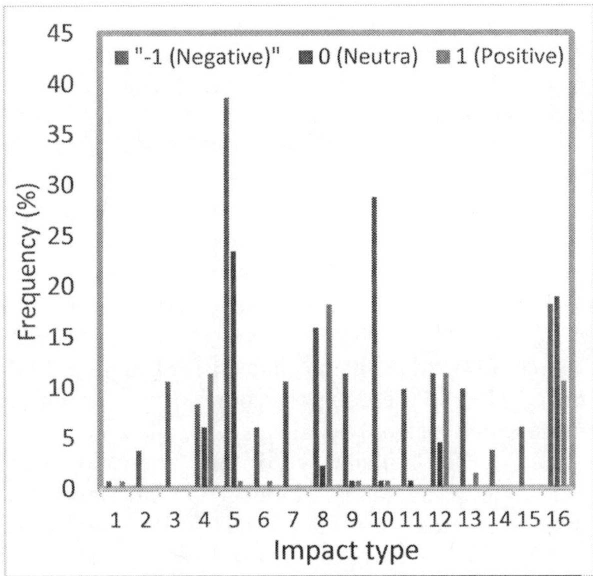


| Impact type | |
|---|---|
| 1=Animal/plant collection | 7=Forestry |
| 2=Animal/plant product | 8=Livestock |
| 3=Crop | 9=Nature Fauna (Biodiversity) |
| 4=Cultural Amenity | 10=Nature Flora (Biodiversity) |
| 5=Environment/ecology | 11=Rare/Protected Species |
| 6=Fisheries | 12=Human health |
| | 13=Tourism |
| | 14=Trade/international relations |
| | 15=Transport/Travel |
| | 16=Economy/Livelihood |

**Figure 5**. Documented impact of QLD horizon weeds on nature and socio-economic factors in their invaded ranges around the globe.

**Table 1.** Top 25 horizon weeds of QLD worthy of management action.

| Species | No. of countries invaded aside from Aust | Life form |
|---|---|---|
| *Gliricidia sepium* | 116 | Tree |
| *Coffea arabica* | 115 | Shrub |
| *Coix lacryma-jobi* | 108 | Grass |
| *Pithecellobium dulce* | 97 | Tree |
| *Jatropha curcas* | 96 | Shrub/Small tree |
| *Arundo donax* | 95 | Grass |
| *Pennisetum purpureum* | 76 | Grass |
| *Syzygium jambos* | 69 | Tree |
| *Spathodea campanulata* | 67 | Tree |
| *Mimosa pigra* | 66 | Shrub |
| *Robinia pseudoacacia* | 55 | Shrub/Small tree |
| *Ipomoea alba* | 53 | Vine |
| *Pennisetum polystachion* | 52 | Grass |
| *Thunbergia fragrans* | 49 | Vine |
| *Dichrostachys cinerea* | 48 | Shrub |
| *Haematoxylum campechianum* | 48 | Tree |
| *Thunbergia grandiflora* | 47 | Vine |
| *Leonotis nepetifolia* | 45 | Shrub |
| *Elephantopus mollis* | 44 | Herb |
| *Cereus uruguayanus* | 42 | Succulent |
| *Chromolaena odorata* | 42 | Shrub |
| *Gmelina arborea* | 42 | Tree |
| *Ulex europaeus* | 39 | Shrub |
| *Clerodendrum chinense* | 38 | Shrub |
| *Caesalpinia decapetala* | 37 | Vine |

DISCUSSION

Much like well-established weed populations in Queensland (Osunkoya *et al.* 2019), most horizon weeds come from the Americas, suggesting that entry of goods (e.g., horticultural products, food crops, machineries etc) from this continent, along with those from the Indian subcontinent require greater scrutiny if we are to avoid further problems of IAS. South Africa, and selected countries in East Africa (Kenya, Uganda, Madagascar) and certain provinces in southern China are also worth noting as common sources of weed incursions into QLD. It is noted that the climates in these regions/countries are similar to those experienced over large parts of QLD.

The horizon weeds are likely to have varying impacts, though the negative influences are often reported, especially on biodiversity and ecology. However, from the analyses there are cases of reported positive impacts (cultural amenity, livestock, human health and livelihood). The positive impacts of alien species are probably underestimated, as there is often a perception bias against alien species (Goodenough 2010, Shackleton *et al.* 2019). Hence management must balance the two impacts in terms of cost-benefit analyses to address the trade-offs in IAS management so that successful implementation of management practices is facilitated.

It is important to note that dominant invasion pathways include the commercial trade in garden ornamentals, forestry/ habitat restoration and mail-order trade. While import restrictions on plants have improved greatly over the past 10-20 years, it is clear that the nursery trade has served as a primary invasion pathway in the past and still needs close monitoring and regulation. In addition, post-border dispersal via water ways, agricultural machines, land vehicles and soil disturbance/ movement appear to be common vectors of spread. Hence, these vectors are worthy of closer surveillance, where feasible, to limit the spread of horizon weeds in QLD. Regional regulation, coupled with improved public awareness for consumers, are also desirable to minimise the spread and impact of invasive plants on the horizon weeds list.

**Conclusion**

No doubt, some of the indices used for assessment need refinement, particularly climate-matching which, for now, is based on 19 variables of rainfall and temperature. We have also modelled habitat matching/ suitability in QLD as a single entity when there are strong regional differences. Hence the modelling work could be improved if key components of these numerous driver variables were narrowed down. Also, the overall index has been based on simple summation of individual indices, which may not necessarily hold, as some indices might be more important than others. As such, varying weightings may be desirable to facilitate a more robust assessment.

ACKNOWLEDGMENTS

We thank The Queensland's Department of Agriculture and Fisheries for providing funding for the project.

REFERENCES

Beaury, E.M., Patrick, M. and Bradley, B.A (2021). Invaders for sale: the ongoing spread of invasive species by the plant trade industry. *Front Ecol Environ* 19(10), 550-556

Csurhes, S. 2021. Risk assessment and prioritization of 229 emerging weed threats detected in Queensland. Technical report. Biosecurity Queensland. 54 pages.

Cuhls, K.E. 2019. Horizon Scanning in Foresight – Why Horizon Scanning is only a part of the game. *Futures & Foresight Science* 2(e23).

Goodenough, A.E. 2010. Are the ecological impacts of alien species misrepresented? A review of the "native good, alien bad" philosophy. *Community Ecology* 11, 13–21

Neve, P., Barney, J.N., Buckley, Y. et al 2018. Reviewing research priorities in weed ecology, evolution and management: a horizon scan. *Weed Res* 58, 250–258.

Randall, R.P. 2007. The introduced flora of Australia and its weed status. CRC for Australian Weed Management, University of Adelaide, Glen Osmond, South Australia.

Shackleton R.T., Shackleton C.M., & Kull C.A. 2019. The role of invasive alien species in shaping local livelihoods and human well-being: A review. *J. Environ Manage* 229, 145-157.

Osunkoya, O.O., Lock C.B., Dhileepan K. & Buru J.C. 2021. Lag times and invasion dynamics of established and emerging weeds: insights from herbarium rec ords of Queensland, Australia. *Biol. Invasions* 23, 3383–408.

Osunkoya O.O., Froese J.G., Nicol S., Perrett C., Moore K., Callander J., & Campbell S. 2019. A risk-based inventory of invasive plant species of Queensland, Australia: Regional, ecological and floristic insights. *Austral Ecol* 44, 1123–1138.

# Making the most out of a sticky situation - acting on Sticky Nightshade as an emerging threat

Marita Sydes[1]
[1]Central Tablelands Local Land Services, Orange, Australia
(marita.sydes@lls.nsw.gov.au)

**Summary** Sticky nightshade (*Solanum sisymbriifolium*) is an invasive species that is native to South America. For a long time in Australia, sticky nightshade was considered a weed of disturbed sites in western Sydney, with scattered records outside this area in NSW, Queensland, Victoria and Western Australia. Sticky nightshade was probably introduced into the central tablelands with fodder in the 1980's drought and since then has been largely confined to the Upper Belubula south of Orange. In 2018, members of the Central Tablelands Regional Weed Committee observed that the sticky nightshade was spreading from this initial infestation area and the weed was subsequently identified as an emerging threat in the region.

In late 2018, the committee moved to investigate further the potential impacts and distribution of sticky nightshade in the central tablelands. Records for sticky nightshade over recent years supported that the core infestation was still in the Upper Belubula, but that new incursions had been recorded in the surrounding council areas. As a result, members of the regional weed committee and landholders impacted by the weed met to conduct a weed risk assessment for sticky nightshade. With the support of the Central Tablelands Local Land Services Board and New South Wales Department of Primary Industry it was determined that sticky nightshade should be listed as a regional priority species in the Central Tablelands Regional Strategic Weed Management Plan – a first for NSW.

This presentation will outline the process to get Sticky Nightshade listed as a regional priority weed species as well as the resources and actions that have been developed to raise awareness of this emerging weed. It is hoped that through listing sticky nightshade in the central tablelands region – other regions will be aware of the threat posed by this weed and be proactive in managing the species.

**Keywords** Sticky nightshade, Local Government, Local Land Services, NSW DPI, weed risk assessment, priority weed species. emerging weeds

# Competition from great brome and barley grass reduces wheat yield

Harmohinder Dhammu[1], Abul Hastem[3], Glen Riethmuller[2], Catherine Borger[1]
[1]DPIRD, Northam, Australia,
[2]DPIRD, Merredin, Australia,
[3]Ex- DPIRD, Northam, Australia
(harmohinder.dhammu@dpird.wa.gov.au)

**Summary** In Western Australia (WA), some minor weeds like brome grass (*Bromus diandrus*) and barley grass (*Hordeum leporinum*) are now emerging as major weeds with greater infestation and frequency. Information on extent of grain yield losses due to these weed species in intensive no-tillage cropping systems of WA is limited. With an aim of determining loss in wheat yield due to competition from these weed species, four field trials were carried out from 2016-2019 on sandy to sandy loam soils at the DPIRD's Wongan Hills Research Facility (-30°47'59.99" S 116°36'59.99" E). The trial sites were selected with very low weed burden and no prior seed bank of the target species. The trials were laid out in completely randomised split plot design with weed species as the main plot factor and weed density as the sub plot factor with four replications. Seed of each weed species was broadcasted at four densities (weed free, low - 125-250 seeds/m2, medium - 250-500 seeds/m2 and high - 500-1000 seeds/m2) over a 5m x 1.1m area of each unit plot, prior to sowing wheat. Wheat cv Mace was sown using 50 kg/ha seed rate on 22 cm row spacing at the end of May-middle of June and machine harvested in November each year. The data was subjected to ANOVA using GENSTAT 19th edition. The results indicate that great brome at low, medium and high densities (averaged over four years) of 74, 141 and 214 plants/m2 respectively, caused 13, 20 and 28% reduction in wheat grain yield compared to the weed free treatment. Similarly, barley grass at an average density of 74, 148 and 245 plants/m2 resulted in 9, 11 and 18% reduction in wheat grain yield. Effect of seasonal conditions (e.g. rainfall) on weeds emergence and their competitiveness was important and will be discussed.

**Keywords** Crop-weed competition, wheat, weeds, grain yield loss

# Predicting profitability of summer weed control timing and impact on crop yield potential: $ummer

Yvette M. Oliver[1] and Rick Llewellyn[2], Therese McBeath[2], Bill Davoren[2] and Andrew Ware[3]
[1]CSIRO, Private Bag 5, Wembley 6913, Western Australia
[2]CSIRO, Locked Bag 2, Glen Osmond SA 5064
[3] 7 Seaton Avenue Port Lincoln SA 5606
(Yvette.oliver@csiro.au)

**Summary**

Summer fallow weed control has been shown to be an integral component of modern crop production systems in a changing Australian climate. At the same time suboptimal control of summer weeds have been shown to have high continuing cost to yield and profitability. Field trial results are used to show the importance of the soil water and nitrogen drivers of the impact of summer weed control timing options under different soil and weed scenarios. A predictive tool ($ummer) designed to inform summer weed control investment decisions including the impact of timing options is presented.

**Keywords** Summer Weed control, Yield benefits, APSIM

## INTRODUCTION

Summer fallow weed control is an increasingly important component of modern cropping systems and has played an important role in how Australian grain growers have profitably adapted to climate challenges (Hunt and Kirkegaard 2011). Although usually shown to increase average profitability, there are many situations where summer weed management and timing decisions are not always clear-cut, and returns from fallow can vary greatly by region, soil type and season (Oliver *et al.* 2010). The cost of control and impact of summer weeds on crop yield is high, with estimates of annual revenue loss due to summer weeds in southern and western Australian cropping regions estimated at $350 M (Llewellyn *et al.* 2016). Sub-optimal summer fallow weed control has been identified as one of the major reasons for Australian wheat yields not reaching yield potential (Hochman and Horan 2018).

To inform summer weed decisions which typically take place under uncertainty of the coming summer and winter crop season conditions, we have applied APSIM-based modelling to produce a tool designed to be applied at the time of specific summer weed control decisions. The $ummer tool produces probabilistic estimates of the two important elements of summer weed impact; soil water, soil nitrogen (Hunt *et al.* 2013) and subsequent crop yield impacts from summer weed populations and timing options.

To support development of the tool we have also conducted field trials to expand the relatively limited range of summer weed control timing field experiments that have measured both soil water and nitrogen impacts in conjunction with simulation modelling on characterized soils.

## MATERIALS AND METHODS

### Trials

In addition to preliminary field validation trials conducted in Western Australia in 2020-21 (data not presented), trials have been conducted in the southern region in 2021.

Trial 1 - Wharminda South Australia: dune location, sand, increasing clay from approximately 0.3 m. Summer rainfall (November-March) 2021/2022 was 205 mm which is significantly higher than the long-term average of 86 mm. There was a 71 mm and 16 mm of rainfall on the 22nd and 24th January respectively.

Trial 2 - Bute South Australia: a) dune site, shallow sand with clay and calcrete increasing with depth and b) flat site, a loamy sand over clay. The 2021/2022 summer rainfall (November-March) was 127 mm, which slightly higher than the 101 mm long-term average. There were large rainfall events of 54.4 mm on 12th November 2021 and 33 mm on 21st January 2022.

Treatments:

1. Full control with follow-up as required (10th -17th February and 4th -17th March),
2. Initial control with no follow-up (10th -17th February)
3. Delayed control (4th -17th March)
4. No control

All treatments were treated with knock-down herbicide in April in preparation for crop seeding. Soil water and nitrogen measurements are presented here, with crop yields to be measured from harvested crops in November-December 2022.

APSIM simulations are being conducted on the trials to validate the modelling and show the range of probable yields with different soil water measured in April from the weed control treatments in the trials with 100 years of season conditions.

**$ummer weed app**

The $ummer tool is currently populated with APSIM modelling output based on:

1) Different locations (7 sites across Western Australia, South Australia, Victoria)
2) Contrasting soils of sand, loam or clay at each site
3) A range of periods in which the target population of weeds can germinate based on rainfall (December, January, February, March) where there are no weeds at other times.
4) Different weed types (deep and shallow rooted)
5) Difference in weed density (from 1 to 50 plants $m^{-2}$)
6) Differences in maturity of the weed population at the time of assessment (days since typical emergence).
7) A range of spray timing options from time of assessment through to various delayed options, compared to no weed control prior to pre-seeding time control.

The modelling is focused on wheat impacts, and the crop was assumed to be sown between 25th April and 30th June when 15 mm rain has fallen over 5 days, with 100 plants $m^{-2}$ of Mace wheat, sown with 50 kg $ha^{-1}$ nitrate and a further 50 kg $ha^{-1}$ 40 days after sowing. The high level of fertiliser was to ensure that only water stored, and rainfall were limiting the yield potential, rather than any difference in N due to weeds. The impact of the weed populations on available soil N is reported separately.

The app is designed for the scenario that the user is standing in a paddock with summer weeds and inputs the location, date, basic soil type and general weed population characteristics (i.e. density and age of deep or shallow rooted weeds) (Fig. 1). The app then asked the user to compare 'control now' and 'control in X days' and/or 'do nothing' scenarios.

From this input the app chooses the set of simulations to determine the expected yield benefit from the control options, determined by the soil water conditions. The gain in starting N available to the subsequent crop is also estimated. The app uses an adjustable wheat price to determine expected returns from control and presents results as the likelihood of achieving an outcome (e.g. probability of a $20 $ha^{-1}$ yield gain) as well as expected value (average).

To demonstrate the $ummer weed app outputs (Fig 4) the Minnipa South Australia site was chosen as the closest site to Wharminda, with inputs of 50 deep rooted summer weeds $m^{-2}$ on the 10th February that had emerged 10 days ago. The $ummer weed app compared scenarios where these could be killed in that week or in the next 30 days compared to no summer weed control prior to pre-seeding control. Both Sand and Loam soil results are shown for comparison.

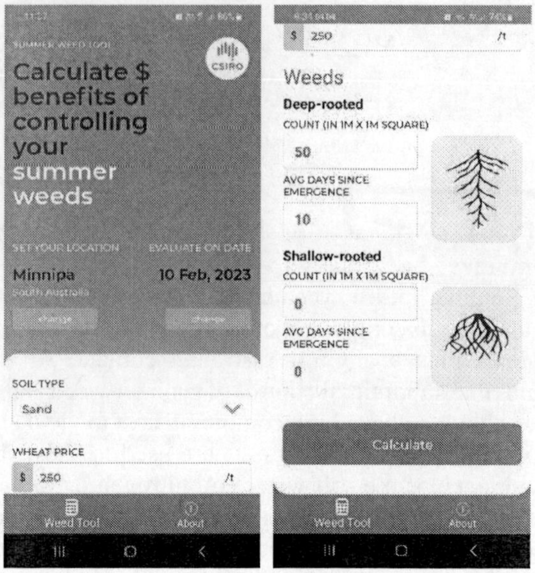


**Figure 1**. Two screen shots of the $ummer weed app input pages

## RESULTS

*Trial results*

The high summer rainfall in November and January caused high weed density and different species at the sites. At Wharminda the average weed density was 206 plants $m^{-2}$ with blanket weed (79 plants $m^{-2}$), volunteer wheat, (79 plants $m^{-2}$), Lovegrass (21 plants $m^{-2}$), capeweed (17 plants $m^{-2}$), medic (13 plants $m^{-2}$), and heliotrope (10 plants $m^{-2}$). At Bute Dune and Flat sites the weed density was 79 and 65 plants $m^{-2}$ with the dominant weed volunteer wheat (78, 49 plants $m^{-2}$) as well as a mixture of large and small heliotrope (1,8 plants $m^{-2}$).

**Table 1.** Additional water (mm) in profile (to 60 cm) compared to no weed control

| Treatment | Wharminda | Bute Dune | Bute Flat |
|---|---|---|---|
| Full ongoing control | 22.2 | 22.6 | 16.9 |
| Initial control with no follow up | 15.9 | 14.2 | 9.5 |
| Control delayed 30 days | 9.3 | 8.8 | 3.7 |

Between February and early April, the uncontrolled weeds used an additional 17-23 mm of stored soil water compared to when the weeds were fully controlled (Table 1). Even when the weed control was delayed by 30 days, the additional water was 4-10mm. At Wharminda, there was additional soil nitrate to 0.9 m of 30, 13 and 4 kg ha$^{-1}$ for the full controlled, initial control with no follow up and delayed control treatments respectively.

However, in April the GSR is unknown, so these April soil water levels were used with APSIM and climate data from 100 seasons to estimate the range of likely yield increases from this stored soil water for 2022 season.

At Wharminda, there was a 50% chance of achieving a 0.31 t ha$^{-1}$ yield increase from managing weeds completely, 0.22 t ha$^{-1}$ for only for the early weed control or 0.19 t ha$^{-1}$ for delaying weed control by a month. In this case, the average and 50$^{th}$ percentile values were similar (Fig 2). There was about 20% of years where there was no yield increase from these differences in soil water, which is likely in years with high growing season rainfall e.g. 13% of the years had greater than 280 mm (Fig 3).

At Bute, the soil water differences resulted in less chance of achieving yield increases, with up to 50% of years having no difference in yield (Fig 2). This lower chance of a yield benefit at Bute may be caused by the higher GSR where 33% of years have growing season rainfall greater than 280 mm (Fig 3). However, due to the abnormal distribution the average yield increase for full weed control was 0.19 at Dune and 0.26 t ha$^{-1}$ at Flat sites and for delayed weed control was 0.12 t ha$^{-1}$ at Dune and 0.08 t ha$^{-1}$ at Flat sites.

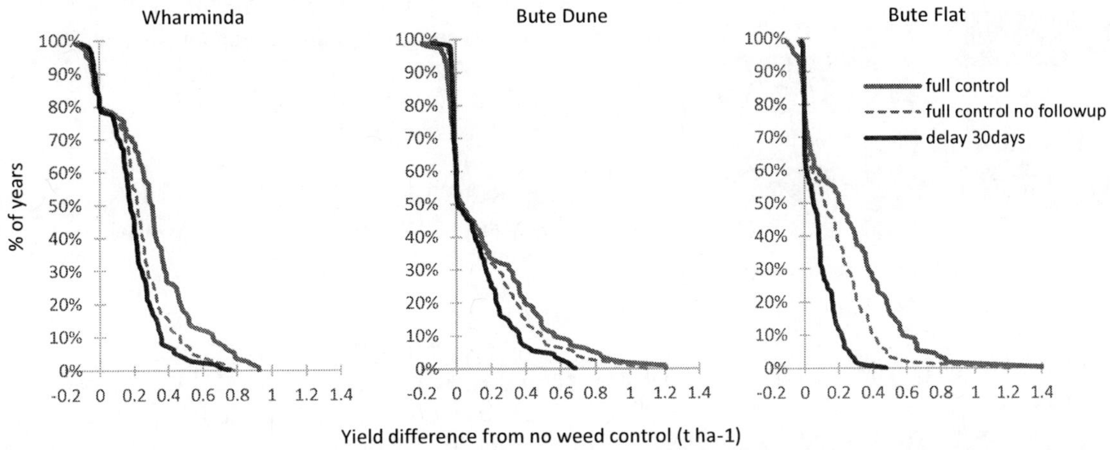


**Figure 2.** The yield increase (kg ha$^{-1}$) estimated using APSIM with the different starting soil water measurements for the treatments and 100 years of climate at Wharminda and Bute.

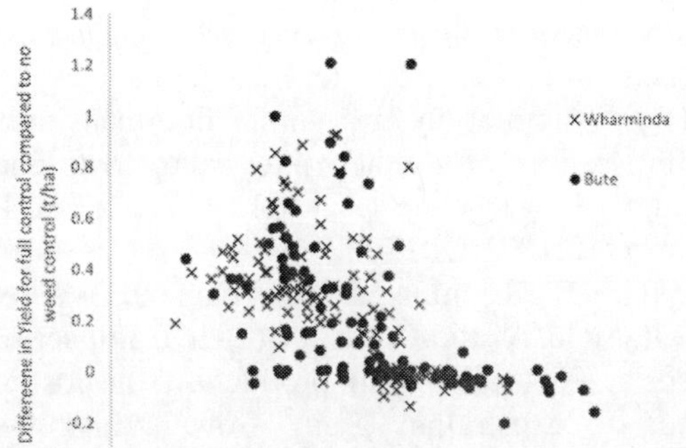


**Figure 3.** The growing season rainfall compared to the APSIM predicted yield increase (kg ha$^{-1}$) using the water content on 4$^{th}$ April 2022 from full weed control and no weed control with 100 season finishes at Wharminda and Bute.

APSIM modelling forms the basis for the $ummer weeds app. However in the simulations the weeds are grown to create the soil water difference, rather than using measured values (Fig 2) and only the years where weeds germinated are used to estimate the yield differences. An example of the output from the app is shown in Figure 4.

At Minnipa, with 50 deep rooted weeds that were controlled now or in 30 days, the main findings are: there are lower yield benefits in sandy compared to loam, which is due to the lower ability to store water in sands (Table 2). Delaying the weed control by 30 days reduced the yield benefit. When you use threshold values of 0.2 t ha-1 (similar to a break-even yield) this was achieved in 78-90% of years even if the weed control was delayed by 30 days.

The Nitrogen in the soil was similar for the two soils at Minnipa with an average 17-21 kg ha$^{-1}$ left in the soil if weeds were managed now, which reduced to 9-10 kg ha$^{-1}$ if weeds were managed in 30 days. A difference of at least 10 kg N ha$^{-1}$, assumed to be an amount potentially influential in nutrient management decisions, occurs less often when weed control is delayed (Table 2).

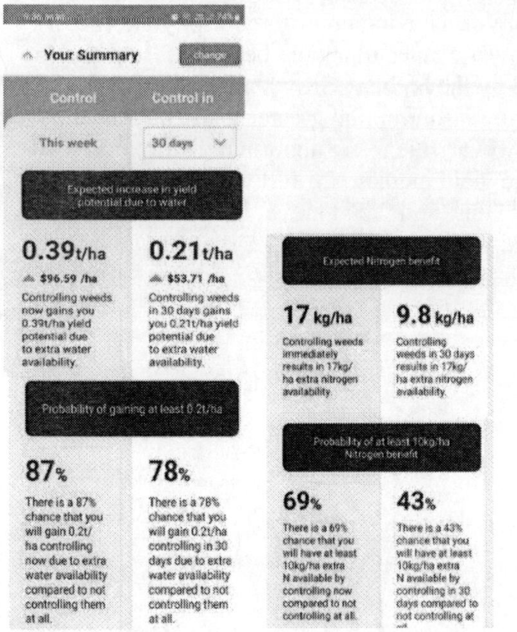


**Figure 4**. $ummer weed app output at Minnipa using a Sand for the Yield and nitrogen benefits

**Table 2.** Yield (t ha$^{-1}$) and Nitrogen (kg ha$^{-1}$) increases compared to no weed control and the probability of achieving yield and nitrogen targets.

| | Weeds controlled | |
|---|---|---|
| | This week | 30 days |
| Sand | | |
| Yield increase (t ha$^{-1}$) | 0.39 | 0.21 |
| Probability of gaining > 0.2 t ha$^{-1}$ | 87% | 78% |
| Nitrogen increase (kg ha$^{-1}$) | 17 | 9.8 |
| Probability of achieving > 10 kg ha$^{-1}$ Nitrogen benefit | 69% | 43% |
| Loam | | |
| Yield increase (t ha$^{-1}$) | 0.52 | 0.26 |
| Probability of gaining > 0.2 t ha$^{-1}$ | 96% | 90% |
| Nitrogen increase (kg ha$^{-1}$) | 21 | 8.6 |
| Probability of achieving > 10 kg ha$^{-1}$ Nitrogen benefit | | |

## DISCUSSION

With the high summer rainfall at the Wharminda and Bute sites, there were large densities of weeds which if left uncontrolled used 17-20 mm of stored soil water. Whether this additional water also increased yield depended on the following GSR. GSR greater than 280 mm led to little yield benefit to stored soil water. There was a large range of likely yield increases, which highlights the risk of weed management when the GSR is unknown, and early season indicators may play a role in summer weed management.

The $ummer weed app can be used with other sites, weed types and densities in addition to those shown here and is designed to be readily applicable at the time of summer weed control decisions to inform expectations of yield benefits and likelihood of profitable yield benefit for different spray timings.

## ACKNOWLEDGMENTS

The authors would like the thank GRDC for funding projects Locally important weeds and Better summer weed management decisions in southern and western Australia (CSP2201-005RTX) and Alex Douglas, Gurjeet Gill, Mark Saunders, Jacob Giles, Jason Emms and SquareV.

## REFERENCES

Hochman, Z. & Horan, H. (2018). Causes of wheat yield gaps and opportunities to advance the water-limited yield frontier in Australia. Field Crops Research. 228, 20–30 (doi.org/10.1016/j.fcr.2018.08.023)

Hunt, J.R., Kirkegaard, J.A. (2011) Re-evaluating the contribution of summer fallow rain to wheat yield in southern Australia. Crop & Pasture Science 62, 915–929. (doi.org10.1071/CP11268).

Hunt J.R., Browne, C., McBeath, T., Verburg, K., Craig, S., Whitbread, A.M. (2013). Summer fallow weed control and residue management impacts on winter crop yield through soil water and N accumulation in a winter dominant, low rainfall region of southern Australia. Crop and Pasture Science 64, 922–934. (doi.org/10.1071/CP13237)

Llewellyn R.S., Ronning D., Ouzman J., Walker S., Mayfield A. and Clark M. (2016). Impact of weeds on Australian grain production: the cost of weeds to Australian grain growers and the adoption of weed management and tillage practices. Report for GRDC: CSIRO (www.grdc.com.au/ImpactOfWeeds)

Oliver, Y.M., Robertson, M.J., Weeks, C. (2010) A new look at an old practice: Benefits from soil water accumulation in long fallows under Mediterranean conditions. Agricultural Water Management 98, 291–300. (doi.org/10.1016/j.agwat.2010.08.024)

# A collaborative and national approach for understanding the distribution of weeds in Australia

Katherina Ng[1], Kirstin Proft[1], David Mitchell[1], Miles Keighley[1], Phil Tennant[1], Margarita Medina[1]
[1]Australian Bureau of Agricultural and Resource Economics and Sciences, Canberra, Australia
(katherina.ng@agriculture.gov.au)

**Summary** Invasive weeds have significant impacts on biodiversity, ecosystem services, social amenity and agricultural industries. However, lack of reliable data is a barrier to quantifying both the distribution of priority weed species and the extent of their impacts. While distribution information is available across the country, an up-to-date, compiled national dataset is not currently available. Further, how distribution data are collected across the country varies in terms of protocol, geographic scale, and temporal frequency.

ABARES is running a project in collaboration with jurisdictions, CSIRO and other partners, to address these fundamental data gaps and challenges to provide a better understanding of weed distributions nationally. Data is being collated from many sources, including formal surveys, control activities, and citizen science programs. The distribution data collated by this project will also inform robust estimation of economic impacts, which allows us to see who benefits from weed management and allocate resources efficiently.

Preliminary results show that the variety of approaches used to collect and store distribution data do not allow for simple comparisons between regions. We are developing modelling approaches to integrate data from diverse sources into national distribution maps. Collaboration through data and information sharing is also essential in creating a nationally consistent dataset, and will enable weed managers to make more effective management decisions. This presentation will cover the opportunities for collaboration and coordination within the weeds community, and identify what is required to build a national understanding of the distributions and impacts of weeds in Australia.

**Keywords** Distribution, map, modelling

# The evolution of environmental weed management on Auckland's regional parks

Holly Cox- Auckland Council Private Bag 92300, Victoria Street West, Auckland 1142, NZ

**Summary** In 2018 Auckland's ratepayers supported an unprecedented investment in the regional natural environment. One of the programs that has benefitted from this investment is the management of environmental weeds on regional parks.

This program has undergone improvements across spatial scale, species targeted, ecological prioritization and information management, while also responding to diverse external drivers including the loss of social license for herbicide use, emerging plant pathogens and increased community appetite for conservation activities.

Local government reform, resulting in the amalgamation of staff from seven local government agencies has also led to different challenges and opportunities to integrate biosecurity and biodiversity planning.

**Keywords** asset based management, environmental weeds, invasive plants, pest plants, prioritisation, site based management

## INTRODUCTION

There are more than 25,000 introduced plant species in NZ, and currently naturalisations are equal to the quantum of indigenous flora (Stanley and Bassett 2015). Once species are beyond eradication, research points towards an asset-based management program.

Tāmaki Makaurau, Auckland, is one of the sixteen regions of New Zealand and contains the country's largest urban area and corresponding population. Auckland Council is the unitary authority that forms the local government for the region. Distinct from other regions, the unitary authority manages twenty eight[1] regional parks encompassing 41,000 hectares. These parks support diverse ecosystems and threatened species and are highly valued by the people of Auckland. One of the most significant pressures on these parks are environmental weeds. Due to Auckland's temperate climate, sprawling urbanization and enthusiastic uptake of exotic garden plants, it has earned the reputation as being the 'weediest city' in the world.

The Natural Environment Targeted Rate will enable a step change for environmental protection and restoration. One of the programs that has benefitted from increased investment is the management of environmental weeds on regional parks. This investment has come at the perfect time for this program with the foundational work to handle increased investment being at a stage to take advantage.

## DOING WHAT ALWAYS WAS DONE

Auckland's regional parks are a mixed-use public asset, with recreational and amenity values historically prevailing over conservation outcomes in park management especially in a weed management context, despite parkland being home to two of the region's largest remaining tracts of indigenous ecosystems, and many other smaller fragments of high value. In 2007 rangers on parks were more likely to be clearing prickly gorse away from tracks rather than worrying about an environmental weed growing in the middle of a forest fragment.

Management of the weed control program was split between sector-based conservation rangers and a regional biosecurity officer. The annual budget was held by the biosecurity team, funded by a biosecurity rate as well as general rates. Decision making on annual work factored in fairness across all parks despite some clearly having greater ecological value than others. Particular species, usually ones that the public recognized and complained about were targeted alongside species that were known forest ecosystem weeds. Rules in the Regional Pest Management Strategy supported this approach[2]. Biodiversity specialists had little or no input. The meagre budget was split into weeks of contractors work and then divvied across the then 22 parks. Often the same contractor was used to do weed control across all parks with emailed instructions to control certain species of weeds. The contractor's report contained qualitative data with small amounts of quantitative data thrown in. All the geospatial data

[1] There is a 29th park in the network- Te Motu a Hiaroa/Puketutu Island is part of the regional parks weed control program however is co-managed by a trust involving Auckland Council, mana whenua and Watercare.

[2] Wild ginger (*Hedychium gardnerianum* Ker Gawl.) and moth plant (*Araujia hortorum* E.Fourn.) rules in the Regional Pest Management Strategy for the Waitakere Ranges and Hunua Ranges, and region wide woolly nightshade (*Solanum mauritianum* Scop.) containment rules and gorse (*Ulex europaeus* L.) boundary rules.

and abundance data was retained by the contractor with council staff having very little ability to track success over time.

## CHANGE THROUGH DATA IMPROVEMENTS

In 2006 the Auckland Regional Council[3] geospatial team won an award for a geospatial database (BioMap) that helped general biosecurity staff manage their complaints work. It was intended to adapt this database to include regional parks work.

To prepare the parks to be included in the database all the regional parks were divided into smaller management units. Geographical features such as tracks, streams, ridges were used and generally of a size that a contractor team could cover in a few days. All data would then be recorded against these management units as data was recorded against properties in the BioMap system.

Data collected against management units included species present, species controlled, methodologies used, hours per methodology, herbicide and additives per methodology, and weather conditions during the control period.

BioMap also enabled the collection of point data, and it was decided to permanently collect location and infestation size of plants that required multiple treatments per season. These plants either spread vegetatively, i.e., were unlikely to be spread geographically across a park, or had other characteristics that meant that they were more likely to found at this point than elsewhere, e.g. moth plant where the majority of seed drop around mature plants and have long lived seed banks. By including this data in BioMap it also resulted in the transfer of data from contractors back to the council.

Annual specifications of work now included instructions and priorities per management unit with some point data information that was gradually added to as work progressed. Due to the improved BioMap system being developed, contractors captured and reported data via reports submitted per control period. Reports were stored until BioMap 2 became available for use.

## CREATION OF AUCKLAND COUNCIL

In November 2010 Auckland Regional Council was amalgamated within the new unitary authority Auckland Council. The Regional Parks and Biosecurity departments remained largely unchanged; however the Information Technology department amalgamation and restructure was delayed and ultimately shelved BioMap.

Amalgamation however brought improvements and presented great opportunities. The small Natural Heritage department was expanded into the Biodiversity team with highly experienced staff from seven agencies across the region. Strategic and operational work was beginning to join up across the landscape from private property, local parks and connecting into regional parks.

## CONTRACT IMPROVEMENTS

The new council brought new staff introducing new ideas. Time based contracts were changed to job size contracts with key performance indicators. Enhanced specifications were developed to support accurate quotations with the eventual work specified to fit the budget. These specifications greatly assisted auditing which had been poorly undertaken to date.

In theory all work was accurately quoted on and delivered. In reality the opposite was true. Contractors were used to not being audited and felt that they would not get work if they did not propose a palatable price. Contract staff were having to rush through areas to meet the hours allocated and missed mature plants, setting back achievements at a site. As this was a rule rather than exception, one on one discussions with suppliers at site led to more accurate quotes, a focus on quality rather than quantity, and a shared recognition of the skill set that experienced weed control contractors held. Improved relationships with suppliers led to increased communication and with mutual understanding of outcomes this resulted in improved delivery across the program.

The operational program was maturing and held up to intense scrutiny generated from the implementation and review of the Auckland Council Weed Management Policy (2013) which was further compounded by concerns around glyphosate use.

What was more concerning for staff was the stagnating area of parks considered under management for pest plants in the region. This aspect was not helped by a budget that had not significantly changed since 2007, combined with the regional park portfolio increasing in size.

[3] Auckland Regional Council was the local body managing the regional parks prior to the 2011 amalgamation of all Auckland local government.

## TURNING THE TIDE

In New Zealand 2016 marked a groundswell of public support for conservation. The Predator Free 2050 goal brought nationwide attention to the biodiversity crisis. This was the perfect time for an equally ambitious project in Auckland. The review of how Auckland manages pests and the prioritisation of management of Auckland's ecosystems lined up with preparation for the long-term plan. [4]

**Biodiversity Focus Areas** Biodiversity Focus Areas (BFAs) are prioritised areas of ecological significance, which were created to achieve the objectives of the Auckland Biodiversity Strategy (2011) and are used to guide the planning and delivery of conservation activity in Auckland. Prioritisation was achieved by using the zonation model (Moilanen et al. 2005) with further refinements from national and local biodiversity specialists. These areas demark a representative range of all indigenous species and ecosystems, with a good number being contained within the regional park network, and a proportion of which were currently not receiving any weed management. All wetlands and gumlands, and most dunelands are considered high priority for management due to their threat status and ongoing decline in habitat quality.

**RPMP review** Previous pest management plans in Auckland were always restricted by budget availability and were mainly species focused. The new proposed Regional Pest Management Plan took advantage of the significant ecological areas under council management and protected them with site led and buffer programs. Other programs within the plan also utilized the site led approach to an extent, e.g., eradication on Aotea Great Barrier Island of weeds that are low incidence on the island despite being widespread elsewhere in the region. The bold programs proposed were supported by a budget bid to the long-term plan in the form of a targeted rate.

**Natural Environment Targeted Rate** In Auckland there has been precedent to use targeted rates to fund biosecurity work. The former Auckland Regional Council had a separate rate to fund biosecurity programs, and more recently Auckland Council charged location specific rates to fund possum control in said areas.

In 2017, as part of the long-term plan process, the council consulted Aucklanders about the option of a Natural Environment targeted rate. Three options, including retaining the status quo, were offered as part of the engagement. Aucklanders overwhelmingly voted in support of the higher of the two options. The rate is expected to raise $311 million over 10 years, which is predicted to protect a minimum of 66% of parks with significant ecological areas, a significant boost for conservation in Auckland.

## "WALKING THE TALK"

Budget from the new rate became available from July 2018. For regional parks this meant over double the budget in that financial year with further increases year on year. Immediate work included contracting an auditor, and surveying 735 hectares of land in the Waitākere Ranges to plan future weed control. Existing work programs were ramped up where it was easy to do so. Where good information existed for BFAs novel weed control on species commenced e.g. tree lupin (*Lupinus arboreus* Sims) and boxthorn (*Lycium ferocissimum* Miers) on duneland at Muriwai to protect three dune ecosystems and the threatened plants pīngao (*Ficinia spiralis* (A.Rich.) Muasya et de Lange) and sand coprosma (*Coprosma acerosa* A.Cunn.). Where limited information was available botanical surveys for non-forest ecosystems commenced to aid weed management.

However a range of operational and logistical steps needed to happen before the full effect of the increased budget could be utilised. The Biodiversity and Biosecurity teams were merged and operational responsibility for the regional parks weed program was handed back to the parks department with the inclusion of new supporting roles. Increased conservation delivery requirements created challenges for a constrained supplier market compounded by the COVID 19 outbreak. This experience highlights the importance of long lead in times and other considerations to support a viable contractor pool.

Also, what gave the regional parks program an advantage to make the most of the targeted rate initially, i.e., good data management, other weed control programs started to benefit from enhanced data management through geospatial apps connected to a centralized database. The simplicity of other council run weed programs meant that they were first to receive the upgrades.

**Ruru our geospatial database.** The restructure established a Bio Information team who are specialists in biodiversity and biosecurity

[4] The 10-year budget sets out the priorities and funding for Council activities that are planned over a 10-year period, for the whole of Auckland Council.

information management. One tool that is under development by this team is Ruru, an award-winning[5] information system in which all conservation work undertaken by Auckland Council will eventually be held and managed through. All the regional parks weed data is in the process of being uploaded, and in time all contractor work will be reported through this system. We have yet to realise the full benefit of having this system in place combined with a 10-year dataset, but one direct benefit will be more clarity of the program to interested parties outside the immediate regional parks team.

## NEXT STEPS

Aside from bedding in the new management team for the regional parks weeds and increasing annual work to meet the key performance indicators of the targeted rate program, additional work is needed.

Ecosystem prescriptions are being developed for all the biodiversity focus areas to address all environmental activity, and where these sites have greater complexity, ecosystem management plans are proposed. Outcome monitoring is being set up to track the success of all the targeted rate programs through measuring changes in ecological integrity within managed sites. Monitoring will also examine management interventions and pressure reduction, such as removal of pest plants, to understand the impacts of different interventions to inform ongoing site based management decisions.

## DISCUSSION

The regional park weeds program now follows management practices better aligned to what is now understood to be best practice management for widespread environmental weeds (Clarkson and Grice 2013). New Zealand's Parliamentary Commissioner for the Environment recently released a review of how New Zealand manages weeds that threaten native ecosystems. The report mentioned an ideal state of effective management of environmental weeds in a regional context. A region would have a biodiversity strategy that clearly identifies which remaining ecosystems are most precious and where they are. There would be a biosecurity strategy that helps identify, require monitoring, and prioritise management of weeds that threaten the identified ecosystems. Finally, a pest management plan with clear rules for the management of these weeds at these sites (Parliamentary Commissioner for the Environment 2021). Although Auckland's current mix of strategies, plans and programs focused through the delivery of the targeted rate does not quite follow this exact recipe, the approach covers broadly similar concepts and will deliver the same outcomes.

This work illustrates the fundamental importance of adequate resourcing and the gains that can be made. Instead of weeds transforming ecosystems the targeted rate program will transform conservation in Auckland. The regional parks weed program also demonstrates continual change and adaptation. The work will never be complete, there will always be pressures from exotic species. Management practices need to evolve, to make the most of technological improvements and continue to use the best information available.

[5] Ruru won the Environment and Sustainability Award at the 2021 New Zealand Spatial Excellence Awards

## ACKNOWLEDGMENTS

Auckland Council: Dr Imogen Bassett, Jonathan Boow, Senior Conservation Rangers (2007-2022)

## REFERENCES

Auckland Council, 2020. Auckland Regional Pest Management Plan 2020-2030. Mahere ā-Rohe Whakahaere Kaupapa Koiora Orotā mō Tāmaki Makaurau

Auckland Regional Council, 2007. Auckland Regional Pest Management Strategy 2007-2012

Clarkson, J.R. and Grice, A.C. 2013 Managing plant invasions: Strategic options defined. Proceedings 12th Queensland Weed Symposium 39-42

Moilanen, A., Franco, A.M.A., Early, R., Fox, R., Wintle, B. and Thomas, C.D. 2005. Prioritising multiple use landscapes for conservation: methods for large multi species planning problems. Proc. R. Soc. Lond. B Biol. Sci. 272: 1885–1891.

Parliamentary Commissioner for the Environment, 2021. Space invaders: A review of how New Zealand manages weeds that threaten native ecosystems

Stanley, M.C. and Bassett, I, E (2015) Environmental weeds in New Zealand: impacts and management. Austral Ark: The State of Wildlife in Australia and New Zealand. 134-161 (http://dx.doi.org/10.1017/CBO9781139519960 )

# Integrated Community and Agency-Led Wilding Conifer Management across the Marlborough Region, New Zealand

Jono Underwood[1]
[1]Marlborough District Council, Blenheim, New Zealand
(jono.underwood@marlborough.govt.nz)

**Summary** In 2016, with new government funding, New Zealand embarked on a nationwide operational programme addressing the threat of wilding conifers. The Marlborough region has no shortage of wilding conifer issues from a legacy of will-intended attempts at erosion control or other plantings of species which turned out to be prolific invaders. With the impending ramping up of wilding confier management programmes in the Marlborough region, as a small unitary authority – Marlborough District Council (MDC) recognised early that it cannot deliver it all on its own. As a result, MDC embarked on establishing a truly collaborative partnerships between all the players actively delivering management programmes in the region. This involved two Community Trusts, the Department of Conservation Wairau/Renwick, St Arnaud & Waitohi/Sounds Area Offices and MDC as the local unitary authority. There was also input by the National Wildings programme team at Biosecurity New Zealand and Land Information New Zealand.

In addition, an early bottom-line principle established was that all programmes were to be tenure-neutral and focused on addressing the threat - not getting hung up on land ownership boundaries. To their credit, all players bought into that notion which likely led to the success being seen. A 'Regional Steering Group' platform was established and chaired by a MDC Environment Committee Member then coordinated and facilitated by MDC staff. With the backbone relationships in place, MDC was then able to coordinate and facilitate the implementation of various programmes receiving NWCCP investment in Marlborough. This including exploring new contracting and project management models, and was all delivered within existing staff resource. MDC was able to demonstrate a focus on achieving the best possible outcome in the most efficient manner.

**Keywords** Wilding, pine, conifer, community, landscape, coordination

# Genetic structure of the invasive coastal weed *Euphorbia paralias* (sea spurge) in Australia

Gavin Hunter[1], Celeste Linde[2], Isabel Zeil-Rolfe[1], Ben Gooden[1]
[1]CSIRO, Health and Biosecurity, Black Mountain Science and Innovation Park, Acton, Australia,
[2]Australian National University (ANU), Research School of Biology, ANU College of Science, Canberra, Australia
(gavin.hunter@csiro.au)

**Summary** Sea spurge (*Euphorbia paralias*) is an herbaceous perennial native to coastal environments of Europe and northern Africa. Sea spurge plants produce buoyant seeds capable of surviving extended periods afloat on ocean currents and seeds are believed to have been accidentally introduced to Australia through ships ballast during the early twentieth century. Since its introduction, sea spurge has progressively invaded many beaches along Australia's southern coastline resulting in dense infestations that reduce the public amenity of beaches. Little is known regarding the genetic complexity of sea spurge in Australia or its possible invasion routes. The aim of this study was to use informative genome-based markers to elucidate the genetic structure and diversity of sea spurge within Australia. To do this, single nucleotide polymorphisms (SNPs) were identified from 374 sea spurge samples from 51 populations across the Australian distribution of the weed using the DArTseq genome complexity reduction technology. Following several filtering steps, several hundred SNP loci were identified for population genetic analyses. Principal component analysis indicated no clear separation or clustering of sea spurge and multi-locus genotypes appeared to be randomly distributed across the Australian range. Genetic structure analysis did however suggest that Australian sea spurge populations were derived from at least two different genetic sources. At a State level, genetic diversity values of sea spurge are low to moderate with diversity of South Australian sea spurge slightly higher than those of other States. Analyses are ongoing but these initial results suggest that ocean currents and possibly anthropogenic movement of propagules have led to a mixing of sea spurge across Australia and that no particular genotype(s) of the weed dominates.

**Keywords** Euphorbiaceae, invasion biology, weed population genetics

# Flying to the Rescue of Shorebird Habitat in the Coorong NP - Turning the Tide on African Boxthorn Invasion

Robbie Andrew[1]
[1]Limestone Coast Landscape Board, Meningie, Australia
(Robert.Andrew@sa.gov.au)

**Summary** African Boxthorn (*Lycium ferocissimum*), a Weed of National Significance and listed as a Declared weed across the Limestone Coast Landscape Board region is able to withstand harsh growing conditions and poses a threat to the fragile Coorong Ramsar listed site and surrounding coastal habitat of beach nesting and migratory shorebirds. 'Our Coorong Our Coast' (OCOC) is a 5 year National Landcare Program funded project addressing terrestrial-based threats to the Coorong RAMSAR site and the habitat of threatened bird species at strategic locations between the Murray River Mouth and the Victoria-South Australia Border.

The OCOC project has been targeting inaccessible Boxthorn infestations in remote coastal dune areas. Operating a Robinson R44 helicopter the pilot hovers low to the ground whilst the passenger applies a measured dose of *Tebuthiuron granules* around the dripline of each boxthorn plant. The herbicide is moisture activated and will remain in the sandy soil until it is taken up following rainfall. Following trials of a range of treatment methods previously, aerial hand broadcasting has been found to be the most efficient and accurate way to apply the herbicide, resulting in the least amount of off-target damage. Additionally this method retains the dead boxthorn in situ, providing shelter for germinating native climbers to use the structure as a trellis. Economically aerial treatment has proven to be the most cost efficient way to treat large amount of boxthorn and in difficult to access land. Limitations with this method include not being able to fly safely in windy conditions and air sickness can be an issue for less experienced passengers tasked with applying the herbicide. Investing $160,000 over 5 years the project has delivered 70,578ha of control (Canunda National Park and Coorong National Park) to date, with 7,288ha remaining for the project to be completed in 2022.

# Six years managing a created Saltmarsh

Geraldene Dalby-Ball[1]
[1]Dragonfly Environmental, Kings Park, Australia
(gm@dragonflyenv.com.au)

**Summary** Dragonfly Environmental has been deeply connected with the first planning then planting and now ongoing maintenance of the constructed salt marsh at Penrhyn Estuary in Botany Bay Sydney. This case study presents weed management that occurred prior to the creation of the salt marsh including the use of machinery, reshaping the dunes, removing bitou bush, and both machine and hand removal of *Juncus acutus* or Spiny Rush from throughout the proposed saltmarsh recreation area. A small area of Mangroves were also removed to improve habitat for migratory birds. With over six years maintaining the site we share the lessons and successes in keeping *Juncus acutus* out of the constructed saltmarsh and managing mangrove so they do not grow within the key migratory bird habitat. This case study includes considerations of working in contaminated environments as well as work health safety practicalities of working in an intertidal zone next to a large port and by the Airport.

**Keywords** Saltmarsh, Coastal, Machinery, *Juncus acutus*, Bitou Bush, Migratory Birds.

# Effect of emergence timing on growth and phenotypic plasticity of feathertop Rhodes grass (*Chloris virgata* Sw.) in Southern New South Wales

Md Asaduzzaman, Hanwen Wu, Eric Koetz, Michael Hopwood and Adam Shepherd
NSW Department of Primary Industries, Wagga Wagga, NSW 2650, Australia
(md.asaduzzaman@dpi.nsw.gov.au)

**Summary** Feathertop Rhodes grass (*Chloris virgata* Sw.) is now a problematic weed in many cropping and non-cropping areas of Southern NSW. This study evaluated the effect of emergence dates on the growth and phenotypic plasticity of this summer weed where cohorts of four different populations were initiated in early-spring (04 Sep), late-spring (04 Nov), mid-summer (04 Jan) and early-autumn (04 Mar) in southern NSW, Australia. Among the four sowing times, the late spring sowing treatment took the longest time from emergence to the first seed head emergence (70-110 days), while it had the shortest seed maturity period (8-16 days). Length of the reproductive and total life period of the four populations differed across the four sowing-time treatments. The plants that emerged in mid-summer had the longest reproductive period (30 days) whereas the early-autumn emerging plants died before the reproductive stage due to cold temperatures during winter. The mid-summer cohort required slightly longer time (63-85 days) to achieve seed head formation and less time (19-24 days) for seed maturity than those plants which emerged in early or late spring. All the reproductive features were varied by sowing times and population numbers. The reproductive biomass allocation pattern and seed production generally increased in the mid-summer emergence cohort. Seed production in the mid-summer (9,942 seeds/plant) cohort was higher than the late spring (8,000 seeds/plant) and early spring (3,240 seeds/plant) cohorts respectively. The ratio of reproductive biomass to vegetative biomass increased in the mid-summer sowing times in all populations, and this species displayed true plasticity in reproductive allocation.

**Keywords** biomass, allocation, reproductive fitness, climate change.

## INTRODUCTION

Weed phenological features including the timing of emergence, growth and sexual reproduction can be used to predict the distribution of a weed species under varying environmental conditions (Ghersa and Holt 2006). Therefore, a biologically meaningful description of phenological events is fundamental for better understanding of the temporal dynamics of a weed species, which would contribute to the well-informed recommendation of suitable timing for effective control (Rafferty and Ives 2011). Properly timed control with suitable tactics could achieve maximum control efficacy, reducing both the cost of managing weeds and risks of new infestations into other areas.

Weed plasticity responses to environmental changes have become one of the most important aspects for evolutionary biology research in weeds and have been accelerated by climate change concerns (Clement *et al.* 2004). Feathertop Rhodes grass (*Chloris virgata* Sw.) is a warm-season, annual grass that is widely distributed globally (Anderson 1974) and was ranked in the top 10 weed species in Australia (Werth *et al.* 2013). In southern New South Wales (NSW), Australia, feathertop Rhodes grass mainly dominates roadsides, fence lines and wasteland areas. Now this small-seeded species has become an issue in the cropping country of southern NSW. Given the large area of feathertop Rhodes grass' expansion to diverse soil and climatic conditions, it is not surprising that feathertop Rhodes grass is very plastic.

The objectives of this research were 1) determine the impact of different emergence dates on the growth and development timing of phenological, morphological and reproductive traits, and 2) compare timing and duration of each phenophase among four different feathertop Rhodes grass populations.

## MATERIALS AND METHODS

**Environmental (emergence timing) and populations treatment** There were four sowing time treatments: early spring (4 Sep 2020), late-spring (04 Nov 2020), mid-summer (04 Jan 2021) and early-autumn (04 Mar 2021). Four feathertop Rhodes grass populations from diverse geographical areas were selected to cover the phenotypic variability of this species. Seeds of four populations including FELT 04/20, GLEN 03/18, STURT 16-17 and PARK 01/20 were collected from Felton, Toowoomba, Wagga Wagga and Parkes respectively. Populations FELT 04/20 and STURT 16-17 were sourced from non-cropping situations.

**Experiment set up and measurements** A total of 400 seeds in each sowing operation were used for each population where sowing trays (32 cm x 40 cm) were prefilled with a commercial potting mix under natural conditions in a net house at Wagga Wagga Agricultural Institute (WWAI), NSW. The emergence time was recorded before seedlings were manually transplanted at 4-5 leaf stage to a plastic pot (18 cm in diameter) pre-filled with the potting mix. Two separate experiments were conducted.

The first experiment was set up for measurements of phenological events. For each sowing time, each population was maintained one seedling/pot to measure the date of: (1) emergence, (2) booting stage (3) first seed head emergence, (4) first mature seed observed in each of the first five emerged seed heads from each plant, and (5) plant senescence. This data was used to calculate: (1) emergence period, (2) vegetative period (number of days between emergence and booting stage), (3) seed maturity period (between seed head emergence and the first mature seed formed), (4) post-reproductive period (between the first mature seed and plant senescence), (5) reproductive period (between the first seed head emergence and plant senescence), and (6) total life period (between emergence and plant senescence).

The second experiment was set up for destructive measurements of plant biomass. There were two seedlings/pot for each population at each sowing time. Each plant at physiological maturity was carefully cut near the soil surface to obtain vegetative (leaf and stem) and reproductive biomass (seed head with seeds) separately. Based on 500-seed weight, we estimated seed production per plant. Other morphological measurements were carried out over time.

## RESULTS

**Effect of emergence timing and population treatments:** Sowing time and population had a significant ($p<0.001$) effects on the timing of key phenological events, and more importantly on the period each population spent within a phenological growth stage (Table 1). Among the four sowing times, the late spring sowing treatment took the longest time from emergence to the first seed head emergence (70-110 days), had the longest post-reproductive period (8-23 days), and the shortest seed maturity period (8-16 days). The average reproductive period was longest in the mid-summer sowing time treatment (30 days) and the shortest in early spring (4 September) time (22 days). Similarly, the mid-summer sowing time treatment resulted in the longest total life period (92-116 days). Feathertop Rhodes grass that emerged in early autumn (04 Mar) did not progress to the reproductive stage in all four populations as a result of low temperatures and frosts in winter which eventually killed the plants.

Populations differed significantly ($p$ = 0.004) in the duration required for vegetative growth and reproduction. Total life period of FELT 04/20 and STURT/16-17 was similar within a given sowing time treatment and responded in a similar manner with sowing time treatments. These two populations were sourced from non-cropping situations and their life period increased from the early spring to late spring sowing times and then decreased. PARK 01/20 and GLEN 03/18 were sourced from cropping situations and tended to have a longer life period when emerged in mid-summer than other sowing times. Across the four feathertop Rhodes grass populations, the variation in reproductive period (seed maturity + post-reproductive period) was greater in the late spring sowing (17-39 days), while it was narrower in both the early spring (20-24 days) and mid-summer sowing (28-32 days).

The total reproductive period (seed maturity + post-reproductive period) for all populations was higher in the mid-summer sowing time than early and late spring sowing times (Table 1). All four populations had a similar trend for reproductive period when they emerged in early spring.

**Plasticity in reproductive effort** Seed production of the four feathertop Rhodes grass populations across the four sowing treatments was significantly ($p$ = 0.003) different (Table 1). On average, the seed production increased from the early spring (3,240 seeds/plant), and late spring (8,000 seeds/plant) to mid-. summer sowing time (9,942 seeds/plant). The ratio between reproductive tissue biomass and vegetative tissue biomass (R-V) increased from the early spring emergence to late spring and summer emergences (Table 1).

**Table 1.** Phenological traits, duration of developmental periods and seed production of four populations of feathertop Rhodes grass emerged at different times (environmental condition).

| Sowing time | Population | Emergence to first seed head emerging (days) | Vegetative period | Seed maturity period | Reproductive period | Life period | Seeds/plant | Slope (R-V) |
|---|---|---|---|---|---|---|---|---|
| 04 Sep | FELT 04/20 | 69-71 | 64. 00 (±0.28) | 13.07 (±0.65) | 22.45 (±0.70) | 86.45 (±0.65) | 6521 (±770) | 0.045c |
| | STURT16-17 | 63-73 | 63.87 (±0.85) | 14.50 (±1.11) | 21.10 (±0.93) | 84.97 (±1.34) | 5951 (±390) | 0.063b |
| | PARK 01/20 | 72-80 | 69.00 (±1.17) | 10.00 (±1.30) | 24.28 (±0.92) | 93.75 (±1.5) | 4390 (±415) | 0.061b |
| | GLEN 03/18 | 85-95 | 65.27 (±4.90) | 11.37 (±0.77) | 20.37 (±0.92) | 85.64 (±5.4) | 1969 (± 600) | 0.008c |
| 04 Nov | FELT 04/20 | 70-72 | 68.10 (±2.00) | 15.28 (±0.60) | 39.07 (±0.85) | 107.17 (±2.34) | 11112 (±1575) | 0.085b |
| | STURT16-17 | 85-95 | 77.07 (±4.00) | 16.00 (±2.15) | 28.72 (±2.76) | 105.79 (±5.00) | 9875 (±1362) | 0.094b |
| | PARK 01/20 | 95-100 | 69.00 (±0.75) | 8.25 (±0.80) | 16.79 (±0.99) | 85.79 (±3.86) | 8071 (±446) | 0.042c |
| | GLEN 03/18 | 100-110 | 90.00 (±2.26) | 8.78 (±0.65) | 17.67 (±0.71) | 107.67 (±3.28) | 2937 (± 1400) | 0.059c |
| 04 Jan | FELT 04/20 | 65-70 | 61.50 (±1.13) | 19.75 (±1.30) | 30.38 (±1.81) | 91.80 (±1.61) | 9930 (±1232) | 0.233a |
| | STURT16-17 | 63-75 | 70.75 (±4.20) | 18.75 (±1.03) | 27.87 (±1.77) | 98.62 (±0.73) | 10867 (± 1362) | 0.194a |
| | PARK 01/20 | 72-78 | 83.18 (±3.03) | 23.75 (±1.04) | 32.75 (±1.04) | 115.93 (±1.07) | 16687 (±670) | 0.002c |
| | GLEN 03/18 | 78-85 | 70.37 (±1.50) | 23.00 (±0.92) | 32.25 (±1.04) | 102.62 (±1.64) | 12285 (± 579) | 0.078b |
| 04 Mar | FELT 04/20 | 67-80 | 69.57 (±1.34) | - | - | - | - | - |
| | STURT16-17 | 67-81 | 64.15 (±0.48) | - | - | - | - | - |
| | PARK 01/20 | 80-90 | 71.25 (±1.11) | - | - | - | - | - |
| | GLEN 03/18 | 80-90 | 79.26 (±1.63) | - | - | - | | |
| - | *p*-value | $p = 0.003$ | $p = 0.004$ | $p = 0.003$ | $p = 0.002$ | $p = 0.002$ | $p = 0.006$ | $p = 0.005$ |

Vegetative period = the interval from emergence to booting stage; seed maturity period = the interval from seed head emergence to the first mature seed formed; post-reproductive period = the interval from the first mature seed to plant senescence; reproductive period = the interval from the first seed head emergence to plant senescence, and life period = the interval from emergence to plant senescence. Slope (R-V) from the reproductive biomass which was regressed with the vegetative biomass using linear functions.

Principal component analysis (PCA) shows that, the proportion of the first and second components were 56.6% and 14.4% respectively (Figure 1). The first component (PC1) mainly consists of reproductive features, and these are plotted in proximity, including reproductive period, reproductive mass, seed head length, number of spikelets, seed head mass, and post-reproductive period. The second component was mainly due to the vegetative period, and total life period.

## DISCUSSION

This study demonstrated that feathertop Rhodes grass can emerge throughout the warmer months (early September to March) of southern NSW. The late-spring emerged plants had the longest vegetative period and the shortest seed maturity period conversely the mid-summer plants produced seed heads earlier in their life span compared to the early- and late- spring sowing times. The results suggest that photoperiod and temperature could be primary environmental factors in determining when the reproductive events occur. The low temperatures during winter (June to August) in southern NSW significantly impacted the growth and development of feathertop Rhodes grass. Therefore, the frost frequency and intensity are an important benchmark for implementing an economic post-emergent control program of feathertop Rhodes grass particularly in southern NSW.

All populations tended to have the highest post-reproductive period at the mid-summer sowing treatment. This could have a positive effect on high seed production and contribute to the overall fitness by replenishing extra seeds in the soil seed banks. The total life period was affected by reproductive period which can correlated to reproductive performance including seeds production per plant. Under favourable conditions, both total life and reproductive periods are maximised for higher seed head production and higher number of seeds, thereby improving the overall fitness.

The slope of the relationship between reproductive tissue biomass and vegetative tissue biomass within a population also varied among sowing time treatments and between populations (Table 1) and this relationship indicated true plasticity in biomass allocation across treatments.

**Figure 1**. Principal component analysis (PCA) of plant morphological and reproductive features of feathertop Rhodes grass. The first axis explains 56.6% and the second axis 14.4% of the variance.

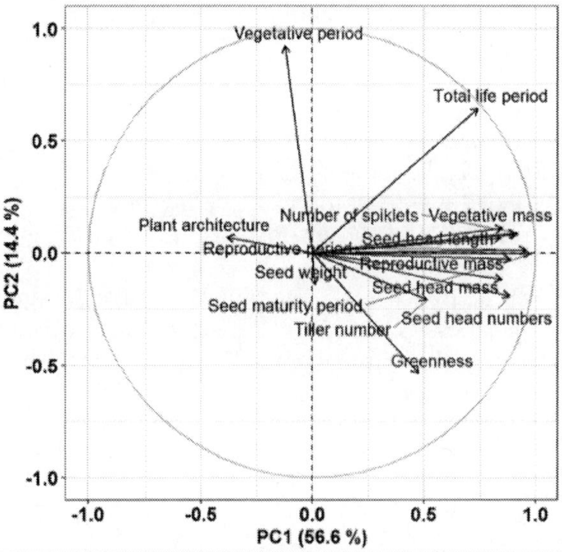


The study confirmed the major role of environmental conditions such as temperature and day length as the driving factors of feathertop Rhodes grass phenological development and confirmed the diversity of the populations and their suitability to different environments. Low temperatures or frosts can slow the rate of plant development and can stop the seed production. The January sowing (mid-summer) produced the greatest number of seeds whereas the early spring resulted in the lowest seed production. The recommendation is that controlling feathertop Rhodes grass seedlings prior to reproduction will reduce populations growth and alleviate their negative effects on crop yield in future generations. The control action should be diverted to control early emergence, especially the mid-summer emergence due to the removal of crop competition after harvest and due to more aggressive growth and seed production.

## ACKNOWLEDGMENTS

The authors acknowledge co-investment from the Grain Research Development Corporation (GRDC) and NSW Department of Primary Industries, Australia

## REFERENCES

Anderson, D. E. (1974). Taxonomy of the Genus Chloris (Gramineae). Provo, UT: Brigham Young University Science Bulletin. p133.

Ghersa, C. M. and Holt, J. (2006)0. Using phenology prediction in weed management: A review. *Weed Research* 35(6), 461-470.

Clements, D.R.,Tommaso, A., Jordan, N., Booth, B.D., Cardina, J., Doohan, D., Mohler, C.L., Murphy, S. D. and Swanton, C. J. (2004). Adaptability of plants invading North American cropland. *Agriculture Ecosystems Environment* 104, 379–398.

Rafferty, N. E. and Ives, A. R. Effects of experimental shifts in flowering phenology on plant–pollinator interactions. (2011). *Ecology Letter* 14, 69– 74.

Werth, J., Boucher, L., Thornby, D., Walker, S. and Charles, G. (2013). Changes in weed species since the introduction of glyphosate resistant cotton. *Crop and Pastures* 64, 791–798.

# The impact of selection and breeding for above-ground vigour on below-ground traits associated with weed competitiveness in wheat

Pieter-Willem Hendriks[1,2,3*], Saliya Gurusinghe[3], Emmanuel Delhaize[4], Greg J. Rebetzke[1], Peter R Ryan[1] and Leslie A. Weston[2,3]
[1] CSIRO, Black Mountain Science and Innovation Park, Canberra, ACT, 2601,
[2] School of Agriculture, Environment and Veterinary Sciences, Charles Sturt University, Wagga Wagga, Locked Bag 588, Wagga Wagga, NSW, 2678
[3] Gulbali Institute for Agriculture Water and the Environment, Charles Sturt University, Locked Bag 588, Wagga Wagga, NSW, 2678
[4] Australian Plant Phenomics Facility, Australian National University node, Canberra, ACT 2601

**Summary** Weeds are a critical pest management issue for wheat producers. The weed competitiveness of wheat has typically been reduced through a century of selection for dwarf genotypes to prevent lodging and increase the harvest index. Many of the cultural and chemical strategies employed to manage weeds are costly and some may lead to the development of herbicide resistance. Therefore, novel and cost-effective integrated approaches for weed management in cereals are needed. One approach to enhance competitiveness without reducing harvest index is through selection for early shoot vigour. However, the impact of above-ground vigour on below-ground root traits and competitiveness has not been well characterised to date.

Competitive root traits in wheat were evaluated against commercial cultivars and parent genotypes using a collection of wheat genotypes generated by recurrent selection at CSIRO. In field and controlled environments, shoot and root traits were evaluated in a series of experiments. Historic cultivars, triticale or cereal rye were included as positive weed suppressive controls for enhanced competitiveness. Genotypic differences in root traits and the allelopathic potential from root exudation were noted among selected early vigour and commercial lines. Their weed suppressiveness was quantified in field and controlled environments.

**Keywords** Root architecture traits, metabolomics, weed suppression, early vigour

## INTRODUCTION

Wheat (*Triticum aestivum* L.) is one of the world's most cultivated crops and is grown across more than 215 million hectares every year (FAO, 2022). As a food crop, wheat accounts for ~20 % of the global human calorie intake (Savary *et al.* 2019). Weeds are considered the leading biotic limitation to wheat production (Oerke 2006), potentially reducing yields by as much as 23%, as they compete for resources that would otherwise be used by the crop (Gallandt and Weiner 2015).

Historically, crops competed with weeds by smothering and shading through increased plant height (Murphy *et al.* 2008). The introduction of the dwarfing genes (*Rht*) during the "Green Revolution" improved yields but reduced weed competitiveness (Vandeleur and Gill 2004), relying on primary tillage and pesticides to control weeds (Evers and Bastiaans 2016). While successful, these practices among other issues caused the emergence of herbicide-resistant weeds through increased selection pressure (Broster *et al.* 2013; Heap 2018). Improving the competitive ability of cereals would be a valuable and cost-effective alternative strategy for suppressing weeds above-ground (Lowry & Smith 2018) without compromising the harvest index (Bertholdsson 2005; Zerner *et al.* 2016).

However, interference occurring between the roots of crop species and weeds may be even more important than above-ground competition (Kiær *et al.* 2013). Traits that increase soil volume occupation (Craine and Dybzinski 2013), nutrient exploitation (Giehl *et al.* 2014), as well as the exudation of secondary phytotoxic metabolites (Weston 2005) generally enhance below-ground competitiveness. In this study, we analysed the effect of enhanced shoot vigour traits on root traits and assessed their contribution to the overall competitiveness of wheat. We measured growth parameters in a set of breeding lines selected for high shoot vigour and compared them with commercial and historic wheat cultivars and triticale, in a series of replicated controlled environment and field experiments.

## MATERIALS AND METHODS

**Genetic material** High shoot vigour and weed-competitive wheat lines (W lines), generated from top-crosses between germplasm from a recurrent selection for increased shoot vigour (Zhang *et al.* 2015) and Australian commercial cultivars were evaluated. Genotypes included

: W400203 and W470201 derived from a cross with cv. Yitpi, and W010709 and W670704 derived from a cross with cv. Wyalkatchem (Zerner et al. 2016). Reference genotypes included commercial cultivars Condo, Yitpi, Wyalkatchem and Mace. Triticale (cv Chopper) was a vigorous and weed-suppressive control (Beres *et al*. 2010). Seeds were sourced from a previous glasshouse experiment and standardised by seed weight assessment.

**Controlled environment experiments** Root growth and competitive interaction were assessed both in hydroponics using root pouches as well as in 35cm tall PVC tubes filled with field soil collected from Wagga Wagga, NSW, Australia, characterised as a red clay-loam kandosol. The experiments were performed in growth chambers at 12 h (day/night) with 20 °C/15 °C and light 600 μmol.m$^{-2}$s$^{-1}$. Six replicates of each genotype were sown per experiment, and the experiments were performed three times.

**Field experiments** Field trials were conducted from May to December in 2018 through 2020 at the Graham Centre (now Gulbali Centre) research field site in Wagga Wagga (35°03S, 147°36 E; 227 m altitude; NSW, Australia). 2018 and 2019 experienced below average rainfall, while in 2020 it was above average. Each cereal genotype was sown in six replicated plots with dimensions 12.0 x 1.8 m, arranged in a randomised complete block design. Above-ground wheat and weed growth was assessed over the growing season, along with ground cover, light interception, leaf area and crop and weed biomass.

## RESULTS

In the controlled environment experiment, the commercial wheat control for high vigour (cv. Condo) had significantly wider leaves than the control for low vigour (cv. Mace). The vigour lines had significantly increased leaf width than Condo (Figure 1).

Early on (at second leaf expansion), total root length differed among genotypes (Figure 2). Triticale cv. Chopper, selected as the control for enhanced vigour and competitiveness, showed the longest total root length of all entries. The more vigorous Condo had a significantly longer root system than the low vigour Mace cultivar. Federation, a historic cultivar, exhibited a longer root system than all the commercial wheat cultivars, as did early vigour genotypes in contrast to their commercial parents (Figure 2). The vigour lines also exhibited more and longer root hairs than did commercial cultivars.

**Figure 1.** Leaf width measured on the third leaf (n=36 for C3 and n=24 for the other entries). Error bars represent standard errors. Letters identify significant differences between means (P≤0.05) as determined by one-way ANOVA

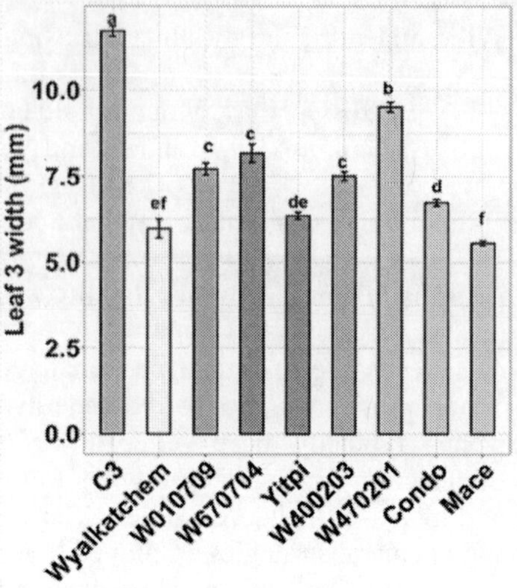


**Figure 2.** Total root length at early growth (two-leaf stage) in selected wheat and triticale genotypes. The error bars represent standard errors (n=24). Letters identify significant differences between means (P≤0.05) as determined by one-way ANOVA.

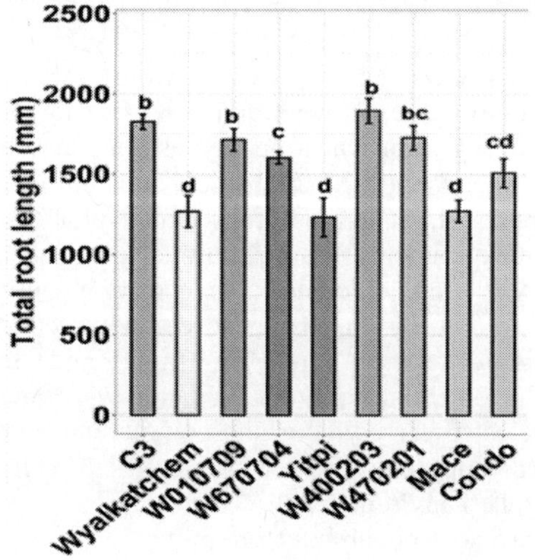


In the field, the high vigour lines closed the canopy earlier and intercepted more photosynthetically active radiation than did the commercial cultivars (P<0.05, Figure 3). A random forest regression performed on the complete set of measured above-ground traits identified that light interception at early tillering and ground cover at the end of tillering were key parameters in the model that explained the variability associated with weed suppression among the wheat entries (data not shown). Indeed, the vigour lines suppressed weeds by up to 200% more than their commercial cultivar parents (P<0.05, Figure 4).

**Figure 3.** Interception of photosynthetically active radiation over the growth of the HV lines and parental lines in the 2020 Wagga Wagga field experiment (Hendriks *et al.*, 2022)

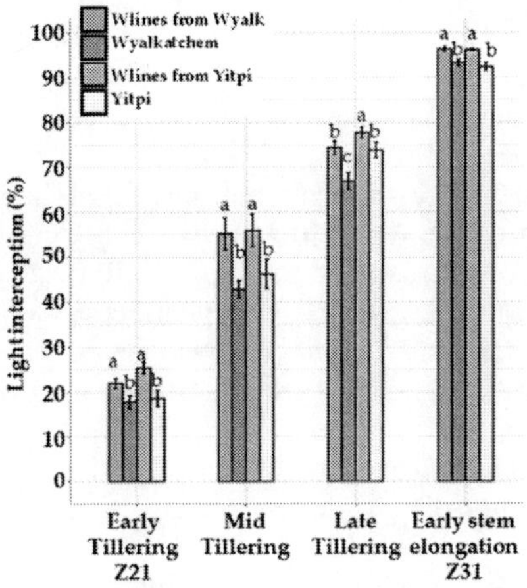


**Figure 4.** Weed biomass measured under each plot at anthesis (n=24). Letters identify significant differences between means (P≤0.05) as determined by one-way ANOVA

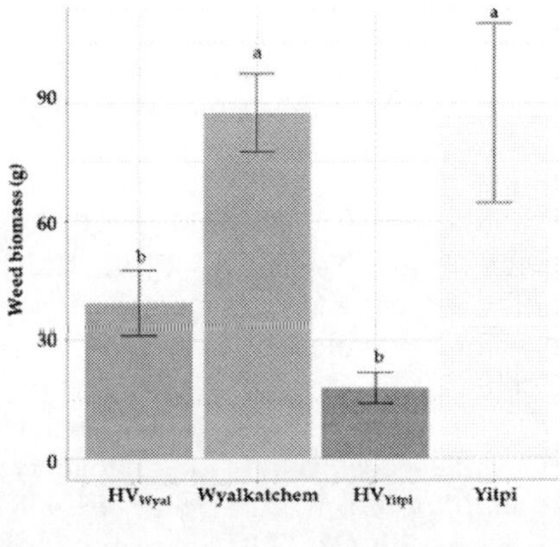


Controlled environment experiments that separated above and below-ground interaction suggest that the competition below-ground was more intensive as there were no significant differences between the weed suppression for plants that were allowed to compete above and below-ground and plants that could only compete below-ground (Figure 5).

## DISCUSSION

Our results demonstrated that the incorporation of shoot vigour into commercial wheat cultivars was associated with significant modifications of root traits related to competitive ability. The progeny of crosses between commercial cultivars and genotypes resulting from recurrent selection for early shoot vigour exhibited longer root systems with longer root hairs than their commercial cultivar parents. Experiments designed to evaluate interference with weeds in hydroponics and in field soil suggest that both weed tolerance and weed suppressive ability of early vigour lines was significantly enhanced. These differences were observed as early as the second leaf formation in the controlled environment experiment, at a developmental stage where competition for light may not yet be limiting. In contrast, differential weed suppression was noted by early tillering in the field, approximatively 30 days after sowing. Our findings clearly demonstrate that early vigour is associated with weed competitiveness at growth stages considerably earlier than previously described (Mwendwa *et al.* 2020).

**Figure 5.** Weed suppression measured as the reduction of ryegrass biomass. Wheat was allowed to interact below ground against four (4Rg) or eight (8Rg) ryegrass plants or to interact above and below-ground (Full) with 4 ryegrass plants

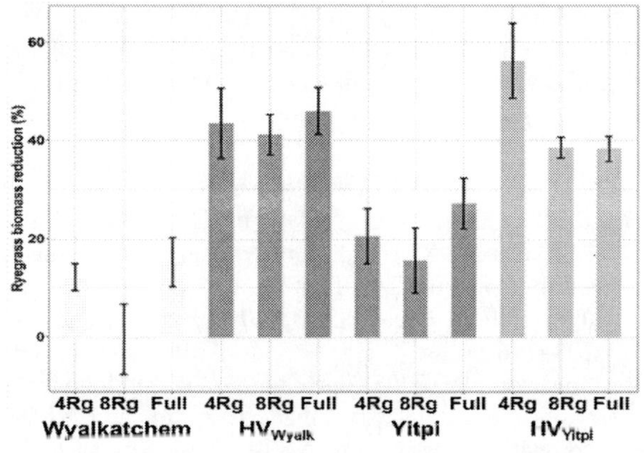


Our findings also demonstrate the importance of below-ground root interaction on the outcomes of crop and weed interference, and are in agreement with results of several studies where above- and below-ground interference were separated. We and others have repeatedly shown that below-ground competition may impact crop growth to a greater degree than above-ground competition (De Lucas and Froud-Williams 1994; Exley and Snaydon 1992; Stone *et al.* 1998).

Soil volume occupation allows plants to access more resources, including soil water and nutrients, and can improve efficiency of their utilisation by crops (Craine *et al.* 2005). We show that high vigour lines exhibit enhanced root growth that allows increased soil volume occupation. Our results also demonstrate that introgressing enhanced early shoot vigour into commercial wheat cultivars resulted in root traits that increased the ability of wheat to successfully interfere with developing populations of weeds under field and controlled environment conditions.

New early vigour lines generated from more advanced recurrent selection and current commercial cultivars are being assessed by private industry. The competition with weeds is a synergy of multiple traits and for a better understanding of the below-ground interactions further studies remain to be conducted.

## ACKNOWLEDGMENTS

This research is supported in part through GRDC project, CSP00182, "Breeding weed competitive wheat", and US00084, "Innovative crop weed control for weed management in the northern grains region". P-W Hendriks was supported by a CSU postgraduate research scholarship and the AE Howard Memorial Trust through a Tim Healey Memorial Scholarship. We acknowledge the support of technical support staff for planting, plot maintenance, and data collection, including Dave Smith, AJE Haddis, Gilbert Permalloo, Tom Mc Lucas, Matt Lynch, Gerard Ryan, Stuart Brown, Graeme Heath and Ian Johnson. We also thank Washington Gapare and Alex Zwart for their assistance with statistical analysis.

## REFERENCES

Beres, BL, Harker, KN, Clayton, GW, Bremer, E, Blackshaw, RE and Graf, RJ. 2010. Weed-Competitive Ability of spring and winter cereals in the Northern Great Plains. Weed Technology, *24*, 108–116

Bertholdsson NO. 2005. Early vigour and allelopathy – two useful traits for enhanced barley and wheat competitiveness against weeds. Weed Research 45, 94-102.

Broster JC, Koetz EA and Wu H. 2013. Herbicide resistance levels in annual ryegrass (*Lolium rigidum* Gaud.) and wild oat (*Avena spp.*) in southwestern New South Wales. Plant Protection Quarterly 28, 126-132.

Craine JM and Dybzinski R. 2013. Mechanisms of plant competition for nutrients, water and light. Functional Ecology 27, 833-840.

Craine JM, Fargione J and Sugita S. 2005. Supply pre-emption, not concentration reduction, is the mechanism of competition for nutrients. The New phytologist **166**, 933-940.

De Lucas CB and Froud-Williams RJ. 1994. The role of varietal selection for enhanced crop competitiveness in winter wheat. Aspects of Applied Biology 40, 343–350.

Evers J, Bastiaans L. 2016. Quantifying the effect of crop spatial arrangement on weed suppression using functional- structural plant modelling. Journal of Plant Research 129, 339-351.

Exley DM and Snaydon RW. 1992. Effects of nitrogen 4ertilizer and emergence date on root and shoot competition between wheat and blackgrass. Weed Research **32**, 175-182

Gallandt ER, Weiner J. 2015. Crop-Weed Competition. *eLS*, 1-9.

FAO (www.fao.org/faostat/en/#data/QV, accessed on 05/02/2022).

Giehl RFH and Gruber BD, von Wirén N. 2014. It's time to make changes: modulation of root system architecture by nutrient signals. Journal Of Experimental Botany 65, 769-778.

Heap I. 2018. The international survey of herbicide resistant weeds. Vol. 2019. www.weedscience.org.

Hendriks P-W, Gurusinghe S, Ryan PR, Rebetzke G, J. and Weston LA. 2022. Competitiveness of early vigour wheat (*Triticum aestivum* l.) genotypes is established at early growth stages. Agronomy (Basel) 12, 377.

Kiær LP, Weisbach AN and Weiner J. 2013. Root and shoot competition: a meta-analysis. Journal of Ecology 101, 1298-1312.

Murphy KM, Dawson JC and Jones SS. 2008. Relationship among phenotypic growth traits, yield and weed suppression in spring wheat landraces and modern cultivars. Field Crops Research 105, 107-115.

Mwendwa JM, Brown WB, Weidenhamer JD, Weston PA, Quinn JC, Wu H and Weston LA. 2020. Evaluation of commercial wheat cultivars for canopy architecture, early vigour, weed suppression, and yield. Agronomy (Basel) **10**, 983.

Oerke EC. 2006. Crop losses to pests. Journal of Agricultural Science 144, 31-43.

Savary S, Willocquet L, Pethybridge SJ, Esker P, McRoberts N and Nelson A. 2019. The global burden of pathogens and pests on major food crops. Nature ecology & evolution.

Stone M, Cralle H, Chandler J, Bovey RW and Carson K. 1998. Above- and below-ground interference of wheat (*Triticum aestivum* L.) by Italian ryegrass (*Lolium multiflorum*). Weed Science. **46**, 438-441.

Vandeleur RK and Gill GS. 2004. The impact of plant breeding on the grain yield and competitive ability of wheat in Australia. Australian Journal of Agricultural Research 55, 855-861.

Weston L. 2005. History and current trends in the use of allelopathy for weed management. Horticulture Technology 15, 529-534.

Zerner MC, Rebetzke GJ and Gill GS. 2016. Genotypic stability of weed competitive ability for bread wheat (*Triticum aestivum* L.) genotypes in multiple environments. Crop and Pasture Science 67, 695-702.

Zhang L, Richards RA, Condon AG, Liu DC and Rebetzke GJ. 2015. Recurrent selection for wider seedling leaves increases early biomass and leaf area in wheat (*Triticum aestivum* L.). Journal Of Experimental Botany **66**, 1215-1226.

# Impacts of Strategic Tillage in a No-Tillage Conservation Cropping System on Wild Oats Population

Mohammad Ali[1], Sijia Xu[1], Michael Widderick[2], Alwyn Williams[1] and Steve Adkins[1]
[1]The University of Queensland, School of Agriculture and Food Sciences, Gatton, QLD 4343, Australia
[2]Department of Agriculture and Fisheries, Leslie Research Facility, Toowoomba, QLD 4350, Australia
(m.ali2@uq.net.au; mohammad.aliuq@gmail.com)

**Summary** Wild oats (*Avena sterilis* ssp. *ludoviciana*) is a major winter weed in no-tillage conservation agriculture (NTCA) systems of Australia's Northern Grains Region. Many wild oat seeds shed before crop harvest and build a persistent seed bank in the 0-2 cm soil layer in a NTCA paddock. Reintroduction of tillage in a strategic way can be useful to manage this weed effectively. To understand the impact of strategic tillage (ST) on the soil seedbank dynamics and emergence pattern of wild oats, four tillage treatments were evaluated in a long-term NTCA paddock. The treatments were NT: no-tillage, $ST_1$: ST applied in the first year of study only, $ST_2$: ST applied in the second year of study only, and CT: conventional tillage. Soil samples were collected from 0-5, 5-10, and 10-15 cm depths to study the seedbank dynamics, and two 1 $m^2$ permanent quadrats were established per plot to study the emergence pattern of wild oats. In NT, 95% of wild oat seeds were found at 0-5 cm depth, whereas 88% of seeds were distributed to a depth of 10 cm under $ST_1$ or $ST_2$, depending on when ST was applied. In CT, 75-85% of seeds were buried below 10 cm depth. In NT, 88% of seedlings emerged during the first 20 days after wheat planting (DAWP), but the rest continued to emerge until 45 DAWP, leading to a wide window of emergence. This produced multiple wild oat cohorts under NT with staggered phenology. In contrast, emergence of seedlings was completed within 20 DAWP from ST or CT and produced a single cohort of wild oat plants with matching phenology. Among the tillage plots, emergence was minimal in ST throughout the season. Strategic use of tillage was found to be effective to minimize wild oat infestation by burying seeds to a depth from where the seedlings could not emerge.

**Keywords** emergence pattern, seed burial, soil seedbank, tillage, wild oats infestation

## INTRODUCTION

There has been a major transformation in broadacre crop production systems in Australia's Northern Grains Region (NGR; which comprises parts of Queensland and all of New South Wales) over the last 50 years (Walsh *et al.* 2019). About 90% of broadacre farmers of this region have already adopted conservation agriculture (CA) i.e., no- or reduced-tillage coupled with stubble retention (Llewellyn *et al.* 2012). However, due to the reduction in tillage operations, farmers are facing weed control challenges. Wild oats (*Avena sterilis* ssp. *ludoviciana* (Durieu) Nyman), a major difficult-to-control winter weed has already increased its abundance in CA systems, due to buildup of a persistent seedbank in the 0-2 cm soil layer (Widderick and McLean 2017).

In recent years, there has been increased interest in the use of strategic tillage (ST) in the no-tillage CA (NTCA) system to address the soil seedbank issue of wild oats (Walsh *et al.* 2019). The aim of ST is to bury wild oat seeds to a depth from where they cannot emerge (Walsh *et al.* 2019). However, application of any tillage operation, whether it is ST or conventional tillage (CT), can lead to a change in weed population dynamics by changing the vertical seed distribution pattern in the soil environment (Cousens and Mortimer 1995). A CT operation involves multiple passes of tillage each year.

The present study, therefore, aimed to evaluate the impact of different tillage treatments on the population dynamics of *A. sterilis* ssp. *ludoviciana* (hereafter *A. ludoviciana*) in a field setting. The study evaluated the seedbank dynamics (seed distribution pattern in 0 to 15 cm soil depth) and the emergence pattern of *A. ludoviciana* when growing in a wheat crop (*Triticum aestivum* L.). Knowing this information will enhance our understanding of the impact of different tillage practices on *A. ludoviciana* population dynamics and will assist in improving the effectiveness of its management program.

## MATERIALS AND METHODS

**Experimental site establishment** The study was conducted at Gatton, Queensland for two consecutive years, 2019 and 2020. The experimental site had a history of growing different winter crops for fodder purposes using NTCA approach for the previous 15 years (2003 to 2018). The crop sequence planted from 2003 to 2018 was 1 year of wheat followed by 1 year of either lupin (*Lupinus albus* L.) or barley (*Hordeum vulgare* L.) before returning to wheat. The soil type at the site was a black Vertosol (48% clay, pH 8.2) and had 2.8% total C, 0.22% total N, 6.9, 2.4 and 5.4 mg $kg^{-1}$ of P, K and S, respectively. An initial seedbank study confirmed the experimental site was free from *Avena* spp.

The field experiment was conducted using a randomized complete block design with four replications. The site with an area of 2,888 $m^2$ (38 m × 76 m; width/length) was first divided into four blocks. Each block was then sub-divided into four plots to carry four tillage treatments (Table 1). Each plot (5 m × 15 m) was separated by 2 m lateral alleys and 2 m block alleys. Each plot was artificially infested with *A. ludoviciana* seed (Westmar biotype) by broadcasting 30 viable spikelets $m^{-2}$ (i.e., 30 viable primary and 30 viable secondary seeds; in total 60 viable seeds $m^{-2}$). The tillage treatments were then imposed randomly to the plots within a block.

**Wheat crop establishment** Immediately after completing tillage operations, wheat (cv. LRPB Lancer) was planted at the rate 60 kg seeds $ha^{-1}$ with 25 cm row spacing and with 3 cm planting depth. To control broadleaf weeds, a tank mixture of selective in-crop herbicides, halauxifen and florasulam at 25 g a.i. $ha^{-1}$ of Paradigm®, and ethylhexyl ester at 440 mL a.i. $ha^{-1}$ MCPA 570 LVE® were applied by boom spray 45 days after wheat planting (DAWP) during the first year of the experiment. Paraffinic oil and alkoxylated alcohol non-ionic surfactants at the rate 0.5% v/v of Uptake® was also used in the tank mixture. Each year, the crop was irrigated four times.

**Soil sample collection for seedbank study** To determine the distribution of *A. ludoviciana* seeds present across different soil depths, 10 intact soil cores (each 5 cm diameter, 0 to 15 cm depth) were extracted across the experimental site at random (immediate after wheat planting) from the centre 3 m × 10 m of each 5 m × 15 m plot in both years. The intact cores were divided into three sub-samples *viz.* 0 to 5, 5 to 10 and 10 to 15 cm. Each sub-sample was spread thinly within a shallow germination tray. The trays were watered daily to field capacity and observed at 2- day intervals to count *A. ludoviciana* seedling emergence. The number of emerged seedlings was recorded replication-wise.

**Seedling emergence of *Avena ludoviciana*** To count the number of in-crop *A. ludoviciana* seedling emergence, two 1 m × 1 m permanent quadrats were established in each plot during May to October in both years. The seedling emergence data were recorded weekly and especially after an irrigation/rainfall event. Emerged seedlings were tagged to avoid double counting and distinguish between cohorts.

**Statistical analysis** The data of all parameters were analyzed separately year-wise, as the year effect was found significant ($P \leq 0.05$) for the parameters used. The data were subjected to analysis of variance (ANOVA) using Minitab software. Means were separated using Fisher's protected least significant difference (LSD) test at $P \leq 0.05$. All graphs were prepared using SigmaPlot software.

## RESULTS

**Soil seedbank dynamics of *Avena ludoviciana*** The effect of tillage treatments was found to significantly affect ($P \leq 0.001$) the seedbank dynamics of *A. ludoviciana* in both years of study for every soil depth except 10-15 cm in 2019 (Figure 1). In 2019, as expected, the seeds broadcasted in the NT and $ST_2$ (which was still NT in first year) were all concentrated in the 0-5 cm soil depth. In contrast, seeds were distributed mainly in the 5-10 cm soil depth (75% of total seeds) followed by the 0-5 cm depth (25%) under $ST_1$. In CT, 50% of seeds were buried in the 10-15 cm depth and the rest were

**Table 1.** Details of tillage treatments on a long-term no-tillage field at Gatton, Queensland to determine the impact of different tillage treatments on the population dynamics of *Avena sterilis* ssp. *ludoviciana*.

| Tillage treatments | Description of tillage operations | Operational depth of tillage (cm) | Code |
|---|---|---|---|
| No-tillage (with crop residue retention) in both years | No pre-sowing tillage | - | NT |
| Strategic tillage in first year, no-tillage (with crop residue retention) in second year | Pre-sowing tillage imposed with one-pass tine implement on two occasions during April, 14-days apart | 10 cm (first year), no-tillage in second year | $ST_1$ |
| No-tillage (with crop residue retention) in first year, strategic tillage in second year | Same tillage operation as described in $ST_1$, but imposed in second year of the trial | No-tillage in first year, 10 cm in second year | $ST_2$ |
| Conventional tillage in both years | Two-passes of disc implement followed by one-pass rotary hoe, and one-pass harrow conducted during April to May; each tillage 14-days apart | 15 cm in both years | CT |

distributed equally to the other two depths. In 2020, 96% of the seeds of NT were retained in the 0-5 cm depth and the rest were buried to the 5-10 cm depth (Figure 1). In the $ST_1$ (which became NT in second year), 86% of seeds were distributed in the 0-5 cm depth, 13% in the 5-10 cm, and only 1% in the 10-15 cm depth. When ST was applied to NT plot in the second year (i.e., $ST_2$), 88% of seeds were buried at 5-10 cm with only 10% of seeds distributed at 0-5 cm, and 2% were distributed at 10-15 cm. In CT, half of the total seeds were buried in the 10-15 cm depth followed by 37% at 5-10 cm whereas only 13% of seeds were distributed at 0-5 cm.

**Seedling emergence** The tillage treatments were found to significantly affect the seedling emergence of *A. ludoviciana* in a cropping environment in both years ($P \leq 0.001$) (Figure 2). In 2019, 60% of seedlings of NT and $ST_2$ emerged during the first 20 DAWP, 30% of seedlings emerged during 20-40 DAWP, and 10% emerged from 40-50 DAWP (Figure 2). In 2020, 88% of seedlings of NT and $ST_1$ emerged during the first 20 DAWP, 8% of seedlings emerged during 20-40 DAWP, and 4% emerged from 40-50 DAWP (Figure 2). Application of ST or CT restricted seedling emergence to within 20 DAWP in both years. Strategic tillage (either $ST_1$ or $ST_2$) was able to reduce seedlings load by almost half compared to NT or CT in both years (Figure 2). Applying ST in year one and then returning to practicing NTCA in the second year resulted in 25% less seedling emergence of *A. ludoviciana* than a continuous NTCA practice.

## DISCUSSION

**Seedbank dynamics** The NT treatment helped to retain > 95% of *A. ludoviciana* seeds on or close to the soil surface in both years (Figure 1). However, 4% of the seeds were found in the 5-10 cm depth in this treatment. Possibly those seeds fell into the cracks of the Vertosol soil during the time of their shedding. Medd (1996) reported that in a Vertosol soil a functional seedbank of wild oats can exist below 5 cm depth. When a ST operation was applied in NT plots, 10% less seeds were retained in the 0-5 cm depth in the following year compared to NT (Figure 1). A ST operation could bury as many as 88% of seeds to a 5-10 cm depth as observed in the second year (Figure 1). On the other hand, several passes of tillage with a disc, rotary hoe and power harrow under CT distributed *A. ludoviciana* seeds at different depths in the soil profile with 75-85% of seeds concentrated to 10 cm or greater depths (Figure 1). A greater number of seeds buried to 10 cm or greater depth means they would germinate but fail to reach the surface and die (Cousens and Mortimer 1995). However, 15-25% seeds were also found to be retained in 0-5 cm depth under CT in both years, which favoured more seedlings to emerge (20 and 50% higher than NT and ST, respectively in both years) from this treatment (Figure 1).

**Seedling emergence** The maximum number of seedlings were able to emerge immediately after planting the wheat crop from all tillage treatments in both years (Figure 2). However, seedlings continued to emerge up to 50 DAWP in no-tillage plots (Figure 2). In contrast, no seedlings were found to emerge after 20 DAWP in CT or ST ($ST_1$ in first year, $ST_2$ in second year; Figure 2). The staggered nature of germination under NT was attributed mainly to poor seed-soil contact (Froud-Williams 1983). This staggered nature of germination was found to stop in the tillage-operated environment (CT or ST; Figure 2). Between the two types of tillage, seedling emergence of *A. ludoviciana* was increased under CT than ST probably due to increased seed-soil contact (Figure 2). By applying a selective in-crop herbicide, seedling emergence during 20 DAWP could be minimized. Applying a ST operation was found to lower the population density of *A. ludoviciana* from the beginning, thereby offering less competition with the crop as well as reducing the production cost related to herbicide application.

## ACKNOWLEDGMENTS

This research was funded by the Grains Research and Development Corporation (project ID 9175899).

## REFERENCES

Cousens, R., and Mortimer, M. (1995) 'Dynamics of weed populations' (Cambridge University Press, Cambridge, UK).

Froud-Williams, R.J. (1983) The influence of straw disposal and cultivation regime on the population dynamics of *Bromus sterilis*. *Annals of Applied Biology* 103,139 –148.

Llewellyn, R.S., D'Emden, F.H. and Kuehne, G. (2012) Extensive use of no-tillage in grain growing regions of Australia. *Field Crops Research* 132, 204 –212.

Medd RW (1996) Ecology of wild oats. *Plant Protection Quarterly* 11:185 –187.

Walsh, M., Broster, J., Chauhan, B., Rebetzke, G. and Pratley, J. (2019) Weed control in cropping systems - past lessons and future opportunities. *In* 'Australian Agriculture in 2020: From Conservation to Automation', eds J. Pratley and J. Kirkegaard. pp. 153 –172.

Widderick, M. and McLean, A. (2017) Tillage and farming system - impacts on weed germination and seedbank longevity. GRDC update papers.

**Figure 1.** Effect of different tillage treatments (NT: no-tillage in both years; $ST_1$: strategic tillage in first year and no-tillage in second year; $ST_2$: no-tillage in first year and strategic tillage in second year; CT: conventional tillage in both years) on *Avena sterilis* ssp. *ludoviciana* seed distribution pattern at soil depths of 0-5, 5-10, and 10-15 cm. Error bars represent standard errors of the mean of four replicate plots.

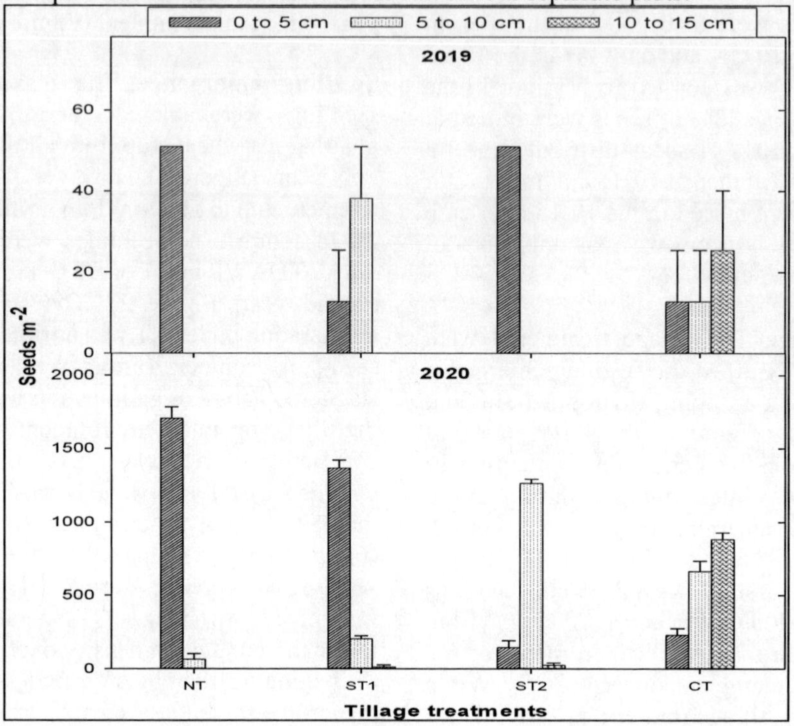


**Figure 2.** Seedling emergence of *Avena sterilis* ssp. *ludoviciana* under different tillage treatments (NT: no-tillage in both years; $ST_1$: strategic tillage in first year and no-tillage in second year; $ST_2$: no-tillage in first year and strategic tillage in second year; CT: conventional tillage in both years). Error bars represent standard errors of the mean of four replicate plots.

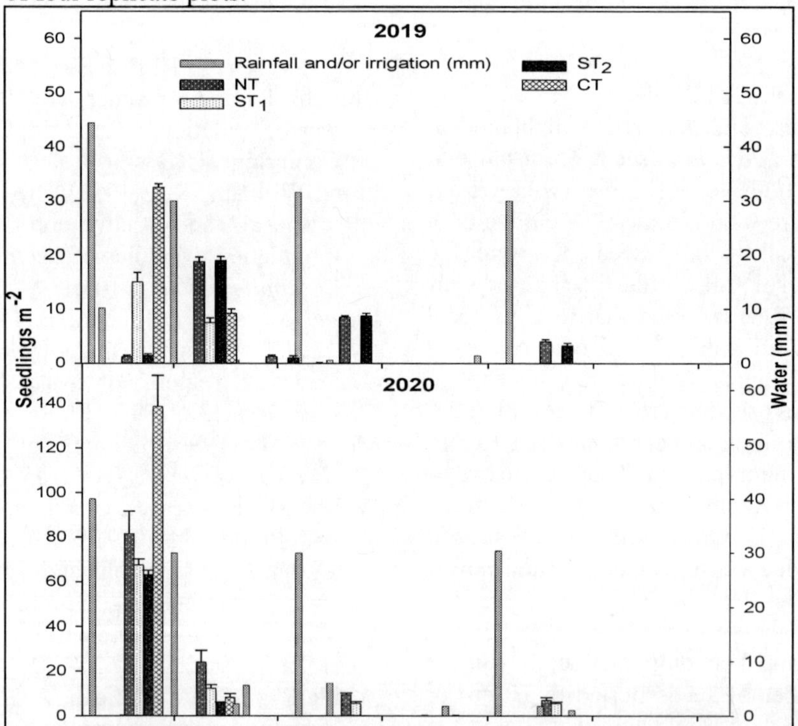

# Six years of crop competition research in the northern grains region – key trends in impact on weed and crop growth

Michael Widderick[1], Kerry Bell[1], Asad Shabbir[2], Hanwen Wu[3], Linda Heuke[2], Greg Harvey[1], Michael Walsh[2]

[1]Queensland Department Of Agriculture And Fisheries, Toowoomba, Australia,
[2]University of Sydney, Camden, Australia,
[3]NSW Department of Primary Industries, Wagga Wagga, Australia
(michael.widderick@daf.qld.gov.au)

**Summary** Resistance to herbicides in weeds is widespread in the northern grains region of Australia. To reduce continued reliance on herbicides and improve the control of herbicide resistant weeds, alternative non-chemical and agronomic approaches for weed management are being explored. One such option is growing a competitive crop by manipulating crop row spacing, density and cultivar. Over a period of five years (2017 – 2022), researchers in the northern grains region have been investigating the effect of growing competitive summer (sorghum and mungbean) and winter (faba bean, chickpea and wheat) crops on key weeds *Echinochloa colona*, *Chloris virgata* (summer crops) and *Sonchus oleraceus* and *Conyza bonariensis* (winter crops). Over the six experimental years, a total of 22 summer and 49 winter field trials were conducted in Wagga Wagga (winter only), Narrabri and Southeast Queensland at either Hermitage, Wellcamp or Kingaroy. Competition effects were assessed through the collection of weed biomass, weed seed production and crop yield data. The large data set is being analysed to identify consistent trends across sites and seasons with each crop by weed combination being analysed independently. While results are still pending, preliminary summation of the data suggests narrow row spacing is a reliable agronomic approach for increasing crop competitiveness and reducing weed growth and seed production. Increasing crop density has also been shown to reduce weed growth and reproduction but results have been more variable. The impact of cultivar is also variable, being greatly influenced by location and seasonal conditions. The impact of narrow row spacing and increased crop density on crop yield has differed between crops, sites and seasons. However, yield responses were generally found to be positive in favourable growing seasons. Results of the meta analysis will help inform decision making on the fit and application of growing a competitive crop.

**Keywords** Crop competition, meta analysis, row spacing, crop density

# Ornamental environmental weeds were marketed earlier and for longer than non-weeds in New Zealand

Jennifer L Bufford[1], Philip E Hulme[1]
[1]Lincoln University, New Zealand
(jennifer.bufford@lincoln.ac.nz)

**Summary** Many plants now considered invasive weeds were originally introduced intentionally as ornamentals. Increasingly, evidence suggests that attributes of the ornamental market may explain the success of environmental weeds but detailed assessments of this assumption are rare. Here we test the hypotheses that compared to non-weeds, environmental weeds will have entered the ornamental market earlier and have been sold for a longer time. We collected data from dozens of nursery catalogues in New Zealand ranging from 1866 to 1992 for 254 environmental weed species in 188 genera and their congeners. We evaluated the first date of record for each species and the range of catalogues in which it appears. Our catalogues have captured at least half of the environmental weed species thought to have been introduced or sold as ornamentals, and of these the majority (80%) are recorded in a nursery catalogue before they are known to have been naturalized. We found that environmental weed species were first sold in nurseries up to two decades earlier on average than other species and were sold in more catalogues across a greater number of years. These results suggest that early introduction and sustained propagule pressure through continued marketing and sale increases the likelihood that a species will become an environmental weed. Our results can be used to assess those species currently sold today that have the potential to become environmental weeds as well as species that pose a low risk. As sale of ornamental species can play an important role in driving which species become environmental weeds, this emphasizes the importance of continued engagement with the horticulture industry to minimize the risk of future invasions.

**Keywords** Alien species, Anthropocene, gardens, historical analysis, plant nurseries, plant invasions, non-invasive species, propagule pressure

# The science of prevention: Risk assessment tools for ornamental plants underpin the Plant Sure Scheme and the Gardening Responsibly Initiative

Aimee Freimanis[1], Michelle R. Leishman[2], Victoria Graham[2], Guyo Duba Gufu[2], Claire Laws[2], Katherine McClellan[2], Samantha Newton[2], Hillary Cherry[3], Chris O'Connor[4]
[1]EcoHort Pty Ltd, Glenhaven, Australia,
[2]School of Natural Sciences, Macquarie University, Sydney, Australia,
[3]NSW Dept of Planning & Environment, Parramatta, Australia,
[4]Nursery & Garden Industry NSW & ACT, Kenthurst, Australia
(hillary.cherry@environment.nsw.gov.au)

**Summary** Historically, the ornamental plant trade has been a significant pathway for weed introductions. For example, over 72% of environmental weeds in Australia are considered garden escapes. Strong desire exists within industry, government and the community to reduce the use of 'weedy' ornamental plants to prevent future weed invasions. The industry-led Gardening Responsibly initiative provides new tools to scientifically assess the invasive risk of ornamental plants, reduce the sale and use of high-risk plants and make it easier for everyone to 'garden responsibly'. Gardening Responsibly is similar to international certification initiatives such as those of the Forest Stewardship Council and Marine Stewardship Council, that use an eco-label to identify environmentally friendly products. The Gardening Responsibly initiative aims to provide access to and increase demand for, low invasive risk ornamental garden plants. To ensure the certification eco-label is trusted and effective, it is underpinned by a robust and transparent plant risk assessment framework and categorisation process that classifies plant species and cultivars according to invasive risk. Species assessed as a low invasive risk are promoted while species assessed as a high invasive risk do not qualify for the eco-label. The Gardening Responsibly website includes a research portal where risk assessments and categorisation protocols are freely accessible.

The Gardening Responsibly initiative is being trialled in New South Wales in 2022 and is designed for national uptake. It is open to all industries across the 'supply chain', including breeders, growers, sellers and plant recommenders, such as local governments and landscape architects. This presentation details: i) collaboration and co-design to build the initiative, ii) plant risk assessment and decision support tools, and iii) the accessibility of the initiative to all audiences, from backyard gardeners to professional horticulturists. The Gardening Responsibly initiative aims to ultimately elicit long-term attitudinal and behavioural change to generate positive environmental outcomes.

**Keywords** Environment, risk assessment, certification, ornamental plants, decision support, behaviour change, prevention

# Late-staged weed-control options are significantly reducing annual ryegrass seed production

Yaseen Khalil[1], Mike Ashworth[1], Anglo Loi[2]
[1]Australian Herbicide Resistance Initiative (AHRI) / School of Agriculture and Environment (SAgE)/ The University of Western Australia (UWA), Perth, Australia,
[2]Livestock - Research & Industry Innovation, Department of Primary Industries and Regional Development, Perth, Australia
(yaseen.khalil@uwa.edu.au)

**Summary** French serradella is a productive annual pasture legume species suited for Australian agriculture. Despite its use, there is little information about the management of these species in the context of weed management, particularly if the weeds are herbicide-resistant, either in terms of directly maximizing pasture productivity or their potential impact on farming systems as part of an integrated weed management package. This study focused on (1) Evaluate the effectiveness of weed management options in reducing weed seed banks in serradella; (2) Evaluate pasture productivity, weed control and its impact on weed seed and serradella seed viability including changes in the soil weed seed bank. Two trials conducted at the UWA-Ridgefield farm/Pingelly. In addition to the control treatments, Trial_1 includes PSPE (propyzamide, imazethapyr, propyzamide+imazethapyr, flumetsulam); Post (Thistrol Gold, imazethapyr, imazamox, flumetsulam); Full canopy (weed wiper, spray-topping, mowing+spray-topping, hay/silage production, green-manuring, brown-manuring). Trial_2 treatments include spray-topping with clethodim at early tillering and late tillering stage; spray-topping with paraquat at heading and early filling stage of ryegrass. Herbicidal and cultural control options applied later in the weed lifecycle, such as weed wiping using glyphosate, spray-topping using paraquat, mowing prior to ryegrass flowering followed by spray-topping, biomass cutting for hay/silage production, incorporation of green biomass into the soil (green manuring), and the non-selective use of glyphosate to kill all plants prior to ryegrass flowering (brown manuring) were effective at reducing annual ryegrass seed production by more than 80% compared to the untreated control. The highest serradella yield was achieved following the application of flumetsulam applied post-sowing/pre-emergent with the second greatest serradella seed production following propyzamide and/or imazethapyr treatment. Paraquat, especially when used at the heading stage of ryegrass, reduced the ryegrass seed production to <2.5% of the untreated control (weedy control). The application of glyphosate or paraquat was effective at reducing ryegrass seed production, however these treatments also greatly reduced serradella seed production, making these treatments unsuitable in self regenerating pastures.

**Keywords** Serradella, herbicide-resistance, spray-topping, seed-bank

# Finding fungi to fight invasive grasses: developing a mycoherbicide for GRT in Australia

Tracey Steinrucken[1], Joe Vitelli[2], David Holdom[2], Yu Pei Tan[3]
[1]CSIRO Health & Biosecurity, Dutton Park, Australia,
[2]Biosecurity Queensland, Dutton Park, Australia,
[3]Queensland Department of Agriculture and Fisheries, Dutton Park, Australia
(tracey.steinrucken@csiro.au)

**Summary** The five weedy *Sporobolus* grasses (a.k.a. the rat's tail grasses, which includes *Sporobolus natalensis*: "GRT") collectively cause a $60 million per year problem for Northern Australia's beef industry. If left unmanaged, these unpalatable, undesirable, and highly competitive grasses have the potential to spread to over 30% of Australia. With seed production of around 80,000 seeds per m² per year and seed viability up to 10 years, managing these grasses in a long term, sustainable and low-cost way is essential. Enter biocontrol. CSIRO, together with Biosecurity Queensland, have spent the last six years finding, identifying, and testing a suite of native fungal pathogens as potential biocontrol agents against the rat's tail grasses. After starting with 110 pathogens, we are now closer to developing a mycoherbicide as we now have a priority list of 8 potential agents. During the process of identifying our pathogens, we discovered and have described at least six novel species, with dozens more to be published, including several new genera. Our priority list includes some of these new fungal species such as *Stagonospora tauntonensis* and *Phaeosphaeria* sp., as well as known species from Alternaria and Fusarium. In this talk we describe the results of virulence bioassays, phylogenetic analyses, glasshouse pathogenicity experiments and host range studies. We also discuss the way forward in the development of a safe, effective and reliable product. We've had continuous consultation with, and input from, landholders, property managers and council representatives. We now have a clear way forward for turning this project into useful, highly adaptable tools, which we hope will reduce the cost and reliance of herbicides for controlling these invasive rat's tail grasses, and which can complement existing management strategies.

**Keywords** Pasture weeds, pathogens, biocontrol, invasive grasses, GRT, Queensland

# Herbicide and fertilizer application trials to improve production in Giant rat's tail grass (GRT) infested pastures.

Shane Campbell[1], Fred Oudyn[2], Rose Campbell[3], Megan Crossing[1], Alana Connolly[1], Karen Harper[1], Wayne Vogler[4], Stephen Martin[5] and Joe Vitelli[3]

[1] School of Agriculture & Food Sciences, The University of Queensland, Gatton Campus, Gatton, QLD 4343, Australia

[2] Chemistry Centre, Department of Environment and Science, GPO Box 5078, Brisbane, QLD 4001, Australia

[3] Biosecurity Queensland, Department of Agriculture and Fisheries, GPO Box 267, Brisbane, QLD 4001, Australia

[4] Biosecurity Queensland, Department of Agriculture and Fisheries, PO Box 976, Charters Towers, QLD 4820, Australia

[5] Powerlink Queensland, PO Box 1193, Virginia, QLD 4014, Australia

**Summary** Giant rat's tail grass (GRT) (*Sporobolus natalensis* (Steud.) T.Durand & Schinz and *S. pyramidalis P.Beauv.*) is an invasive weed of pastures. Conventional control efforts for GRT centre on pasture management, the use of chemical and mechanical control and plant competition. To improve management options, recent studies in south-east Queensland have focused on (a) better understanding the residual effects of the most widely used herbicide (flupropanate) and (b) fertilization to determine if it can enhance forage quality and utilization of GRT, particularly in high rainfall environments. In the herbicide trial, granular or liquid flupropanate were applied at label recommendation of 1500 g a.i. $ha^{-1}$, to mature GRT plants growing in one of five soil types and to pots containing soil only. Residue levels were monitored annually in both soil and in GRT for four years. In an initial ungrazed fertiliser trial eight rates of nitrogen (0 – 300 kg N $ha^{-1}$) were applied to a GRT infested setaria (*Setaria sphacelata (Schumach.) Stapf & C.E.Hubb.*) pasture. A second integrated trial was testing the combination of four fertiliser (0, 50, 100 and 200 kg N $ha^{-1}$) and two herbicide applications (± herbicide) under grazed conditions.

Irrespective of soil type, GRT plants in the herbicide trial contained 22±0.3% (granular) and 31±1.3% (liquid) of the applied flupropanate after 12 months, with levels dropping to <5% after 24 months. Flupropanate in the corresponding soil pots were 20±1.7% (granular) and 7±1.3% (liquid) after 12 months, with similar levels recorded after 24 months. No significant difference was observed between flupropanate formulations when applied to bare soil at 12 (83±3.3%) and 24 (73±1.8%) months after application. Whilst a range of plant response measurements are being undertaken in the fertilizer trials, in this paper we focus on changes in leaf tensile strength and differences in grazing patterns. GRT leaf material was found to have a much higher tensile strength than setaria, and it increased with maturity for GRT but not setaria. Increased fertilisation had a weak negative correlation ($P=0.065$) with leaf tensile strength. In the grazed trial, irrespective of fertilizer regimes, cattle introduced to 5-week-old regrowth tended to heavily graze both GRT and setaria over the first 2 weeks, particularly setaria which was grazed lower (24.1 cm) than GRT (38 cm). This has allowed wick wiper applications of a flupropanate + glyphosate based mixture to be applied to the taller GRT plants, with efficacy and non-target damage assessments the focus of on-going monitoring.

**Keywords** *Sporobolus natalensis*, GRT, fertilising, nutrition, flupropanate, tensile strength.

## INTRODUCTION

*Sporobolus natalensis* and *Sporobolus pyramidalis*, commonly known as Giant rat's tail grass (GRT) is an invasive weed of pastures, natural reserves, forestry and utilities with the capacity to reduce the productivity of agricultural land, decrease land value, reduce the biodiversity of natural ecosystems and increase control expenses to non-sustainable levels (Simon and Jacobs 1999). The grasses are of extremely low palatability and high tensile strength, and when tussocks are mature, livestock generally avoid utilising the plant. GRT was introduced into Australia through contaminated seed, with *S. pyramidalis* now widespread from Cooktown in north Queensland and south to the New South Wales Central Coast, whilst *S. natalensis* is found widespread from Rockhampton in central Queensland to Port Macquarie on the mid north coast of NSW (AVH 2017). Populations of both species are present in the Northern Territory (AVH

2017). The importance of these species is reflected in both being Weeds of National Significance with estimated annual losses of $60 million per annum to the cattle industry in northern Australia.

Current control efforts for GRT center on the use of chemical and mechanical control, plant competition and pasture management. Despite the production of a best practice manual for GRT management, control has not been achieved and GRT continues to spread into new areas. This paper reports on recent studies in south-east Queensland aimed at better understanding the residual effects of the most widely used herbicide (flupropanate) and determining if fertilisation can enhance forage quality and utilization of GRT, as part of an integrated management approach.

## MATERIALS AND METHODS

**Flupropanate potted trial** A 5 × 2 × 3 factorial experiment was undertaken using a complete randomised design and four replications. Factor A was five contrasting soil types (chromosol, dermosol, ferrosol, kurosol and vertosol) assigned to main plots, factor B was two pasture treatments (GRT, bare ground) assigned to subplots, and factor C was three herbicide treatments (nil, liquid, granular) assigned to sub-subplots.

The different agricultural soils were collected from locations in south-east Queensland known to sustain GRT populations (AVH 2017). At each site ~950 kg of soil was mechanically removed by scraping the top 10 cm of soil from a 5 m × 5 m area. The soil was transported to the Ecoscience Precinct (ESP) at Dutton Park where it was sieved through a 2 mm mesh.

Mature GRT plants (*S. natalensis*) were collected from a cattle property near Conondale, Queensland (26°42'53"S; 152°41'51"E) and transported to ESP. There they were separated into single tillers containing an established root system and placed in 4 L pots filled with 4000 g of oven dry equivalent soil from each of the five selected soil types. Plants were then grown for two months in a glasshouse prior to herbicide application. No inflorescences were present at the time of spraying. Throughout the entire experiment each pot was maintained at 40% soil moisture content, which provided sufficient water for plant growth and microbial activity without the leaching of any herbicide from the pot.

The liquid herbicide application of flupropanate (1564 g a.i. ha$^{-1}$) (Grow Choice Tussock™ Herbicide) was applied using a 12 V electric powered fine air compressor unit (Iwata Studio Series) with 0.35 mm nozzle and operating pressure of 1 mPa. Each plant was sprayed just prior to the point of run-off (~400 L ha$^{-1}$), with the fine nozzle and a spray guard attachment ensuring the solution was applied directly to the plant without any contamination to the soil. For granular application, a 5 ml vial containing a perforated lid was used to uniformly apply granular flupropanate (1564 g a.i. ha$^{-1}$) (Granular Products GP Flupropanate Granular Herbicide) to the soil surface. Bare ground pots were also treated to a uniform application of both liquid and granular formulation of flupropanate. The concentration rate used in this experiment was based on the recommended application rate of flupropanate for GRT control given on the label.

At three, six, 12, 24 and 48 months post-herbicide treatment, 24 samples for each soil type were randomly selected for flupropanate determination. Data for 12 and 24 months only are shown in this paper. The soil from each pot was removed, passed through a 2 mm sieve and mixed thoroughly to ensure that the sample was uniform. A 200 g subsample was then removed from each pot for flupropanate soil analysis and delivered to the Department of Environment and Science, Chemistry Centre at ESP.

At the designated sampling times, GRT plants were also removed from each pot and their fresh weight recorded before placing the samples in a drying chamber set at 25°C. A lower temperature was selected to avoid potential heat impacts with the herbicide. The plant samples remained in the drying chamber for 10-14 days. Once the samples were dry, they were processed through a 200 V electric plant grinder (Culatti Type MFC), using a 0.5 mm mesh and delivered to the Chemistry Centre at ESP for flupropanate residue determination.

Soil and plant data was statistically analysed using ANOVA, but beforehand it was transformed using an arcsine transformation. If significant treatment differences were detected ($P<0.05$), the means were separated using Fishers' Protected Least Significant Difference (LSD) test. Data was back transformed for presentation.

**Fertiliser trials** The field site was located near Mapleton (26°62'S; 152°87'E) and comprised a dense infestation of GRT (average of 2 ± 0.07 plants m$^{-2}$) within a setaria based pasture (*Setaria sphacelata (Schumach.) Stapf & C.E.Hubb.*). In February 2022 a randomized complete block experiment was established with eight treatments each replicated three times. Experimental units were 5 m × 4 m plots with a 2 m buffer between blocks. Treatments comprised eight rates of nitrogen (0, 25, 50, 75, 100, 150, 200 and 300 kg N $ha^{-1}$). Initially the trial site was fenced (to exclude grazing), slashed and all cut material removed from each plot. Plots were then fertilized, which entailed an initial base application of Diammonium phosphate (DAP; 138.89 kg $ha^{-1}$) followed by the addition of Urea to achieve the designated rates of nitrogen. Fertiliser was applied using a handheld Ozito® spreader. Whilst a range of plant response measurements are being undertaken, in this paper we focus on changes in leaf tensile strength between GRT and setaria at 5 and 9 weeks after slashing and fertilization. This was achieved by testing a minimum of five leaves of each species per plot using a device developed by Dr. Marcelo Benvenutti (Queensland Department of Agriculture and Fisheries, Gatton, QLD, Australia) to replicate the way cattle grip and tear grass material during grazing.

A second integrated trial is testing the combination of four fertiliser (0, 50, 100 and 200 kg N $ha^{-1}$) and two herbicide applications (±herbicide) under grazed conditions. It comprises a split plot design, with fertilizer treatments allocated to main plots and herbicide applications to sub plots. Each treatment is replicated three times and experimental units comprise 3 m × 5 m plots, with a 1-2 m buffer between blocks. In February 2022 the site was slashed and fertilized using a similar procedure to trial 1. Cattle were excluded for the first five weeks, but then given access to graze the trial, which was located in a 52.6 ha paddock stocked with 70 animals, giving an overall stocking rate of 1 animal 0.75 $ha^{-1}$. Herbicide treatments were implemented on 19 April 2022 using a customised wick wiper device to apply a herbicide mixture containing 150 g a.i. glyphosate plus 81 g a.i. flupropanate $L^{-1}$. In this paper we focus on whether there were differential grazing responses between fertilizer treatments prior to the application of herbicides. This was achieved by measuring weekly changes in height of five GRT and five setaria plants permanently located in each plot, except for week 1 when flooding prevented access to the site. Data from both trials was statistically analysed using ANOVA and if significant differences ($P<0.05$) occurred the means were separated using Fishers' Protected Least Significant Difference (LSD) test. Regression analysis was also undertaken to determine the response of fertilization on leaf tensile strength of GRT and setaria.

## RESULTS

**Flupropanate potted trial** No significant difference was observed between soil types for the 12- and 24-month residue analysis. However, the decline in flupropanate irrespective of formulation between 12 (61.2%) and 24 (46.2%) months was significant ($P<0.0005$). Irrespective of time, more than double the amount of flupropanate was recovered in bare soil only pots (77.6%) compared to planted pots (soil plus plant material) (29.8%) Figure 1). The percentage of total flupropanate recovered when applied as a granular application (58.7%) was significantly higher ($P<0.021$) when compared to a liquid application (48.8%), irrespective of soil type, time and pot treatment (Figure 1).

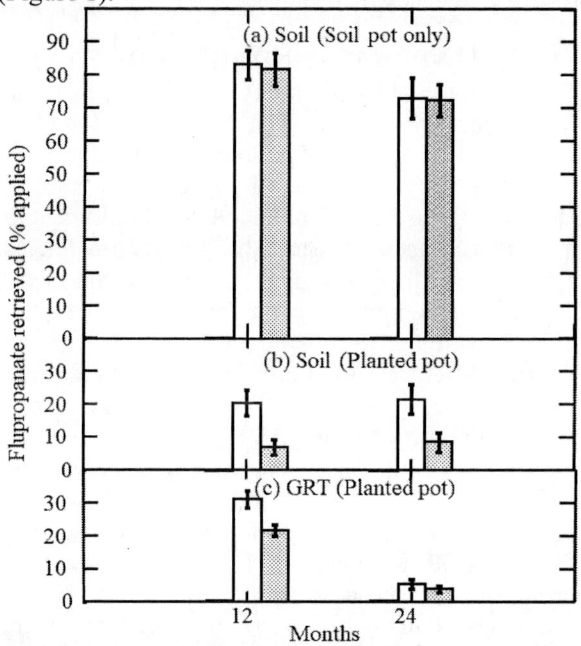


**Figure 1.** Percentage of applied flupropanate retrieved in soil where plants were excluded (a), and planted pots (b) in soil and (c) within the GRT plant when treated with granular (☐) and liquid (▩) flupropanate, at 12 and 24 months post-herbicide application, irrespective of soil type. Error bars represent standard error of the mean.

**Fertiliser trials** Overall, GRT leaf material was found to have a significantly higher tensile strength ($P<0.05$) than setaria (Figure 2). Age of regrowth ($P<0.05$) also had a significant effect for GRT but not setaria, with 5-week-old regrowth recording lower leaf tensile strengths than the mature 9-week regrowth (Figure 2). Increased fertilisation had a weak negative correlation with leaf tensile strength ($P=0.065$) (Figure 2).

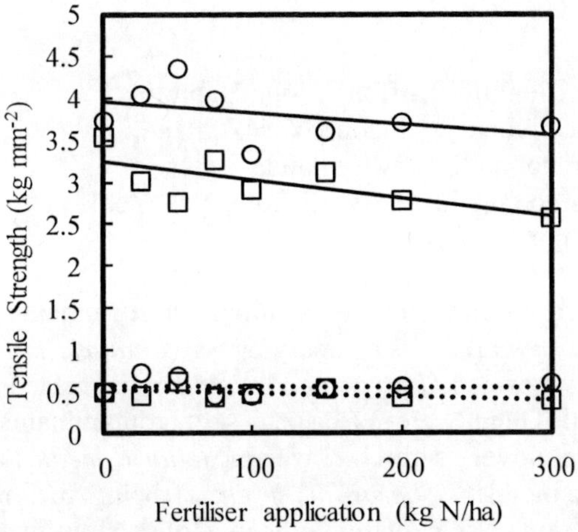


**Figure 2.** Relationship between leaf tensile strength and fertilizer application of 5 (☐) and 9 (O) week old regrowth of GRT (solid line) and setaria (dotted line).

In the grazed trial, initially there was a significant plant height difference (*P<0.05*) between fertilizer treatments for setaria (Figure 3), but not GRT (*P>0.05*). The unfertilized control treatment tended to have the shortest setaria plants (54.2 cm), while the two highest fertilizer treatments had the tallest plants (71.7 to 77.3 cm). GRT averaged 67.9 cm across all fertiliser treatments. Following the introduction of cattle, both GRT and setaria were grazed heavily over the first 2 weeks, particularly setaria which was grazed lower than GRT. At this stage, GRT and setaria averaged 38 and 24.1 cm respectively, with no significant differences (*P>0.05*) between fertilizer treatments. Only small reductions in plant height occurred over the following four-week period, with GRT and Setaria averaging 34.2 and 16.7 cm after six weeks grazing, respectively (Figure 3).

## DISCUSSION

While the integrated fertilizer and herbicide trial under grazed conditions is ongoing, the results reported above provide some valuable insights for the integrated management of GRT. Flupropanate is thought to be mostly absorbed through the root system, and when applied to foliage is reliant on rain to be washed onto the soil for root uptake. Results here suggest flupropanate can also be absorbed through the green or actively growing foliage of GRT, translocated to the roots and exudated into the soil, albeit <10% of applied is found in the soil. Despite having higher leaf tensile strengths than setaria, cattle readily consumed 5-week-old regrowth of both setaria and GRT. However, they had a tendency to graze setaria lower to the ground (i.e. 16.7 versus 34.2 cm after 6 weeks grazing) which provides a point of differentiation for subsequent herbicide applications to control GRT using wick wiper style equipment. If flupropanate is included in wick wiper herbicide mixtures it has the potential to provide some residual control of GRT. Despite a slight reduction in the leaf tensile strength of GRT ongoing measurements (e.g. nutritional analysis) and monitoring is required to determine any benefits of fertilization in high rainfall environments.

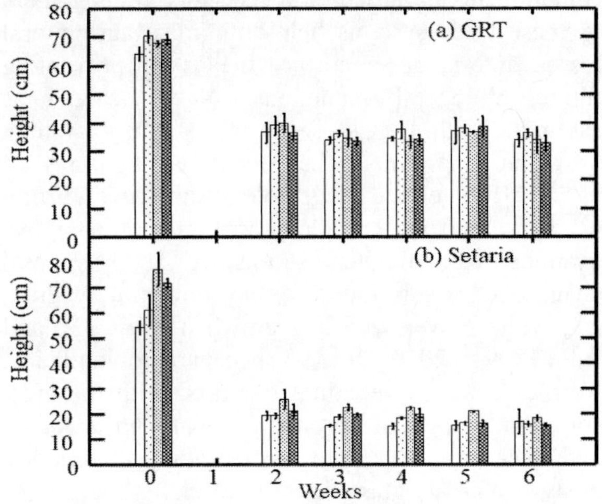


**Figure 3.** Influence of different nitrogen rates [unfertilised control (☐), 50 (☐), 100 (☐), and 200 (▩) kg N ha$^{-1}$] to weekly height (cm) measurements of (a) GRT and (b) Setaria. Error bars represent standard error of the mean.

## ACKNOWLEDGMENTS


The authors acknowledge the financial assistance provided by Powerlink Queensland, the Queensland Government, The University of Queensland, and the A.W. Howard Memorial Trust; land owner David Ross for his patience and understanding in dealing with researchers; Dr. Vincent Mellor for assistance with statistical analysis and Dr. Marcelo Benvenutti for providing equipment to measure leaf tensile strength.


## REFERENCES


AVH 2017. Atlas of Living Australia (AVH). Australia's Virtual Herbarium, Council of Heads of Australasian Herbaria. Available at: http://avh.chah.org.au

Simon, B.K and Jacobs, S.W.L. (1999). Revision of the Genus *Sporobolus* (Poaceae, Chloridoideae) in Australia. *Australian Systematic Botany* 12,375-448.

# A century of weed change in New Zealand's forage seed multiplication industry

Jesse Rubenstein[1], Philip Rolston[2], Philip Hulme[3], Jennifer Bufford[3], John Hampton[1]
[1]Department of Agricultural Sciences, Lincoln University, Lincoln, New Zealand,
[2]Foundation for Arable Research, Christchurch, New Zealand,
[3]Department of Pest Management and Conservation, Lincoln University, Lincoln, New Zealand
(jesse.rubenstein@lincolnuni.ac.nz)

**Summary** International seed trading provides a significant pathway for seed contaminants, and many globally established weeds originated as contaminants in agricultural seed lots. Management of these trade systems helps minimise agricultural losses and is an important means of preventing future biological incursions. New Zealand is essential within this seed for sowing system, providing one-third to half the world's supply of various forage and vegetable crop seeds. Using historical and current plant contaminant data, we examined the frequency, identity and temporal changes of weeds found within perennial ryegrass and white clover seed lots grown in New Zealand from 1909 to 2020. Over 95 species of contaminants were detected in perennial ryegrass, with the most common being soft brome (*Bromus hordaceus*) and hair grass (*Vulpia bromoides*). Correlation analysis for ryegrass identified eight species of contaminants that significantly decreased over the 110-year study period and five that increased. Catsear (*Hypochaeris radicata*), hawkbit (*Leontodon* sp.) and sorrel (*Rumex acetosella*) decreased the most over time, while annual poa (*Poa annua*), lesser canary grass (*Phalaris minor*) and wireweed (*Polygonum aviculare*) increased the most. There were 115 species of contaminants in white clover, with chickweed (*Stellaria media*) and field madder (*Sherardia arvensis*) being the most common. No contaminants in clover significantly increased over time, but cress (*Barbarea* sp.), dodder (*Cuscuta* sp.) and mouse-ear chickweed (*Cerastium* sp.) decreased the most. Considering New Zealand trades crop seed with approximately half of the world's countries and contributes substantially to the global supply of forage seed, our study provides a unique insight into changes in the weed spectrum in New Zealand and throughout the seed for sowing system over the last century. Information provided by this study is also useful for biosecurity agencies and land managers trying to identify problematic weed species on which to focus resources.

**Keywords** Biosecurity, agricultural weeds, ryegrass, clover, temporal changes

# Does awareness of invasive freshwater plants mitigate the dispersal risk posed by lake users?

D Clements[2], Philip Hulme[1], P Champion[2]
[1]Bio-Protection Research Centre, Lincoln University, Canterbury, New Zealand,
[2]National Institute of Water and Atmospheric Research (NIWA), Hamilton, New Zealand
(philip.hulme@lincoln.ac.nz)

**Summary** Recreational users of freshwaters (e.g. fishers, boat users etc.) are leading vectors of alien plant spread among lakes. To date no study has integrated information on the associations between awareness, mitigation and residual risk of different lake users, that might provide insights into more effective management of this introduction pathway. Using data from over one thousand face-to-face interviews of lake users across New Zealand to capture details of more than 1700 lake visits, we present the first comprehensive analysis of this pathway. Interviews captured data on the main activity, location of residence, visit frequency, other lakes visited in the last fortnight, awareness of alien freshwater species and any actions they might take to prevent their spread. The dominant lake users were water-skiers, swimmers, boat fishers, jetskiers, kayakers and lakeside fishers with other users including jetboater, sailors and hikers, less frequent. Awareness of alien plant species was high overall but with marked variation among user groups. While almost all jetboaters knew the name of at least one alien plant species, this was true for only half of all swimmers. In general, awareness was higher in users who had been directly affected by alien plants, particularly those whose equipment or boat engines were fouled causing negative associations. As a result, it was these users who were most likely to take mitigating actions such as cleaning and/or drying their equipment to prevent further spread. To derive an overall assessment of the risk posed by different users, data on distances travelled, likelihood of visiting invaded lakes, willingness to take action to prevent spread and the relative abundance of users were integrated. As a result, this study recommends that awareness raising should better target boat users, particularly water-skiers, focusing on the impacts upon their leisure activity rather than biodiversity.

**Keywords** Aquatic, behaviour change, pathway, propagule pressure

# Control of the emerging aquatic weed Amazon frogbit with flumioxazin

Tobias O. Bickel[1], Bahareh Shahrabi Farahani[1], Christine Perrett[1], Junfeng Xu[2], Joseph Vitelli[1]
[1] Invasive Plant Science, Department of Agriculture and Fisheries, Brisbane
[2] School of Agriculture and Food Sciences, The University of Queensland, Gatton

**Summary** Amazon frogbit is an emerging aquatic weed that causes significant impacts to freshwater systems in Australia. Currently there are limited control options available, with only one herbicide (flumioxazin) registered for its control.

Outdoor pond trials demonstrated that foliar and subsurface flumioxazin application provide excellent Amazon frogbit control (95-100% biomass reduction) at intermediate label rates. Subsurface application provided slightly better control than foliar spray. However, in deeper waterbodies foliar application will be more economical. In a field trial, a subsurface injection of flumioxazin (200 ppb) in a farm dam covered by a dense Amazon frogbit infestation provided ~99% control with a single application within three months.

Overall, our work demonstrated that flumioxazin is an excellent control tool to manage Amazon frogbit and will greatly enhance the management of this invasive aquatic weed.

**Keywords** floating macrophytes, *Limnobium laevigatum*, aquatic herbicide, aquatic weed management.

## INTRODUCTION

Invasive aquatic weeds cause significant environmental and socio-economic impacts worldwide. Amazon frogbit *Limnobium laevigatum* (Humb. & Bonpl. ex Willd.) Heine (Hydrocharitaceae; also known as sponge plant) is a free-floating aquatic plant originating from freshwater habitats of tropical and subtropical South America, Central America, and the Caribbean (Cook and Urmi-König 1983). The plant also produces inconspicuous white flowers, forming a fleshy capsule with up to 100 seeds per capsule. A popular aquarium species, it has been introduced in multiple locations around the world, including California, Japan, southern Africa and Australia (Howard *et al.* 2016, Anderson and Akers 2011, Kadono 2004). Amazon frogbit is a fairly recent arrival in Australia, first detected in 2003, it is now present in multiple states (QLD, NSW, WA) and is rapidly expanding its range (Atlas of Living Australia 2022).

Amazon frogbit has morphologically distinct growth forms. Starting as small seedlings that resemble duckweed, the plant develops larger spongy (aerenchymatic) floating leaves that lay flat on the water surface. Once the water surface is covered, it extends its leaves vertically and can become up to 50 cm tall, resembling water hyacinth (Cook and Urmi-König 1983). Amazon frogbit readily reproduces asexually which allows it to rapidly overgrow entire water bodies. But the plant also produces inconspicuous white flowers, forming seed pots that can contain 20-30 seeds. Seeds can germinate immediately or persist in the environment for at least three years (Weerasinghe 2020).

The impact of Amazon frogbit species is not fully documented. However, frogbit can form dense mats with up to 2000-2500 plants per square meter (Weerasinghe 2020) and the similarity of its growth habit to water hyacinth indicate that it can cause significant environmental and socio-economic impacts. Amazon frogbit readily outcompetes other aquatic plants (Perryman 2013) and the rapid growth results in a thick cover of the water surface, affecting water quality and interfering with recreational and commercial use of freshwater systems. Like other free floating aquatic weeds, the thick floating mats prevent gas exchange and light to penetrate the underlaying water column, thereby modifying aquatic habitats and making them unsuitable for native flora and fauna (Perna and Burrows 2005).

Currently there are limited options available to manage this highly invasive aquatic weed (Anderson and Akers 2011). The herbicides imazamox and penoxulam were found to be effective in controlling Amazon frogbit in the USA (Willis *et al.* 2018), but these herbicides are currently not registered for aquatic use in Australia. The congeneric *Limnobium spongia* (Bosc) Steudel native to the USA can be controlled with diquat, triclopyr and 2,4-D; glyphosate was not effective for control (Madsen *et al.* 1998). Flumioxazin is a new herbicide registered in Australia for control of aquatic weeds (Clipper, Sumitomo Inc.). Preliminary field trials showed that flumioxazin effectively controls Amazon frogbit (authors' observations), but there is no published data on the effect of application techniques (foliar vs. subsurface) and application rate on control efficacy. To address this knowledge gap, we conducted a mesocosm (pond) trial to analyse the efficacy of flumioxazin in controlling Amazon frogbit and carried out a small-scale field trial to measure control efficacy in a real-world scenario.

## MATERIALS AND METHODS

**Mesocosm trial** The experiment was conducted in an outdoor area at the Ecosciences Precinct, Dutton Park, Queensland, Australia, in February- April 2017 (late summer to autumn). Frogbit was cultured in 35 plastic crates (60 x 35 x 37 cm; ~70 L) filled with de-chlorinated tap water, aerated to prevent stratification. Mesocosms were fertilized monthly with 5 g of soluble fertiliser (Thrive, Yates, Australia) containing macro nutrients (NPK 25:5:8.8) and trace elements to support healthy plant growth. Once Amazon frogbit plants covered the entire water surface, crates were randomly assigned to five treatments (seven replicates each): control (no herbicide), low subsurface, high subsurface, low foliar, and high foliar (see Table 1 for rates). The dosages represent low and medium application rates listed on the of the flumioxazin Clipper (Sumitomo) label.

The plants were treated with flumioxazin (Valor, Sumitomo Inc) in late summer, 22 February. For subsurface application, aliquots (10 - 20 mL) of a flumioxazin stock solution were injected into the water column. For foliar treatments, plants were sprayed with a paint gun (8 - 16 mL stock solution) with a plastic shroud in place to prevent spray drift; water was added to the low foliar treatment dose (8 mL) to keep spray volume consistent (equivalent to 750L $ha^{-1}$). The paint gun produced small droplets that achieved even coverage with the applied volume. On the treatment day, the water temperature was 27.01 °C and the water was acidic (pH 5.1) and had a specific conductance of 152 µS $cm^{-1}$.

Plant health was visually assessed twice weekly for the remainder of the experiment. At the end of the experiment (20 April; eight weeks after treatment), all remaining frogbit was harvested from each crate to measure wet mass (WM). We tested for differences between treatments with ANOVA and Tukey's HSD after transforming data (LN) to meet the requirements of parametric tests. Statistical analysis was carried out in R (v4.0.5, The R Foundation).

**Field trial** A field experiment was performed in a small dam (27°48'27.3"S ,153°01'47.4"E; 495 $m^2$ surface area, 520 $m^3$ Volume) on a private property in Jimboomba, QLD. The dam was partially shaded by tall eucalypt trees and the entire dam surface area was covered with a thick mat of Amazon frogbit. To determine pre-treatment biomass, four samples of Amazon frogbit (0.25 $m^2$) were collected from random locations in the dam. Plant samples were dried for 72 hours at 50 (°C) and weighed to determine dry mass (DM). On 11 September 2019 (early spring), 103 g flumioxazin a.i. (Valor, Sumitomo Inc) was mixed in 200 L of water in a truck-mounted quick spray unit. The herbicide solution was injected 50 cm below the water surface in 15 evenly spaced treatment spots (~ three meters distance between injection spots) to achieve a target concentration of 200 ppb flumioxazin in the water column. Plant health and water chemistry parameters were monitored regularly for the next three months and at the end of the trial the entire remaining frogbit biomass was harvested to determine dry mass.

## RESULTS

**Mesocosm trial** On the day of treatment, frogbit covered the entire water surface of the experimental crates forming a tall canopy (mean plant height 15.1 cm ± 3.3 SD); the average wet mass was 15.4 kg $m^{-2}$ (± 2.5 SD). Frogbit plants exhibited herbicide damage (blackened leaf veins) in all treatments (except control) within 24 hours. Plant health deteriorated rapidly over the next 7 days. Visual damage was most severe in the high subsurface application and frogbit was severely compromised, disintegrated and began to sink within 2 weeks. From DAT40, most plants in the high subsurface were dead and there was no further change in conditions until week 8 (harvest). The decline in health over time was similar in all herbicide treatments. Plant damage was lower in low subsurface and high foliar treatments and the low foliar application caused the least amount of visual damage.

At the end of the experiment, there was a significant difference in canopy height (ANOVA: df=4, $F = 39.0$, $p<0.0001$) and biomass (ANOVA: df=4, $F = 27.97$, $p<0.0001$) between treatments (Fig. 1, Table 1). Final canopy height in the control ponds was similar to starting conditions and the biomass had increased by 22% (Fig. 1, Table 1). Herbicide application significantly reduced plant height and biomass in all treatments compared to the control (Fig. 1, Table 1). High subsurface application was the most efficient treatment and achieved complete control in all but one crate. Control efficacy in terms of biomass and plant height reduction declined in order from high subsurface to high foliar, low subsurface with least control achieved with the low foliar treatment (Fig. 1, Table 1). However, the differences in final biomass between herbicide treatments were not statistically significant except for the low foliar treatment (Fig. 1).

**Field trial** Before treatment, Amazon frogbit covered the entire water surface of the dam with an average dry mass of 611 g $m^{-2}$ ± 72 SD, providing an estimate of around 5 t of plant wet mass for the entire dam. After subsurface flumioxazin application, Amazon frogbit leaves started darkening within 48 hours of exposure. However, the thick plant mat took considerable time to break down. Two months after treatment, some of the plants had decayed and sunk and with about half of the dam area had having an open water surface. After three months, only a small amount of decaying and fragmented frogbit remained in the dam and was harvested; the total biomass was 1.3 kg DM of frogbit (~21.1 kg wet mass) for the entire dam, giving a control efficacy of 99.6% with the single herbicide application.

Dam water physico-chemical variables changed over the period of the experiment from DAT0 to three months after treatment. Water temperature increased over time from 14.6 to 27.6 °C following the seasonal warming of air temperature from spring to summer. Initially, the water in the dam was acidic (pH 5.5) but the pH started to increase in the last month of the trial, coinciding with the opening of the water surface, until it was neutral (pH 7.0) at the end of the trial. Specific conductance increased steadily over time from 47 to 493 µS $cm^{-1}$.

**Table 1.** Flumioxazin application rates, final Amazon frogbit canopy height (maximum plant height), wet biomass and % control in terms of biomass reduction compared to before treatment. Values are means ± SD. Lettering indicates a statistically significant difference at $p<0.05$ (Tukey's HSD).

| treatment | flumioxazin a.i. rate | canopy height cm | wet mass kg $m^{-2}$ | % control |
|---|---|---|---|---|
| control | 0 | 14.3 ± 3.6a | 19.7 ± 2.3 a | -26 |
| low foliar | 105 g $ha^{-1}$ | 2.4 ± 2.2b | 4.1 ± 3.7 | 74 |
| high foliar | 210 g $ha^{-1}$ | 0.5 ± 1.1c | 0.7 ± 1.3 | 96 |
| low subsurface | 100 ppb (µg $L^{-1}$) | 0.8 ± 1.2bc | 1.7 ± 2.9 | 87 |
| high subsurface | 200 ppb (µg $L^{-1}$) | 0.0 ± 0.0c | 0.0 ± 0.0 | 100 |

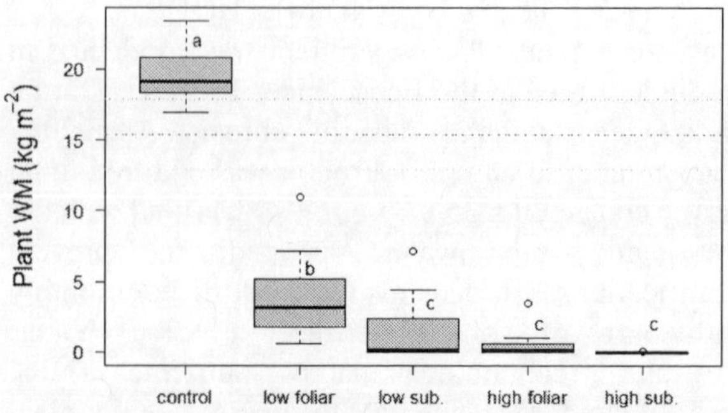


**Figure 1.** Final Amazon frogbit biomass eight weeks after herbicide application with different application techniques and dosage. Lettering indicates a statistically significant difference at $p<0.05$ (Tukey's HSD); 'low/high sub.' stands for low/high subsurface application, respectively. Horizontal lines indicate mean biomass (dashed) before treatment ± SD (dotted).

## DISCUSSION

Our research demonstrated that flumioxazin is an excellent herbicide tool to manage Amazon frogbit in Australian freshwater systems at low to intermediate label rate applications. In outdoor mesocosm trials we found little difference in control efficacy (87 – 100%) between foliar and subsurface application. Only the low foliar spray provided significantly less control efficacy than the other treatments. But even at this low 105 g $ha^{-1}$ foliar spray a 74% reduction in biomass was achieved with a single application. While foliar application was slightly less efficient than subsurface application, it will still be more economical in deeper water bodies as only the surface area is treated instead of dosing the entire water body volume. Therefore, a smaller amount of product is needed with foliar application. Despite taking care to achieve even herbicide coverage when spraying the Amazon frogbit, the slightly lower foliar efficacy could be result of uneven coverage when applying the herbicide, a common issue with foliar spraying of aquatic weeds (Willis *et al.* 2018, Mudge *et al.* 2012). Additionally, it is possible that uptake of flumioxazin by Amazon frogbit is more efficient through the root system and the leaves that are in contact with the water surface, than through the tougher ticker cuticle of emergent leaves. Lastly, the total amount of flumioxazin applied to the crates in the subsurface treatment was ~ three times higher, so plants can take up more product.

The field trial further demonstrated that flumioxazin is an efficient control tool for managing Amazon frogbit in a real-world scenario, removing more than 99% of the biomass with a single subsurface application of 200 ppb. However, it took three months to achieve control in the field site compared to a few weeks in the mesocosm experiment. We hypothesize that the shading through trees in the field site slowed down the control. Nevertheless, final control was the same as in the small-scale experiment. The removal of the dense frogbit cover on the water surface dramatically improved the water quality. Before herbicide application the water in the dam was highly acidic (low pH) and would have been unsuitable for a wide range of aquatic organisms. After the removal of frogbit the pH became neutral, greatly improving habitat quality.

Flumioxazin provides similar or better control than other aquatic herbicides reported in the literature. While imazamox provided around 90% biomass control at intermediate application rates (Willis *et al.* 2018) the application rate (280g $ha^{-1}$) was still more than twice that of the flumioxazin dose (105 g $ha^{-1}$) from the current study; imazamox is not registered for aquatic use in Australia. The congeneric *L. spongia* was controlled well with diquat in the USA (Madsen *et al.* 1998). While some diquat products are registered for aquatic use in Australia, diquat is a broad spectrum herbicide that potentially can cause considerable non-target damage to native macrophytes. From the authors' experience, flumioxazin is far more specific and carries a lower risk of damaging other aquatic plants. Flumioxazin also hydrolyses rapidly once applied to the water (Mudge *et al.* 2010, Katagi 2003), therefore, no long-term non-target damage should not be expected. Glyphosate products are registered for aquatic weed control in Australia. However, the literature suggests that it only provides poor control of the congeneric *L. spongia* (Madsen *et al.* 1998) and therefore similar poor control of Amazon frogbit is anticipated, suggesting that flumioxazin will provide far better control.

Future research should investigate dose-response relationships for foliar and subsurface flumioxazin application to control Amazon frogbit in more detail and determine minimum contact times compared to breakdown rates.

## ACKNOWLEDGMENTS

Sumitomo Inc. supplied the Valor herbicide used in the trials. The Logan City Council facilitated access to the Jimboomba field site and provided operational support with manpower and spray equipment. The Queensland Government and Queensland local governments funded the research.

## REFERENCES

Anderson, L. and Akers, P. (2011). Spongeplant: a new aquatic weed threat in delta. Cal-IPC News, 19(1), 4-5.

Atlas of living Australia. (2022). *Limnobium laevigatum* (Humb. & Bonpl. ex Willd.) Heine. https://bie.ala.org.au/species/https://id.biodiversity.org.au/node/apni/2913443

Cook, C.D.K. and Urmi-König, K. (1983). A revision of the genus *Limnobium* including *Hydromystria* (Hydrocharitaceae). Aquatic Botany, 17(1), 1-27.

Howard, G.W., Hyde, M.A. and Bingham, M.G. (2016). Alien *Limnobium laevigatum* (Humb. & Bonpl. ex Willd.) Heine (Hydrocharitaceae) becoming prevalent in Zimbabwe and Zambia. BioInvasions Records, 5(4), 221-5.

Kadono, Y. (2004). Alien aquatic plants naturalized in Japan: history and present status. Global Environmental Research, 8(2), 163-9.

Katagi, T., (2003). Hydrolysis of N-Phenylimide herbicide flumioxazin and its anilic acid derivative in aqueous solutions. Journal of Pesticide Science, 28(1), 44-50.

Madsen, J.D., Owens, C.S., and Getsinger, K.D. (1998). Evaluation of four herbicides for management of American frogbit (*Limnobium spongia*). Journal of Aquatic Plant Management, 36, 148-50.

Mudge, C.R., Haller, W., Netherland, M. and Kowalsky, J. (2010). Evaluating the influence of pH-dependent hydrolysis on the efficacy of flumioxazin for Hydrilla control. Journal of Aquatic Plant Management, 48, 25-30.

Mudge, C.R., Heilman, M., Theel, H. and Getsinger, K. (2012). Efficacy of subsurface and foliar penoxsulam and fluridone applications on giant salvinia. Journal of Aquatic Plant Management, 50, 116-24.

Perna, C. and Burrows, D. (2005). Improved dissolved oxygen status following removal of exotic weed mats in important fish habitat lagoons of the tropical Burdekin River floodplain, Australia. Marine Pollution Bulletin, 51(1-4), 138-48.

Perryman, M. J. (2013). Evaluating the invasive potential of South American spongeplant, *Limnobium laevigatum* (Humboldt and Bonpland ex Willdenow) Heine, in California's Sacramento-San Joaquin Delta. (Environmental Sciences Major). University of California at Berkeley, Berkeley, California.

Weerasinghe, R. (2020). Amazon frogbit - Limnobium laevigatum ex-situ experiments and field investigations. South East Regional Centre for Urban Landcare (SERCUL), Beckenham, WA.

Willis, B. E., Heilman, M. A., Bishop, W. M. and Shuler, S. W. (2018). Evaluation of multiple herbicides for control of sponge plant (*Limnobium laevigatum*). Journal of Geoscience and Environment Protection, 6(6), 56-64.

# Implications of seedbank dynamics in managing aquatic weeds

Hoang Nguyen[1], Tobias Bickel[2], Christine Perrett[2], Bahar Farahani[2], Steve Adkins[1]
[1]The University Of Queensland, Brisbane, Australia,
[2]Queensland Department of Agriculture and Fisheries (QDAF), Brisbane, Australia
(hoang.t.nguyen@uqconnect.edu.au)

**Summary** Most aquatic invasive weeds predominantly reproduce vegetatively which allows them to quickly take over invaded habitats. However, even after successful removal of vegetive materials, some water weeds can re-establish from the seedbank. Additionally, substrates also contain seeds of native macrophytes that can aid restoration after weed removal. Thus, seedbank dynamics in wetlands is an important aspect of long term weed management. We extracted seeds from soil cores and assessed seedling emergence to determine seedbank dynamics in lake Kurwongbah, a Brisbane drinking water reservoir, southeast Queensland, Australia. *Cabomba caroliniana* A.Gray (cabomba) is a serious invasive aquatic plant that has infested lake Kurwongbah littoral. The project will investigate the ability of native species to recruit from the soil seedbank after removal of the invasive weed and determine the potential of cabomba to re-establish from seeds. The outcomes of this project will contribute to improving aquatic invasive weed management and restoration of aquatic ecosystems and to provide a better understanding of the soil seedbanks of aquatic plants.

**Keywords** Seedbank dynamics, aquatic weed management, restoration, freshwater ecosystem

# *Cabomba caroliniana* eradication - integrated weed control success in the NT

Christopher Collins[1], Tony Dugdale[2] and Thomas Price[1]
[1]Northern Territory Government - Department of Environment, Parks and Water Security, Palmerston, NT, Australia
[2]Victorian State Government – Agriculture Victoria Research, Department of Jobs, Precincts and Regions, Victoria, Australia
(christopher.collins@nt.gov.au)

**Summary** Cabomba (*Cabomba caroliniana* A. Gray) is a submerged aquatic plant from the southern United States, is one of the world's most serious aquatic weed species. It is recognised in Australia as a Weed of National Significance. The species was first detected in the Northern Territory (NT) in 1996 and since that time the Northern Territory Government's Weed Management Branch has successfully eradicated three separate naturalised infestations. A fourth site, along a 2.2km stretch of the Darwin River, is presently on track for eradication. The success of the NT cabomba eradication effort is multi-faceted and predominantly sits with the integration and timing of multiple infestation stressors over time. These main stressors were natural annual flooding of the infestation area during the wet season, broad scale application of the herbicide carfentrazone and restriction of propagule spread. Importantly, cabomba programs in the NT have had consistent resourcing for over 20 years and the support of multiple agencies which enabled the NT Weed Management Branch to move past failures to achieve eradication.

**Keywords** aquatic weeds, carfentrazone-ethyl, eradication, integrated weed management.

## INTRODUCTION

Eradication of *Cabomba caroliniana* (cabomba), a Weed of National Significance (WONS), from the wet dry tropics of Australia's Northern Territory (NT) has been the aim of the Northern Territory Government's Weed Management Branch (WMB) since first detection in 1996. Between 1996 and 2018 WMB has locally eradicated three naturalised cabomba infestations located in Palmerston's Marlow Lagoon, Pine Creek and upper Darwin River. Predominately combinations of hand pulling and applications of the herbicide 2,4–D n-butyl ester (2,4-D) resulted in local eradication of cabomba from these water bodies (Price and Collins, 2016). Although WMB applies a stringent requirement of ten years of no detection before eradication is declared, it was observed in each of the cases above that where a 12 month period of no detection of cabomba is achieved, cabomba does not return (14 plus years of subsequent nil detection at these sites has followed, as of 2021).

A fourth site that contains cabomba is the lower Darwin River (Lok Landji). It is currently in its fifth year of no detection, after an extensive eradication program, which began in 2004 (when the upper Darwin River eradication program began). It represents the only other known cabomba infestation in the NT.

This paper outlines the integrated weed management measures implemented by WMB in Lok Landji since 2016 that have resulted in this fourth site being on track for cabomba to be declared eradicated from the NT in 2027.

## FIELD SITE AND INFESTATION

**Darwin River Cabomba Infestation NT** On 21 October 2004 cabomba was reported and positively identified in Darwin River. Subsequently cabomba was identified at multiple locations along an 11 km reach (Department of Natural Resources Environment and the Arts, 2006). The river itself stretches for a total of 16 km draining into Darwin Harbour. The management of the infestation was split between the top 8 km (Upper Darwin River) and a downstream billabong, Lok Landji, by shallow rocky anabranches. Lok Landji is a perennial water body located in the lower reaches of Darwin River. It is a 2.25 km long billabong with a variable width up to 40 m and depth up to 10 m (average 3 to 4 m). Its volume is ~170 ML and surface area ~6 ha.

## METHOD

**Cabomba Eradication Methods Lok Landji** The active control program, 2016 to 2019, was modified with the following activities undertaken:

- Intensive surveillance.
- Booms installed.
- Four broad scale applications of Shark® Aquatic Herbicide (240 g $L^{-1}$ carfentrazone-ethyl).
- Physical removal of plants (carried out as detected during surveillance).
- Quarantine area declared.

**Intensive Surveillance Techniques** Surveillance has been an integral part of the Program since 2004. From 2016 to January 2019 about 100 separate survey events have been carried out in Lok Landji, with surface surveys making up the bulk of the events. In the NT's Top End, entering the water is not recommended as the probability of a salt water crocodile encounter is greater than zero. Five salt water crocodiles have been found and removed from Lok Landji over the last eight years, the largest being a 3.25 m animal in September 2017.

<u>Surface Surveys</u> Surface surveys for cabomba in Darwin River were conducted from a vessel or from the bank on foot (where the river is unnavigable by vessel). The vessel is moved at idle speed adjacent to the bank with one observer surveying forward and down and the other looking between the vessel and the bank. It takes 28 days from flowering for cabomba to produce a seed (Tarver and Sanders, 1977). Weekly surveys were deemed necessary as surface surveys are not 100% effective, thus providing multiple chances for individual plants to be detected and removed before setting seed.

<u>Benthic raking surveys</u> Benthic raking involves a grapnel (two rake heads, fastened back to back) on the end of a length of metal chain which is attached to over 10 meters of rope. The grapnel is thrown into the water, allowed to sink to the bottom and then dragged along the river bed some few meters and then raised up. Cabomba fragments, if present, are readily snagged by the tines on the rake heads and able to be brought to the surface for inspection.
Benthic raking was used by Weed Officers in an ad hoc fashion at random and historic cabomba infestation sites only.

<u>Boom Inspection Surveys</u> Floating booms with curtains were used to reduce the ability of cabomba fragments to disperse up or downstream of infestations. The booms were strung up across the river at strategic locations to catch fragments that moved with the current or were windblown along the surface in any direction. In 2016 two booms were installed across Lok Landji. One was installed upstream of the most upstream cabomba location and the other at the junction of the carfentrazone treated and untreated areas. The booms were inspected weekly for the presence of fragments. The side of the boom the fragments were found on gave an indication of whether an infestation was present up or down stream of the boom. The upstream boom also reduced potential for movement of windblown floating fragments entering cabomba free areas of the billabong. Booms were typically removed from the river once a significant flow was experienced as a result of wet season rainfall.

<u>Underwater camera survey</u> The Director of Marine Ecosystems (NTG Flora and Fauna Division) developed a cabomba detection protocol for WMB after a pilot survey was conducted. The aim was to assess the effectiveness of the aquatic herbicide treatment in Lok Landji with a monitoring program using an underwater video camera and occupancy modelling framework.

The surface area of Lok Landji was divided up into 612 sites averaging about 10 m by 10 m (~100 $m^2$). To determine survey sensitivity prior to application of carfentrazone, 97 randomly selected sites were surveyed in August 2016 across Lok Landji. At each site a video camera was dropped to the river bed six times to determine the presence or absence of cabomba. Cabomba was found in 15 out of the 97 sites surveyed (naïve occupancy = 15.4%). The detectability was quite low (0.26) which means it often only turned up once or twice in the 6 camera drops per site. Based on this, we were 85-90% confident that we would find cabomba by dropping the camera 6 times per site, if it is present. To be 95% sure of detection it would require 10 camera drops per site (Griffiths, 2016).

<u>eDNA Sampling</u> In 2018 the Centre for Tropical Water and Aquatic Ecosystem Research (TropWATER) at James Cook University Queensland developed an environmental DNA (eDNA) assay for the detection of cabomba (Edmunds and Burrows, 2019). In late 2018 WMB collected surface and benthic water samples from Darwin River for Cabomba eDNA assay.

**Herbicide application** Two key factors drove the herbicide application technique. Firstly, eradication requires that all plants are controlled, and no surveillance program can detect every stem of cabomba in a 6 ha waterbody, even if it weren't inhabited by crocodiles. Secondly, submersed aquatic weeds such as cabomba require hours of exposure to herbicide to be killed (FMC Corporation, 2012). Broadscale application of the herbicide carfentrazone to the entire infested area (i.e. the sections of the billabong that were known to contain some cabomba) results in exposure at the target concentration (2 ppm a.i.) of the herbicide for a much greater period (in contrast to spot applications, which result in rapid dilution and were used in the past with 2,4–D). Furthermore, it results in all plants in the infested area being exposed, even if they had not been detected by the surveillance program. This circumvents the necessity of needing to detect every plant of the target species to achieve control, a limitation known to cause eradication programs to fail.

Cabomba infestations were known to occur in the downstream section of Lok Landji and occupy 65% of the billabong volume. Given a maximum of 50% of the volume could be treated under the APVMA permit, the upstream-most infested part of this area, equating to 50% of the billabong volume, was treated. Upstream areas must be treated first to completely remove the risk of upstream reestablishment. The remaining 15%, at the most downstream part of the billabong, was left untreated.

Carfentrazone applications were carried out to coincide with ideal treatment conditions, those being healthy and actively growing cabomba, high light conditions with clear water and low water flow (FMC Corporation, 2012).

**Seasonal Flooding – Impact on Cabomba** Wet season flooding is not a direct action of the program managers but is a significant annual event that warrants consideration. Major flooding was experienced in Lok Landji between January and May 2017. The 2016/17 wet season for Darwin and surrounds was the third biggest on record. The resultant high flow rate of Darwin River coincided with the Darwin River dam reaching capacity and the dam spillway flowing for an estimated 68 days, with high river flows well into the early dry season. This high river flow and subsequent high turbidity resulted in a period of around five months where cabomba growth would have been supressed due to suboptimal growing conditions (low light caused by high turbidity and high flow).

## RESULTS

**Application and monitoring** In total four 50% by volume treatments of cabomba with carfentrazone were carried out annually in Lok Landji in September or October between 2016 and 2019. A significant water quality monitoring protocol was implemented for each application as a requirement of the Northern Territory Environment Protection Authority.

**Cabomba abundance** Between 2004 and 2009 cabomba had been present in an estimated 340 out of the 612 camera survey sites and in August 2016, before carfentrazone was applied in October 2016, cabomba was present in 86 of the 612 sites.

Three months post-first application (Jan 2017) no floating cabomba fragments could be located in either the treated or untreated infestation areas, however, in the untreated area a single, healthy 10 cm fragment was retrieved via benthic raking. This healthy stem indicated that targeted management would be required once the river stopped flowing in the 2017 dry season. In May 2017, seven months after application, no cabomba was detected in any of the camera survey sites. The herbicide application had effectively controlled the whole infestation. This result supports the findings in Glenbrook Lagoon, NSW (Day, *et al.* 2014) where half of the waterbody was treated but very effective control was achieved over the entire waterbody.

Multiple integrated surveys were undertaken throughout the 2017 dry season months with nil cabomba detected. In mid-September 2017, the second 50% by volume treatment for cabomba using carfentrazone was completed as a follow up application relative to the detection of the fragment in January 2017 and to treat any other undetected cabomba.

In 2018 and 2019 multiple integrated surveys were undertaken throughout the dry season months, cabomba was unable to be detected. However, cabomba DNA was detected in water samples collected in 2018, mostly from benthic samples (Edmunds, *et al.* 2019). The positive detection is, in itself, not a direct indication of live of viable cabomba plants or propagules being present. TropWATER advised that the eDNA assay cannot distinguish between eDNA from viable cabomba and eDNA from dead or decaying cabomba, legacy eDNA. It is believed that it is possible for eDNA of aquatic vegetation to be present and detectable some years post the death of the last viable plant or propagule (Edmunds, *et al.* 2019). This information was reason enough for program managers to carry out a third (September 2018) and fourth (September 2019) 50% by volume treatment for cabomba using carfentrazone.

Not a single cabomba plant or fragment has been detected since January 2017, as of December 2021. Cabomba is likely to be declared eradicated from Lok Landji and the NT generally in 2027.

## DISCUSSION

The NT cabomba eradication program has adapted and overcome the challenges that had seen the progression towards eradication falter in Lok Landji prior to 2016.

Reintroduction of cabomba or establishment of new infestations in the NT cannot be ruled out. The aquarium industry trade cannot wholly be prevented and poses a risk. It is also probable that cabomba is currently present in ponds and aquariums in the NT with release into the environment possible. It is also possible that viable cabomba seed is present in Lok Landji, although this seems unlikely given that regeneration has not occurred in any of the other three NT sites where eradication has been declared.

With detection of dormant cabomba propagules being almost impossible, a lot of weight has been placed in the supporting evidence that cabomba has not returned at any other location in the NT where a nil detection period of 12 months has occurred. It is noted, for Lok Landji, that it has been over four years since last detection but only 30 months since active control ended though natural flooding continues unaided.

Annual wet season flooding is likely to have contributed to the long-term control observed after the 2016 application. Similar long-term control has been observed in Lake Benalla, Victoria, where drawdowns were used to control cabomba. Sampling of the cabomba at the end of the drawdowns demonstrated that cabomba was still viable and capable of rapid regeneration, however, severe flooding in the lake after the drawdown (and associated turbid water) is thought to have resulted in long term control, although cabomba returned several years later (Author's pers. obs.).

Another risk to the NT cabomba eradication goal is the potential decline in political and departmental will to continue investment in weed eradication programs. Having seen, across all NT cabomba infestation, 25 years (and counting) and almost $5 million invested so far, there is a possibility of support being withdrawn given competing priorities for government resources. The opportunity cost of not eradicating cabomba is the realisation of the threats the weed possess to the NT lifestyle, tourism and economy. Rough costings to build a water treatment plant, should cabomba enter Darwin River Dam (Darwin's main potable water supply), was estimated to be $40 million in 2004 and up to $100 million in 2016.

ACKNOWLEDGMENTS

The authors would like to thank the Kungarakany People of the Northern Territory upon whose land and water these activities were undertaken. Thanks also to the 30 plus individual Weed Management Branch staff that have been tasked with devising and implementing programs to eradicate cabomba from the NT and the NTG senior managers/ED's/CEO's and Ministers that have championed those programs ensuring ongoing support for cabomba eradication for over 25 years. Cabomba eradication in the NT was funded by the Northern Territory Government.

REFERENCES

Day, C., Petroeschevsky, A., Pellow, B., Bevan, J., O'Dwyer, T., Lawrence, A. S., et.al. (2014). Managing a priority outlier infestation of Cabomba caroliniana in a natural wetland in the Blue Mountains: could this be eradication? Ninteenth Australasian Weeds Conference, (pp. 147-152). Hobart.

Edmunds, R.C. and Burrows, D. (2019). Development of Cabomba caroliniana (fanwort) eDNA Assay: Technical Report. Centre for Tropical Water and Aquatic Ecosystem Research (TropWATER) Report 19/09, James Cook University Press, Townsville.

Edmunds, R.C., Villacorta-Rath, C., Collins, C., and Burrows, D. (2019). Cabomba caroliniana eDNA detection in two Darwin, NT billabongs following eradication attempts: Technical Report. Centre for Tropical Water and Aquatic Ecosystem Research (TropWATER), James Cook University Press, Townsville.

FMC Corporation (2012). Shark Aquatic Herbicide Technical Bulletin.

Griffiths, T. (2016) Summary of Cabomba monitoring Darwin River. Northern Territory Government unpublished report.

Department of Natural Resources Environment and the Arts . (2006). Cabomba Eradication Program Report for 2004/05. Darwin: Northern Territory Government Department of Natural Resources Environment and the Arts.

Price, T. and Collins, C. (2016). Silt, snags and snapping handbags - managing Cabomba caroliniana in the tropics of the Northern Territory.

Tarver, D.P., Sanders, D.R., (1977). Selected life cycle features of fanwort. J. Aquat. Plant Manag. 15, 18–22.

# Interactions between the native *Azolla filiculoides* and exotic *Salvinia molesta*

Guyo Gufu[1], Anthony Manea[1], Michelle Leishman[1]
[1]Macquarie University, Sydney, Australia
(guyo.gufu@bct.nsw.gov.au)

**Summary** Competition between freshwater plant species that occupy similar positions in the water column tends to be more intense relative to that between plants which grow in different parts of the water column. However, the ongoing environmental and climate change might modify the competitive interactions between such species. Using an environmentally controlled glasshouse experiment, we investigated the effect of CO2 and nutrient enrichment on competition between two free-floating fern species that co-occur in south-eastern Australia - native *Azolla filiculoides* and invasive exotic *Salvinia molesta*. The species were grown in monoculture and competition in nutrient cultures that were replaced weekly to simulate a dynamic system. We hypothesised that resource enrichment will enhance relative growth rates (RGR) of both species. We further hypothesised that although RGR of both species will be suppressed under competition relative to their counterparts grown in monoculture, *S. molesta* will be the dominant species in the competition treatment. We found that the relative growth rate (RGR) of both species was greater under high resource conditions as hypothesised. Surprisingly, competition did not result in suppression growth in either of the species. On the contrary, *A. filiculoides* had a facilitative effect on *S. molesta*. In addition, *A. filiculoides* gained more biomass under high resource conditions relative to *S. molesta* and the opposite was true under low resource conditions. We conclude that CO2 and nutrient concentration did not mediate competition between the species but instead influenced RGR independent of competition. These findings suggest that species composition in dynamic water bodies under future environmental conditions may be determined by the species' responses to environmental changes rather than by changes in competitive interactions.

**Keywords** Atmospheric CO2, competition, nutrient enrichment, relative growth rate

# Using a 25 W diode laser to control annual ryegrass and turnipweed

Guy Coleman[1], Christopher Betters, Sergio Leon-Saval, Michael Walsh
[1]University Of Sydney, Brownlow Hill, Australia
(guy.coleman@sydney.edu.au)

**Summary** The development of machine learning algorithms for precise weed detection is creating an opportunity to selectively apply non-selective physical weed control options. Recent research suggests energy intensive approaches can be used in large-scale cropping systems when applied as a site-specific treatment. Laser weeding is one such opportunity, whereby energy is tightly focused in a beam of light directed onto the weeds resulting in cell heating, rupture and death from the incident energy. Lasers are flexible in deployment with opportunities to adjust treatment length, beam qualities (width and intensity) and light wavelength through the type of optics and laser emission method. This precision targeting by laser weeding treatments provides a substantial advantage in energy use efficiency over other thermal methods such as flaming and microwaves. A 25 W, 942 nm diode laser was evaluated for control efficacy on annual ryegrass (*Lolium rigidum*) and turnipweed (*Rapistrum rugosum*) two representative winter weeds for the northern production region. A 15 and 60 second treatment provided control of both weeds at the 2-leaf and 4 to 6-leaf stages. When laser weeding did not result in plant death, plant biomass was severely diminished by longer treatments in early tillering ryegrass and 8 to 12-leaf turnipweed. An evaluation of larger beam diameters suggested larger diameters improved the ease of targeting the growth point with a lower requirement for precision placement. Using energy density to extrapolate from these results indicates that a laser of at least 400 W is needed for sub second control with a 10 mm beam diameter.

**Keywords** Site-specific weed control, weed detection, computer vision

# Biological control of priority weeds of cropping systems in Australia: A viable proposition?

Raghu Sathyamurthy[1,2]
[1]CSIRO Health and Biosecurity, Canberra, Australia,
[2]CSIRO Health and Biosecurity, Brisbane, Australia
(raghu.sathyamurthy@csiro.au)

**Summary** Classical biological control (biocontrol) is a management technique that could contribute to landscape scale reduction of invasive weeds that pose a continuous treat to the grain industry. The suppression of weed performance by biocontrol agent(s) in unmanaged contexts, beyond crop fields and fallows, could limit the rate of incursion of weed seeds within cropping systems. Since 2016, CSIRO has led a major research initiative to find and assess candidate biocontrol agents for two herbicide-resistant weeds affecting the grain industry in Australia; fleabane (primarily *Conyza bonariensis*) and common sowthistle (*Sonchus oleraceus*). This initiative has involved several research scientists, technical support staff and students from CSIRO and collaborating organisations in France, Colombia, Brazil and United States. Several promising candidate agents have been found on fleabane in the native range, with host-specificity testing well advanced for a microcyclic rust fungus and a tephritid gall fly. In contrast, despite extensive field surveys in Europe and northern Africa over several years, all potential biocontrol agents found on common sowthistle have been capable of attacking two key Australian native species closely related to the weed target during initial testing. Based on these results, we have concluded that there would be limited value in pursuing further classical biocontrol of common sowthistle. To guide future investments in biocontrol, we have assessed important weeds of relevance to the grain industry in Australia through a transparent prioritisation framework, adapted from previous projects.

**Keywords** Biological control, crop weeds, host-specificity testing, target prioritisation

# Performance and Weed Suppressive Potential of Winter Cover Crops Established as Monocultures and Multispecies Mixtures in Southern Australia

Saliya Gurusinghe [1], Shamsul Haque [1,2], Asad Shabbir [3], Michael J. Walsh [3] and Leslie A. Weston[1]
[1] Gulbali Institute of Agriculture, Water and the Environment, Charles Sturt University, Locked Bag 588, Wagga Wagga, NSW 2678, Australia.
[2] School of Agriculture and Wine Sciences, Charles Sturt University, Wagga Wagga, NSW 2678, Australia.
[3] Sydney Institute of Agriculture, School of Life and Environmental Sciences, University of Sydney, Narrabri, NSW 2390, Australia
(sgurusinghe@csu.edu.au)

**Summary** Cover crops provide rotational diversity that frequently results in reduced erosion and improved soil health, weed control, and moisture retention in mixed farming and cropping systems. They have also been shown to suppress weeds due to their physical abilities and/or chemical properties. Field experiments were established in winter 2021 in Wagga Wagga and Narrabri NSW to evaluate the establishment of selected cover crop species potentially well adapted to each region and assess their ability to provide winter annual weed suppression through weed interference. Cover crop species evaluated in these experiments included conventional grass, legume and non-legume broadleaf established either as monocultures or multi-species binary mixtures. Crop and weed competitive traits were estimated by assessment of crop competitive traits including canopy light interception (LI), leaf area index (LAI) and crop and weed biomass. Grazing oats and tillage radish strongly reduced weed biomass accumulation at both Wagga Wagga and Narrabri sites. Multispecies binary mixtures of cover crops performed similarly to monocultures with respect to formation of suppressive ground covers by shading the soil surface and thereby reducing weed biomass. Total biomass accumulation in winter cover crop treatments was generally similar between sites.

**Keywords** Weed suppression, interference, competition, legumes, multi-species mixtures

## INTRODUCTION

Cover crops provide important agro-ecological functions by improving soil tilth and moisture availability for subsequent crops while suppressing annual weeds and reducing soil erosion (Schipanski et al., 2014). Cover crops incorporated into integrated weed management strategies have resulted in reduced herbicide use and the incidence of herbicide resistance, both of which are important considerations for Australian grain producers (Peterson et al. 2018). The weed suppressive potential of cover crops typically manifests either through competition for resources (Lawley et al., 2012) or through the release of phytotoxic secondary metabolites from crop residues and root exudates (Bhadoria 2011). However, the latter has not always been well demonstrated under field conditions. In addition, residues remaining on the soil surface over time can create a mulch-like effect, forming a physical barrier to weed seedling establishment and seed germination (Galloway and Weston, 1996).

In southern grains region farming systems, the decision to incorporate cover crops in rotations is driven by factors including cost of establishment as well as water use requirements of cover crops during vegetative growth, and their ability to conserve soil moisture post-termination and before establishment of the subsequent grain crops (Bell et al., 2012). The net water benefit associated with establishment of common cover crop species on subsequent winter crops was previously evaluated (Erbacher et al., 2019). However, performance when established as monocultures or multispecies mixtures, in various soil types, requires further investigation.

Therefore, a series of field, glasshouse and laboratory experiments were designed to assess the performance of a diverse collection of summer and winter cover crops with respect to weed suppression, biomass accumulation and water usage over time. Studies were performed in both northern and southern NSW and addressed the following objectives: 1) evaluation of cover crop establishment at each field site 2) comparison of the competitive ability of diverse covers established as monocultures or multispecies mixtures contributing to early season weed suppression and 4) impact of cover crops on plant-available soil water over time.

## MATERIALS AND METHODS

**Cover crop establishment** Field experiments were established in 2021 at the Graham Centre field site, Wagga Wagga NSW and the University of Sydney research site in Narrabri. All experiments were arranged in a randomised complete block design with four replications per treatment. Soil types were characterised as a fine red kandosol at Wagga Wagga with pH of 5.9 and grey vertosol with a pH of 8.7 at Narrabri. Monocultures were established in separate split blocks from the multispecies mixtures to reduce the influence of spatial variation on soil type, tilth and cover crop establishment. Monocultures were sown at commercially recommended sowing rates for New South Wales. Multispecies binary mixtures were also sown at various ratios within the recommended sowing rate ranges (ie. 1:0, 0.75:0.25; 0.5:0.5, 0.25, 0.75, and 0:1).

Crops were sown on 23rd April 2021 in Narrabri, and 13th May 2021 and 19th May 2021 for multispecies mixtures and monocultures in Wagga Wagga, respectively. Cover crops were seeded using a precision cone planter with plot dimensions of 10 m × 1.6 m, at 20 cm row spacing at Wagga Wagga. At Narrabri, the row spacing was 25 cm. Row spacings utilised reflect the preferred recommendations at each location based on resource availability. Plots were established in a randomised complete block design at each location, with 4 replications. Experiments were fertilised with 100 kg ha–1 of Croplift 12 (NPS 12: 18: 6; Incitec Pivot, Melbourne, VIC) at the time of sowing at Wagga Wagga and with 65 kg ha–1 Cotton Sustain (NPS 6:12:22; Incitec Pivot, Melbourne, VIC) in Narrabri. To facilitate soil moisture assessment over time in various treatments, two 50 mm diameter soil cores were removed to a depth of 20 cm and 70 cm from the centre of monoculture treatments and multispecies treatments containing equal (50/50) ratios of crop mixtures in Narrabri. Polyvinyl chloride (PVC) pipes (50 mm diameter) were installed where soil cores were removed, and later capped. Fortnightly, soil moisture measurements were obtained using a moisture sensor (MP406; ICT International Ltd, VIC) and reported as volumetric water content. Crops were subsequently terminated using glyphosate at a rate of 2 L $ha^{-1}$ at 120 days after planting at late vegetative growth stage, or at maturity at 150 days after planting.

**Crop assessments** Crop emergence was recorded in 2 × 0.25 $m^2$ quadrats per treatment, at 25 days after sowing. Aboveground biomass samples were collected at 80 days after sowing (DAS) for the crop and at 80, and 120 DAS for weeds within each plot by cutting at the soil surface in two 0.25 $m^2$ quadrants and total biomass of both the crop and weed species was determined. Plant material was collected and sorted immediately after harvest before drying at 40 °C for 120 h, and dry weights of crop and weeds were determined. Aboveground crop competition was also assessed within each treatment sub-plot at ~30, 50, 70 and 90 days after sowing. Canopy cover was assessed by determination of percent interception (LI) of photosynthetically active radiation at the base of the crop using a light ceptometer (AccuPAR LP-80 Ceptometer, Decagon Devices®), typically performed on a cloudless day.

**Statistical analysis** Trial randomization, design and data analysis were performed using Agricultural Research Manager (ARM) version 9.0, a statistical software package by GDM (Gylling Data Management Inc., 2014). Statistical analysis of data was performed by one-way or two-way analysis of variance for randomized experiments with four replicates using the statistical software, R (R Core Team, 2017). Significant differences were separated using Tukey's HSD test for multiple comparisons, with significance declared at $P<0.05$.

## RESULTS

**Performance of winter cover crop monocultures**
All winter crops established successfully at Wagga Wagga and Narrabri. Maximal crop biomass in Wagga Wagga was recorded in tillage radish at ~ 8 t $ha^{-1}$ and was higher than grazing oats at~ 6.5 t $ha^{-1}$ ($P<0.05$) (Figure 1A). Maximum biomass accumulation at Narrabri was observed in field pea and grazing oats treatments at ~ 8 t $ha^{-1}$ and was significantly higher than that of other cover crop species assessed (Figure 1B) ($P<0.05$).

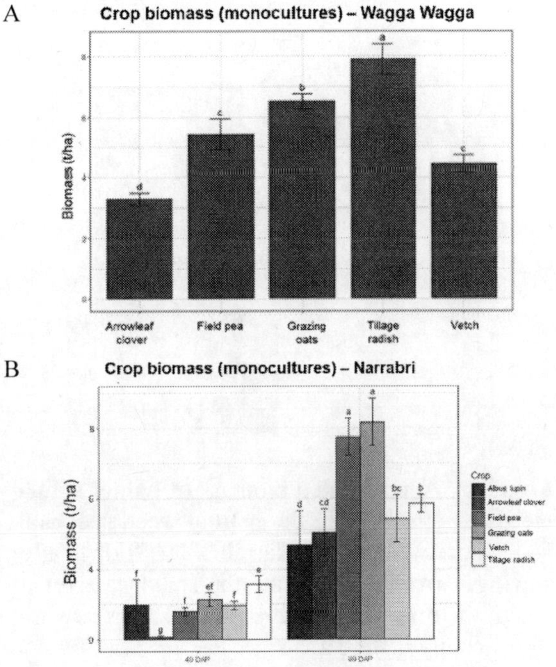


**Figure 1**. Accumulated biomass of winter cover crop monocultures at (A) Wagga Wagga at maturity and (B) Narrabri at 40 and 80 days after planting. DAP; days after planting. Error bars indicate the standard error of means. Means

sharing the same letters are not significantly different.

**The weed suppressive potential of winter cover crop monocultures** The key winter annual weed species present in the natural weed seedbank at Wagga Wagga included annual ryegrass (*Lolium rigidum* Gaudin), barley grass (*Hordeum* spp.), sowthistle (*Sonchus* spp.), poppy (*Papaver* spp.) and fumitory (*Fumaria* spp.). At Narrabri, the natural weed seedbank consisted mainly of annual ryegrass, brome grass (*Bromus* spp.) and sowthistle. The accumulated weed biomass was recorded at maturity (80 DAS) and following cover crop desiccation at 120 DAS at Wagga Wagga, and at 40 and 80 days after sowing at Narrabri, as crops matured early at this site.

Weed biomass accumulation was significantly reduced by all treatments at maturity, when compared to the untreated control (Figure 2) (P<0.05), except for Albus lupin and arrowleaf clover at Narrabri due to late emergence and slow growth of these crops at that location. At crop maturity, all other cover crop treatments reduced the accumulated weed biomass similarly, with reductions in weed biomass ranging from 30% - 92% when compared to the untreated control (P<0.05).

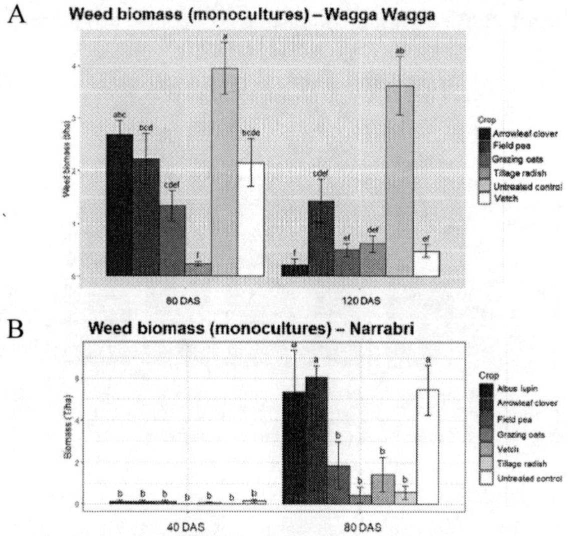


**Figure 2.** Accumulated biomass of winter annual weeds emerging from the natural weed seedbank (A) Wagga Wagga (B) Narrabri. DAS; days after sowing. Error bars indicate the standard error of means. Means sharing the same letters are not significantly different.

**Performance of multispecies mixtures** Crop biomass accumulation in the multispecies mixtures vetch/ oats and field pea/ oats was similar among the companion species (data not shown). However, in the tillage radish/ field pea mixture, all sowing ratios containing the radish treatment accumulated more crop biomass when compared to a monoculture of field pea at Wagga Wagga (P<0.05) (Figure 3A). At Wagga Wagga, all ratios of cover crop mixtures suppressed weed biomass accumulation by approximately 50% (P<0.05) when compared to the untreated control, except for the monoculture treatment of the tillage radish/ field pea mixture, which impacted the weed biomass accumulation less than the other sowing ratios in the mixture (P<0.05) (Figure 3B).

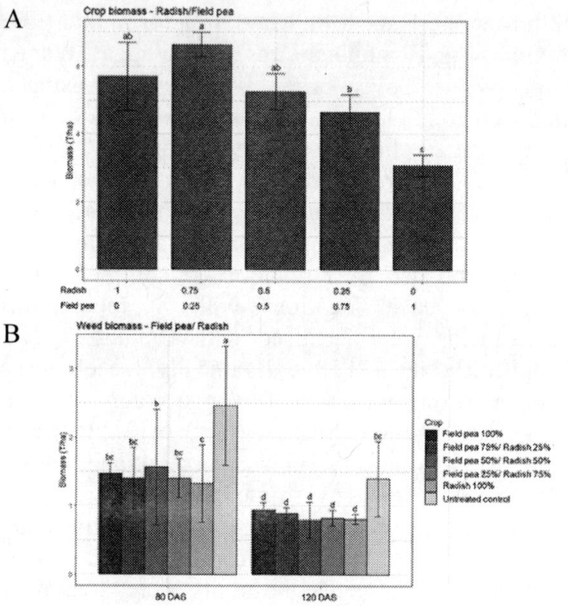


**Figure 3.** (A) Crop and (B) weed biomass accumulation in the multispecies binary mixture field pea/ tillage radish, established at Wagga Wagga. DAS; Days after sowing. Error bars indicate the standard error of means. Means sharing the same letters are not significantly different.

Multispecies mixtures established at Narrabri also performed similarly with no difference in crop biomass and weed biomass accumulation observed between treatments (data not shown).

**Soil moisture usage** Moisture measurements were performed fortnightly at soil depths of 20 cm and 70 cm in all monocultures established at Narrabri. Soil moisture content at the initiation of sampling 20 days after sowing ranged between approximately 26-32% across the site at both 20 and 70cm below the soil surface (data from 20cm depth shown; Figure 4). Differences in soil moisture among monocultures were more pronounced between 120 and 150 days after sowing at both depths, with the early terminated treatments, except for arrowleaf clover conserving between 10- 17.5% volumetric soil water compared to the late termination time ($P<0.05$).

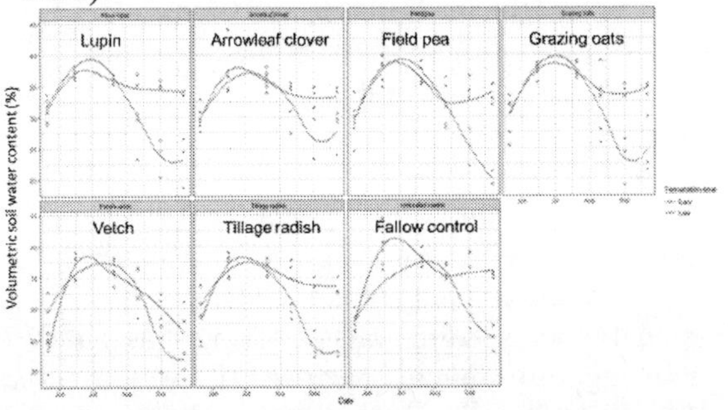


**Figure 4.** Volumetric soil water content measured in monocultures established at Narrabri NSW, 20 cm below the soil surface. Red: early termination; Blue: late termination.

## DISCUSSION

Field experimentation performed over the winter growing season in northern and southern NSW identified cover crop species appropriate for establishment in the low to medium rainfall zones in diverse soil types and climatic conditions. Favourable environments experienced at both field locations also enabled the evaluation of the weed suppressive potential of both monocultures and binary mixtures of those cover crop accessions. Strongly competitive crops were those which exhibited early vigour and biomass accumulation, which included grazing oats and tillage radish in the winter growing season These treatments provided excellent suppression of weed growth and establishment. Most binary multispecies mixtures of winter cover crops reduced the accumulation of weed biomass similarly to that in their respective monocultures. This suggests that multispecies mixtures could provide ample biomass and crop biodiversity in the cover cropping phase of crop rotations, while reducing annual weeds in broadacre cropping systems.

The selection of the appropriate cover crop termination time is critical, particularly in the case of grazing oats, a species that conserved soil water when terminated early, but showed significant soil water depletion when terminated at crop maturity. Results from this experiment describing suppressive, biodiverse cover crops as useful rotational crops will better inform farmers interested in regimes for enhancement of crop biomass, soil moisture conservation and productivity on-farm and reducing weed seedbank density over time.

## ACKNOWLEDGMENTS

The authors acknowledge investment from GRDC (US000084) and technical assistance at from field research teams at Charles Sturt University, University of Sydney and Queensland Department of Agriculture and Fisheries.

## REFERENCES

Adler, M.J. and Chase, C.A. (2007). Comparison of the allelopathic potential of leguminous summer cover crops: cowpea, sunn hemp, and velvet bean. *Horticulture science* 42, 289–293.

Bell, L. W., and Moore, A. D. (2012). Integrated crop–livestock systems in Australian agriculture: Trends, drivers and implications. *Agricultural Systems*, *111*, 1-12.

Bhadoria, P.B.S. (2011). Allelopathy: A natural way towards weed management. *American Journal of Experimental Agriculture*. 1, 7.

Dear, B.S. and Ewing, M.A. (2008). The search for new pasture plants to achieve more sustainable production systems in southern Australia. *Australian Journal of Experimental Agriculture*. 48, 387–396.

Erbacher, A., Lawrence, D., Freebairn, D., Huth, N., Anderson, B., and Harris, G. (2019). Net water benefit of cover crops in Northern grains production. Farming water with ground cover. In Proceedings of the 2019 Australasian Agronomy Conference.

Galloway, B. A., and Weston, L. A. (1996). Influence of cover crop and herbicide treatment on weed control and yield in no-till sweet corn (*Zea mays* L.) and pumpkin (*Cucurbita maxima* Duch.). Weed Technology, 10(2), 341-346.

Latif, S., Gurusinghe. S., Weston, P.A., Brown, B., Quinn, J.C., Piltz, J.W. and Weston, L.A. (2019) Performance and weed suppressive potential of selected pasture legumes against annual weeds in southeastern Australia. Crop and Pasture Science. 70.2 (2019): 147-158

Lawley, Y.E., Teasdale, J.R. and Weil RR (2012). The mechanism for weed suppression by a forage radish cover crop. Agronomy Journal. 104, 205–214.

Peterson, M.A., Collavo, A., Ovejero, R., Shivrain, V. and Walsh, M.J. (2018). The challenge of herbicide resistance around the world: a current summary. Pest management Science. 0.

R core team (2017). R: A language and environment for statistical computing. R foundation for statistical computing, Vienna, Austria.

Schipanski, M.E., Barbercheck, M., Douglas, M.R., et al. (2014). A framework for evaluating ecosystem services provided by cover crops in agroecosystems. Agricultural Systems. 125, 12–22.

# Integrated Weed Management – alternative strategies for weed control in pulses

Navneet Aggarwal[1,2], Penny Roberts[1,2], Helena Oakey[2] and Larn McMurray[3]
[1]Agronomy, South Australian Research and Development Institute, 70 Farrell Flat Road, Clare-5453, SA.
[2]The University of Adelaide, Waite Campus, Urrbrae-5064, South Australia.
[3]Global Grain Genetics Pty Ltd, Clare-5453, South Australia.
(navneet.aggarwal@sa.gov.au)

**Summary:** The increased frequency of Group 2 tolerant pulses in crop rotations is reducing herbicide diversity due to limited safe options particularly for in-crop broadleaf weed control. Therefore, developing alternative weed control methods is important for delaying imidazolinone herbicide resistance in broadleaf weeds. Experiments were established to test integrated weed management (IWM) strategies including, crop-weed competition in faba bean at Turretfield (2017-2019) and Salter Springs (2019), and wick-wiping and clipping brassica weeds at different growth stages in a lentil crop at Turretfield (2019-2020) and Tickera (2021). Increasing faba bean densities from the standard grower practice of 24 plants $m^{-2}$ to 30 plants $m^{-2}$ reduced vetch seed set by up to 45%, due to increased crop competition. Wick-wiping brassica weeds with Glyphosate + LVE MCPA up to two weeks after its pod initiation reduced pod set by 62-100% compared to no wick-wiping. Further, clipping brassica weeds three weeks after pod initiation (with green and squashy weed seeds) reduced pod set by 94% compared to the no clipping treatment.

**Keywords:** Crop competition, plant density, wick-wiping, clipping.

## INTRODUCTION

With the increasing frequency of pulses in crop rotations, broadleaf weed control has become challenging due to limited safe herbicide options. While there is increasing availability of herbicide tolerance traits in new crop varieties, alternative herbicide strategies are crucial to maintaining weed control options. The development of new IWM strategies will allow greater in-crop broadleaf control and reduce in-crop weed seed set and build-up of the soil weed seed bank in pulse-based rotations. Growers are becoming more accepting of IWM programs that would help to maximise the heterogeneity of selection pressures, minimise resistance evolution, achieve satisfactory weed control and allow sustainable long-term herbicide use.

Crop competition through early closure of the crop canopy is a vital cultural strategy that plays an important role in IWM by reducing the weed biomass and their fecundity and leads to increased crop yields (Lemerle *et al.* 2004). In a survey of 130 Western Australian growers, 61% were using higher seeding rates of wheat as an IWM tactic against annual ryegrass (Llewellyn *et al.* 2004). The agronomic tactic of increasing seeding rates to increase in-crop competition over weeds can be effective especially in pulse crops that have low plant densities and slow initial growth, such as lupin and faba bean. Until now, a major effort has been made to control grass weeds with increasing cereal crop densities, but limited literature is available citing the effect of increasing plant density on increasing crop competitiveness of pulses over problematic broadleaf weeds and needs exploring. Additionally, novel weed management practices, such as wick-wiping and clipping have a role to play in controlling resistant weed seed set from plants surviving early weed control strategies, especially when weeds set seed before crop maturity. Application timing of the wick-wiping and clipping treatments is crucial for reducing broadleaf weed seed set and needs investigation.

Therefore, the present studies were carried out to develop IWM strategies for the pulse phase of the crop rotation, reducing broadleaf weed seed set by improving crop competition in faba bean, and with wick-wiping and clipping broadleaf weeds in lentil.

## MATERIALS AND METHODS

Four field experiments focused on crop competition were established in the Lower North region of South Australia, with three at Turretfield Research Centre (TRC) (34°.32'38" S, 138°.50'49" E, at 116 m above sea level in medium rainfall zone with 471 mm average annual rainfall) over the growing seasons of 2017, 2018 and 2019, and one experiment at Salter Springs (34°.12'39.70" S, 138.37'50.89" E at 229 m above sea level in high rainfall zone with 501 mm average rainfall) in 2019. The soil at the TRC field sites was a light clay over medium clay (2017), clay (2018 and 2019) in texture with organic matter content of 1.1-2.0% and a pH (water) of 7.1-8.5 in 0-20 cm layer. Soil at the Salter Springs site was very heavy clay in texture with organic matter content of 1.5-2.1% and a pH (water) of 8.4 in 0-20 cm layer. Rainfall received at TRC site in 2017 and 2018 was 278 and 188 mm, respectively.

The experiments were established in a factorial randomized complete block design with three faba bean densities 12, 24 and 36 plants $m^{-2}$ (2017), five faba bean densities 12, 18, 24, 30 and 36 plants $m^{-2}$ (2018 and 2019), and three herbicide treatments Simazine 1100 (PSPE), Simazine 1100 (PSPE) + imazapyr + imazamox 750 (POST at 5-6 crop-node stage), and unsprayed control with three replicates. Group 2 imidazolinone resistant faba bean PBA Bendoc was sown at a depth of 5-6 cm using a no-till plot seeder fitted with knife-point openers and press wheels. Plots were 10 m long and contained six crop rows spaced 22.5 cm apart. The seed of faba bean for all experiments was obtained from the same source (Faba bean Breeder, The University of Adelaide) in all the three years, to avoid any potential influence of seed source on early vigour. Vetch seeds were broadcast prior to sowing @ 50 seeds $m^{-2}$ to contribute to the existing background medic weed population at Turretfield. There was a background population of vetch and bifora at Salter Springs. Herbicides were applied by using a tractor mounted sprayer delivering 100 L $ha^{-1}$ water at a pressure of 200 kPa. Crop plant emergence counts were recorded when faba bean was at approximately the 2-3 node stage. Crop biomass was sampled at faba bean flowering stage with cuts made from four rows × 1-m length in each plot starting 1-m inside from the plot end. The crop biomass was dried at 60° C for 48 hours till constant weight, and weighed.

Additionally, the potential benefits from wick-wiping and clipping for reducing the seed set of brassica weeds in lentil were studied at Turretfield in 2019 and 2020, and at Tickera in 2021. These experiments were established in randomised complete block design with three replicates by using lentil cultivar PBA Hurricane XT sown at a density of 120 plants $m^{-2}$. Experiments tested the response of brassica weeds including wild turnip, wild radish and Indian hedge mustard to wick-wiping with Glyphosate + LVE MCPA + water mixed as 1:1:1 and the application of weed clipping just above the lentil canopy at different growth stages. The wick-wiping and clipping treatments were applied at weekly intervals, starting from pod initiation stage. A gravity-based wick-wiper was used, and clipping of weed growing parts above the crop canopy was done manually. Seed/pod set of broadleaf weeds was determined by counting the pods and seeds obtained from plants sampled in a 0.25 $m^2$ quadrat placed at three random locations in each plot.

**Statistical Analysis.** Weed and crop data were analysed with ANOVA. A square-root variance-stabilizing transformation was used for vetch and medic plant density, vetch pod and seed set, medic pod set and bifora seed set data before analysis. Least squared means were used to determine significant differences ($P < 0.05$) between herbicide application, faba bean densities, and the interaction between herbicides and faba bean densities. The interaction effects were non-significant, therefore main treatment effects are presented in this paper.

## RESULTS AND DISCUSSION

### A. Effect of clipping and wick-wiping (lentil)

The timing of clipping treatments was an important factor in reducing pod set of brassica weeds (wild turnip, Wild radish and Indian hedge mustard), with later clipping treatments (at two and three weeks after wild turnip pod initiation) reducing pod set compared to the earliest treatment (at pod initiation) (Table 1). However, the opposite effect was observed with wick-wiping in 2019, where earlier treatments (up to two weeks after pod initiation) resulted in reduced turnip weed pod set compared to the late wick-wiping (three weeks after pod initiation). Wick-wiping at weed pod initiation and at later stages before embryo development proved equally effective in 2020 and 2021, reducing wild radish and Indian hedge mustard pod set by up to 100% and 78%, respectively, compared to no wick wiping. Using a combination of wick-wiping and clipping resulted in reduced weed pod set compared to the control (no wick-wiping/clipping) of up to 96%. When combining the two treatments of clipping and wick-wiping, earlier timing (at pod initiation) was the most effective in 2019. All of the combined treatments resulted in reduced weed pod set compared to the control, however the combination treatments were not significantly different to the singular treatments of either delayed clipping or early wick-wiping (Table 1).

### B. Crop competition studies (faba bean)

*Effect of increasing faba bean density*

*(i) On crop growth*

Crop biomass increased with increasing seeding rates in all the years and at all sites (Table 2). Plant height also increased with increasing faba bean densities, except in 2018 at Turretfield. The increase in plant height with increasing densities is an adaptive response due to the close proximity of other plants, known as shade avoidance syndrome, and is triggered by plant hormones and photoreceptor proteins (Ballaré and Pierik 2017). Increasing faba bean plant densities from standard growers' practice of 22-24 plants $m^{-2}$ to 30-36 plants m$^{-2}$ resulted in a 13-18% increase in grain yield at Turretfield in all three years, however, no yield advantage was seen at Salter Springs in 2019. Earlier sowing at Salter Springs (31/05/2019) might explain this result, due to the longer crop season as compared to the Turretfield site that was sown comparatively later in all three years (19/6/2017, 06/06/2018 and 18/6/2019).

*(ii) On broadleaf weed seed set control*

Vetch seed set was reduced by 45-88% with increasing faba bean plant density from the standard grower practice of 22-24 plants m$^{-2}$ to 30-36 plants m$^{-2}$ (Table 2), resulting from increased competition on weeds due to greater crop biomass and plant height. This helped faba bean in smothering vetch plants and seed set was reduced with increasing seeding rates. However, increasing faba bean density from 22-24 plants m$^{-2}$ to 30-36 plants m$^{-2}$ was not as effective for reducing seed set of medics in two out of three years. The medic grew in a prostrate manner between the crop rows and set the same number of seed at both high and standard crop densities. Similarly, bifora set the same number of seed in both the standard and increased faba bean densities at Salter Springs in 2019, by growing as tall as the faba bean crop in the denser canopies. Therefore, the benefits for reducing broadleaf weed seed set are dependent on the adaptations of associated weed species to differences in crop canopy structure.

Effective weed management strategies should not only focus on killing weeds when they have emerged in crop, they should also target control of weed seed set and the reduction of weed seed banks. In-crop competition has a potential application in faba bean, which is sown at low plant densities and has slow initial growth. Novel approaches to integrated weed management such as wick-wiping with low volume concentrated herbicides and mechanical clipping of weed plants growing above the lentil crop canopy, improves control of brassica weed pod set. Integrated weed management strategies incorporating the use of these agronomic tactics and other novel approaches, in combination with rotating chemistries, will reduce the overall selection pressure on weeds and delay herbicide resistance build up.

ACKNOWLEDGEMENTS

The research undertaken as part of this project is made possible by the significant contributions of growers through both trial cooperation and the support of the GRDC (DAS00168-BA, DAV00150 and UOA2105-013RTX) the authors would like to thank them for their continued support.

REFERENCES

Ballaré C.L. and Pierick, R. (2017). The shade-avoidance syndrome: multiple signals and ecological consequences. *Plant, Cell and Environment* 40, 2530-43.

Lemerle, D., Cousens, R.D., Gill, G.S., Peltzer, S.J., Moerkerk, M., Murphy, C.E., Collins, D. and Cullis, B.R. (2004). Reliability of higher seeding rates of wheat for increased competitiveness with weeds in low rainfall environments. *The Journal of Agricultural Science* 142, 395-409.

Llewellyn, R.S., Lindner, R.K., Pannell, D.J. and Powles, S.B. (2004). Grain grower perceptions and use of integrated weed management. *Animal Production Science* 44, 993-1001

**Table 1.** Brassica weed pod set at Turretfield (in 2019 and 2020) and at Tickera (in 2021) as affected by clipping and wick wiping. Bars labelled with the same letter are not significantly different ($P \leq 0.05$).

| Treatment | Turretfield 2019 | Turretfield 2020 | Tickera 2021 |
|---|---|---|---|
| | Wild turnip pods (m$^{-2}$) | Wild radish pods (m$^{-2}$) | Indian hedge mustard pods (m$^{-2}$) |
| Clipping at early pod initiation stage | 170$^{ab}$ | 145$^{b}$ | 1156$^{ab}$ |
| Clipping after one week of early pod initiation | 121$^{bc}$ | 15$^{c}$ | 660$^{bc}$ |
| Clipping after two weeks of early pod initiation | 48$^{cdef}$ | 10$^{c}$ | 182$^{d}$ |
| Clipping after three weeks of early pod initiation | 15$^{ef}$ | 4$^{c}$ | - |
| Clipping + wick wiping at early pod initiation stage | 11$^{f}$ | - | - |
| Clipping + wick wiping after one week of early pod initiation | 66$^{cde}$ | - | - |
| Wick wiping at early pod initiation stage | 56$^{cdef}$ | 0$^{c}$ | 853$^{bc}$ |
| Wick wiping after one week of early pod initiation | 34$^{def}$ | 15$^{c}$ | 384$^{cd}$ |
| Wick wiping after two weeks of early pod initiation | 91$^{bcd}$ | 10$^{c}$ | 660$^{bc}$ |
| Wick wiping after three weeks of early pod initiation | - | 11$^{c}$ | - |
| No weed wiping/clipping | 243$^{a}$ | 454$^{a}$ | 1772$^{a}$ |

**Table 2.** Crop growth, yield, and broadleaf weed seed set as affected by increase in faba bean densities at Turretfield in 2017-2019 and at Salter Springs in 2019. Bars labelled with the same letter are not significantly different ($P \leq 0.05$).

| Location | Target faba bean density ($m^{-2}$) | Faba bean density achieved ($m^{-2}$) | Crop biomass at flowering ($t\ ha^{-1}$) | Plant height at flowering (cm) | Grain yield ($t\ ha^{-1}$) | Vetch seed set ($m^{-2}$) | Medic pod set ($m^{-2}$) | Bifora seed set ($m^{-2}$) |
|---|---|---|---|---|---|---|---|---|
| A. Turretfield 2017 | 12 plants | 12 plants | $1.13^{c}$ | $52.2^{c}$ | $1.92^{c}$ | $778^{a}$ | $732^{a}$ | - |
| | 24 plants | 24 plants | $2.55^{b}$ | $59.1^{b}$ | $2.83^{b}$ | $636^{a}$ | $356^{b}$ | - |
| | 36 plants | 36 plants | $3.72^{a}$ | $68.4^{a}$ | $3.25^{a}$ | $381^{b}$ | $159^{c}$ | - |
| B. Turretfield 2018 | 12 plants | 12 plants | $1.52^{c}$ | $44.1^{a}$ | $0.77^{c}$ | $558^{a}$ | $146^{a}$ | - |
| | 18 plants | 18 plants | $2.21^{b}$ | $45.4^{a}$ | $0.95^{b}$ | $369^{ab}$ | $102^{ab}$ | - |
| | 24 plants | 24 plants | $2.49^{ab}$ | $45.5^{a}$ | $0.98^{b}$ | $404^{a}$ | $40^{bc}$ | - |
| | 30 plants | 30 plants | $2.94^{a}$ | $47.0^{a}$ | $1.15^{a}$ | $219^{c}$ | $40^{bc}$ | - |
| | 36 plants | 36 plants | $2.9^{a}$ | $45.7^{a}$ | $1.18^{a}$ | $178^{c}$ | $32^{c}$ | - |
| C. Turretfield 2019 | 12 plants | 11 plants | $2.67^{c}$ | $55.0^{d}$ | $1.15^{e}$ | $142^{a}$ | $9^{a}$ | - |
| | 18 plants | 16 plants | $2.96^{c}$ | $57.4^{c}$ | $1.39^{d}$ | $160^{a}$ | $5^{ab}$ | - |
| | 24 plants | 22 plants | $3.71^{b}$ | $60.5^{b}$ | $1.61^{c}$ | $152^{a}$ | $3^{ab}$ | - |
| | 30 plants | 27 plants | $3.46^{b}$ | $63.2^{a}$ | $1.72^{b}$ | $65^{b}$ | $1^{b}$ | - |
| | 36 plants | 32 plants | $4.29^{a}$ | $63.3^{a}$ | $1.86^{a}$ | $84^{b}$ | $0^{b}$ | - |
| D. Salter Springs 2019 | 12 plants | 12 plants | $2.7^{c}$ | $82.8^{c}$ | $2.49^{c}$ | $153^{a}$ | - | $237^{a}$ |
| | 18 plants | 14 plants | $3.2^{c}$ | $84.9^{c}$ | $2.77^{b}$ | $113^{ab}$ | - | $139^{ab}$ |
| | 24 plants | 22 plants | $4.1^{b}$ | $88.2^{b}$ | $2.88^{ab}$ | $75^{ab}$ | - | $61^{bc}$ |
| | 30 plants | 25 plants | $4.3^{b}$ | $88.5^{b}$ | $2.89^{ab}$ | $28^{bc}$ | - | $22^{c}$ |
| | 36 plants | 30 plants | $5.0^{a}$ | $92.0^{a}$ | $2.94^{a}$ | $9^{c}$ | - | $20^{c}$ |

# Developing strategies to mitigate and manage resistance to key herbicides: A project overview

Jenna Malone[1], Navneet Aggarwal[2], Michael Widderick[3], Danica Goggin[4], Roberto Busi[4], Hugh Beckie[4], Christopher Preston[1]
[1]University of Adelaide, South Australia, Australia,
[2]South Australian Research and Development Institute, Clare, Australia,
[3]Department of Agriculture and Fisheries, Toowoomba, Australia,
[4]University of Western Australia, Perth, Australia
(jenna.malone@adelaide.edu.au)

**Summary** The objective of this project is to provide new knowledge on regionally effective strategies for the management of herbicide resistance, particularly to the pre-emergent herbicides, glyphosate and the imidazolinone herbicides, in key grain cropping weeds. With new pre-emergent herbicides arriving, for some of which there is existing resistance, it is important to understand the patterns of pre-emergent herbicide resistance that occur in annual ryegrass to develop strategies to delay resistance. This includes understanding the genetic relationship of resistance to existing herbicides and new modes of action. In addition, field trails have been established in Western and Southern Australia to investigate management of resistance to pre-emergent herbicides in annual ryegrass to Groups 15, 3, 13 and 30 herbicides. The increasing number of imidazolinone-tolerant crops being grown in rotations has increased the risk of resistant weeds evolving. In some species, such as annual ryegrass, resistance has evolved quickly, while in other species, such as brome grass, only a few cases of resistance have occurred. Research is being conducted to determine whether ploidy, and how much ploidy, plays a role in selection for resistance to imidazolinone herbicides. Glyphosate resistance continues to evolve in both summer and winter weeds, and several different mechanisms of resistance have been identified in Australia. Glyphosate resistance in a number of different species is being investigated to gain a better understanding of the extent and variation of different resistance mechanisms present in individual weed species. Field trials have been established to investigate which management strategies may be most effective for different species and different resistance mechanisms.

**Keywords** Resistance mechanisms, resistance management, glyphosate, pre-emergent herbicides, imidazolinone herbicides

# Winter cover crops and their weed suppressive abilities

Asad Shabbir, Campbell Parish and Michael Walsh
School of Life and Environmental Sciences, The University of Sydney Camden 2570 New South Wales, Australia
(asad.shabbir@sydney.edu.au)

**Summary** The management of annual weeds in cropping systems is a primary issue for growers in the northern grains region of Australia. Cover crops are a non-chemical weed management option that can be used in conservation cropping systems to reduce the reliance on herbicides. The aim of this study was to determine the effectiveness of three winter cover crop species (forage oats, purple vetch, and tillage radish) in suppressing emergence and growth of annual winter weeds at two field sites near Camden, NSW. On average forage oats and purple vetch provided 42% greater suppression of early weed emergence than tillage radish at Bringelly (heavy soil site). At Lansdowne (sandy soil site), forage oats and tillage radish suppressed weed biomass by 67% and 88% more than vetch, at 80- and 120-days post crop emergence, respectively. At Bringelly, forage oats suppressed weed biomass on average 70% more compared to tillage radish and purple vetch at 80 days post crop emergence. There were no differences in weed suppression among the three crop species 120 days post crop emergence at Bringelly. The higher weed suppressive ability of different cover species was related to their ability to produce more biomass during early growth phase. This study has demonstrated that cover crops can be used to supress the emergence and growth of annual winter weeds.

**Keywords** cover cropping, northern grains region, oats, vetch, tillage radish, weeds.

## INTRODUCTION

The management of annual weeds in cropping systems is a primary issue for growers in the northern grain production region of Australia. It is estimated that the total cost of weeds in terms of revenue loss and weed management expenditure to Australian grain growers is approximately $3,318 million, equivalent to $146/ha (Llewellyn, 2016).

As a result of a major shift from conventional farming to conservation agriculture, weed management practices have changed to focus on the use of herbicides. Conservation agriculture, based on minimal soil disturbance and residue retention, has resulted in a reliance on herbicides for weed control in Australian cropping systems (Llewellyn and D'Emden, 2010). Whilst herbicides are the most effective weed management practice, the frequent application of herbicides has led to the widespread evolution of herbicide resistance in several major weed species (Walsh and Powles, 2007).

Cover crops are grown during non-grain crop phases (e.g., fallows) primarily to increase the retention of soil moisture and nutrients for subsequent grain crop production, however the resulting biomass production can restrict the growth of weeds. Cover crops are now being considered as new non-chemical alternative weed management option that through competition can suppress the growth of weeds (Reeves, 1994). Studies have shown that competitive winter cover crops, fodder radish (*Raphanus sativus* cv. Brutus), winter oilseed rape (*Brassica napus* cv. Emerald) and winter rye (*Secale cereale* cv. Protector) can suppress weed growth by more than 70% at the experimental farm of Wageningen University, the Netherlands (Kruidhof *et al*. 2008).

Cover crops may have potential for addressing weed issues in the northern grains region of Australia; however, it is currently unclear what cover crop species are suitable for production in this region and their subsequent impact on weed emergence and growth. The general objective of this research was to evaluate the weed control potential of winter cover species suited to the northern grains region of Australia. Specifically, the aims were to determine the effectiveness of three different winter cover crop species, forage oats (grass), purple vetch (legume) and tillage radish (brassica) in supressing weed emergence and growth.

## MATERIALS AND METHODS

To determine the weed suppressive abilities of winter cover crops, field trials were conducted in winter 2021 at two locations with contrasting soil types near Camden, NSW. The first site, Bringelly, is a sloping location with a dark loam soil while the second site, Lansdowne, is a grey sand soil type. At both sites, three winter cover crops, forage oats (*Avena sativa*), purple vetch (*Vicia benghalensis*) and tillage radish (*Raphanus sativus*), and a control with no crop (fallow) treatments included. The cover crops were sown in six rows, 25cm apart in plots of 12×2 m dimensions (24 $m^2$) on the 8$^{th}$ of March 2021. The treatment plots were laid out in a completely randomised block design with 4 replicates. Standard planting rates of 8, 35 and 40 kg $ha^{-1}$ were used for tillage radish, purple vetch and forage oats, respectively. The field sites were monitored and observed on a fortnightly basis, with the option of supplementary irrigation when required.

Weed plant emergence counts were taken 42 days after crop sowing (DAS) by counting number of plants along 1 $m^2$ transect randomly placed at three randomly selected positions in each plot. Cover crop and weed plants were harvested at two different times (80- and 120-days post crop emergence) for shoot dry biomass determination. To achieve this, all plants along a 1 $m^2$ transect were cut at soil surface level and packed into paper bags. The freshly harvested plants were then kept inside an oven set at 70°C for 72 hours.

Analysis of variance (ANOVA) was performed on all data using a statistical package GenStat Ver. 19.1 (VSN International – UK). The means of crop and weed biomasses at both times and locations were compared using Tukey's 95% confidence intervals to determine the significant difference among treatments.

RESULTS

**Weed emergence** The Lansdowne site was dominated grass weeds (annual ryegrass, broom grass) while more broad-leaved species (*Fumaria* sp., capeweed, common sowthistle, chickweed etc.,) were present at Bringelly. At 42 DAS weed densities were substantially higher (85%) at Bringelly than at Lansdowne as indicated by the fallow plot weed count data (Table 1). Weed emergence was 42% lower in oats and vetch treatments compared with tillage radish at Bringelly, however weed emergence was similar in all three cover crops at Lansdowne (Table 1).

**Table 1.** Weed emergence 42 days after crop sowing at Lansdowne and Bringelly sites.

| Treatments | weed emergence (plants $m^{-2}$) | |
|---|---|---|
| | Lansdowne | Bringelly |
| tillage radish | 18 ±1.5 | 125 ±8.3 |
| forage oats | 15 ±1.2 | 66 ±3.3 |
| purple vetch | 15 ±1.3 | 77 ±2.7 |
| fallow | 19.6 ±5.8 | 134.6 ±11.9 |

**Crop and weed biomass – Lansdowne** At 80 DAS tillage radish produced the highest biomass, averaging 3.2 t $ha^{-1}$ followed by forage oats (2.5 t $ha^{-1}$) and vetch (1.6 t $ha^{-1}$) at Lansdowne (Fig. 1 A). Tillage radish and forage oats both reduced weed biomass by 82% more than purple vetch (Fig. 1 B).

At 120 DAS tillage radish and forage oats produced the highest crop biomass of over 4 t $ha^{-1}$, which was 44% greater than vetch (1.6 t $ha^{-1}$) at Lansdowne (Fig. 2 A). Tillage radish and forage oats both reduced the weed biomass 90% more compared to purple vetch (Fig. 2 B).

**Crop and weed biomass – Bringelly** At 80 DAS tillage radish produced 64% higher biomass compared to forage oats and vetch at Bringelly (Fig. 3 A). Forage oats showed the highest weed suppression, reducing the weed biomass by over 90% compared to vetch and tillage radish (Fig. 3 B).

At 120 DAS tillage radish on average produced 41% biomass compared to forage oats and purple vetch, averaging 3.2 t $ha^{-1}$ at Bringelly (Fig. 4 A). There were no significant differences between oats, tillage radish and vetch in terms their weed suppressive ability.

**Figure 1**: Biomass of cover crops (A) and weeds (B) at 80 days post crop sowing at Lansdowne site, Camden New South Wales. The different small letters above bars show significant differences ($P \leq 0.05$). In the absence of cover crops, weeds produced 1.8 t $ha^{-1}$ biomass at 80 days post crop sowing.

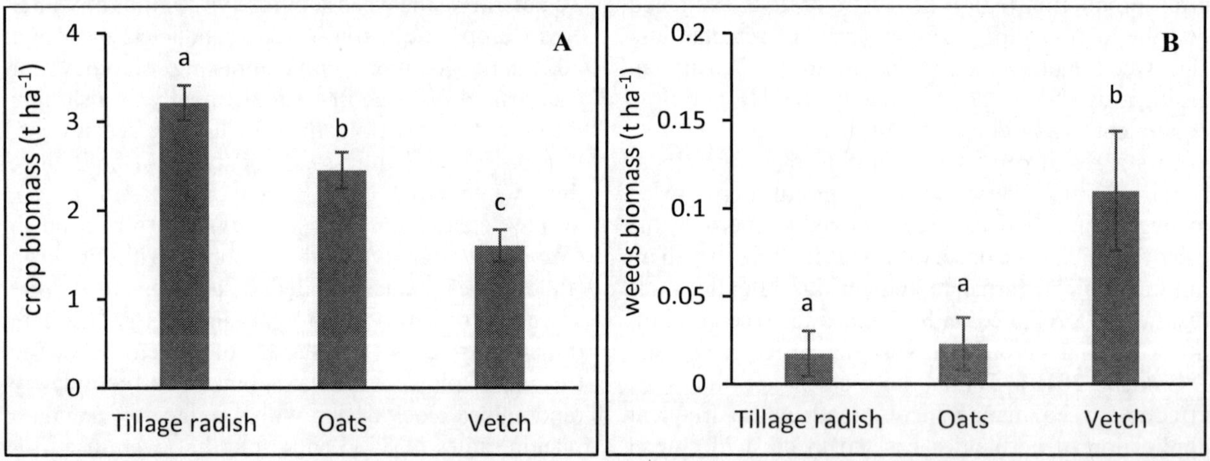

**Figure 2**: Biomass of cover crops (A) and weeds (B) at 120 days post crop sowing at Lansdowne site, Camden NSW. The different small letters above bars show significant differences ($P \leq 0.05$). In the absence of cover crops, weeds produced 2.9 t $ha^{-1}$ biomass at 80 days post crop sowing.

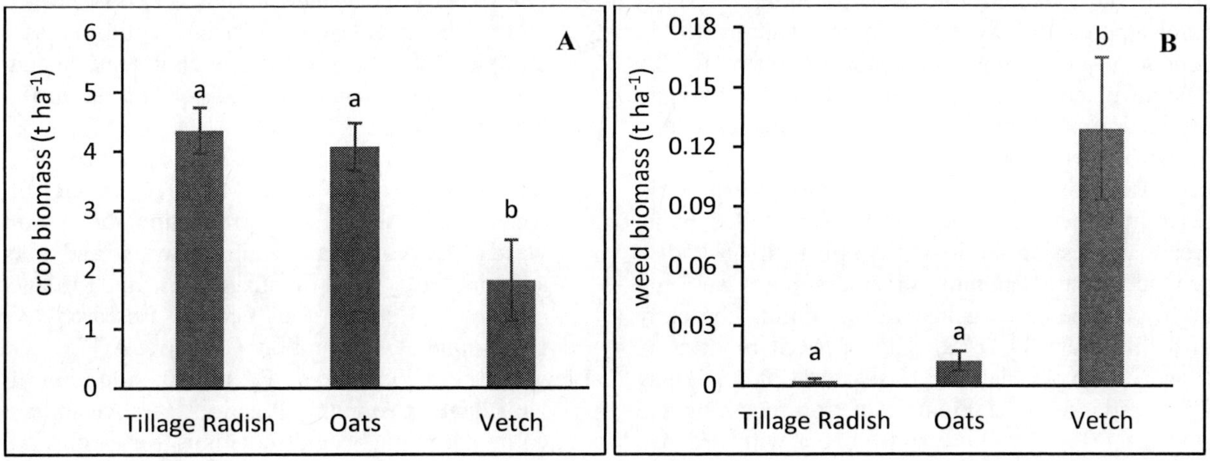


**Figure 3**: Biomass of cover crops (A) and weeds (B) at 80 days post crop sowing at Bringelly site, Camden NSW. The different small letters above bars show significant differences ($P \leq 0.05$). In the absence of cover crops, weeds produced 1.2 t $ha^{-1}$ biomass at 80 days post crop sowing.

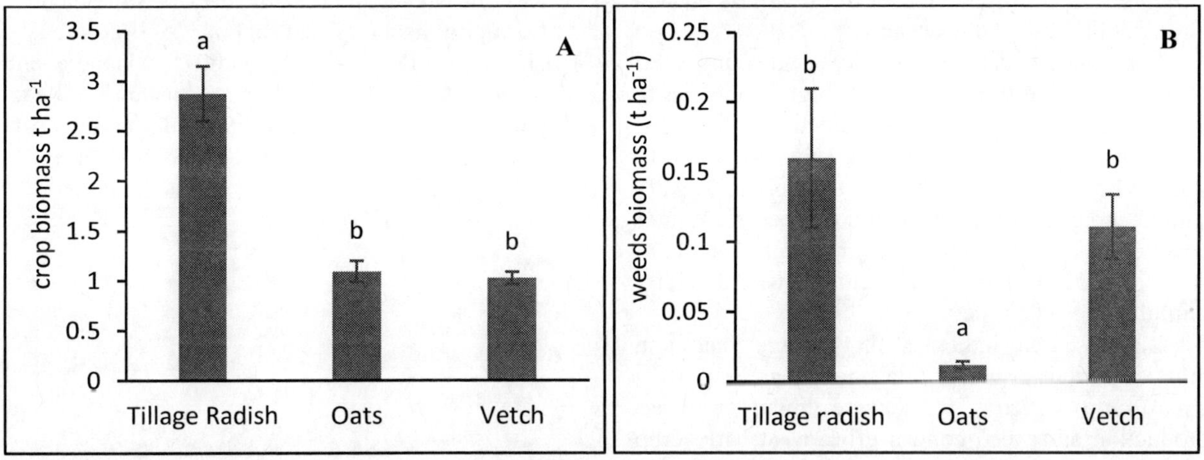


**Figure 4**: Biomass of cover crops (A) and weeds (B) at 120 days post crop sowing at Bringelly site, Camden NSW. The different small letters above bars show significant differences ($P \leq 0.05$). In the absence of cover crops, weeds produced 2.1 t $ha^{-1}$ biomass at 120 days post crop sowing.

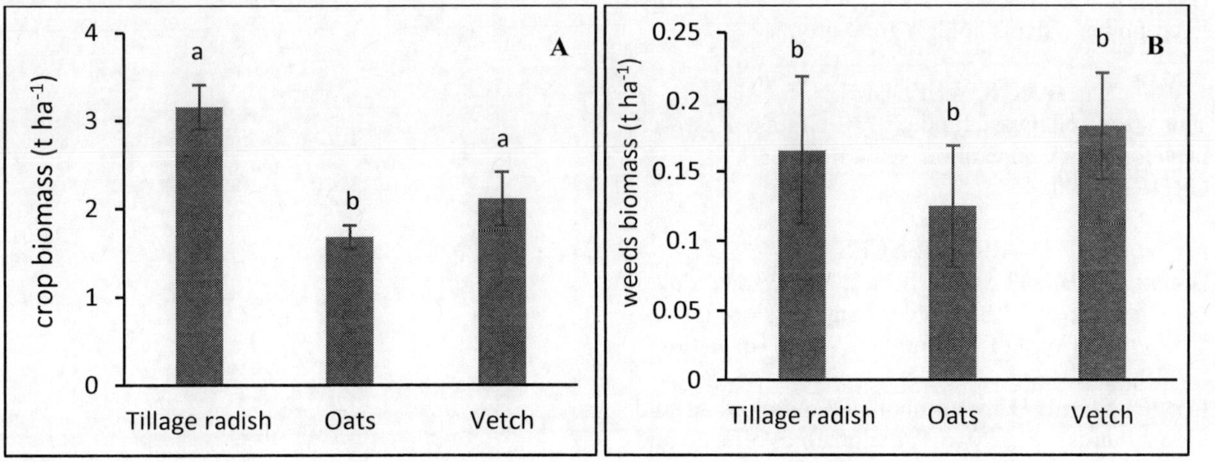

## DISCUSSION

The three cover crop species evaluated in these studies have demonstrated the potential for high-level suppression of weed emergence and growth. In general, weed suppression corresponded to the amount of cover crop growth with higher biomass production resulting in lower weed emergence and growth. Even though the biomass of forage oats decreased over time, this species consistently performed better at both locations in terms of its weed suppression ability. In contrast, tillage radish produced a large amount of biomass, but it was only highly competitive against weeds, during the early growth stages (80 DAS). Higher weed biomass in tillage radish plots later in the season (120 DAS) may be in response to the initiation of crop senescing and flowering (Fig. 5). Our results concur with Brennan *et al.* (2005) who reported that mustard (brassica) produced higher early season biomass than oats and legumes. In another study, Ch, *et al.* (2016) found tillage radish, purple vetch and white mustard (*Sinapis alba* L.) suppressed weed growth by 60% at three field locations in Germany.

The Bringelly had much higher background weed densities than Lansdowne. Despite higher densities at Bringelly, forage oats maintained high weed suppression capability during the early crop growth stage at both sites. This indicates that forage oats are a good choice as a winter cover crop species. Winter cover crops are relatively more important in southern parts of the northern grains regions where winter rainfall pattern dominates.

Our research indicates that forage oats and brassica varieties, such as tillage radish are important cover crop varieties to achieve higher biomass production and weed control efficacy at early crop growth stage. However, purple vetch is also effective, but it took longer to establish and has weaker effect on weed growth. This information is useful for grain growers in the northern cropping region of Australia where integration of cover crops in cropping systems holds a great promise.

## ACKNOWLEDGMENTS

This research was funded by Grains Research Development Cooperation, grant number GRDC US00084

## REFERENCES

Brennan, E.B. and Smith, R.F., 2005. Winter Cover Crop Growth and Weed Suppression on the Central Coast of Californial. Weed Technology, 19, 1017-1024.

Ch, K., Sturm, D.J., Varnholt, D., Walker, F. and Gerhards, R., 2016. Allelopathic effects and weed suppressive ability of cover crops. Plant soil and environment, 62, 60-66.

Kruidhof, H.M., Bastiaans, L. and Kropff, M.J., 2008. Ecological weed management by cover cropping: effects on weed growth in autumn and weed establishment in spring. Weed research, 48, 492-502.

Llewellyn, R., Ronning, D., Ouzman, J., Walker, S., Mayfield, A. and Clarke, M., 2016. Impact of weeds on Australian grain production: the cost of weeds to Australian grain growers and the adoption of weed management and tillage practices. Report for Grains Research & Development Corporation, Canberra, ACT.

Llewellyn, R.S., D'Emden, F.H. 2010. Adoption of No-tillage Cropping Practices in Australian Grain Growing Regions. Grains Research and Development Corporation, Kingston, ACT, pp. 1–31. www.grdc.com.au/notill_adoption.

Reeves, D.W. 1994. Cover Crops and Rotations. In: Hatfield, J.L. and Stewart, B.A., Eds., Crops Residue Management. Advances in Soil Science, Lewis Publishers, Boca Raton, 125-172.

Walsh M.J., Powles S.B. 2007. Management strategies for herbicide-resistant weed populations in Australian dryland crop production systems Weed Technology, 21, 332-338

# WeedScan - a website and smartphone app for identifying, reporting and managing priority weeds in Australia

Claire Lock[1], Andrew Mitchell[2], Alexander Schmidt-Lebuhn[2], Matthew Shillam[1], Ron Li[2], Hanwen Wu[1]
[1]New South Wales Department of Primary Industries, Wollongbar, Australia,
[2]CSIRO, Black Mountain, Australia
(claire.lock@dpi.nsw.gov.au)

**Summary** Mobile apps like PlantSnap and PictureThis have revolutionised plant identification through artificial intelligence, which can analyse a plant photo and instantly suggest what the plant may be. The potential of accessible, rapid diagnostic tools to aid the early detection of new weeds is clear; however, existing plant identification apps are often paywalled and typically do not tell users whether their plant is a weed, link them to locally-relevant weed management information or facilitate the reporting of priority weeds to government weeds staff. To bridge this gap, WeedScan is being developed by the Centre for Invasive Species Solutions, CSIRO and the NSW Department of Primary Industries (DPI) with input from other states and stakeholders. The first release of the WeedScan website and smartphone app is scheduled for mid-2023. WeedScan's artificial intelligence model is being trained by CSIRO to recognise approximately 300 priority weed species across Australia and will help users to identify weeds from photos, with this aspect of the smartphone app working offline. Identification suggestions will include links to existing weed profiles, which will be filtered according to the user's state or territory if known. Additionally, users will be prompted to make a record if the weed is a priority in their state or territory. If a user records their weed observation, alerts will be sent to government weeds staff who have set up notifications for the weed in that state or local government area. Public WeedScan records will be visible on a map which can be viewed and searched by users. NSW DPI is continuing to scope WeedScan's functionality through user workshops across Australia. Once fully developed, WeedScan will be a valuable tool for farmers, agronomists, landholders, weeds officers and NRM groups to improve weed identification and management.

**Keywords** App, AI, identification, reporting

# Digital 3D weed models - an innovative identification tool for early detection.

Rachel K. Klyve[1], Wendy L. Gibney[2], Ian K. Turnbull[2]
[1] 3/17 Guna Street, Charlestown, New South Wales 2290, Australia
[2] Department of Primary Industries, Locked Bag 6006, Orange 2800 New South Wales, Australia
(rachelklyveart@gmail.com)

**Summary** In a world-first initiative, NSW Department of Primary Industries (DPI) has developed a comprehensive and accessible tool that assists users to identify weeds that pose the highest risk to NSW. Interactive digital three-dimensional (3D) models of weed species that are prohibited in NSW due to the risk they pose, and some animal species, have been produced by scientific artists using digital 3D modelling and texturing software. These models, which form an integral component of biosecurity field personnel training, have been rated by learners as highly useful.

**Keywords** Prohibited matter, biosecurity, weeds, plant identification, weed officers, digital 3D, enabling tools, technology, Blender, Adobe Substance 3D Painter, Adobe Photoshop, Sketchfab.

## INTRODUCTION

In contemporary biosecurity management, government investment is targeted toward prevention and early eradication activities that provide the greatest return on investment (NSW DPI 2018). Eradication programs are more likely to be successful when early detection of new incursions occurs. Early detection is only possible if biosecurity field personnel are trained to identify species that do not, or rarely, occur in their state or country. Unfortunately, restricted access to live specimens and the fragmentation of alternate resources such as photos, botanical line drawings, and textual descriptions can limit training opportunities and outcomes.

To address these barriers NSW DPI has developed a comprehensive and accessible tool in the form of realistic interactive digital 3D models of the highest risk weed, and animal species. Scientific artists have produced these models using digital 3D modelling and texturing software. The models are accurate to 1 mm. Entire species or parts of species are depicted including all diagnostic features.

To maximise early detection of high-risk weed species NSW DPI has designed and implemented a training program for NSW biosecurity field personnel which utilise the digital 3D models.

## BACKGROUND

**Regulation of Prohibited Matter, NSW** Schedule 2 of the NSW *Biosecurity Act (2015*) lists 28 entries of terrestrial and freshwater weeds as prohibited matter in the state. These plants are identified as posing the highest risk of adversely impacting the economy, environment, and people of NSW. These species (and any of their parts) are not to be imported, kept, grown, moved, sold or dealt with in any other way. Of the 28 entries in the schedule: three of the species are not known to be found in Australia; one species was previously found in Australia and was eradicated; and seven of the species are present in NSW and under active eradication programs. The remainder are present, or have previously been present, in other states and territories.

Local government is responsible for weed control functions in NSW under the NSW *Biosecurity Act 2015* (s370). Councils and county councils appoint authorised officers to prevent, eliminate, minimise and manage weed biosecurity risks on their behalf. These authorised officers are referred to broadly as weeds officers. There are approximately 180 authorised officers appointed under the NSW *Biosecurity Act 2015* across NSW.

In a training needs survey of weeds officers at the NSW Weeds Conference held in Newcastle and online in 2019, 94% of respondents stated they wanted training on prohibited matter.

**Digital 3D models in education** Observing and interacting with live specimens is generally accepted to be the best way to gain plant identification skills that are transferable to field work. Due to the biosecurity risk associated with prohibited matter plants, including importing them and transporting them around the state, it was determined that the use of live specimens was not appropriate.

The most used alternative is photographs, which are an essential tool for visualising the habitat and habit of plants. But when photographs are used to identify other diagnostic features, it can lead to confusion due to multiple factors including, variable lighting, complex and competing backgrounds, poor focus and narrow depth of field. Botanical illustration is an historically established method for the comprehensible depiction of plants that benefits from the skill of the illustrator to simplify extraneous detail and highlight diagnostic features (Hickman *et al.* 2017). While illustrations are still used in many botanical contexts, their use for people who have not been trained in the visual language of botanical illustration may be challenging. Also, understanding a 3D object (a plant) from 2D resources is inherently difficult.

Other biological fields, such as medicine, face similar issues when educating students. Access to cadavers is considered the best method for gaining an accurate understanding of human anatomy but issues exist with access and cost (Triepels *et al* 2020). Therefore, new technologies are being embraced and tested. Studies show that the use of digital 3D models in human anatomy education is an effective method for improving student understanding (Triepels *et al* 2020; Yammine and Violato 2015). The use of digital 3D models compared to traditional resources result in greater objective and subjective spatial understanding, memory retention, and identification of critical features, plus an increase in the motivation and interest of students to engage (Triepels *et al* 2020).

The use of digital 3D visualisation in botanical identification is in its infancy. This project represents a test case for the usefulness of this technology in botanical education and species identification.

**Model making process** Scientific artists with tertiary training in botany and scientific illustration have produced the weed models discussed in this paper. The models are created using several digital 3D modelling and texturing programs; Blender, Adobe Substance 3D Painter, Adobe Photoshop, and Sketchfab. The plant species are modelled by hand, rather than laser scanned or produced by photogrammetry as is common with non-biological digital 3D model making. The trained scientific artists possess subtle skills that are better suited to accurately represent the complexity and delicacy of plants in an economical, timely, and aesthetically engaging fashion.

The production of the model begins with understanding each species, ideally by viewing a live specimen. When live specimens are unavailable, photographs are utilised in conjunction with written botanical descriptions and botanical line drawings, when available. All parts of the plant are modelled to scale in Blender using a mesh of digital polygons. Each part of the plant is accurate to one millimetre. Accuracy is achieved with Blender's in-built scale feature and bespoke rulers created by the artists. The plant parts are positioned accurately to describe the habit of the species. Colour and texture details, including venation, hairs, lenticels, glossiness or roughness etc., are added by painting in Adobe Substance 3D Painter and through the use of some photographic textures in Adobe Photoshop. Lifelike lighting of the model environment enables realistic visualisation of the glossiness and opacity of the plant surfaces giving further diagnostic information.

The draft models are assessed by NSW and interstate biosecurity experts with first-hand knowledge of each species. Recommended changes are incorporated into the final models which are uploaded to the interactive online platform Sketchfab.

Within Sketchfab the user can move the models in 'space' to view from 360 degrees and zoom into key features. This presents a realistic sense of the size and position of the features relative to the whole. Independent scale indicators (rulers or scale bars) are included in the final models. The diagnostic features of each species are further highlighted on the models through written annotations.

**Weed officer training** In 2022 NSW DPI have commenced 14 'Getting to know Prohibited Matter' training sessions for weeds officers. The digital 3D weed models form an interactive component of this training alongside other training tools, such as look-alike species, physical models, and access to a range of species-specific information (e.g. WeedWise species' descriptions).

The annotation function in Sketchfab is used as one learning activity where learners complete blank annotations, enabling greater retention of what they see as the key diagnostic features of each species. Many weeds officers already have a sound understanding of botany and plant features so the training has been designed to be somewhat self-directed.

An additional 12 models have been created for incorporation into NSW DPI's established Water Weeds training package.

**Table 1.** Weed species modeled

| Scientific name | Common name |
|---|---|
| *Alternanthera philoxeroides* | Alligator weed |
| **Andropogon gayanus* | Gamba grass |

| | |
|---|---|
| **Annona glabra* | Pond apple |
| **Asparagus declinatus* | Bridal veil creeper |
| **Bassia scoparia* | Kochia |
| *Cabomba caroliniana* | Cabomba |
| **Centaurea stoebe* subsp. *micranthos* | Spotted knapweed |
| **Centaurea X moncktonii* | Black knapweed |
| **Chromolaena odorata* | Siam weed |
| **Clidemia hirta* | Koster's curse |
| **Cryptostegia grandiflora* | Rubber vine |
| *Eichhornia crassipes* | Water hyacinth |
| **Eichhornia azurea* | Anchored water hyacinth |
| *Egeria densa* | Leafy elodea |
| *Equisetum arvense* | Horsetails |
| *Gymnocoronis spilanthoides* | Senegal tea plant |
| *Heteranthera reniformis* | Kidney-leaf mud plantain |
| **Hydrocotyle ranunculoides* | Hydrocotyl |
| *Hygrophila costata* | Hygrophila |
| **Lagarosiphon major* | Lagarosiphon |
| **Limnobium laevigatum* | Frogbit |
| **Limnobium spongia* | Spongeplant |
| **Limnocharis flava* | Yellow burrhead |
| *Ludwigia longifolia* | Long-leaf willow primrose |
| *Ludwigia peruviana* | Ludwigia |
| **Miconia calvescens* | Miconia |
| **Mikania micrantha* | Mikania vine |
| **Mimosa pigra* | Mimosa |
| **Myriophyllum spicatum* | Eurasian water milfoil |
| **Nassella tenuissima* | Mexican feather grass |
| *Orobanche minor* | Common Broomrape |
| **Parthenium hysterophorus* | Parthenium |
| **Pilosella aurantiaca* | Orange Hawkweed |
| *Sagittaria platyphylla* | Sagittaria |
| *Salvinia molesta* | Salvinia |
| **Stratiotes aloides* | Water soldier |
| **Striga asiatica* | Witchweed |
| **Trapa natans* | Water caltrop |
| **Vachellia nilotica* | Prickly acacia |
| **Vachellia karroo* | Karoo acacia |

* denotes prohibited matter species

## DISCUSSION

**Assessment** The usefulness of the digital 3D weed models is being assessed in ongoing subjective surveys of the 'Getting to know Prohibited Matter' training participants. To date 58 participants have completed the training evaluation. Participants are asked to respond to the statement 'The 3D digital models helped me learn about the weeds.' 83% 'strongly agreed' and 17% 'agreed.' Feedback on the training program is generally very positive with 100% of respondents strongly agreeing or agreeing that they *"feel more confident about being able to identify prohibited matter in the field*."

Engagement with Prohibited Matter species on the NSW DPI WeedWise website has increased since the addition of the digital 3D models. Over 4000 views of the Prohibited Matter 3D models were registered in the first four months of their presence on NSW WeedWise. Furthermore, activity on the DPI internal and public social media pages highlighting the presence of the 3D models on WeedWise has generated engagement with the species and very positive feedback.

**Universal access to models** In the spirit of national and international collaboration, the digital models are publicly available on the NSW DPI WeedWise website. This enables access to the models for weed professionals from other state agencies plus the general public. This accessibility enables greater possibility of early detection and eradication of high-risk species across Australia and the world.

## ACKNOWLEDGMENTS

We thank the many experts who gave their time to review the models and offered feedback on their accuracy. Your help was invaluable. The models were reviewed by experts from NSW Department of Primary Industries, NSW National Parks and Wildlife Service, QLD Department of Agriculture and Fisheries, and Parks Victoria. Jemma Gillard and Lila Raymond also created models for the project, plus gave technical support. And many thanks to Dr Bernadette Drabsch from the University of Newcastle for her tireless encouragement and inspiration.

## REFERENCES

Hickman, E.J., Yates, C.J., and Hopper, S.D.

(2017). Botanical illustration and photography: a southern hemisphere perspective. *Australian Systematic Botany* 30(4), 291-325.

NSW DPI, NSW Department of Primary Industries (2018). New South Wales invasive species plan 2018-2021. https://www.dpi.nsw.gov.au/__data/assets/pdf_file/0003/807753/InvasiveSpeciesPlan2018.pdf (accessed 25 March 2022).

Triepels, C., Smeets, C., Notten, K., Kruitwagen, R., Futterer, J. J., Vergeldt, T., and Van Kuijk, S. (2020). Does three-dimensional anatomy improve student understanding? *Clinical anatomy 33*(1), 25–33.

Yammine, K. and Violato, C. (2015), A meta-analysis of the educational effectiveness of three-dimensional visualization technologies in teaching anatomy. *American Association of Anatomists* 8, 525-538.

# Implementation of a rapid response biosecurity program to eradicate a novel invasive species (*Bocconia frutescens*) in NSW

Terry Inkson[1], Matt Ansley[1], Alysha Wynan[1]
[1]MidCoast Council, TAREE, Australia
(Terry.Inkson@MidCoast.nsw.gov.au)

**Summary** As illustrated by the generalised invasion curve, a conceptual framework widely adopted within the biosecurity community, early detection and rapid response programs are necessary for the timely and cost-effective eradication of new invasive species. In January 2021 a specimen of the potentially invasive Plume Poppy (*Bocconia frutescens*) was detected near Taree on the mid-north coast of New South Wales. The incursion appears to be the result of seed bank dispersed from a horticultural plant that has germinated after a significant bushfire disturbance in 2019. It has been detected in rural residential land, as well as adjacent high-value conservation assets including National Park and Council bushland reserve. Plume Poppy has been recorded as being a highly invasive weed of disturbed areas in tropical and subtropical regions throughout the world, notably in Hawaii. This paper describes the response of the local control authority (MidCoast Council) in carrying out an identification, eradication and education program in the area where the incursion was detected, in accordance with the Hunter Regional Weeds Committees' 'New Weed Incursion and Rapid Response Plan (2017-2022)'. A cross-tenure collaborative management approach is described, involving cooperation with stakeholders including NPWS and volunteer groups working on Council land. The implementation of a rapid response process is recorded in the context of managing the Plume Poppy incursion, and the efficacy of the initial control is documented through ongoing monitoring of the distribution of the target species. This case study illustrates the importance of community engagement in enhancing the early detection capabilities of biosecurity agencies, as well as the importance of community cooperation in the ongoing detection and control of an invasive species during an eradication program.

**Keywords** Post-fire weed management, Biosecurity Rapid Response, Invasive Species Management, Novel Infestation, *Bocconia frutescens*, Plume Poppy, Priority Weed, midcoast Council, Hunter Regional Weeds

# The threat of black knapweed (*Centaurea* x *moncktonii*) on the Northern Tablelands of NSW

Graham W. Charles[1], Kerinne Harvey[2] and Claire Lock[3]
[1]Weed Biosecurity Research, NSW Department of Primary Industries, Locked Bag 1000, Narrabri 2390, New South Wales, Australia
[2]Weed Biosecurity Research, NSW Department of Primary Industries, PMB 2, Grafton 2462, New South Wales, Australia
[3]Weed Biosecurity, NSW Department of Primary Industries, Wollongbar 2477, New South Wales, Australia
(graham.charles@dpi.nsw.gov.au)

**Summary** Black knapweed (*Centaurea* x *moncktonii*) is Prohibited Matter under the NSW *Biosecurity Act 2015*. It is a member of the Asteraceae family, in the genus *Centaurea*, and is closely related to some common weeds of Northern NSW, including Maltese cockspur (*Centaurea melitensis*) and St. Barnaby's thistle (*Centaurea solstitialis*). Plants in this family are known to be difficult to control with herbicides when treated as flowering plants. However, plants are hard to detect prior to flowering, making targeted control at this earlier growth stage very challenging to achieve. No infestations of black knapweed, also known as meadow knapweed, were known to exist in NSW, until a 150ha infestation was discovered on the Northern Tablelands of NSW in early 2019. The discovery followed several years of drought and at the time this weed was one of the few green plants in the infested paddocks and was being heavily grazed by cattle. Very high plant numbers were present in a 2 ha section of the infested area, with densities exceeding ten small plants $m^{-2}$, and most plants were present in just one paddock of approximately 74 ha. Interestingly, there is evidence that black knapweed had been deliberately introduced to the infected property over 100 years earlier. While there is no direct evidence that this introduction is related to the present infestation, investigations have not revealed any other likely source. Since the find in 2019, the most heavily infested paddock received a boom application of Grazon® Extra Herbicide and escapes have been managed with spot-spraying. Survey transects established prior to treatment in 2019 show there has been a large reduction in the black knapweed population over time, following treatment and heavy competition from other pasture species. This weed remains an eradication target and evaluation continues.

**Keywords** eradication, infestation, prohibited matter.

## INTRODUCTION

Black knapweed (*Centaurea* x *moncktonii* C.E.Britton) is Prohibited Matter under the New South Wales *Biosecurity Act 2015*. It is a member of the Asteraceae family, in the genus *Centaurea*, and closely related to some common weeds of Northern NSW, including Maltese cockspur (*Centaurea melitensis*) and St. Barnaby's thistle (*Centaurea solstitialis*). Black knapweed, also known as meadow knapweed, is an invasive weed that can be problematic in pastures and sensitive ecosystems. It is a thornless, rhizomatous, perennial thistle and is considered a serious weed in many states of the United States of America.

Black knapweed is a fertile hybrid between two European knapweeds, *Centaurea nigra* and *Centaurea jacea* (brown knapweed), and can cross back with either parent species. The lineage of the black knapweed found at Tenterfield is unknown.

Black knapweed identification has in the past been confused with *C. nigra* in Australia. *C. nigra* doesn't occur in NSW, but populations have established in Victoria and South Australia. Some plants of brown knapweed have also been found in Victoria where it may have been introduced as an ornamental. Black knapweed can also itself hybridise with *C. solstitialis* (St. Barnaby's thistle), a thorned thistle and a common weed throughout much of eastern Australia (Roche and Susanna 2010).

A black knapweed plant at Tenterfield, 2019. Photo: Josh Biddle.

Black knapweed is not established in NSW, but plants have been found in Queensland, South

Australia, Victoria, and Tasmania (Anon 2022). Black knapweed is a widespread weed in Europe, New Zealand, and the US, so many potential pathways for infestation exist. Old reports indicate that black knapweed was introduced as a potential pasture species in the Tenterfield area of NSW in the late 1800s, and a herbarium specimen was collected from the area in 1903. Black knapweed had not been detected in NSW since that time.

**Black knapweed at Tenterfield** An infestation of black knapweed was discovered on a roadside near Tenterfield on the northern NSW Tablelands in March, 2019. Given that neither *C. nigra* nor *C. jacea* (the two parent species) occur in NSW, it is unlikely that the infestation resulted from hybridisation between the parent species, but is more likely from a direct introduction of *C.* x *moncktonii*.

Mature black knapweed plant at Tenterfield, 2019. Photo: Josh Biddle.

Surveys of the surrounding area detected a heavy infestation of black knapweed in an adjoining paddock, with additional plants detected along the road and in a few surrounding paddocks. The total infestation covered an area of around 150 ha. The core area of infestation was approximately 74 ha, with a very heavy population of knapweed along 2 ha of creek line, exceeding ten small plants $m^{-2}$ in this area. Tenterfield was experiencing severe drought conditions at the time and the plants were easily seen as they were almost the only green in the paddock. They were being heavily grazed by cattle and few reproductive plants were obvious in the grazed area. Plants along the roadside were generally larger and more mature.

**Characteristics of black knapweed from Tenterfield** Seed heads were collected from the black knapweed plants at Tenterfield to test the seed viability, time to emergence etc. Viability testing under glasshouse conditions showed that most of the seed was not viable. Most of the seeds collected were small, white, and immature. Heads contained a small number of darker, apparently mature seeds. The first collection of seed heads contained 0.9 mature seeds per head. Just 16% of seeds germinated after planting, with emergence in 10 to 34 days. This test was repeated, with similar results from a second batch of seed heads. Viability was higher at 64% on an additional batch of seed heads, but the heads contained only 0.4 mature seeds per head. A further three plants (2%) emerged from these pots in the following year. Hence, it seems that the black knapweed from Tenterfield produces few viable seeds per head, which may have limited the spread of this weed. The seed number and viability were too poor to allow us to accurately assess the seedbank longevity of this weed.

There was concern that the cattle grazing in the infested paddock may have been ingesting knapweed seed heads and spreading the seed via their manure, particularly as knapweed was almost the only "pasture" plant alive in autumn 2019 in the infested paddock. Fresh manure from a series of cowpats was collected in autumn 2019 to test this possibility. Dry cow manure was added to the surface of 30 pots in a glasshouse, with 40 g of manure applied per pot. No knapweed seedlings emerged from the manure over a 59-day period, although 4.4 grass seedlings/pot and some other broadleaf seedlings did establish from the manure. This result is in line with the low viability of seed observed earlier and suggests that ongoing grazing of the infected paddock is not a major issue for the control of knapweed.

The main area of infestation was a heavily grazed paddock at Tenterfield, detected in the 2019 drought.

**Controlling black knapweed** Seed heads were removed from black knapweed plants on the roadside and from mature plants found in the paddock. Cattle were grazing the infested paddock at the time of the initial discovery and this practice has been allowed to continue.

Information from the US indicates that black knapweed can be controlled using a range of herbicides, including: 2,4-D, clopyralid, dicamba, picloram, and glyphosate (Duncan *et al.* 2022). None of these herbicides is registered for controlling black knapweed in NSW, but we were able to use Grazon® Extra herbicide (300 g/L triclopyr + 100 g/L picloram + 8 g/L aminopyralid) under permit. Grazon Extra was applied by boom spray to the most heavily infested areas in April and May 2019, and escapes have been spot sprayed in the paddock and on the roadside.

To determine the effectiveness of treatment, we established ten permanent transects in the main paddock prior to treatment (Transects 1-10). The transects were each 10 m in length, and the groundcover of knapweed (cm presence) was recorded in each transect. A 30 m transect was also established at a tangent to the creek line (Transect 11). Groundcover was recorded rather than plant number, as this is a rhizomatous, perennial plant, where the presence of above-ground plant parts does not necessarily indicate individual plants, as multiple plant parts could arise from a single rhizome. An additional twelve transects were established on the fence line and roadside in October 2019 (Transects 12 – 23). Results are shown in Table 1.

Table 1. Percentage ground cover of knapweed over five observations. Values are averages and standard errors from sets of transects.

| Date | T 1-10 | T 11 | T 12-23 |
|---|---|---|---|
| 17 April 2019 | 18.2% | 23% | |
| s.e. | ± 3.9% | | |
| 31 Oct 2019 | 0.8% | 2.2% | 1.7% |
| s.e. | ± 0.5% | | ± 0.7% |
| 25 Feb 2020 | 0.5% | 0.2% | 0.6% |
| s.e. | ± 0.3% | | ± 0.7% |
| 9 Jun 2020 | 0% | 0% | 0% |
| 3 Dec 2020 | 0% | 0% | 0% |

Results from the fixed transects indicate a rapid decline in knapweed density along fixed transects in response to boom applications and spot spraying of Grazon Extra Herbicide. Initial levels of knapweed presence occupied around 20% of the ground cover in April 2019, to no knapweed occurrence by December 2020 (Table 1). Heavy competition from perennially grasses was observed after the drought broke in early 2020 which likely contributed to the decline in knapweed presence.

No knapweed plants have been observed in the transects since June 2020, although occasional plants remain in the paddock. These plants continue to be treated by spot-spraying but are difficult to find in the tall and heavy grass sward, primarily African lovegrass (*Eragrostis curvula*), that established in the paddock following rain.

The main area of knapweed infestation covered by competitive perennial grass following rain (photo 26th February 2020).

Two knapweed flowerheads (center of photo) among tall perennial grasses demonstrate the difficulty in observing knapweed presence.

**Future management for black knapweed at Tenterfield** The aim of the current work at Tenterfield is to contain the infestation and over time to eradicate this weed. The initial program has proven to be successful, with few plants now apparent even in what were previously the most heavily infested areas.

An issue for the eradication program has been that small black knapweed plants are difficult to distinguish from many of the other broadleaf plants growing in the infected paddock. Plants can be distinguished by close examination, but this level of scrutiny is difficult to achieve in the very dense grass sward that has established post-drought and impractical to undertake on an area of over 100 ha.

Plants are readily identified by their flower heads once these become obvious through the grass sward, but mature seed may have already developed on these plants by the time they are detected, such that the weed seed bank is replenished, and the weed problem is perpetuated.

To ensure the success of the eradication program, the level of scrutiny and targeted treatment may need to be increased to ensure plants do not produce viable seed. This increased level of scrutiny may need to be maintained over the proof-of-freedom period, which will be a number of years after the last plant is found and destroyed.

A more technically challenging, but in the long-term, potentially cheaper, and more effective alternative for detecting and eradicating black knapweed might be the use of drones and robot sprayers that are able to identify black knapweed flowers (by drone) and target these plants for control (by robot). This option is becoming increasingly feasible with the development of robot and artificial intelligence (AI) technologies. A single person should be able to manage such a system, surveying the infected area several times a week and managing the robot/s with a minimal manpower requirement. The main cost of the approach would be the technology used, but this cost will decline with time and the approach should have great value for future weed eradication programs, should these occur. At Tenterfield, this approach may have to be augmented by some hand-spraying of areas such as the creek line where it may not be feasible for a robot to access.

This approach of using drones to locate knapweed plants would be particularly attractive during the years of the proof of freedom phase, when no active spraying would be required.

The black knapweed flower head appears to the human eye to be distinctively different to anything else in the infected areas and we anticipate that it would be easily distinguished by AI. The head is relatively large, around 2 cm in diameter and pinkish-purple in color. Flowering heads are present from spring through to autumn.

The flowers could be confused with other thistles, such as spear thistle (*Cirisium vulgare*) or variegated thistle (*Silybum marianum*). However, examples of these thistle have not been commonly observed in the infected areas, and their inadvertent inclusion in the initial eradication effect (should this happen), would be of minimal negative consequence.

**Conclusion** We consider that the current program to eradicate black knapweed from NSW is viable but will require a continuing investment in time and resources. Augmenting the eradication program with AI may be a better option for this weed and would be a biosecurity investment that could be invaluable for dealing with future prohibited matter events.

Black knapweed flower head.

## ACKNOWLEDGMENTS

We acknowledge the efforts of Chris Battersby and Tenterfield Shire Council in controlling this weed, the owners of the infected properties for their cooperation, and the inputs from the DPI teams under Dr. Pete Turner and Dr. Elissa Van Oosterhout.

## REFERENCES

Anon (2022). Atlas of Living Australia. https://bie.ala.org.au/species/https://id.biodiversity.org.au/node/apni/2907391. Accessed: 4 May 2022.

Duncan, C., Story, J., Sheley, R., Parkinson, H. and Mangold, J. (2022). Montana Knapweeds. http://msuinvasiveplants.org/documents/mt_noxious_weeds/Montana%20knapweeds.pdf. Accessed: 21 April, 2022.

Roche, C.T. and Susanna, A. (2010). New habitats, new menaces: *Centaurea* x *kleinii* (*C. moncktonii* x *C. solstitialis*), a new hybrid species between two alien species. *Collectanea Botanica* (*Barcelona*) 29, 17-23.

# Significant New Weed Detections in the Northern Territory 2017-2021 – consequences for management

Louis Elliott[1]
[1]NT Department of Environment, Parks and Water Security, Darwin, Australia
(louis.elliott@nt.gov.au)

**Summary** Seven detections of significant new weeds in the Northern Territory (NT) occurred during the five year period 2017-2021. These include: (1) two species currently or formerly targeted for national eradication – *Limnocharis flava* and *Chromolaena odorata*; (2) two new incursions of Weeds of National Significance previously detected and eradicated in the NT – *Parthenium hysterophorus* and *Cryptostegia grandiflora*; and (3) the first Australian records for three species previously identified as potential weed threats to northern Australia – *Boerhavia erecta*, *Spigelia anthelmia*, *Ludwigia leptocarpa*. While responses to four of these incursions have strained limited government resources and required prioritisation over other weed management issues, there have also been positive consequences. First, they have stimulated more effective collaborations between a wide range of stakeholders (including land managers, industry, other parts of the NT government, the States and the Commonwealth) to improve the detection and response to weeds across a vast area (1.4 million km2) with very limited resources. Second, they have required the development, revision and implementation of emergency weed response processes and procedures informed by other national eradication programs, and consistent with the Biosecurity Incident Management System (BIMS). Third, they underscore the importance of jurisdictional and national weed risk assessment systems and committees to provide evidence-based evaluations of the significance of new incursions. During these five years, we have found that (1) close working relationships, (2) generalised incident response procedures and (3) evidence-based assessment have been central to the implementation of strategic responses to these incursions by both government and industry.

**Keywords** New detections, weed risk assessment, emergency response, eradication, collaboration

# Low-volume high-concentration applications of glyphosate to control gamba grass (*Andropogon gayanus*)

Melissa Setter[1], Stephen Setter[1], Melinda Perkins[2,3], Hayley McMillan[2,4] and Shane Campbell[2]
[1] Biosecurity Queensland, Department of Agriculture and Fisheries, Centre for Wet Tropics Agriculture, South Johnstone, Queensland, Australia
[2] School of Agriculture and Food Science, the University of Queensland, Gatton, Queensland, Australia
[3] Agri-Science Queensland, Department of Agriculture and Fisheries, Gatton Research Facility, Gatton, Queensland, Australia
[4] Agri-Science Queensland, Department of Agriculture and Fisheries, Ecosciences Precinct, Boggo Road, Queensland, Australia

(Melissa.setter@daf.qld.gov.au)

**Summary** Gamba grass (*Andropogon gayanus* Kunth.) is a tussock-forming perennial species capable of out-competing other pasture grasses to form dense stands up to 4 m tall. Infestations occur in the Northern Territory, Queensland, and Western Australia, but its current distribution is only a small proportion of its potential range. Once established, gamba grass impacts on the biodiversity and ecosystem function of an area, whilst also imposing a significant fire hazard due to the large biomass that it produces. Controlling invasive grasses amongst other desirable grass species is a challenge, particularly in difficult to access areas where movement of vehicles and equipment is impaired.

To overcome some of these challenges, we undertook a rate response trial on gamba grass to test the efficacy of low-volume high-concentration applications of glyphosate. At a field site near Mt Garnet in North Queensland, a dense stand of gamba grass was slashed in December 2017 and allowed to regrow until April 2018. A randomised complete block experiment comprising seven treatments, three replicates and clusters of 15 gamba grass plants as experimental units was then established. Using a gas operated splatter gun attached to backpack style equipment, six rates of glyphosate (0, 9, 18, 27, 36, 45 and 54 g a.i. $L^{-1}$) were applied, with each plant directly receiving 4 mL of herbicide mixture per half metre of plant height. An untreated control was also included for comparison. After 3 months, gamba grass showed a strong dose-dependent response ranging from 30% mortality at 9 g a.i. $L^{-1}$ to 100% mortality at rates of 36 g a.i. $L^{-1}$ or higher. Regrowth of any surviving plants was also adversely affected at rates above 9 g a.i. $L^{-1}$ with plants taking longer to reshoot. While promising, a follow up trial has been undertaken on more mature gamba grass plants to determine if similar results can be achieved.

**Keywords** Herbicide, invasive grasses, splatter gun, weed control.

## INTRODUCTION

Gamba grass is native to the tropical and sub-tropical savannas of Africa (Biosecurity Queensland, 2016). Infestations occur in the Northern Territory, Queensland and Western Australia, but its current distribution is only a small proportion of its potential range (e.g. Setterfield *et al.* 2014, Csurhes & Hannan-Jones 2016,). Once established, gamba grass negatively impacts on the biodiversity and ecosystem function of an area, whilst also imposing a significant fire hazard due to the large biomass it produces (Rossiter-Rachor *et al.* 2009, Setterfield *et al.* 2014).

Reducing gamba grass populations is problematic as it is well adapted to the northern wet dry tropics, is a prolific producer of wind dispersed seeds and responds well to periodic burning (Rossiter *et al.* 2003, Bebawi *et al.* 2018). It is highly competitive due to its rapid growth, high biomass and its soil nitrogen harvesting process which limits nitrogen availability (e.g. Setterfield *et al.* 2005, Rossiter-Rachor *et al.* 2009,) to other plants within the ecosystem it invades.

Another challenge is controlling gamba grass in difficult to access areas. For some other weeds like lantana (*Lantana camara*) and Siam weed (*Chromolaena odorata*), the use of low volume high concentration applications using backpack style equipment has proven effective (e.g. Somerville *et al.* 2011, Brooks *et al.* 2014). In this study we tested whether this approach could be used to effectively control gamba grass. A rate response trial was implemented whereby six rates of glyphosate were applied to gamba grass using splatter gun style equipment, and efficacy compared against an untreated control.

MATERIALS AND METHODS

**Site details** A cattle property approximately 50 km south-west of Mount Garnet in far north Queensland (18°02'23.9"S 144°52'32.2"E) was used for this trial. It had a patch of gamba grass growing at a density of about 15, 000 plants ha$^{-1}$ in a cleared area, with a predominately red earth soil type. To provide a uniform trial area, the experimental site was slashed 140 days prior to treatment (13 December 2017), and cattle were excluded thereafter with electric fencing. The trial was established in April 2018, and treatments implemented on 3 May 2018. In the year preceding, and the year of the experiment, the area received average (723 mm) and above average (996 mm) rainfall, respectively, compared to the 10-year annual average of 722 mm (Queensland Government 2020).

(a)

(b)

(c)

(d)

**Figure 1.** Experimental trial site (a) prior to and (b) after slashing in December 2017. The area was (c) fenced at the time of slashing to exclude livestock and (d) individual plants designated for treatment were marked with white pegs in early May 2018.

**Experimental design** The experiment was undertaken using a randomized complete block design, and compared the efficacy of six application rates of glyphosate against an untreated control treatment. Glyphosate rates used were 9, 18, 27, 36, 45 and 54 g a.i. L$^{-1}$, corresponding to Roundup® herbicide (360 g a.i. L$^{-1}$) mixture rates of 25, 50, 75, 100, 125 and 150 mL L$^{-1}$, respectively. There were three replicates per treatment. Each replicate plot contained a cluster of 15 tagged gamba grass plants, and 2 m buffers surrounded each plot.

Water was used as a carrier and each mixture also contained 2 mL L$^{-1}$ Pulse® Penetrant (1020g/L Polyether modified polysiloxane; Nufarm Australia, Laverton North, Vic.). Treatments were implemented on 3 May 2018 between 13:00 and 16:00 using a gas powered 'Forestmaster' applicator (N.J. Phillips®) set to deliver 4 mL shots via a fan nozzle adjusted to spray to a width of c.a. 30 cm. Plants received 4 mL of mixture per half metre of plant height. Shots were applied to individual plants in a single strip beginning at the top of the plant and ending at the bottom of the plant. Each plant received an average of 15 ± 2 mL of herbicide mixture. Environmental conditions during herbicide application ranged from 23—32°C, 41—70% relative humidity, 20—80% cloud cover and 0.6—1.0 km h$^{-1}$ wind speed.

**Data collection** A 1 m tall wooden peg was inserted at the base of each of 315 monitored plants, which were assigned a unique identification number. Visual monitoring of leaf injury (brownout), plant health, regrowth height and presence/absence of crown moisture was conducted monthly for 13 months after treatment (MAT). Brownout (%) was estimated as the proportion of leaves exhibiting necrosis, expressed as a percentage of all leaves on the plant. Plant health was assessed using a 10-point rating scale: (1) biomass erect, 100% green; (2) biomass erect 50–99% green; (3) biomass semi-erect, 5–49% green; (4) biomass semi-erect, yellow-green; (5) biomass semi-erect, 0% green; (6) biomass erect, 100% brownout; (7) biomass collapsing, 100% brownout; (8) biomass near flat, 100% brownout; (9) biomass flat to ground, 100% brownout; and (10) no biomass remaining. Regrowth height of surviving plants was measured as the standing height (i.e., not pulled straight prior to measurement). Plant mortality (%) was calculated as the number of dead plants expressed as a percentage of the total number of plants. Plants were classified as dead once they displayed 100% brownout (health rating of 6 or higher) with no subsequent regrowth.

**Statistical analysis** All data analysis was conducted using Minitab®, Version 17.3.1 (Minitab Pty Ltd, Sydney, Australia). Data expressed as percentages (i.e. brownout and plant mortality) were arcsine transformed prior to analysis and then back transformed for presentation within tables or graphs. All data were subjected to an analysis of variance using a general linear model that included herbicide treatment as a fixed effect and block as a random effect. Means were compared using Fisher's least significant difference (LSD) test at a significance level of 0.05. Binary logistic regression with a confidence interval of 95% was used to model the relationship between glyphosate concentration in the herbicide mixture and plant mortality 3 MAT.

## RESULTS

The Gamba grass plants were all flowering at the time of treatment, and had an average height of 1.9 ± 0.3 m, and basal circumference of 1.0 ± 0.3 m.

The three highest rates of splatter gun application (36, 45 and 54 g a.i. L-1) produced 100% mortality at 3 MAT (Table 1). The lower rates of glyphosate achieved 29–80% mortality 3 MAT with no further mortality over time. By 13 MAT, regrowth had occurred in 99% of all surviving plants. Those treated with 9 g a.i. $L^{-1}$ glyphosate displayed regrowth 2 months earlier than plants treated with higher glyphosate rates.

Surviving glyphosate-treated plants showed stunted regrowth compared to untreated plants (Table 1). This difference was most pronounced 10 MAT, as regrowth heights began to plateau or fall after that due to stem lodging. Furthermore, a strong ($R^2$ = 0.93; P <0.001) sigmoidal, dose-dependent relationship was observed for plant mortality (Figure 2).

**Table 1.** Responses of gamba grass plants to splatter gun application of different rates of glyphosate.

| Glyphosate rate g a.i. L[1] | Plant health score 13 MAT[1] | Mortality (%) 3 MAT[1] | Regrowth Height (cm) 10 MAT[1] |
|---|---|---|---|
| 0(control)[2] | 2.0 d | 0 e | 73 a |
| 9 | 3.5 c | 29 d | 63 b |
| 18 | 5.7 b | 69 c | 47 c |
| 27 | 6.5 b | 80 b | 41 c |
| 36 | 7.5 a | 100 a | - |
| 45 | 7.3 a | 100 a | - |
| 54 | 7.5 a | 100 a | - |

[1] Means within a column that do not share a letter are significantly different (P < 0.05) according to Fisher's LSD test. [2] Untreated plants served as a control treatment.

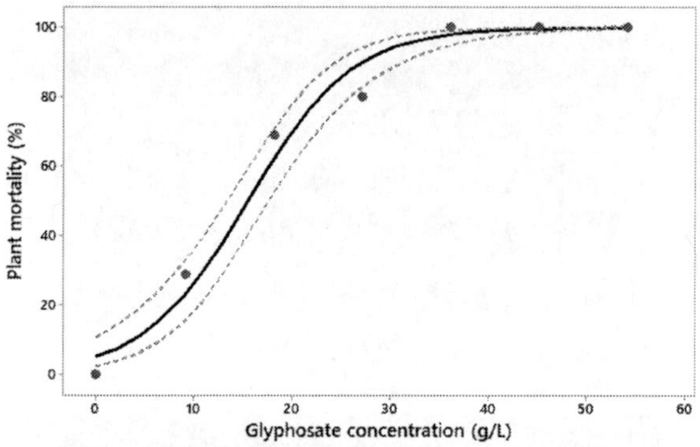


**Figure 2.** Relationship between gamba grass plant mortality 3 MAT and glyphosate concentration in splatter gun applied herbicide mixture. Data points represent treatment means. Solid line represents a binary logistic regression equation (mortality rate = exp(-2.905 + 0.1887 glyphosate concentration)/(1 + exp(-2.905 + 0.1887 × glyphosate concentration); $R^2$ = 0.93; P <0.001), in which mortality rate as a proportion has been converted to a percentage. Dotted lines represent 95% confidence interval.

## DISCUSSION

The findings from this experiment suggest that 36 g a.i. $L^{-1}$ glyphosate is the optimal rate for controlling gamba grass via splatter gun application, as it is the lowest rate capable of achieving 100% mortality within 3 MAT. It is important to note however that this trial was undertaken on gamba grass regrowth and a follow up trial has been undertaken on more mature gamba grass plants to determine if similar results can be achieved. If effective, low volume high concentration herbicide application is advantageous in that it requires a smaller volume of herbicide, so can be useful where access is difficult, e.g., on hillsides. Gas-powered applicators are also available and very suitable for larger infestations. The weight of herbicide mix, as well as the amount of water required to create the mix, can be greatly reduced when using splatter application. Another advantage is that the gamba grass clumps are specifically targeted, resulting in less wastage and off-target damage. Brooks *et al* (2014) noted that "Low-volume high concentration applications of herbicide provide an additional treatment option for areas not accessible to high-volume ground-based spray equipment."

Although splatter application of glyphosate was highly effective, further research into alternative herbicides is warranted, due to potential future issues with glyphosate resistance, and possible regulatory restrictions. Any herbicide treatment that displayed a pre-emergent effect, e.g. possibly flupropionate, could also be highly valuable.

Campbell *et al* (2019) used the splatter technique on rubber vine *(Cryptostegia grandiflora)*, prickly acacia (*Vachellia nilotica*), and Chinee apple (*Ziziphus mauritiana*), and noted some variability in results, indicating that "many factors may affect efficacy, including the health, size and density of plants, herbicide choice and mixture/application rate, presence/absence of biological control agents and climatic conditions".

## ACKNOWLEDGMENTS

This research would not have been possible without Commonwealth and State Government funding and support. We are grateful to staff from the Tropical Weeds Research Centre, particularly Kyle Barton, Carl Andersen and Rodney Stevenson. We thank Nadine and Ron Atkinson for the use of their land.

## REFERENCES

Bebawi, F.F., Campbell, S.D., Mayer, R.J., 2018. Gamba grass (*Andropogon gayanus* Kunth.) seed persistence and germination temperature tolerance. Rangeland Journal. 40, 463–472.

Brooks, S. J., Gough, K. L. and Campbell, S. D. (2014) Refining low-volume, high-concentration herbicide applications to control *Chromolaena odorata* (L.) King & Robinson (Siam weed) in remote areas. Plant Protection Quarterly, 29 (2). pp. 71-77. ISSN 0815-2195

Campbell, S. D., McMillan, H., Brazier, D. A., Setter, M. J., & Setter, S. D. (2019). Advancing splatter gun technology for rangeland weeds. In 1st Queensland Pest Animal and Weed Symposium.

Biosecurity Queensland, 2016. Restricted invasive plant fact sheet: Gamba grass, *Andropogon gayanus.* https://www.daf.qld.gov.au/__data/assets/pdf_file/0011/67466/IPA-Gamba-Grass-PP147.pdf. Accessed 22 February 2020.

Csurhes, S., Hannan-Jones, M., 2016. Invasive plant risk assessment: Gamba grass, *Andropogon gayanus*. Biosecurity Queensland, Department of Agriculture and Fisheries, Queensland Government, Brisbane.

Queensland Government, 2020. SILO. https://www.longpaddock.qld.gov.au/Queensland Government/. Accessed 21 February 2020.

Rossiter, N. A., Setterfield, S. A., Douglas, M. M., & Hutley, L. B. (2003). Testing the grass-fire cycle: alien grass invasion in the tropical savannas of northern Australia. Diversity and distributions, 9(3), 169-176.

Rossiter-Rachor, N.A., Setterfield, S.A., Douglas, M.M., Hutley, L.B., Cook, G.D. and Schmidt, S. (2009), Invasive *Andropogon gayanus* (gamba grass) is an ecosystem transformer of nitrogen relations in Australian savanna. Ecological Applications, 19: 1546-1560. https://doi.org/10.1890/08-0265.1

Setterfield, S., Douglas, M., Hutley, L., Welch, M., 2005. Effects of canopy cover and ground disturbance on establishment of an invasive grass in an Australia savanna. Biotropica 37, 25–31.

Setterfield, S., Rossiter-Rachor, N., Douglas, M., Wainger, L., Petty, A., Barrow, P., Shepherd, I., Ferdinands, K., 2013. Adding fuel to the fire: The impacts of non-native grass invasion on fire management at a regional scale. PLoS ONE 8, e59144.

Setterfield, S. A., N. A. Rossiter-Rachor, M. M. Douglas, D. McMaster, V. M. Adams, and K. Ferdinands. 2014. The impacts of *Andropogon gayanus* (gamba grass) invasion on the fire danger index and fire management at a landscape scale. Pages 125– 128 in 19th Australasian Weeds Conference. Weed Society of Tasmania, Hobart, Tasmania, Australia.

Somerville, S., Somerville, W., & Coyle, R. (2011). Regenerating native forest using splatter gun techniques to remove lantana. Ecological Management & Restoration, 12(3), 164-174.

# The advancing front of invasive lovegrasses across Australia's rangelands

Shane Blakely[1], Marie Vitelli[2], Maree Tully[3], Ann-Maree Johnson[2] and Jaime Colley[2]
[1]Precision Beef, PO Box 1910, Toowoomba, Queensland 4350, Australia.shane@rpsys.com.au
[2]AgForce, PO Box 13186, Northbank Plaza, George St, Brisbane, Queensland 4003, Australia
[3]Mitchell and District Landcare Association, 48 Cambridge St, Mitchell, Queensland 4465, Australia.
(vitellim@agforceqld.org.au)

**Summary** Four species of invasive *Eragrostis* lovegrasses threaten the integrity of Australia's rangelands and reduce pasture production for livestock. Concerned rangeland graziers are calling for increased national awareness, development of useful identification guides and further adaptive research into cost-effective management options for invasive lovegrasses.

**Keywords** *Eragrostis*, lovegrass, rangelands.

## INTRODUCTION

Rangeland graziers across northern, inland Australia are calling on four species of invasive lovegrasses to be considered as priority national weeds to help achieve effective prevention and management.

The invasive lovegrass complex of four introduced species has invaded vast pastoral areas of Australian rangelands (Figure 1). African lovegrass, *Eragrostis curvula* (Schrad.) Nees is well documented throughout southern and eastern Australia as an unpalatable, unproductive, invasive, introduced grass species, which is resistant to control measures. Trichophora, *Eragrostis trichophora* Coss. & Durieu, is a related introduced species, first identified in Australia in 1971, which has progressively invaded pastures across southern inland and western Queensland, particularly in the Western Downs and Maranoa districts. The third invasive species is soft lovegrass, *Eragrostis pilosa* (L.) P.Beauv., which occurs across most Australian states, including inland Queensland. Stink grass, *Eragrostis cilianensis* (All.) Vignolo ex Janch., has invaded all Australian states, since the early 1900's. These four introduced grass species are environmental weeds which thrive on low phosphorus, sandy soils with low ground cover or after ground disturbance. Major risk pathways include invasion from roadsides and vehicles into adjacent paddocks.

Once these weedy grasses become established, control options are very limited, and eradication is virtually impossible. High density grazing of young, immature plants, ploughing, mulching or slashing are partially effective at managing infestations. The only herbicide options in pasture situations are non-selective glyphosate and/or flupropanate. Flupropanate has a 14 day to four month grazing withholding period, depending if spot or broadacre applied. Herbicide resistance to flupropanate has been confirmed in African lovegrass from the Southern Tablelands of New South Wales, Australia (Powells 2022).

Rangeland graziers are calling for four actions:-

1. The four invasive lovegrass species are assessed under the new National Established Weed Priorities NEWP framework and all declared as Restricted Biosecurity Matter at state government levels.
2. Initiate research into ecology of trichophora lovegrass to refine prevention and management measures, especially under extensive pastoral conditions, drought impacts and drawing on landowner experiences.
3. Local governments consider listing the complex as a declared pest under local law, include management measures in biosecurity plans and roadside maintenance.
4. As per General Biosecurity Obligation requirements for all biosecurity matter, awareness and education initiatives are developed for all land users and farm visitors outlining control measures and actions to mitigate risk of lovegrass seed spread.

Farmers and graziers are faced with developing and implementing their own farm biosecurity protocols for invasive weeds that are not declared locally, state-wide or recognised nationally. Issues often arise when managing non-declared weeds near boundaries with transport corridors, conservation areas and floodplains. For example, active communication and voluntary collaboration are required with local government and contractors to implement equipment clean down when slashing roadsides infested with a non-declared weed. Co-existing land users such as resource companies, contractors, utility providers, small miners and tourists also pose biosecurity risks, especially when there is no regulatory requirement for them to manage non-declared weeds.

## HISTORY

### African lovegrass

African lovegrass was introduced from south Africa to Australia during the late 1800s or early 1900s, probably accidently as a contaminant of other pasture seeds. Subsequently, other types or cultivars were introduced up until the 1980's for pasture trials and environmental remediation (Parsons 2001). A more leafy and less weedy cultivar is 'Consol' which is used in parts of NSW. African lovegrass is a pernnial, drought tolerant tussock grass, which favours sandy soils and is a prolific seeder. It is widespread across eastern Australia and readily establishes along roadsides, disturbed soils and is a major weed of remnant, native grass regional ecosystems (Weeds Australia 2021).

Meat and Livestock Australia (2009) developed the 3D weed management best practice manual and documented four NSW producer case studies.

### Trichophora lovegrass

The literature for trichophora lovegrass is scant. It is also native to southern Africa (Hosking *et al.* 2007) and is morphologically similar to African lovegrass. The first record from Australia is from Alice Springs in 1971 (Friedel 2020), as a possible contaminant of African lovegrass pasture seed.

Australian Virtual Herbarium records (2021) indicate the distribution of African and trichophora lovegrasses overlaps in the south-east quadrant of Queensland and northern parts of New South Wales.

### Soft lovegrass

An annual environmental weed that is common across eastern Australia and north west Western Australia. It prefers disturbed areas and roadsides, with no shade.

### Stink grass

A common annual environmental weed throughout Australia invading disturbed areas, pastures, roadsides and arid wetlands. The grass produces a distinct odour when wet. Mature flowering plants are not palatable to livestock. Stink grass is native to north Africa and the Mediterranean.

## INVASIVE LOVEGRASS WEEDINESS

There are numerous examples in the literature citing African lovegrass as a pasture weed (Fensham 1998, Batianoff and Butler 2002).

Although not in the top 20, it was one of 71 weed species nominated by state and territory governments for assessment as Weeds of National Significance (WONS) and still remains a weed of potential national significance (Weeds Australia 2021).

African lovegrass is a declared weed in the ACT, New South Wales, South Australia, Tasmania and Victoria (Weeds Australia 2021) and is listed as a weed in Chile, South Africa, Lebanon, Colombia, parts of the United States, New Zealand and Japan (Csurhes *et al.* 2016). Although not declared in Queensland, Reardon-Smith (2009) identified African lovegrass as a significant risk to the Condamine catchment. Firn (2009) called for urgent action to reduce further spread across Australia.

Twelve years down the track and there has been no holistic, multi-stakeholder action to include invasive lovegrass species into national action plans to protect Australia's rangelands. Local grazier and landholder knowledge is rarely embedded into ecological research to protect grassy woodlands (Firn *et al* 2018). In contrast, concerned environmental scientists listed five high biomass pasture grasses as threatening process to northern Australia's biodiversity in 2009, under the *Environmental Protection and Biodiversity Conservation Act 1999.* National coordination under the Australian Weeds Strategy and the Threat Abatement Plan enables awareness and cost-effective management strategies to be developed for the five listed grass species. The five listed species are gamba grass *Andropogon gayanus* Kunth, para grass *Urochloa mutica* (Forssk.) T.Q.Nguyen, olive hymenachne *Hymenachne amplexicaulis* (Rudge) Nees, mission grass *Cenchrus polystachios* (L.) Morrone, and annual mission grass *Cenchrus pedicellatus* (Trin.) Morrone.

Rangeland graziers are concerned about the current and increasing impacts and costs of invasive lovegrasses across inland Australia. This has been the trigger for grass-roots action and requested solutions from graziers across southern inland Queensland and western New South Wales. There is value in sharing grazier knowledge and experiences in developing cost-effective integrated management strategies.

## FUTURE MANAGEMENT

National awareness about the impact of invasive lovegrasses needs to increase. Ideally, the four invasive lovegrasses are assessed for declaration as Weeds of National Significance or Weed Issues of National Significance under the new National Established Weeds Priority Framework.

Resourcing is required to prepare and publish a practical ute guide for identifying invasive lovegrasses, including how to differentiate from beneficial native lovegrass species which are useful fodder. Of the 40 common native *Eragrostis* species, the AusGrass2 database (2015) lists four useful fodder species. These are *E. lacunaria* F.Muell. ex Benth., *E. laniflora* Benth., *E. setifolia* Nees and *E.*

*xerophila* Domin. The emerging technology of DNA taxonomic identification needs to be applied to the native and introduced *Eragrostis* species complex.

Best management strategies require collation and sharing amongst land managers. Further research into alternative herbicide options to glyphosate and flupropanate is required, especially since herbicide resistance is confirmed within populations of African lovegrass across New South Wales. The return on investment for agricultural chemical registrants to invest into Australian pasture weed herbicide research is low. Graziers hold grave concerns about future effective herbicide options, with less entities and research organisations investing in herbicide and integrated management research across Australia's rangelands.

Most importantly, in the current absence of regulated weed declaration processes and national weed status, this is a call from rangeland graziers to weed science networks for collaboration across multiple land users, agribusinesses and organisations to prevent new infestations and manage existing fronts of invasive lovegrass outbreaks which threaten Australia's rangelands and pasture productivity.

## REFERENCES

AusGrass2 Grasses of Australia (2015) https://ausgrass2.myspecies.info/category/australian-grasses/australian-grasses/eragrostis (accessed 23 June 2022).

Batianoff G.N and Butler D.W. (2002). Assessment of invasive naturalized plants in south-east Queensland. *Plant Protection Quarterly* 17(1), 27-34.

Csurhes, S., Leigh, C. and Walton, C. (2016). Invasive plant risk assessment: African lovegrass. Queensland Department of Agriculture and Fisheries.

Fensham, R.J. (1998). The grassy vegetation of the Darling Downs, south-eastern Queensland, Australia. Floristics and grazing effects. *Biological Conservation* 84(3), 301–10. (DOI 10.1016/S0006-3207(97)00105-5).

Firn, J. (2009). African lovegrass in Australia: a valuable pasture species or embarrassing invader? *Tropical Grasslands* 43, 86-97.

Firn, J., Ladouceur, E. and Dorrough, J. (2018). Integrating local knowledge and research to refine the management of an invasive non-native grass in critically endangered grassy woodlands. *Journal of Applied Ecology* 55, 321–30 (DOI 10.1111/1365-2664.12928).

Friedel, M.H. (2020). Unwelcome guests: a selective history of weed introductions to arid and semi-arid Australia. *Australian Journal of Botany*, 68(2), 75–99. (DOI doi.org/10.1071/BT20030).

Hosking, J.R., Conn B.J., Lepschi, B.J.. and Barker, C.H. (2007). Plant species first recognised as naturalised for New South Wales in 2002 and 2003, with additional comments on species recognised as naturalised in 2000–2001. *Cunninghamia* 10(1), 139-66. https://www.biodiversitylibrary.org/item/269996#page/143/mode/1up (accessed 3 May 2022).

Meat and Livestock Australia. (2009). '3D weed management African lovegrass' and 'African lovegrass case studies' https://publications.mla.com.au/login/redirectFrame (accessed 3 May 2022).

Parsons, W. T. and Cuthbertson, E.G. (2001). 'Noxious weeds of Australia'. (CSIRO Publishing, Collingwood).

Powells, J. (2022). Herbicide resistance in perennial pasture systems – The ostrich has bolted. Proceedings of the 20th Australian Agronomy Conference, Toowoomba, Qld. ed. Bell, L. http://agronomyaustraliaproceedings.org/images/sampledata/2022/Weeds/ASApowells_j_510s.pdf. (accessed 12 May 2022).

Weeds Australia – Profiles (2021). *Eragrostis curvula* (Schrad.) Nees. https://profiles.ala.org.au/opus/weeds-australia/profile/Eragrostis%20curvula (accessed 12 May 2022).

**Figure 1**. Australian distribution of four invasive lovegrass species (2022).
Source: The Australasian Virtual Herbarium (https://avh.ala.org.au/).

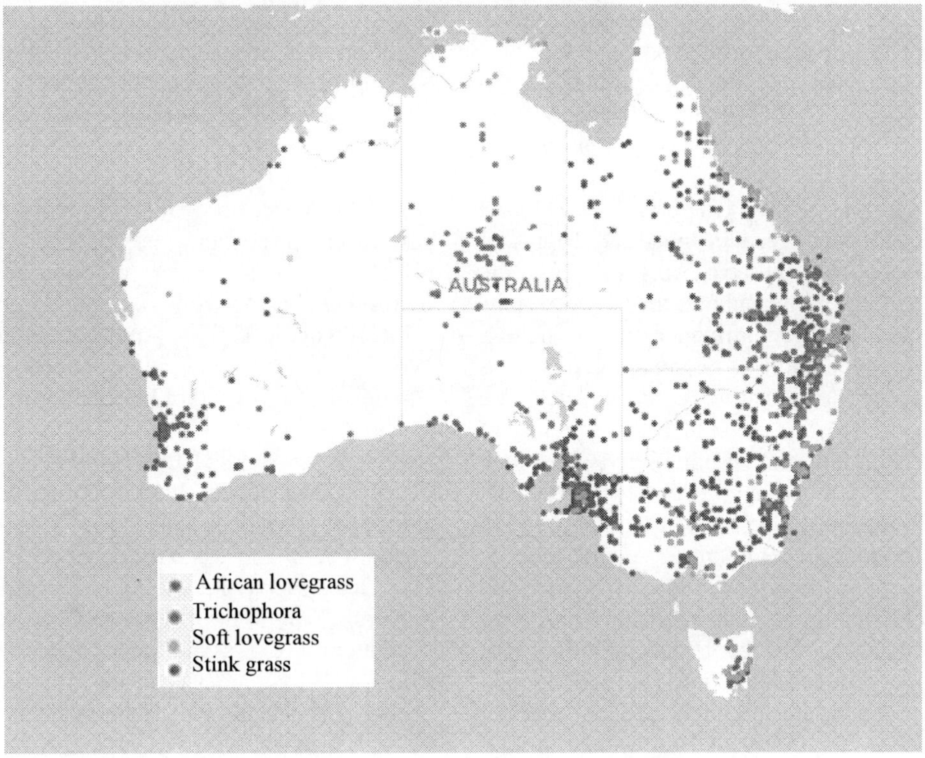

# Evaluation of Di-Bak® Herbicide Capsule System for Control of Chinee Apple (*Ziziphus mauritiana*) in North Queensland

Ciara J. O'Brien[1], Shane Campbell[1], Wayne Vogler[2] and Victor J. Galea[1]
[1] The University of Queensland, Gatton Campus, Gatton, Queensland, 4343, Australia
[2] Department of Agriculture and Fisheries, Tropical Weeds Research Centre, PO Box 976, Charters Towers, Queensland, 4820, Australia
(ciara.obrien@uqconnect.edu.au)

**Summary** Chinee apple (*Ziziphus mauritiana* Lam.) is a significant weed in the drier tropics of northern Queensland, Western Australia, and the Northern Territory. Throughout these regions its densely formed thickets influence the structure, function, and composition of rangeland ecosystems thereby outcompeting the native pasture species. The subsequent loss of pasture cover affects the quality of services (i.e., agronomic productivity, livestock carrying capacity, mustering, ecosystem services) obtainable from this diverse, natural resource. In Australia, the management of *Z. mauritiana* is limited to the application of synthetic herbicides and mechanical clearing operations. Whilst their efficacy is undisputed, there are concerns regarding the suitability of synthetic herbicides in ecologically sensitive or low-value habitats. This greater appreciation for environmental stewardship has promoted significant developments in the field of woody weed management.

This study investigates the effectiveness of a novel stem-implantation system for controlling woody weed species in grassland and rangeland environments. A pair of replicated trials were established among a naturally occurring population of *Z. mauritiana* at Alligator Creek, North Queensland (rural locality south of Townsville). The trials differed in dosage level by adjusting the application spacing (10 cm and 15 cm) of four encapsulated synthetic herbicides. An untreated control and benchmark treatment (drill-and-fill application of Tordon® RegrowthMaster) were also assessed for performance comparison. A significant effect ($p < 0.05$) on plant vigor was discerned for both dosage levels within eight months of trial establishment. The highest incidence of mortality was observed among the individuals treated with aminopyralid + metsulfuron-methyl (37.5 mg/capsule and 30 mg/capsule), metsulfuron-methyl (330 mg/capsule) and picloram (1000 mg/capsule), achieving a similar response to the drill-and-fill application of Tordon® RegrowthMaster (200 g/L triclopyr, 100 g/L picloram and 25 g/L aminopyralid). We predict total mortality (~100%) with these preeminent treatments by the next assessment period (15 months), as well as equally effective control under a reduced dosage (by increasing spacing from 10 cm to 15 cm).

**Keywords** Chinee Apple, *Ziziphus mauritiana*, Woody Weed, Integrated Weed Management, Chemical Control, Stem Implantation

## INTRODUCTION

*Ziziphus mauritiana* Lam. (Chinee Apple, Indian Jujube, Ber) is a deciduous thorny tree or shrub native to south Asia and eastern Africa (Grice 1996, Grice *et al*. 2000, Bebawi *et al*. 2016, Liang *et al.* 2019). In Australia, it was introduced to early mining settlements (e.g., Charters Towers, Ravenswood, Mingela, Hughenden) during the late nineteenth century (1863) for its ornamental and horticultural value (Grice 1996, Grice *et al.* 2000, Bebawi *et al.* 2016). Its current distribution is densest in the northern parts of Queensland (Townsville-Charters Tower region), Western Australia (Northern Kimberley, Pilbara, Dampierland) and the Northern Territory (near Katherine) (Grice 1996, Bebawi *et al.* 2016). Throughout these regions, its densely formed thickets can alter the structure and ecological integrity of rangeland ecosystems by outcompeting native pasture species (Grice 2004, Bebawi *et al.* 2016). This affects the quality of services (agronomic productivity, livestock carrying capacity, water accessibility, mustering) obtainable from this diverse, natural resource (Bebawi *et al.* 2016, Dhileepan 2017, Ani *et al*. 2018).

The management of *Z. mauritiana* is limited to the application of synthetic compounds or mechanical clearing operations (Bebawi *et al.* 2016, Dhileepan 2017). The manual removal of higher density (>150 plants/ha) or isolated infestations can be achieved through stick-raking, blade ploughing or bulldozing of individual trees (terrain and soil type permitting). However, these attempts are often deemed inefficient and cost prohibitive. The basal or cut-stump application of synthetic auxin herbicides (triclopyr, fluroxypyr or picloram carried in diesel) is the most effective approach for the aggressive eradication of lower density (<50 plants/ha) populations. Whilst their efficacy is undisputed, there are concerns regarding the suitability of synthetic herbicides applied with diesel in ecologically sensitive (i.e., riparian zones, woodlands) or low-value habitats (O'Brien *et al.* 2022). Their excessive or imprecise application may result in non-target damage through herbicidal and diesel drift, runoff or leaching into adjacent habitats (O'Brien *et al.* 2022). This greater appreciation for environmental stewardship has promoted significant developments in the field of woody weed management by reducing dosage levels or improving application methods (O'Brien *et al.* 2022).

This study investigates the effectiveness of BioHerbicides Australia's (BHA Pty Ltd) proprietary stem-implantation system and Di-Bak® range of synthetic herbicides for controlling *Z. mauritiana* in rangeland environments. This novel technology was initially developed for the encapsulated delivery of an endophytic bioherbicide in parkinsonia (*Parkinsonia aculeata* L.) (Galea 2021a, b). It has since been expanded to the application of other endophytic organisms, as well as synthetic compounds (herbicides, fungicides, and insecticides) available in dry formulations (Goulter *et al.* 2018, Galea 2021a, Limbongan *et al.* 2021). Unlike its industry counterparts, this device provides a targeted, readily calibrated herbicide application thereby minimising environmental and operator exposure (Goulter *et al.* 2018, Galea 2021a, Limbongan *et al.* 2021, O'Brien *et al.* 2022).

## MATERIALS AND METHODS

**Experimental Design** A pair of replicated trials were established (23 to 25 February, 2021) among a naturally occurring population of *Z. mauritiana* at a cattle property in Alligator Creek, North Queensland (rural locality near the south of Townsville) (19°24'10"S 146°56'24"E). These parallel trials involved the mapping, measurement, and treatment of individual plants with four encapsulated synthetic herbicides sourced from BioHerbicides Australia's (BHA Pty Ltd) Di-Bak® range of registered and developmental products (Table 1). An untreated control and benchmark treatment (drill-and-fill application of Tordon® RegrowthMaster) were also assessed for performance comparison.

The first trial investigated a commercially recommended dosage of one capsule for up to (<) every 10 cm in stem circumference (Goulter et al. 2018, Galea 2021b) by following a randomised complete block design (RCBD) with four blocks. Within each block, the six treatments were randomly assigned to a group of fifteen plants (uniform age and growth). All plants were appropriately tagged, and GPS waypoints recorded for ease of re-location.

**Table 1.** Active constituents (g/kg) and dosage (mg/capsule) of registered and developmental encapsulated synthetic herbicides sourced from BioHerbicides Australia (BHA Pty Ltd).

| | Treatment | Active Constituents (g/kg) | Dose (mg/ capsule) |
|---|---|---|---|
| 1 | Glyphosate | 700 | 261 |
| 2 | Aminopyralid + | 375 | 37.5 |
| | Metsulfuron-Methyl | 300 | 30 |
| 3 | Metsulfuron-Methyl | 600 | 330 |
| 4 | Picloram | 100 | 1000 |

The population dynamics of this trial site provided an opportunity to assess the error tolerance (i.e., resilience to human error) of the stem-implantation system by lowering the dosage level (one capsule per 15 cm in stem circumference). This second trial adopted a similar experimental design with three blocks. The same untreated control and benchmark treatment (drill-and-fill application of Tordon® RegrowthMaster) was evaluated given its proximity to the first trial.

**Treatment Application** The synthetic capsules were delivered via the InJecta® handheld device developed by BioHerbicides Australia (BHA Pty Ltd). The mechanics of this unit are described in-depth by Limbongan *et al*. 2021.

The drill-and-fill application of Tordon® RegrowthMaster (200 g/L triclopyr, 100 g/L picloram and 25 g/L aminopyralid) was achieved with a calibrated drenching syringe (NJ Phillips 5 mL Metal Tree Injector). A series of holes (10 cm spacing) were drilled (10 mm drill diameter) into the sapwood (15 mm to 20 mm depth) of the woody stem at a slight downward angle (45°). The label recommended dose (1 mL at 1:4 dilution with water) of herbicide solution was injected into the drilled holes.

**Trial Assessment** These trials were assessed at establishment (23 February 2021), three months (24 May 2021) and eight months (26 October 2021) by recording the overall vigour (1 = healthy, 2 = slightly distressed, 3 = moderately distressed, 4 = severely distressed, 5 = dead) of each plant. The 'stress score' was discerned by removing the outermost layer of the bark with a rasp, or with severely affected plants, bark removal with a claw hammer to reveal the colour of the tissue beneath (O'Brien *et al.* 2022). A plant was deemed 'dead' if the stem tissue was dry, brown, or bleached. Whilst, there was underlying moisture and colour (often dull, pale green) observed in the stem tissue of moderately or severely distressed plants. Additionally, an auditory assessment of the internal hydration (i.e., vascular fluids) was conducted by hammering the primary stem to confirm mortality (O'Brien *et al.* 2022).

**Data Analysis** The treatment effects on 'stress score' were analysed in RStudio® (RStudio Inc, Boston, Massachusetts, United States). A one-way analysis of variance (ANOVA) was performed by taking the mean (μ) of each replication. All pairwise comparisons among treatment means (μ) were estimated with the emmeans (estimated marginal means, also known as least-squares means) package. The compact letter displays (CLD) were corrected with Tukey's Honest Significant Difference post-hoc test.

## RESULTS AND DISCUSSION

A significant effect ($p < 0.05$) on plant vigour was discerned for both dosage levels within eight months (i.e., 35 weeks) of trial establishment (Table 2). The highest incidence of mortality was observed among the individuals treated with aminopyralid and metsulfuron-methyl (37.5 mg/capsule and 30 mg/capsule), metsulfuron-methyl (330 mg/capsule) and picloram (1000 mg/capsule), achieving a similar deterioration in plant health to the 'drill-and-fill' application of Tordon® RegrowthMaster (200 g/L triclopyr, 100 g/L picloram and 25 g/L aminopyralid) (Table 2, Table 3). This is evidenced by their rapidly increasing 'stress score' values whereby most plants (>85%) were deemed 'dead' (stress score of five) at thirty-five weeks (Table 3). However, there was no significant difference ($p > 0.05$) individually between these four treatments (Table 2) or under different dosage levels (i.e., treatment-trial interaction). This suggests that the recommended dosage has an error tolerance of at least (≤) 5 cm per capsule for the respective synthetic compounds. We predict total mortality (100%) with these preeminent treatments by the next assessment period (15 months).

The performance of glyphosate (261 mg/capsule) is underwhelming relative to the other synthetic treatments. The reduction in plant vigour was mostly 'moderate' (stress score of three) for both dosage levels (Table 2). The untreated plants (i.e., control) were also slightly distressed at thirty-five weeks (Table 2). However, a degree of seasonal defoliation is expected given the deciduous nature of this species due to moisture stress.

**Table 2.** The *p*-value, mean (μ) stress score (SS) and standard error (SE) for trial one (10 cm spacing) and two (15 cm spacing) at each assessment period (week 0, 13. 35). The superscript letters (i.e., compact letter displays) denote all pairwise comparisons among treatment means (μ).

| | Week | | | | | |
|---|---|---|---|---|---|---|
| | 0 | | 13 | | 35 | |
| | Trial 1 | | | | | |
| *p*-Value | 0.451 | | $9.24\times10^{-10}$ | | $2.45\times10^{-09}$ | |
| | SS | SE | SS | SE | SS | SE |
| Control | 1.0[a] | 0 | 1.0[c] | 0 | 2.68[c] | 0.073 |
| Tordon® Regrowth Master | 1.0[a] | 0 | 3.92[a] | 0.043 | 4.80[a] | 0.062 |
| Glyphosate | 1.0[a] | 0 | 2.52[b] | 0.113 | 3.37[b] | 0.099 |
| Aminopyralid + Metsulfuron-Methyl | 1.0[a] | 0 | 3.67[a] | 0.061 | 4.98[a] | 0.017 |
| Metsulfuron-Methyl | 1.0[a] | 0 | 3.63[a] | 0.063 | 5.0[a] | 0 |
| Picloram | 1.0[a] | 0 | 3.88[a] | 0.041 | 4.97[a] | 0.023 |
| | Trial 2 | | | | | |
| *p*-Value | 0.465 | | $2.80\times10^{-10}$ | | $1.36\times10^{-08}$ | |
| Control | 1.0[a] | 0 | 1.0[c] | 0 | 2.62[c] | 0.092 |
| Tordon® Regrowth Master | 1.0[a] | 0 | 3.89[a] | 0.047 | 4.89[a] | 0.047 |
| Glyphosate | 1.0[a] | 0 | 2.22[b] | 0.077 | 3.38[b] | 0.143 |
| Aminopyralid + Metsulfuron-Methyl | 1.0[a] | 0 | 3.93[a] | 0.038 | 5.0[a] | 0 |
| Metsulfuron-Methyl | 1.0[a] | 0 | 3.98[a] | 0.022 | 5.0[a] | 0 |
| Picloram | 1.0[a] | 0 | 3.73[a] | 0.067 | 4.91[a] | 0.043 |

**Table 3.** The percentage (%) mortality among each treatment for trial one and two at the most recent assessment period (Week 35).

| Treatment | Percentage (%) Mortality | |
|---|---|---|
| | Trial 1 | Trial 2 |
| Control | 0.0 | 0.0 |
| Tordon® Regrowth Master | 83.33 | 88.89 |
| Glyphosate | 6.67 | 13.33 |
| Aminopyralid + Metsulfuron-Methyl | 98.33 | 100 |
| Metsulfuron Methyl | 100 | 100 |
| Picloram | 96.67 | 91.11 |

This novel technology has been proven successful for the management of other woody weed species: prickly acacia (*Vachellia nilotica*), leucaena (*Leucaena leucocephala*), calotrope (*Calotropis procera*), camphor laurel (*Cinnamomum camphora*), (Goulter *et al.* 2018), mimosa bush (*Vachellia farnesiana*) (Limbongan *et al.* 2021), Chinese elm (*Celtis sinensis*) (O'Brien *et al.* 2022, Galea 2021a). This study has demonstrated that the encapsulated delivery of synthetic compounds is also highly effective against *Z. mauritiana* in grassland and rangeland environments. Unlike its industry counterparts, a minimum recommended lethal dose is delivered directly into the vascular system of the target species whereby the active agent is fully captured internally (Goulter *et al.* 2018, Galea 2021a). This targeted, readily calibrated herbicide application has the potential to (i) lower active ingredient concentrations (~20% to 30%), (ii) minimise the likelihood of environmental exposure to plant protection compounds and (iii) improve operator safety (Galea 2021a). This technology is a possible replacement for foliar or stem spraying, stem-injection or canopy application (Galea 2021a)

Although not presented in this paper, these trials were also replicated in Mulgrave, Northern Queensland during the dry season (late-May 2021). This research is still underway, and the findings will be reported upon trial conclusion.

ACKNOWLEDGMENTS

The authors would like to thank Mr Wayne Krogh for provision of the study site. We thank Mr Simon Brooks and Ms Clare Warren for their assistance with trial assessment, as well as Dr Vincent Mellor for his guidance with the biometric analysis.

REFERENCES

Ani, O., Onu, O., Okoro, G. and Uguru, M. (2018). Overview of Biological Methods of Weed Control. In 'Biological Approaches for Controlling Weeds', eds R. Radhakrishnan, pp. 5-16. (InTechOpen, London, UK).

Bebawi, F.F., Campbell, S.D. and Mayer, R.J. (2016). Seed Bank Persistence and Germination of Chinee Apple (*Ziziphus mauritiana* Lam.). *The Rangeland Journal* 38, 17-25 (DOI 10.1071/RJ15104).

Dhileepan, K. (2017). Biological Control of *Ziziphus mauritiana* (Rhamnaceae): Feasibility, Prospective Agents and Research Gaps. *Annals of Applied Biology* 170, 287-300. (DOI 10.1111/aab.12338).

Grice, A.C. (1996). Seed Production, Dispersal and Germination in *Cryptostegia grandiflora* and *Ziziphus mauritiana*, Two Invasive Shrubs in Tropical Woodlands of Northern Australia. *Australian Journal of Ecology* 22, 49-55. (DOI 10.1111/j.1442-9993.1996.tb00615.x).

Grice, A.C., Radford, I.J. and Abbott, B.N. (2000). Regional and Landscape-Scale Patterns of Shrub Invasion in Tropical Savannas. *Biological Invasions* 2, 187-205.

Grice, A.C. (2004). Weeds and the Monitoring of Biodiversity in Australian Rangelands. *Austral Ecology* 29, 51-58. (DOI 10.1111/j.1442-9993.2004.01364.x).

Galea, V.J. (2021). Stem Implanted Capsules – A Target, Efficient and Sustainable Method for Delivery of Synthetic and Biological Agents to Plantation Crops. Proceedings of the 2nd International Plantation Conference (Universiti Teknologi MARA, Selangor, Malaysia).

Galea, V.J. (2021). Use of Stem Implanted Bioherbicide Capsules to Manage an Infestation of *Parkinsonia aculeata* in Northern Australia. *Plants* 10 (DOI 10.3390/plants10091909).

Goulter, K.C., Galea, V.J. and Riikonen, P. (2018). Encapsulated Dry Herbicides: A Novel Approach for Control of Trees. Proceedings of the 21st Australasian Weeds Conference, eds S. Johnson, L. Weston, H. Wu and B. Auld, pp. 247-250. (The Weed Society of New South Wales Inc., Sydney).

Limbongan, A.A., Campbell, S.D. and Galea, V.J. (2021). Novel Encapsulated Herbicide Delivery Mechanism: Its Efficacy in Mimosa Bush (*Vachellia farnesiana*) Control. *Plants* 10 (DOI 0.3390/plants10112505).

O'Brien, C.J., Mellor, V. and Galea, V.J. (2022). Controlling Woody Weed Chinese Elm (*Celtis sinensis* Pers.) with Stem-Implanted Herbicide Capsules. *Plants* 11 (DOI 10.3390/plants11030444).

# Monitoring *Striga asiatica* (Orobanchaceae) seedbank for eradication success

Anna M. Williams[1] Natasha Riding[1] and Joseph S. Vitelli[1]
[1]Biosecurity Queensland, Dutton Park, QLD, Australia
(anna.williams@daf.qld.gov.au)

**Summary** To enable evaluation of eradication efforts for the parasitic plant *Striga asiatica* (L.) Kuntze, seed sachets were buried in perforated PVC canisters at 25 sites across the infested area in July 2016. The canister sites were selected based on the proximity to previous *S. asiatica* detections and cover the range of soil types and topographies associated within the eradication program treatment area. At each site the canisters received the same eradication treatments that were applied to the surrounding area. Site characteristics necessitated different combinations of false host (soybean), true host (corn), fumigant (dazomet), stimulant (ethylene) and herbicide treatments, to achieve a rapid decline in the soil seed bank.

The viability of retrieved seeds shows an accelerated decline (92.5 % to 6.5 % in four years) across all sites in response to the treatments irrespective of treatment combination used. The viability of seeds was found to vary across sites, burial depth and time. The information collected has provided valuable data for the eradication of *S. asiatica* in Australia and for evaluating the timing for release of paddocks from active treatment.

**Keywords** eradication, red witchweed, RWW, *Striga asiatica*, quantitative performance measures, persistent seedbank.

## INTRODUCTION

*Striga asiatica* (red witchweed (RWW)) was first recorded in Australia in 2013 and by 2015 the Red Witchweed Eradication Program (RWWEP) was declared (Smith *et al.* 2019). RWW is an annual obligate root hemiparasite, dependent on attachment to a host plant to complete its lifecycle (Joel *et al.* 2013). Host plants for RWW include several important crop species such as rice (*Oryza sativa* L.), sugarcane (*Saccharum officinarum* L.), sorghum (*Sorgum bicolor* (L.) Moench) and corn (*Zea mays* L.) (Shaw *et al.* 1962). Wheat (*Triticum aestivum* L.) and barley (*Hordeum vulgare* L.) have also been confirmed as host plants at the Ecosciences Precinct, Brisbane (J.S. Vitelli, pers com 2017). Individual RWW plants can produce up to 500,000 seeds (Shaw *et al.* 1962). The dust-like seeds are approximately 250 µm long and 120 µm wide and can remain viable in a soil seed bank for 14 years under field conditions (Bebawi *et al.* 1984). Soil cores collected from an infested site at Habana, prior to the commencement of the eradication program, indicated the presence of a soil seed bank of approximately 1,000,000 / m$^2$ (J.S. Vitelli, pers com 2015).

Successful eradication of an invasive plant is dependent on delimitation, containment and extirpation (Panetta and Lawes 2005). Progress towards achieving the goal of eradication success in a weed eradication program is usually evaluated based on the presence or absence of plants detected during surveillance (Panetta and Lawes 2007). An ongoing decline in area and the number of detected plants, in particular adult reproductive plants, can be used to evaluate the effectiveness of different control strategies used (Panetta and Lawes 2007). However, for species with a persistent seedbank or where control methods vary in efficacy, the use of more quantitative performance measures may be required. This is the case for RWW which not only has long lived seed but the control methods available also vary in effectiveness depending on the location of the seed in the soil profile.

The size of RWW seed make traditional methods for monitoring soil seedbank decline extremely difficult. The methodology required to collect soil samples and separate the seeds from the soil is complicated and time consuming. An alternative approach to collecting an extremely large number of soil samples would be to bury seed in retrievable canisters located throughout the infested area. Seed viability data can then be relayed back within the eradication program and management regimes adjusted accordingly. A shortcoming of this method is that there is a finite number of retrievable seeds at any one site. Despite the efforts to predict the likely timeframe to achieve successful eradication and cover the potential permutations of treatments used, there is a risk that the final retrieval may contain viable seeds. Without additional canisters available for continued monitoring, only predictions can be made based on the already collected results.

If RWW was eradicated from Australia, it would be the first country in the world to achieve this. The United States of America has spent more than US$260 million over 66 years trying to eradicate RWW (Iverson *et al.* 2011), and in Africa it has been estimated that losses to farmers from *Striga* species amounts to US$7 billion annually (Berner *et al.* 1995). Eradication activities within the RWWEP center on regular surveillance for the presence of

RWW plants, optimization of control strategies that prevent the emergence and establishment of mature reproductive plants and treatments that target and accelerate the soil seedbank decline. A 100-point system has been implemented as the program moves from eradication to the release of previously infested areas from quarantine (Smith *et al.* 2019). All treated areas accumulate points according to both treatment activities and surveys employed. A total of 100 points is required for an area to become eligible for release from quarantine following the date of the last RWW detection (Smith *et al.* 2019).

This paper reports on the RWW soil seedbank viability over a 4-year eradication program across 25 sites known to be infested with RWW. Control efforts include the use of post-emergent herbicides, catch crops, trap crops and fumigants.

## MATERIALS AND METHODS

An eradication monitoring study was established across 25 sites in July 2016. The sites were spread over eight infested properties near Habana (21°3'48"S, 149°4'21"E), Mackay, Queensland and transected the active eradication management zone of the RWWEP. The 25 sites varied in soil type and topography (Table 1). Soil types included friable non-cracking clay and clay loam (Dermosol), cracking clay (Vertosol) and sand or loam over sodic clay (Sodosol) (Holz and Shields 1985). All sites were located within zones where RWW had previously been detected. A total of 300 canisters (12 per location) each containing three RWW seed sachets placed at 10, 30 and 50 cm depth were used in this study. Five sites amenable to a soil fumigant application contained a fourth sachet placed on the soil surface.

All RWW seeds used in this experiment were collected from a nursery stock of plants grown within a QC2 quarantine facility at the Ecosciences Precinct, Brisbane between November 2013 to June 2014. Seed pods were stored at low humidity at 35°C until sufficient stock had been collected. Only seeds > 150 μm were used to maximize the germination viability (92.5 %) of the seed cohort. Approximately 100 seeds were measured by volume using a miniature scoop. The measured seeds were photographed for counting, then transferred into 40 mm x 65 mm, 62 μm aperture, sachets constructed from precision woven polyamide (nylon) mesh tubes (SAATIFIL PA 62/40 PW WH) that were sealed using self-adhesive nylon tape (PSP Spinnaker repair Tape). The sachets were installed into canisters on site in the field.

The canisters were 550 mm in length and constructed from 100 mm PVC pipe perforated with 10 mm holes to create 20 to 25 % open space. The seed sachets were placed at 10, 30 and 50 cm depths within the canisters and filled with 10 mm sieved soil collected from the study sites. Sixty canisters had an additional seed sachet placed at the soil surface.

Canisters were buried in July 2016 and capped with a Mozzie Stoppa™ Original Screen at the soil surface to prevent loss of seed sachets. Eradication treatments commenced immediately.

At each site the canisters received the same eradication treatments that were applied during the

**Table 1:** Soil type and land uses of each study location site.

| *Site* | *P* | *T* | *Soil type* [a] | *Land use* |
|---|---|---|---|---|
| 1 | 1 | 1 | Dermosol | Cropping |
| 2 | 1 | 1 | Vertosol | Cropping |
| 3 | 1 | 1 | Dermosol | Cropping |
| 4 | 2 | 3 | Dermosol | Cropping |
| 5 | 2 | 3 | Dermosol | Grazing |
| 6 | 2 | 3 | Dermosol | Cropping |
| 7 | 3 | 1 | Dermosol | Cropping |
| 8 | 4 | 1 | Dermosol | Cropping |
| 9 | 4 | 5 | Sodosol | Cropping |
| 10 | 4 | 1 | Vertosol | Cropping |
| 11 | 4 | 5 | Vertosol | Cropping |
| 12 | 5 | 1 | Dermosol | Grazing |
| 13 | 5 | 1 | Dermosol | Grazing |
| 14 | 5 | 5 | Dermosol | Grazing |
| 15 | 8 | 1 | Vertosol | Cropping |
| 16 | 8 | 3 | Dermosol | Headland / Cropping |
| 17 | 34 | 3 | Dermosol | Cropping |
| 18 | 34 | 3 | Dermosol | Cropping |
| 19 | 80 | 5 | Vertosol | Cropping |
| 20 | 80 | 3 | Sodosol | Cropping |
| 21[b] | 2 | 3 | Dermosol | Fenceline / Grazing |
| 22[b] | 2 | 3 | Dermosol | Fenceline / Grazing |
| 23[b] | 4 | 1 | Dermosol | Fenceline / Cropping |
| 24[b] | 4 | 5 | Dermosol | Headland / Cropping |
| 25[b] | 5 | 5 | Dermosol | Fenceline / Grazing |

P = Property T = Topography

[a] Soil type and topography were determined from information from Holz and Shields (1985) held within the Queensland Soil and Land Information database. The topography rating reflects the undulation of the land from low (1) to severe (5). As undulation increases the crests and slopes are steeper and more complex, gullies are narrower and more frequent and farming land is more fragmented and less accessible.

[b] Sites with surface seed sachets included.

management regime. Treatment selection included soybean (*Glycine max* (L.) Merr.) sown at 50 t ha$^{-1}$ as a false host crop; corn (*Zea mays* L.) sown at 30 t ha$^{-1}$ as a sentinel true host crop; ethylene gas was used as a stimulant and injected into the soil using a tractor mounted with a custom built injection system pulling a tyne through the soil and releasing ethylene at a rate of 2.0 kg ha$^{-1}$ (1675 L ha$^{-1}$) applied at a depth of 15 to 30 cm and the fumigant dazomet (Basamid®) applied at 330 kg a.i. ha$^{-1}$ to bare ground along headlands and fence lines. A range of herbicides (glyphosate, haloxyfop, imazapic and halsulfuron) were also applied to maintain weed control and minimise the presence of unwanted host plants. The combination and frequency of treatment application varied both over time and between sites.

Sampling occurred on eight occasions between November 2016 and November 2020. At each sampling event only 4 to 13 sites were randomly selected for canister exhumation with a maximum of six canisters removed at any one site. The remaining six canisters were left in-situ until the treated areas had reached the desired 100 points, as determined by the RWW program point system (Smith *et al.* 2019). At which point three canisters would be removed at these sites and RWW seed viability tested to help determine if these areas can progress from active eradication treatment to a monitoring phase.

All retrieved sachets were stored at low humidity at 35°C until processed. At processing, each sachet was washed to remove exterior dirt. The contents of the sachet were then flushed with RO water onto a Whatman® number 1 filter paper. The opened sachets were checked under a microscope for remaining seeds and seed coats. All seeds and seed coats were counted and intact seeds tested for J Ecol. viability. Viability testing was performed using 1 % 2,3,5-triphenyl tetrazolium chloride (TTC) as per Moore (1976). Seeds were immersed in 80 μl TTC and stored in the dark at 35°C for seven to ten days. Embryos turning red to pink in colour were considered viable.

## RESULTS

The treatments applied during the 4-year period significantly reduced the seed viability across all sites. The viability of seeds started at 92.5 % at the time of field burial and declined to 3.5, 8.9 and 7.2 % at 10, 30 and 50 cm depths respectively by 2020 (Figure 1). No surface sachets were present in canisters retrieved in 2020.

Differences in the combination of eradication treatments applied, seed sachet burial depth and individual site conditions contributed to the variability observed for the seed viability decline. For example, in 2016, the viability of surface seed declined to 8.8 % after a single application of dazomet, while in the same canisters, buried seed viability was 19.5 %, 38.2 % and 46 % at 10, 30 and 50 cm. The viability of seeds collected at the same retrieval event from sites not treated with dazomet was 47.7 %, 49.8 % and 50.1 % at 10, 30 and 50 cm.

Viability differed across seed burial depth. The viability of seeds from sites 13 and 20 decreased more rapidly at 10 cm (39.8 %) than 50 cm (55.8 %), whereas the seed viability from sites 2, 11 and 19 was lowest at 50 cm (20.8 %) and remained higher at 10 cm (48.5 %).

Irrespective of burial depth, seed buried at site 10 declined most rapidly, dropping to 8.8 % viability in 2017. Seeds at site 5 retained the highest viability at 56.7 %, 40 % and 52.8 % at 10, 30 and 50 cm in 2018.

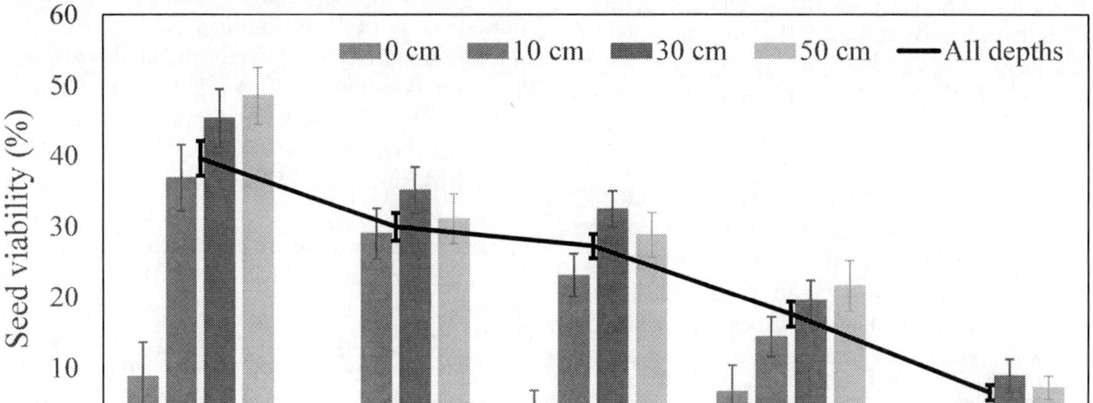


**Figure 1.** Effect of ongoing paddock applied treatments on seed viability at different seed burial depth. Each sample is the mean seed viability from all seed sachets retrieved during the year (standard error bars shown).

## DISCUSSION

Seed viability declined with time as the number, frequency and range of treatments were applied across all 25 sites. Despite seed viability declining to 6.5 % viability across all sites within 4 years, this value could potentially equate to 65,000 seeds / $m^2$ remaining in core RWW infested areas.

Differences in soil type and topography appear to strongly influence seed longevity. Heavy clay soil combined with prolonged wet conditions are likely to have caused the accelerated seed bank decline (20.8 %) observed at 50 cm depth for sites 2, 11 and 19. Whereas at site 5 the highest retained viability (49.8 %) across all depths was likely attributed to its location on the side of an eastern facing, gently sloping paddock, with good drainage and friable clay loam soil.

The current data indicates that an accelerated seedbank decline of *Striga asiatica* in Australia could be achievable within five to eight years compared with the expected timeframe of $14^+$ years for natural attrition alone. Successful extirpation of the seed bank depends on preventing seed recruitment. This is especially true for short lived annual species such as *S. asiatica* which can reach reproductive maturity within two to four weeks following emergence. If mature plants are detected, the timely intervention of treatments to prevent incorporation of new seeds into the soil profile is essential. Fortunately, for the RWWEP a relatively straightforward method is available for treating newly emerged RWW plants that is both effective at destroying the plant and any seed present on the soil surface. In these instances, the immediate area (2 m x 2 m) is incinerated and then followed with a tarped dazomet treatment. The scheduled frequency of surveillance of every 5–10 days undertaken within the Eradication Program (Smith *et al.* 2019) also minimises the possibility that seeds have time to reach maturity.

Determining when an area is free of RWW based on the number of treatments applied can be extremely challenging as evident in the USA where an eradication program in North and South Carolina was established in 1956. Despite its success in reducing the quarantined area from 175,000 ha to approximately 445 ha on the Carolina Coastal Plain the eradication program is still active after 66 years (Iverson *et al.* 2011). Currently, treatments and surveillance activities in the US determine an annual hand back of approximately 40 ha to land holders but there is also a return of a similar area back into the program as RWW is subsequently detected in released areas. It is estimated that it will take an additional 10–15 years to fully eradicate RWW from the USA (R.G. Westbrooks, pers com 2019).

The eradication monitoring data collected in this study has been critical for assessing progress towards the goal of eradication of RWW in Australia. Insights gained from monitoring seed bank decline has aided the decision for timely hand back of paddocks based on evidence of a declining seed bank. However, caution also needs to be exercised that the intrinsic variability of each site may not always lend itself to a hand back based solely on a scoring system reliant on the accumulation of treatments and surveys.

## ACKNOWLEDGMENTS

The RWWEP is a nationally cost shared response, with funding provided from government jurisdictions and industry. The authors thank the RWWEP field staff for their assistance with burial and retrieval of seed canisters and the patience and support of the infested landholders.

## REFERENCES

Bebawi, F.F., Eplee, R.E., Harris, C.E. and Norris, R.S. (1984). Longevity of witchweed (*Striga asiatica*) seed. *Weed Science* 32, 494–497.

Berner, D.K., Kling, J.G. and Singh, B.B. (1995). Striga Research and Control – A Perspective from Africa. *Plant Disease* 79, 652–660.

Holz, G.K. and Shields, P.G. (1985). 'Mackay Sugar Cane Land Suitablity Study Part 2. Land Suitability' (QLD DPI, Brisbane).

Iverson, R.D., Westbrooks, R.G., Eplee, R.E. and Tasker A.V. (2011) Overview and Status of the Witchweed (*Striga asiatica*) Eradication Program in the Carolinas. *In* 'Invasive Plant Management Issues and Challenges in the United States: 2011 Overview', eds A.R. Leslie, and R.G. Westbrooks, pp. 51 68. (American Chemical Society, Washington, DC).

Joel, D.M., Gressel, J., and Musselman, L.J. eds. (2013). 'Parasitic Orobanchaceae: Parasitic Mechanisms and Control Strategies' (Springer, Heidelberg).

Moore, R.P. (1976). Tetrazolium seed testing developments in North America. *Journal of Seed Technology* 1, 17–30.

Panetta, F.D. and Lawes, R. (2005). Evaluation of Weed Eradication Programs: the delimitation of extent. *Diversity and Distributions* 11, 435–442.

Panetta, F.D. and Lawes, R. (2007). Evaluation of the Australian Branched Broomrape (*Orobanche ramosa*) Eradication Program. *Weed Science* 55, 644–651.

Shaw, W.C., Shepherd, D.R., Robinson, E.L. and Sand, P.F. (1962). Advances in Witchweed Control. *Weeds* 10, 182–192.

Smith, M., Vitelli, J.S. and Bowditch, T. (2019). 'Red Witchweed Eradiction Response Plan 2019–2025'. (DAF, Brisbane).

# Interaction between wheat establishment timing and pre-emergent herbicide choice on growth and competition of annual ryegrass

Mike Ashworth[1], Roberto Lujan Rocha[1] and Hugh Beckie[1]

[1] Australian Herbicide Resistance Initiative, School of Agriculture and Environment, The University of Western Australia, Crawley, WA 6009, Australia.

**Summary:** Between 2018-2021, nine field trials were conducted across the Western Australian grain belt to assess the interaction between wheat time of sowing (either dry sowed with no pre sowing weed germination or delayed sowing with emerged weeds killed with glyphosate), wheat seeding rate and pre-emergent herbicide choice. With three new pre-emergent herbicides, cynmethalyn (Luximax®), bixalone (Overwatch®) and the mix of aclinofen-pyroxasulfone and diflufenican (Mateno Complete®) released in 2021, this report presents findings from the Dandaragan site in 2021. This study found that early (dry) sowing outyielded delayed sowing, however delayed sowing reduced annual ryegrass seed production. The seed production of annual ryegrass correlated with soil persistence of pre-emergent herbicide choice. This study measured that trifluralin(Triflur X®) and prosulfocarb + metolachlor (Boxer Gold®) degraded the fastest resulting in increased annual ryegrass seed production. Pyroxasulfone (Sakura®), Mateno Complete and Overwatch were highly residual resulting in reduced ryegrass seed production. Increasing wheat seeding rates consistently reduced annual ryegrass seed production. This study demonstrated the need to compensate for the lack of a pre-sowing knockdown herbicide application with increased crop competition and a more residual pre-emergent herbicide choice.

**Key words:** wheat, weeds, annual ryegrass, pre-emergent herbicides, time of seeding, crop competition, herbicide degradation

## INTRODUCTION

In the southern grainbelt of Australia, dry sowing has become popular as it enables growers to plant larger areas with limited machinery, within or before the optimum planting time to maximise yield potentials. At the same time, there has been an increased prevalence of grass weed populations with increased seed dormancy that emerge later to evade knockdown (glyphosate/paraquat) herbicide applications. To control these late emerging individuals there are several pre-emergent 'residual' herbicides that can be safely used within no tillage farming systems to provide an extended period of herbicidal activity. These herbicides are often applied directly to the soil prior to planting.

To control these late germinating populations, it has long been advised that growers should delay sowing of weedy paddocks to maximise the weed control effectiveness of knockdown applications. However, any delay in sowing results in a sharp decline in crop yield potential. Previously, dry sowing techniques have relied upon low weed seed banks as they place significant reliance on longevity and efficacy of soil applied herbicides that are often applied a long time before crop and weed germinating rains.

It has, however, been identified that with some pre-emergent residual herbicides, early sowing may now be the optimum weed control strategy as crops sown early into higher soil temperatures grow at a faster rate and have a competitive advantage against later emerging weed cohorts. Crops that are sowed late generally grow more slowly and take longer to close their canopy, giving weeds a greater opportunity to establish and grow. Earlier sowing, when soil temperatures are generally warmer, provides an opportunity to increase the crop's competitive advantage against weeds while maximising crop yield potentials. However, the early use of pre-emergent herbicides leads to increased rates of herbicide dissipation and microbial degradation. Past research by Minkey (2017) demonstrated that pre-emergent herbicides decayed more rapidly in warm soil conditions; with Sakura® (pyroxasulfone) decaying at the slowest rate and Boxer Gold® (prosulfocarb + s-metolachlor) and trifluralin decaying faster. The potential degradation of pre-emergent herbicides may therefore offset the value of increased crop competitiveness from earlier sowing.

## MATERIALS AND METHODS

Experiments were conducted at Dandaragan in the Western Australian grain belt. The first time of sowing (TOS1) took place in the first week of May and the second time of sowing (TOS2) in the first week of June. Each trial was direct sowed into cereal stubble. A factorial combination of wheat seeding rate, pre-emergent herbicide and time of sowing (TOS1 plus TOS2 (four-week delay) was randomised in complete blocks with four replicates (Table ). The wheat variety used was Scepter (Intergrain Australia), which is a high yielding, mid-late maturing variety, sowed at 25cm row spacing, with the seeding rate treatments outlined in Table 2. The site was sown with no tillage tine openers with press wheels to provide sufficient seed–soil packing and promote good weed germination. All plots were planted at one sowing depth (approx. 2cm) to minimise the confounding effects of emergence rate and seeding depth differences on biomass and grain yield. Immediately prior to sowing, the whole experimental area was treated with 1.5L/ha Roundup Ultramax (Glyphosate 540g/L, Sinochem Australia), 100ml/ha Lontrel (Clopyralid 750g/L, DowAgrosciences Australia), to control all germinated weeds; followed by the application of each individual plot's pre-emergent herbicide treatment (Table 2).

To control dicotyledonous species such as wild radish (*Raphanus raphanistrum* L.), all plots had a post-emergent application of 670ml/ha Velocity (210g/L Bromoxynil + 37.5g/L Pyrasulfotole, Bayer Australia). For the duration of this study, no additional annual ryegrass control was applied. To ensure optimal wheat growth, 100kg/ha Gusto Gold (Summit Fertilisers Australia) (N – 10.2%, P – 13.1%, K – 12%, S – 7.6%, Cu – 0.07%, Zn – 0.14% and Mn – 0.01%) was drilled 3cm below the seed to minimise contact with the germinating wheat seed. To optimise crop growth, supplementary nitrogen fertiliser in the form of urea (Summit fertilisers Australia) (N – 32%) was applied to all plots.

**Table 1.** Factorial combinations of wheat density, pre-emergent herbicide treatment and TOS of wheat at Dandaragan site in 2021.

| Treatments | Comments |
|---|---|
| **Factor 1 - Crop density treatment description** | |
| Low | 100 plants/m$^2$ (sowed at 45kg/ha) |
| Optimum | 150 plants/m$^2$ (sowed at 68 kg/ha) |
| High | 200 plants/m$^2$ (sowed at 90 kg/ha) |
| **Factor 2 - Time of sowing treatment description** | |
| TOS1 | Dry sowed (4 May 2021) |
| TOS2 | Later sowing after opening rainfall (3 June 2021) |
| **Factor 3 - Pre-emergent herbicide treatment description (knockdown plus)** | |
| Nil (knockdown only) | Nil herbicide applied control (knockdown glyphosate only) |
| Triflur X® 2.0L/ha | Trifluralin 480 gai/L |
| Boxer Gold® 2.5L/ha | s-Metolachlor 120 gai/L + Prosulfocarb 800 gai/L |
| Sakura® 118g/ha | Pyroxasulfone 850 g ai/kg |
| Overwatch ® 1.25L/ha | Bixlozone (Isoxazolidinone) 400 g ai/L |
| Luximax® 500mL/ha | 750g ai/L Cinmethylin |
| Mateno Complete® 1L/ha | 400 g ai/L Aclinofen + 100 g ai/L Pyroxasulfone and 66 g ai/L Diflufenican |

At 10 weeks after emergence (WAE), wheat establishment and ryegrass density was assessed. Above-ground biomass samples of annual ryegrass were removed 27 WAE in three 0.25m$^2$ quadrats per plot. From these samples, the number of ryegrass panicles were counted. To estimate annual ryegrass seed production collected panicles from each plot and thrashed to extract seed. The number of seeds extracted was counted using an S-25 optical seed counter (Data Technologies, Kibbutz Tzora, Israel). At 28 WAS, the whole plot was machine harvested to determine grain yield. Grain samples (400g) were analysed for moisture and protein.

**Herbicide bioassay.** Starting at the time of pre-emergent herbicide application (week 0) soil samples were collected from each plot at 14-day intervals. Soil samples were collected by sampling six 30mm diameter cores per plot to a depth of 6cm form the interrow. Soil samples were immediately transferred into sealed plastic trays with no holes and stored at <15°C for no more than 24 hours. Upon receipt at the University of Western Australia, all soil samples were moistened within the sampling trays using 75ml deionised water containing TWEEN 20 ionic surfactant (Polyethylene glycol sorbitan monolaurate, Sigma Adrich Australia). Fifty seeds from the known herbicide susceptible annual ryegrass biotype (VLR1) were sowed at 1cm depth of the moistened soil in each tray before being placed in a temperature-controlled naturally lit glasshouse (15°C night 25°C day). To ensure adequate seed germination, the containers were sealed for 24 hours before lids were removed for the remainder of the growth period. All trays were watered daily to maintain field capacity. The above-ground shoot length was measured 21 days after sowing, with the percentage shoot length inhibition calculated as per (Khalil et al., 2018b) using the following formula:

Inhibition (%) = 1-(Lt/Lo) x 100% [1]

where: Lt is the shoot length measured in the herbicide-treated soil or crop residue and L0 is the shoot or root length in the untreated soil or crop residue as per (Khalil et al 2018a; Khalil et al 2019b)

## RESULTS

At the Dandaragan site, the first time of sowing (TOS1) was 4 May and the second time of sowing (TOS2) was on 3 June. The soil in the top 10cm was a yellow grey sandy loam with a pH 5.6 $CaCl_2$ and organic total carbon content of 1.56%. The first TOS was sowed into moist soil with no pre seeding ryegrass germination. The second TOS was sowed into similar soil moisture following a significant ryegrass germination. Following sowing, subsequent rainfall was average with acceptable soil moisture for the rest of the season.

**Pre-emergent herbicide persistence bioassay .** Immediately following sowing and at two weekly intervals thereafter, soil samples were taken from the top 10cm of soil from the inter-row region where herbicide would have concentrated following seeding. Using the herbicide susceptible annual ryegrass population VLR1, it was demonstrated that for both TOS1 and TOS2, Sakura, Overwatch and Mateno Complete were the most persistent herbicides, limiting ryegrass shoot length to <42% of the untreated control (%UTC) in both TOS1 and TOS2 at 14 WAS. Triflur X degraded at a fastest rate in TOS1 and TOS2 with ryegrass growth of 86% of the UTC for TOS1 and 88% for TOS2 at 14 WAS. Among the herbicides tested Boxer Gold provided the lowest efficacy of control from the first week after application, with ryegrass growth of 17 and 28% of the UTC respectively. In TOS1, Overwatch was the most persistent herbicide, however in TOS2, Mateno Complete maintained its persistence resulting in a shoot length of <12% of the UTC at 14 WAS (Figure 1).

**Effect of pre-emergent herbicide efficacy, time of crop seeding and wheat seeding rate on ryegrass seed production.** The application of herbicides reduced ryegrass seed production with Triflur X, Boxer Gold and Luximax providing the least ryegrass control. When these herbicides were used, delaying sowing (TOS2) and increasing wheat seeding rates further decreased ryegrass seed production. The use of more residual herbicides (Sakura, Overwatch and Mateno Complete) greatly reduced ryegrass seed production, however in these treatments, delayed TOS (P>0.05) and increased seeding rates (P>0.05) did not further reduce ryegrass seed production.

**Wheat yield.** A significant effect of TOS was found (p<0.001), with early TOS increasing yields (Figure 2). The choice of pre-emergent herbicide or seeding rate did not have a significant effect on yield owing to the high rainfall throughout 2021 and lower ryegrass densities leading to high yields.

## CONCLUSION

This study demonstrated the need to compensate for the lack of a pre-seeding knockdown herbicide application with increased crop competition and a more residual pre-emergent herbicide choice. Dry sowing (TOS 1) outyielded delayed sowing (TOS 2) at the Dandaragan site, however the number of annual ryegrass seeds produced at the end of the season was consistently greater in the TOS1 where less residual herbicides (Triflur X, Boxer Gold and Luximax) were used. Bioassay assessments concurred with the ryegrass seed production data demonstrating that herbicides such as Triflur X and Boxer Gold were not highly residual, whereas, Sakura, Mateno Complete and Overwatch were more residual resulting in reduced ryegrass seed production and often higher wheat yields. While early sowing with an excellent pre-emergent herbicide is considered important for maximising wheat yield, this study found that wheat densities should be practically increased to maximise wheat competitiveness and further reduce ryegrass seed production.

## ACKNOWLEDGMENTS

The research was undertaken with investment support of the GRDC.

## REFERENCES

Khalil, Y., Siddique, K., Ward, P., Piggin, C., Bong, S., Nambiar, S., Trengove, R., Flower, K. (2018) A bioassay for prosulfocarb, pyroxasulfone and trifluralin detection and quantification in soil and crop residues. Crop and Pasture Science 69, 606-616.

Khalil, Y., Flower, K., Siddique, K., & Ward, P. (2019). Rainfall affects leaching of pre-emergent herbicide from wheat residue into the soil. PloS one, 14(2)

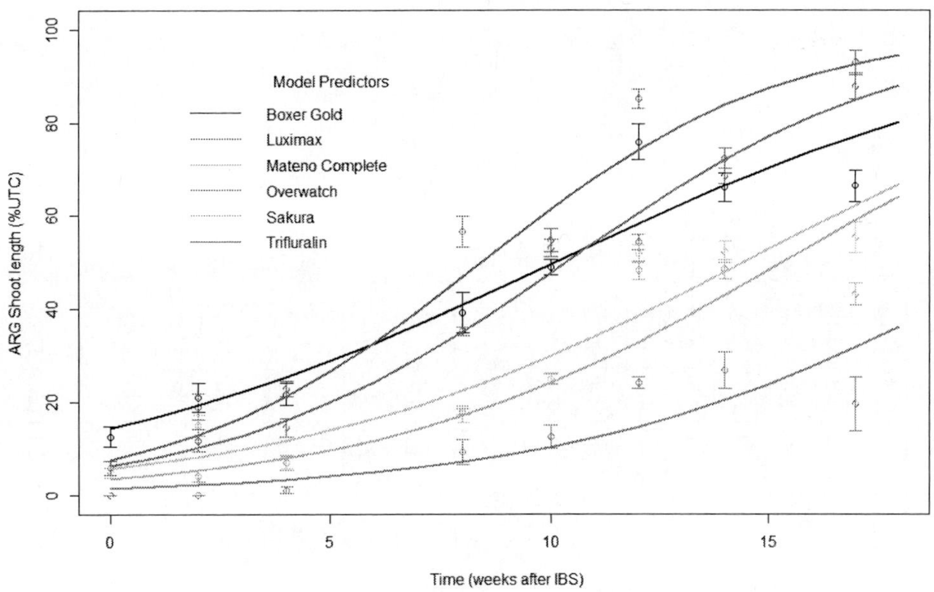


**A.**

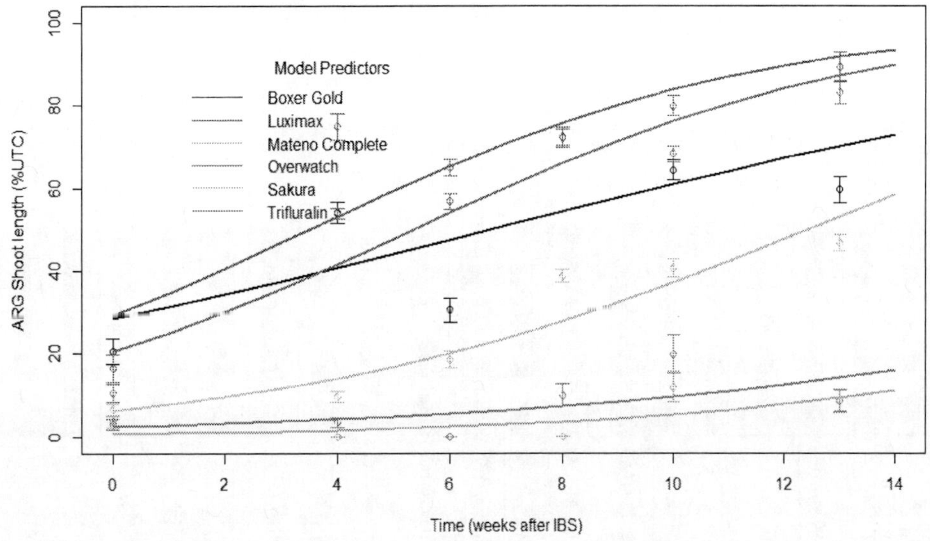


**B.**

**Figure 1:** Herbicide bioassay results. A (TOS1) and B (TOS2) Model predictors for shoot length inhibition of annual ryegrass (VLR1) grown for 21 days in soil sampled at 14-day sampling intervals from plots treated with 1. Nil herbicide (not shown as it is 100%), 2. Trifluralin 480 gai/L, 3. Boxer Gold (S-metolachlor 120 gai/L + prosulfocarb 800 gai/L), 4. Sakura (pyroxasulfone 850 gai/kg) (n=4), 5. Mateno Complete® (aclinofen 400gai/L+ diflufenican 66gai/L + pyroxasulfone 100gai/L), 6. Overwatch ® (bixlozone (isoxazolidinone) 400g/L), and 7. Luximax® (750g/L cinmethylin) (n=4).

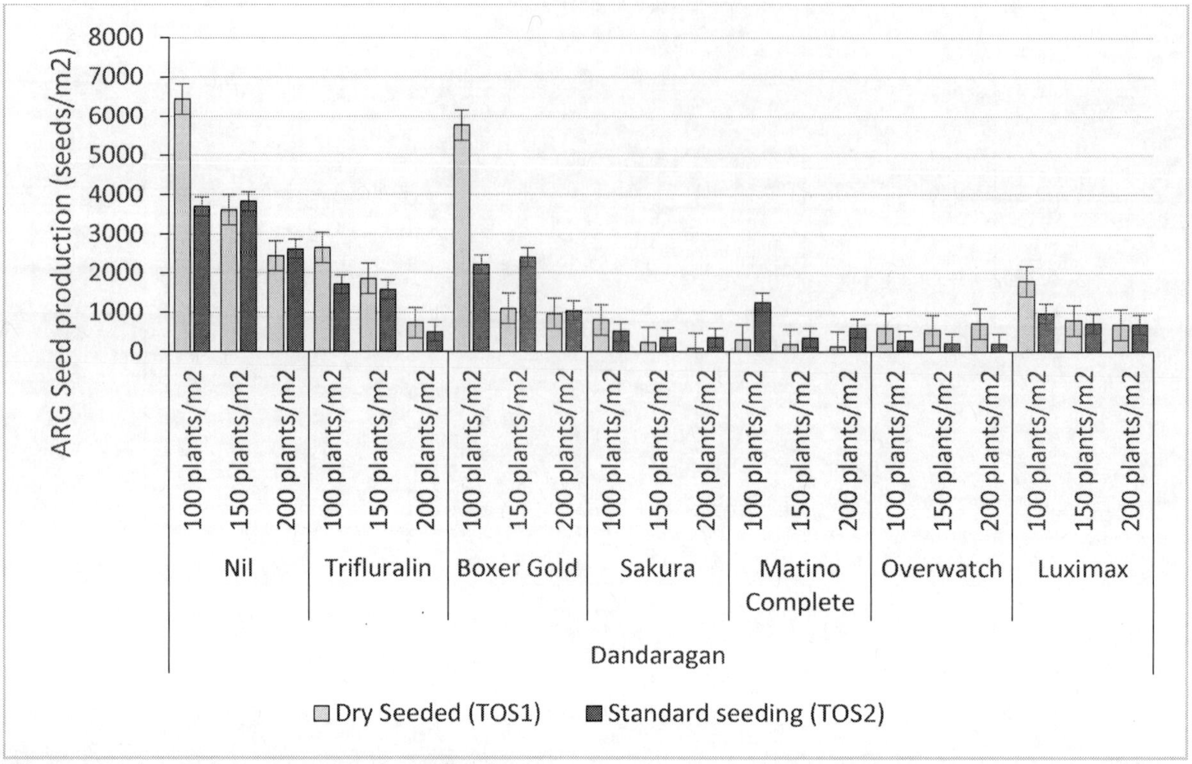


**Figure 2.** Annual ryegrass seed production

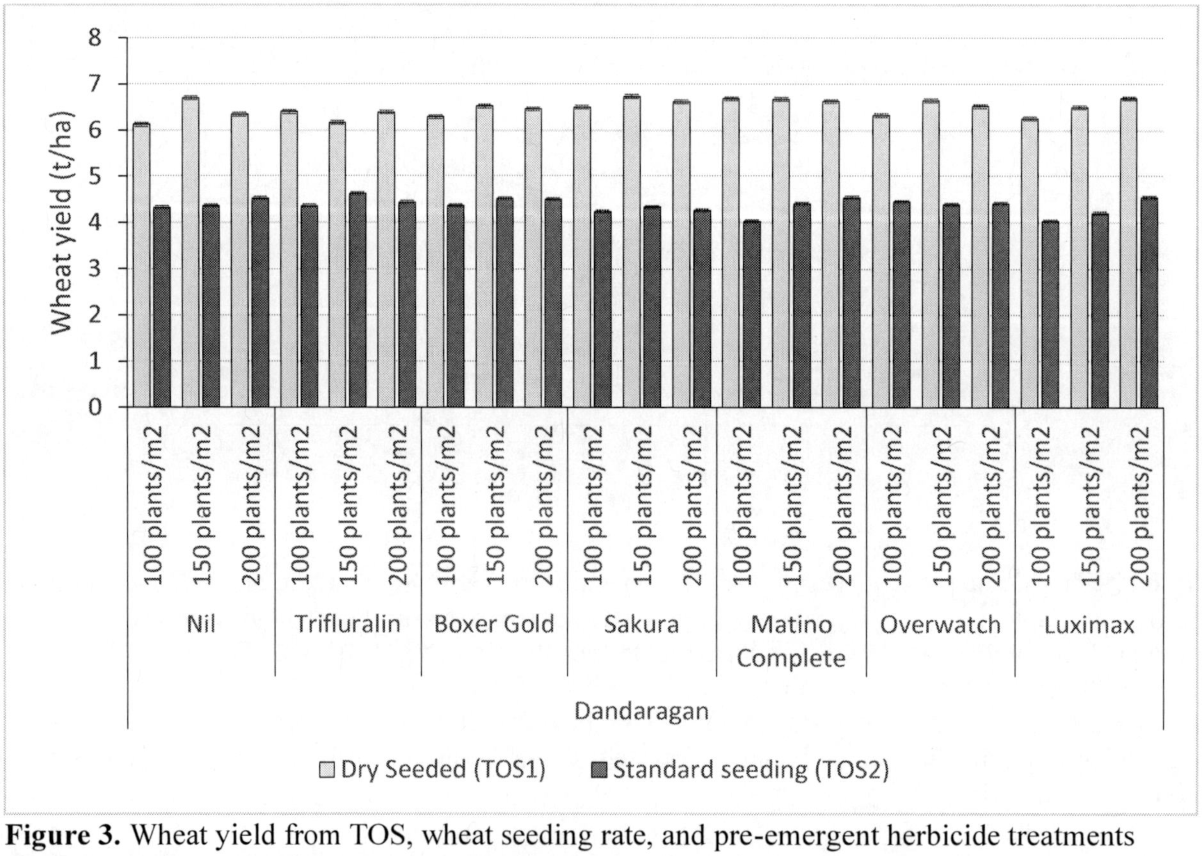


**Figure 3.** Wheat yield from TOS, wheat seeding rate, and pre-emergent herbicide treatments

# Phytotoxicity thresholds for crop seedlings exposed to soilborne residues of diuron and imazapic are regulated by soil type

Michael Rose[1], Annie Ruttledge[2], Jesse Muller[2], Kelvin Spann[1], Lukas Van Zwieten[1], Michael Widderick[2]
[1]NSW Department of Primary Industries, Wollongbar, Australia,
[2]Queensland Department of Agriculture and Fisheries, Toowoomba, Australia
(michael.widderick@daf.qld.gov.au)

**Summary** In the northern grains region of Australia, up to 45% of the cropped area routinely receives a pre-emergent herbicide application. This is largely in response to observed increase in resistance to glyphosate in summer weeds including feathertop Rhodes grass, flaxleaf fleabane, common sowthistle and awnless barnyard grass. In particular, the pre-emergent herbicides imazapic and diuron are now commonly used in summer fallows for the control of these weeds. Although residual herbicides provide a longer control period for certain weeds, their persistence in the soil for lengthy periods (12 – 24 months) may impede the growth of subsequent winter (e.g. barley, wheat, chickpea, lentils, field peas, lupins) or summer (e.g. maize, sorghum, mungbean) crops.

Soil analysis for herbicide residues can be performed through commercial laboratories, however, interpretation of results can be challenging for several reasons. First, there are very few publicly available crop toxicity thresholds that can be used to assess the soil residue concentrations. Second, herbicide bioavailability in different soils depends on the physicochemical properties of both the soil and herbicide of interest, and toxicity thresholds will vary depending on these relationships. We conducted numerous dose-response bioassays for several summer and winter grain crops exposed to residues of diuron or imazapic in different soil types. We report the toxicity threshold values for these crop-herbicide combinations and demonstrate that the soil-specific threshold can be suitably predicted through an understanding of the sorption behaviour of each herbicide in the different soils. The framework used here can be used to derive toxicity thresholds for other priority herbicide-crop combinations prone to carryover damage and the information can be used in decision making to minimise crop loss.

**Keywords** Herbicide residues, bioavailability, toxicity thresholds, imazapic, diuron

# Demonstrating integrated weed management strategies to control barley grass in low rainfall zone farming systems

Amanda Cook[1,2] and Gurjeet Gill[2]
[1]South Australian Research and Development Institute (SARDI) Minnipa Agricultural Centre, 226 McKenzie Road Minnipa, South Australia
[2]School of Agriculture, Food and Wine, University of Adelaide, Glen Osmond SA 5064
(Amanda.cook@sa.gov.au)

**Summary** Barley grass (*Hordeum glaucum*) continues to be the major grass weed in cereal cropping regions on the upper Eyre Peninsula (EP). A three-year field trial was undertaken to investigate barley grass management strategies. The traditional management of pasture systems resulted in an increase in barley grass set and its infestation next year. In contrast, the use of a desiccated late hay freeze in the pasture phase reduced barley grass seed set by 75%. Use of simazine and clethodim in triazine tolerant canola reduced barley grass seed set. Imidazolinone herbicide worked well in the year of application (2019) but barley grass was able to infest the sown pasture system in the following season.

Despite achieving effective control in one season barley grass has the ability to germinate from the weed seed bank the following season and still set high weed seed numbers. Therefore barley grass management and lowering weed seed set needs to be a focus in all seasons in low rainfall farming systems.

**Keywords** Barley grass, weed management, low rainfall farming systems.

## INTRODUCTION

Barley grasses (*Hordeum glaucum* and *H. leporinum*) possesses several biological traits that make it difficult to manage in the low rainfall zone, so it is becoming a more prevalent weed species in field crops in southern and western cropping regions. A survey by Llewellyn, *et al.* (2016) showed that barley grass is now in the top 10 weeds of Australian cropping in terms of area infested, crop yield loss and revenue loss.

The biological traits that make barley grass difficult to manage in low rainfall zones include the early onset of seed production, which reduces effectiveness of crop-topping or spray-topping in pastures; shedding seeds well before crop harvest which reduces the ability of harvest weed seed control effectiveness compared to weeds such as ryegrass which have a much higher seed retention; increased seed dormancy due to a cold vernalisation requirement (Fleet *et al.*, 2012) which reduces weed control from pre-seeding knockdown herbicides due to delayed emergence; and increasing herbicide resistance, especially to Group 1 herbicides commonly used to control grass weeds in pasture phase and legume crops.

Barley grass management is more challenging in the low rainfall zone because the growing seasons tend to be more variable in terms of rainfall, which can affect the performance of the pre-emergent herbicides. Furthermore, many growers in these areas tend to have lower budgets for management tactics, and break crops are generally perceived as a higher risk rotation strategy than cereals. Therefore, wheat and barley tend to be the dominant crops in the low rainfall zone. This research was undertaken at the SARDI Minnipa Agricultural Centre (MAC) as part of a coordinated GRDC research project with farming systems groups across the southern and western cropping regions to demonstrate tactics that can be reliably used to improve the management of barley grass.

## MATERIALS AND METHODS

In 2019 low rainfall growers and weeds researchers met to discuss the issue of barley grass in low rainfall upper Eyre Peninsula farming systems. A three-year broad acre management plan was developed to be implemented with five different strategies to test and compare barley grass management in a replicated broad acre farm trial on the SARDI Minnipa Agricultural Centre from 2019 to 2021 (Table 1). The strategies implemented were System 1 (S1) - District Practice, S2 - Strategic control, S3 - Continuous Cereals, S4 - Two Year Break and S5 - Cultural Control system. The Minnipa Agricultural Centre has an average annual seasonal rainfall of 324 mm, with an average growing season rainfall of 241 mm.

The five management strategies were tested over the three years of rotation with the focus on barley grass weed management and weed seed set. The trial was composed of three replicated broad acre strips of three seeder widths of 27 metres wide of each treatment in MAC paddock S3. Crop establishment, dry matter, barley grass numbers pre-sowing, in crop and at barley grass seed set, grain yield and quality were assessed during the growing seasons. Stubbles and pastures were grazed by sheep over the summer period. The barley grass population present at the trial site has been confirmed to be resistant to group

1 herbicides. For this reason, the use of group 1 herbicides were reduced as management strategies in the broadacre demonstration.

## RESULTS

### 2019

In 2019 the barley grass plant numbers in June/August ranged from 0 to 130 plants/m$^2$. However, treatments with between 3 to 8 plants/m$^2$ (District Practice and Cultural Control) still produced over 300 seeds/m$^2$. In contrast the use of imidazolinone in Scope CL barley in System 2 (S2 - Strategic control) had no barley grass weed seed set in 2019. Compass barley in the District Practice (S1) and Cultural Control systems (S5) had very similar barley grass seed set. Compass barley crop-topped before cutting for hay (S3) reduced barley grass seed set in 2019. The Two Year Break (S4) with self-regenerating pasture in 2019 had higher barley grass plant numbers during the season, but late paraquat application in early September in the pasture phase lowered weed seed set.

| System | 2019 | 2020 | 2021 |
|---|---|---|---|
| 1. District practice | 17 May: Compass barley sown @ 68 kg/ha<br>Glyphosate 1.2 L/ha + Trifluralin 1.5 L/ha | Self-regenerating medic pasture<br>Clethodim 330 mL/ha POST | 2 June: Scepter wheat sown @ 75 kg/ha<br>Glyphosate 1.2 L/ha + Trifluralin 1.5 L/ha |
| 2. Strategic control | 17 May: Scope CL barley sown @ 68 kg/ha<br>Glyphosate 1.2 L/ha + Trifluralin 1.5 L/ha<br>16 July: POST Intervix 700 mL/ha | 26 April: Sultan medic<br>3 June: POST Clethodim 330 mL/ha | 2 June: Scepter wheat sown @ 75 kg/ha<br>Glyphosate 1.2 L/ha + Trifluralin 1.5 L/ha |
| 3. Continuous cereals | 17 May: Compass barley sown @ 95kg/ha<br>Glyphosate 1.2 L/ha + Trifluralin 1.5 L/ha<br>3 Sep: hay freeze with Weedmaster DST @ 1.8 L/ha | 12 May: Scepter wheat sown @ 70 kg/ha<br>PRE Trifluralin 1.5 L/ha | 10 June: Spartacus CL barley sown @ 70 kg/ha<br>6 August: POST Intervix 700 mL/ha |
| 4. Two year break | Self-regenerating grass free pasture<br>17 May: Propyzamide 1 L/ha<br>2 July: Targa Bolt 190 mL/ha + Clethodim 250 mL/ha<br>3 Sep: spray-topping with Paraquat 1.2 L/ha | 26 April: Trident TT canola sown @ 1.8 kg/ha<br>Glyphosate 1.5 L/ha + Hammer 50 mL/ha + Trifluralin 0.8 L/ha + Simazine 0.8 L/ha<br>3 June: Clethodim 330 mL/ha<br>11 June: Atrazine 800 g/ha | 2 June: Scepter wheat sown @ 75 kg/ha<br>Glyphosate 1.2 L/ha + Trifluralin 1.5 L/ha |
| 5. Cultural control | 17 May: Compass barley double seeded @ 120 kg/ha<br>17 May: Glyphosate @ 1.2 L/ha | Self-regenerating grass-free pasture<br>3 June: Clethodim @ 330 mL/ha<br>6 Sep: hay freeze with Paraquat @ 1.2 L/ha | 2 June: Scepter wheat double sown<br>2 June: Glyphosate 1.2 L/ha + Trifluralin 1.5 L/ha |

**Table 1.** The five different management strategies, crops, pastures and herbicide treatments for each season (2019-2021) at Minnipa Agricultural Centre, paddock S3.

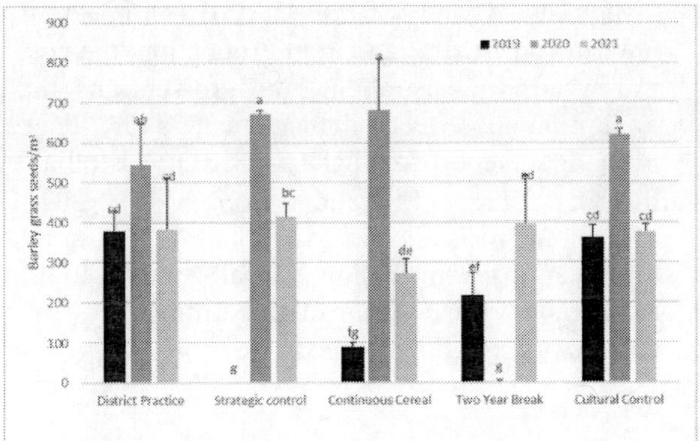


**Figure 1.** Barley grass weed seed set in five different management strategies over three years (2019-2021) at Minnipa Agricultural Centre, paddock S3. Treatments with different letters are significantly different at P=0.05 (LSD =138). Error bars represent standard deviation of the treatment. Refer to Table 1 for information regarding the management strategies investigated.

### 2020

With a late break to the season in 2020 most of the barley grass germinated in mid-July to August thereby avoiding the early weed control with pre-sowing herbicide applications. All crops established well but below average rainfall in May, June and July resulted in very slow crop growth until August and September. The 2020 herbicide applications to the break crop systems of the canola and medic crops reduced barley grass plant numbers, with the triazine tolerant canola system (S4) giving the best late barley grass weed management.

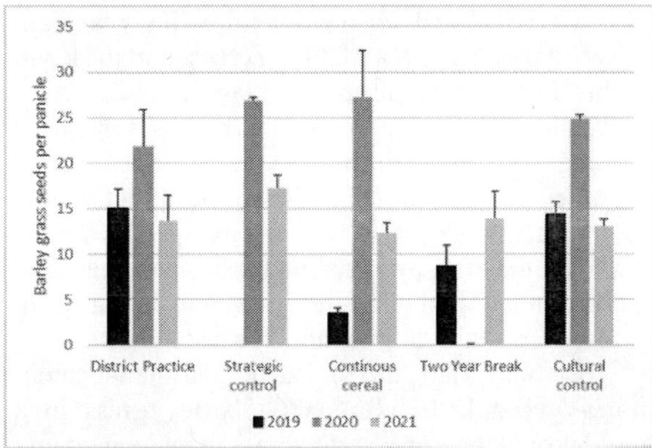


**Figure 2.** Barley grass seeds per panicle for weed seed set in five different management strategies over three years (2019-2021) at Minnipa Agricultural Centre, paddock S3. Treatments with different letters are significantly different at P=0.05 (LSD =4.6). Error bars represent standard deviation of the treatment. Refer to Table 1 for information regarding the management strategies investigated.

Despite excellent weed control in 2019 by the imidazolinone herbicide in S2, barley grass plants and seeds set in 2020 were as high as the other three systems (Figure 1). It is highly likely barley grass was able to establish in this system from the residual seedbank.

### 2021

Some barley grass plants started to germinate by early July in 2021, but like previous years, most barley grass germinated in mid-July to August, which was reflected in the higher late barley grass numbers in September. The Continuous Cereal system sown with Spartacus CL barley had high early barley grass numbers, but imidazolinone applied in early August reduced the barley grass density and lowered the seed set (Figure 1). All other management strategies which were sown to Scepter wheat had a similar barley grass seed set of greater than 370 seeds/m$^2$. There were no differences in grain yield between weed management strategies in 2021.

The barley grass seed set per panicle weed seed set per panicle varied between seasons with 2020 having greater number of weed seeds being set and returned to the weed seed bank than the other seasons (Figure 2).

## DISCUSSION

During the three years of this trial, management tactics found to be effective on barley grass included imidazolinone herbicides, the use of triazine tolerant (TT) canola with simazine and a late hay freeze in pasture with paraquat. Even though imidazolinone worked well in the year of application (2019), barley grass was able to establish in the following year from the seedbank and its population increased in the sown pasture system in the following season.

While the imidazolinone herbicide system is working well in low rainfall farming systems, it must be strategically used to maximise the effectiveness and long-term use of this system. Growers need to be aware of the risk of herbicide resistance and also imidazolinone herbicide residues and plant back periods, especially in low rainfall seasons.

This field trial failed to identify a management strategy capable of eliminating barley grass in a single year. Therefore, barley grass management and lowering weed seed set needs to be a focus of growers in all seasons in low rainfall farming systems.

With confirmed resistance to 'fop' herbicides in barley grass populations at the Minnipa Agricultural Centre, switching to clethodim could be effective in the short term. Generally, a higher rate of clethodim (500 mL ha$^{-1}$) appears to be effective on most populations. Recent work has shown butroxydim was highly effective against most of 'fop' and clethodim resistant populations of barley grass (Gill *et.al.*, 2021). However, resistance to the higher rate is likely to evolve with sustained use over the next few years.

With group 1 herbicide resistance becoming more common and widespread within the upper EP low rainfall zone there needs to be less reliance on their use in the pasture phase and alternative weed control strategies implemented across the rotation are required. If barley grass herbicide resistance is suspected, the barley grass population needs to be tested to know which herbicides can be used effectively.

To ensure group 1 resistance is kept in check, growers should ensure any suspected resistant plants are dealt with in pasture systems by following up with a knockdown herbicide as early as possible to prevent seed set. Growers should always have follow up options to control any survivors and to preserve group 1 herbicides for as long as possible.

The loss of group 1 herbicides within the pasture break system has the potential to change farming systems. Currently farmers on the upper Eyre Peninsula rely on self-regenerating medic-based systems with a profitable livestock enterprise, with grass control applied to prevent weed seed set in spring. Inability to control barley grass with group 1 herbicides will result in medic pasture having to be sprayed out using glyphosate in spring. This will reduce the feed base and livestock carrying capacity, delayed crop sowing times in the cropping phase to gain early weed control or more cropping dominant systems with other break crops (canola, vetch, lentils) and alternative herbicide groups which will increase risk and could impact on profitability.

## ACKNOWLEDGMENTS

This research was made possible through GRDC investment in the project Demonstrating and validating the implementation of integrated weed management strategies to control barley grass in low rainfall zone farming systems. The author would like to acknowledge the SARDI Minnipa Agricultural Centre staff involved in the research delivery especially Ian Richter, Wade Shepperd, John Kelsh, Katrina Brands, Craig Standley, and the previous MAC farm manager, Jake Hull.

## REFERENCES

Fleet B and Gill G (2012) Seed Dormancy and Seedling Recruitment in Smooth Barley (*Hordeum murinum* ssp. *glaucum*) Populations in Southern Australia. *Weed Science* 60(3):394-400. 2012

Llewellyn RS, Ronning D, Ouzman J, Walker S, Mayfield A and Clarke M (2016) Impact of Weeds on Australian Grain Production: the cost of weeds to Australian grain growers and the adoption of weed management and tillage practices Report for GRDC. CSIRO, Australia.

Gill G, Fleet B and Cook A (2021) Herbicide resistance in barley grass populations from the low rainfall zones in South Australia. Eyre Peninsula Farming Systems Summary, Section 7 Weeds: 167-170.

# WeedSmart - changing the face of communications & learning in Australian agriculture

Lisa Mayer[1], Jessica Strauss[1]
[1]Australian Herbicide Resistance Initiative, The University of WA, Crawley, Australia
(lisa.mayer@uwa.edu.au)

**Summary** Sustainable broad-acre crop production in an ever-changing environment is challenging. To understand these challenges and to improve decision making we need knowledge. The way we attain knowledge has changed over the past decade and we can expect to see greater changes in the future with increasing innovations in digital communications. Online courses, podcasts, video, socials and online publications are platforms that are increasingly utilised to share knowledge. This is the new frontier in communications in agriculture and WeedSmart continues to be a pioneer in this evolving space. WeedSmart is the industry voice that delivers science-backed weed control solutions to growers and advisors for long-term profitability in Australian broad-acre cropping.

WeedSmart is focused on engaging its audience by communicating knowledge through a variety of novel digital formats as well as with face-to-face interactions from extension agronomists located in the regions. The aim is to influence attitudes and actions to minimise crop weeds and sustain herbicides. Farmers and their advisors are constantly challenged to combat weeds and herbicide resistance to maximise profitability in their cropping system. WeedSmart promotes the adoption of a diverse range of agronomic tools grouped into six categories. 'The Big 6' offers tips on essential agronomic practices to reduce the crop weed seed bank. By transducing the Australian Herbicide Resistance Initiative's and other organisations' research, WeedSmart effectively delivers the science of resistance along with chemical, cultural and mechanical weed control advice. Australian farmers are recognized as early adopters and innovators and we anticipate that this audience will continue to seek solutions to weed control through new platforms. WeedSmart is backed by GRDC and industry and is committed to ensuring key weed control messages resonate and engage farmers and advisors to secure profitability into the future.

**Keywords** Communication, resistance, crops, weeds, Big 6, WeedSmart

# Weed Wide Web: Surveillance of Online Trade in Declared Plants

Jacob Maher[1], Lisa Wood[1], Stephanie Moncayo[1], Oliver Stringham[1], Phill Cassey[1]
[1]The University Of Adelaide, Adelaide, Australia
(jacob.maher@adelaide.edu.au)

**Summary** The trade of plants is a major source of invasive species introductions and establishment. In Australia, a large-scale informal plant trade is facilitated by publicly accessible e-commerce websites. This unregulated trade has enabled a rapidly emerging pathway for the spread of invasive plant species (weeds). The trade of declared weeds is illegal and presents biosecurity professionals with a unique set of problems and significant challenges to address. Thus far, investigations of the online weed trade are done on an ad hoc basis by biosecurity professionals manually searching websites. In response, we have developed a semi-automated method utilising webscrapers to systematically capture the online plant trade. We match search terms based on scientific, common and trade names with text from online plant advertisements to detect weeds being traded. Visual identification with listing images is used to verify detections. Our aim is to quantify and describe the trade to better understand trade dynamics and participant motivations. Preliminary results show the online trade of weed species is frequent and widespread, with illegal trade present in all states and territories. To date, we have detected more than 100 different species of declared weeds traded online. Cacti, aquatic plants and horticulturally popular, yet invasive, species are the most traded weeds. Misidentification of plants and the use of generic (non-scientific) names by traders is common. This behaviour suggests an overall lack of awareness of the species being traded, their legal status, and the potential consequences of trading a declared weed. Culinary, medicinal and other uses for plants are purported by sellers, providing insight into the desire for certain weeds. These insights and the methodology developed for this project will provide biosecurity professionals with the information and tools required to detect illegal sales and therefore prevent future weed incursions.

**Keywords** E-commerce, webscraping, online trade, internet, detection, surveillance

# Management of the online trade of *Salvinia molesta* in Victoria: detection, identification and eradication

Victoria Byrne[1], Zachariah Munakamwe[1]
[1]Agriculture Victoria, Australia
(victoria.byrne@agriculture.vic.gov.au)

**Summary** The State prohibited weed (SPW) salvinia (*Salvinia molesta* D.S. Mitch.) is one of the Australia's worst weeds due to its rapid growth and ability to spread via vegetative means. Salvina is managed by Agriculture Victoria with the aim of eradicating it from the State. Online trade of salvinia is increasing and is identified as a major pathway for dispersal. Agriculture Victoria actively searches for listings of SPWs being sold on major online platforms. This online surveillance has recently detected a surge of salvinia advertisements, with many labelled as the related but undeclared species *Salvinia minima* (Baker). However, as *Salvinia minima* has not been confirmed as present in Australia, it is suspected that salvinia species advertised for sale in Victoria are *Salvinia molesta*. This potential mislabelling increases the difficulty of online detections and the ability to undertake compliance activities. In addition, correctly identifying salvinia is problematic due to the genus being poorly described at the species level. These hurdles may have a detrimental impact on the eradication of salvinia from Victoria. Definitive identification of *Salvinia molesta* has not been possible through morphology and can only be achieved through deoxyribonucleic acid (DNA) testing. Agriculture Victoria has established a system involving independent laboratory services, utilising DNA barcoding, to identify salvinia plants to the species level, providing a useful tool to accurately identify *Salvinia molesta*. To date, all seventeen tested samples that were advertised for sale as *Salvinia minima* have been positively identified as *Salvinia molesta*. It is therefore highly likely that despite being labelled as an alternative *Salvinia* species, the majority of salvinia species detected online are the SPW *Salvinia molesta* and will be investigated accordingly to ensure timely detection, identification and treatment to provide the best chance at eradication.

**Keywords** DNA barcoding, dispersal, eradication, noxious weeds, surveillance

# Collaborating to combat Illegal Online Trade of Noxious Weeds: 'Bridging the borders'

Zachariah Munakamwe[1] and Angela Constantine[2]
[1]Agriculture Victoria, PO Box 3100, Bendigo, VIC 3552, Australia.
[2] Agriculture Victoria, PO Box 5646, Cranbourne, VIC 3977, Australia.
(zachariah.munakamwe@agriculture.vic.gov.au)

**Summary** Early detection and rapid response, the key elements of effective pest eradication programs, require collective action and interstate collaboration. This is even more important when trade is involved as there is the potential for plants to be quickly spread over very large distances. Illegal online trade is an emerging pathway of spread of noxious weeds and Agriculture Victoria collaborates with other State and Territory authorities to combat trade of State prohibited weeds (SPWs). In the past five years, Agriculture Victoria detected a total of 143 online sales of water hyacinth (*Eichhornia crassipes* (Mart.) Solms) and salvinia (*Salvinia molesta* D.S. Mitch.). Of these, 48 were located in New South Wales, 36 in Victoria and the rest were in other States and Territories except Tasmania.

The detections outside Victoria were forwarded for action to the respective jurisdictions through a special collaboration arrangement. Feedback from interstate indicated that all the forwarded reports were followed up resulting in the seizure of these species before they established and spread. This collaboration has also increased communication, intelligence sharing and following-up on online detections between the jurisdictions.

**Keywords** Biosecurity, eradication, salvinia, State prohibited weeds, water hyacinth.

## INTRODUCTION

State prohibited weeds (SPWs) are the highest category of declared noxious weeds in Victoria and are managed with the aim of eradicating them from the State. They are either not yet present in Victoria or are only present in small infestations that are considered eradicable from the State. It is illegal to buy, sell or transport SPWs in Victoria (*Catchment and Land Protection Act 1994* (Vic)). When a SPW is detected in Victoria, including in online trade, Agriculture Victoria responds by removing or destroying all plants, and following up with regular monitoring until the site is extirpated. Tracing operations are also undertaken to find any linked sites.

Contrary to our efforts to eradicate SPWs from Victoria, Agriculture Victoria has found a number of these species advertised for sale on online trading platforms around the country. Such sales severely undermine eradication efforts as they present pathways whereby these species can readily spread over long distances, creating new infestations. Interstate infestations are outside the jurisdiction of Agriculture Victoria, but still pose a serious threat to the State. To combat this risk, Agriculture Victoria initiated a program to reduce the illegal online trade of noxious weeds in Australia and this has so far produced positive results.

This paper aims to:

- Evaluate the program to date, highlighting key achievements.
- Assess the major program deliverables as indicators of success.
- Highlight the effectiveness of the national collaboration between key stakeholders.

## MATERIALS AND METHODS

The program and the online search methodology are fully described in Munakamwe and Constantine (2017). An initial industry analysis was done to determine the online sites where SPWs were most often detected in trade, and to understand how these websites operated.

The analysis narrowed the program scope to focus on three sites; eBay (www.ebay.com.au), Gumtree (www.gumtree.com.au) and Facebook (www.facebook.com), and explored how to effectively influence these priority platforms. This approach was considered the most pragmatic way to effectively detect noxious weed trade and influence the very diverse (Hinsley et al. 2016) and variable online marketplace.

Alerts were set on the eBay.com.au and gumtree.com.au websites to trigger an email message whenever an SPW was listed for sale. Key words were set for each SPW species, as were other phrases that are commonly found associated with SPW sales, such as 'pond plants', 'water lilies' and 'water plants'. Daily manual searches were also conducted on the websites in spring through to autumn (peak online sales period) and on an ad hoc basis in winter (when online sales of target plants are less common). This primary data was collected for five years and some secondary data extracted from the Agriculture Victoria Departmental internal database, Bioweb.

## RESULTS AND DISCUSSION

### Online detections

Over five years there were a total of 143 online detections (from March 2016 to June 2021) of the SPWs water hyacinth (*Eichhornia crassipes*) (97) and salvinia (*Salvinia molesta*) (46) in all states and territories except Tasmania (Table 1). The highest number of detections were of plants located in New South Wales, 48 (34%), followed by Victoria with 36 (25%) and Queensland 32 (22%). Only one (0.7%) detection was of plants in the Northern Territory. About 71% (102) of the detections were made on Gumtree, 24% (34) on Facebook and only 5% (7) on eBay. Figure 1 shows the total number of detections in each state and territory over the five years of this program.

All states and territories successfully followed up on water hyacinth and salvinia online trade cases detected by Agriculture Victoria for sale online in their jurisdictions (Table 1), resulting in the seizure of these species before they established and spread. Although not SPWs, New South Wales and Western Australia reported several declared cacti plants advertised for sale in Victoria, which Agriculture Victoria successfully followed up.

### National collaboration

Agriculture Victoria actively promotes collaborations with all States and Territories in tackling the online trade of noxious weeds, including presenting a research paper (Munakamwe and Constantine, 2017) at the 19th NSW Weeds Conference to promote this integrated approach. The conference was also an opportunity to interact and network with interstate counterparts. This resulted in increased communication, intelligence sharing and following-up on online detections between the States.

### Salvinia case study

The following case study demonstrates how effective collaboration can lead to early detection and rapid response; the key elements of all pest eradication programs. An Agriculture Victoria Biosecurity Officer detected salvinia (*Salvinia molesta*) located in New South Wales being advertised for sale on Gumtree.com.au. Salvinia is a State prohibited weed (SPW) in Victoria and a weed of national significance. The detecting officer reported the advertisement to the NSW biosecurity authorities who promptly instigated compliance activities resulting in seizing of the plants on the very same day of the detection.

This salvinia seizure avoided the potential huge economic and environmental impact imposed by this highly invasive weed not only to NSW but to the whole country. This is just one example of several similar detections that Agriculture Victoria passed on to respective jurisdiction authorities for action. Feedback from interstate indicated that all the forwarded reports were followed up.

### Collaboration with major online platforms

Through this program, Agriculture Victoria successfully collaborated with and influenced eBay and Gumtree to develop actions that aim to stop noxious weed advertisements on their platforms. EBay added the SPWs most likely to be traded to their website filters on 9 July 2016. Sellers are stopped and notified when trying to list filtered items, preventing their advertisement. There is also a general warning on eBay that pops up when traders list plants warning that some plants are illegal to sell.

Declared noxious weeds are occasionally advertised from overseas with sellers offering to post the items into Australia. A relationship was built with eBay where such detections can be reported to the website's key contacts, who remove the advertisements instantly. Ebay has a 'plants and seeds policy' which includes information about Biosecurity, including a specific link to Agriculture Victoria's SPW webpage.

Gumtree also published a government announcement on declared noxious weeds and includes links to the relevant state and territory government organisations on their website. On 5 December 2017, Gumtree also offered Agriculture Victoria and other government departments the opportunity to place more articles on their website warning traders of the illegality of trading noxious weeds. As for eBay, a relationship has been built whereby international advertisements of SPWs detected on Gumtree can be reported to a site contact and be immediately removed from the site.

**Table 1: State prohibited weed detections (March 2016 to June 2021).**

| Species | Website | Vic | NSW | QLD | WA | SA | ACT | NT | TAS | **Total** |
|---|---|---|---|---|---|---|---|---|---|---|
| | Gumtree | 11 | 28 | 11 | 8 | 4 | 4 | 0 | 0 | 66 |
| Water hyacinth | Facebook | 20 | 5 | 1 | 0 | 0 | 0 | 0 | 0 | 26 |
| | eBay | 1 | 3 | 1 | 0 | 0 | 0 | 0 | 0 | 5 |
| | Gumtree | 2 | 6 | 19 | 4 | 4 | 0 | 1 | 0 | 36 |
| Salvinia | Facebook | 2 | 5 | 0 | 0 | 1 | 0 | 0 | 0 | 8 |
| | eBay | 0 | 1 | 0 | 0 | 1 | 0 | 0 | 0 | 2 |
| **Total** | | **36** | **48** | **32** | **12** | **10** | **4** | **1** | **0** | **143** |

**Key**: Vic - Victoria, NSW - New South Wales, QLD - Queensland, WA - Western Australia, ACT - Australia Capital Territory, NT - Northern Territory.

**Online trade standard operating procedure**

An online trade Standard Operating Procedure (SOP) has also been developed as part of this initiative. The SOP was developed to provide a standardised guide for authorised officers involved in the program in Victoria to ensure all relevant activities and processes are carried out in a consistent manner and in accordance with the *Catchment and Land Protection Act 1994* (CaLP Act). The SOP has been an effective tool and particularly useful for new authorised officers, who are not familiar with the program. The SOP was shared with interstate colleagues on request to assist in the development of their own programs and this collaboration reflects the integrated approach that this program promotes.

**Online trade webpage**

Agriculture Victoria has now developed an online trade webpage: https://agriculture.vic.gov.au/biosecurity/weeds/illegal-online-trade-of-noxious-weeds-in-victoria. The webpage highlights online trade of noxious weeds as an emerging biosecurity pathway and focusses on the two most common SPW species detected in online trade, water hyacinth and salvinia. The webpage provides an engagement tool to raise awareness that online trade of noxious weeds is illegal and poses a significant biosecurity risk. This webpage has been translated into Vietnamese, Khmer and Simplified Chinese to communicate with Culturally and Linguistically Diverse communities who have historically been found to be likely to have water hyacinth.

**Passive surveillance (Weed Spotters)**

Agriculture Victoria maintains a passive surveillance network of trained volunteers, known as Weed Spotters, strategically recruited from various agencies identified as most likely to encounter, correctly identify and report SPWs as they go about their daily activities. Weed Spotters are trained to identify and report all SPWs, whether detected in person or online. Trained Weed Spotters are more likely to make accurate reports of target species than the general public (Munakamwe *et al* 2018) and provide an important source for detecting and reporting trade in SPWs. There are currently 2,723 registered Weed Spotters in Victoria.

Figure 1. State prohibited weed detections (March 2016 to June 2021)

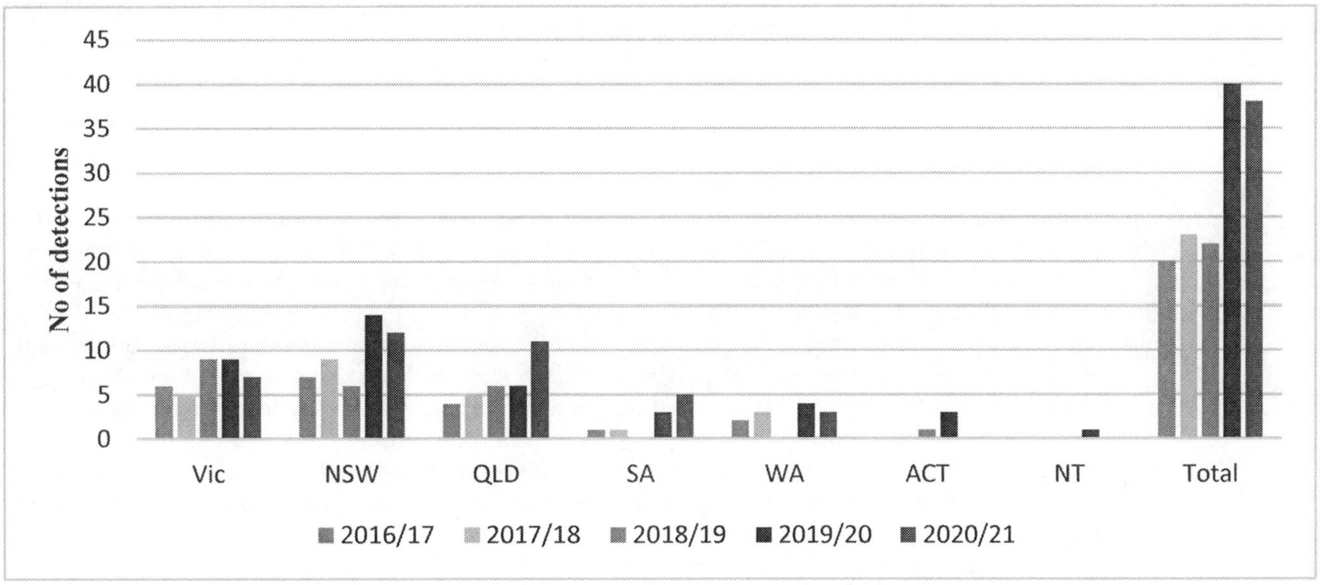


## CONCLUSION

The online detections from this Victorian initiative led to removal and treatment of high-risk invasive noxious weeds in Victoria and interstate. The national collaboration contributed immensely towards the success of this high-impact program.

## ACKNOWLEDGMENTS

Special thanks to Sharyn Williams, Mark Watt, High Risk Invasive Plants Team and all biosecurity regional teams in the Plants, Chemicals and Invasives program – Agriculture Victoria.

## REFERENCES

*Catchment and Land Protection Act 1994* (Vic) s60. Anstat Pty Ltd (Government Printer for the State of Victoria), Melbourne 3002, Victoria Australia.

Hinsley, A., Lee, T.E., Harrison, J.R. and Roberts, D.L. (2016). Estimating the extent and structure of trade in horticultural orchids via social media. *Conservation biology* 30 (5), 1039 – 1047.

Munakamwe, Z. and Constantine, A. (2017). Illegal online trade of noxious weeds in Australia: monitoring and regulating e-commerce. *Proceedings of the 19th NSW Weed Society Conference* 16 – 19 October 2017, University of New England, Armidale, Australia, pp 58-63.

Munakamwe, Z., Constantine, A. and McInerney C. (2018). The Victorian Weed Spotters: Recruitment and training of citizen scientists. Proceedings of the 21st Australasian Weeds Conference 9 – 13 September 2018, Sydney, Australia. eds S. Johnson, L. Weston, H. Wu and B. Auld. pp. 193-196.

# New threats, old foes: problem weeds for farmers

Nyree Stenekes[1], R Kancans[1]
[1]Australian Bureau of Agricultural and Resource Economics and Sciences, Department of Agriculture, Water and the Environment, Canberra, Australia
(nyree.stenekes@agriculture.gov.au)

**Summary** Weeds pose significant management concerns in farming systems across Australia. To better understand this national problem, the Department of Agriculture commissioned ABARES in 2016 and 2019 to collect a national dataset of landholders' views on managing weed species, including all the Weeds of National Significance and other current and emerging weeds. This was the largest single data collection ever run by ABARES, with responses from over 8,000 landholders across 53 natural resource management regions in Australia in the 2019 data collection. The survey provided a national picture of the extent of weed problems: which weeds are a problem for landholders, the impacts on production systems, the effort and cost landholders incur in managing weeds on their land, the types of control actions conducted by landholders and local management groups, and how effective these actions are in addressing weed problems. This paper highlights key results from the 2019 survey of landholders and looks at long-term trends in weed management by comparing results from a previous data collection conducted in 2016. The potential for this dataset to be integrated with other national weed datasets, and how this integration can add value to research and policy, will be discussed.

**Keywords** Weed management, Weeds of National Significance, agricultural land managers, farmers attitudes and practices, national pest and weed survey

# Novel chemistries in Eucalyptus essential oils – a nature's gift for herbicide discovery

Hanwen Wu[1]

[1]EH Graham Centre for Agricultural Innovation (NSW Department of Primary Industries and Charles Sturt University, Wagga Wagga, Australia

(hanwen.wu@dpi.nsw.gov.au)

**Summary** The heavy reliance on herbicides for weed management in conservation farming systems has resulted in the rapid development of herbicide resistance, thereby restricting herbicide options and threatening the continued success of conservation farming systems. More than 45 weed species have been confirmed to have evolved resistance in Australia and some species such as annual ryegrass and wild radish have developed multiple resistance to herbicides with different modes of action. It is therefore imperative to discover new compounds with novel modes of action to combat this ever-evolving herbicide resistance in weeds. It has long been recognised that some eucalyptus species are capable of suppressing understorey vegetation growth via allelopathy. Eucalyptus contains a rich source of bioactive constituents, which have been reported to cause phytotoxicity to a number of weed species. The bioactive compounds in eucalyptus oils could possess potential commercial value for further exploitation as natural herbicides. Eucalyptus is a member of Myraceae family and it is a native to Australia. There are about 800 eucalyptus species readily available in Australia. Leaf materials of more than 300 Eucalyptus species were kindly provided by Dr Dean Nicolle of Currency Creek Arboretum. The essential oils were extracted and evaluated for their herbicidal activities and their potential in weed management. Results showed that eucalyptus species differed significantly in oil contents from 0% to 4.49% (fw, leaf) and in herbicidal activities against the germination and growth of annual ryegrass. Some *Eucalyptus* essential oils at 2.5μl/petri dish promoted the germination up to 19%, while others inhibited the germination up to 96%. The bioactive compounds associated with Eucalyptus allelopathy will also be discussed.

**Keywords** Annual ryegrass, herbicide resistance, chemical control, allelopathy

# Herbicidal Activity of *Digera muricata* against *Melilotus indicus* and identification of Allelochemicals

Ali Raza[1], Muhammad Akbar[1], Tayyaba Khalil[1], Nasim Ahmad Yasin[2], Abdul Nazir[3], Aqeel Ahmad[4]
[1]Department of Botany, University of Gujrat, Gujrat, Pakistan,
[2]University of the Punjab, Lahore, Pakistan,
[3]Department of Environmental Sciences, COMSATS University Islamabad, Abbottabad, Pakistan,
[4]University of Chinese Academy of Sciences (UCAS), Beijing, China
(muhammad.akbar@uog.edu.pk)

**Summary** Synthetic herbicides are in use to combat weeds in crops but all are associated with numerous ill effects. In recent years, search for ecofriendly herbicides is very active, but reports regarding herbicidal activity of *Digera muricata* against a weed, *Melilotus indicus*, under in vivo conditions are missing. So, in the present study, herbicidal activity of *D. muricata* was evaluated against *M. indicus*. Herbicidal activity was checked at two concentrations (50% and 100%) of *D. muricata* aqueous extracts in repeated sets of pot bioassays. Herbicidal activity was evaluated by growing *M. indicus* either alone or in combination with wheat. There were 5 treatments (dH2O, half dose herbicide, full dose herbicide, 50% plant extract, and 100% plant extract). 50 and 100% plant extracts of *D. muricata* decreased the shoot length, fresh weight, and dry weight of *M. indicus* by 69 and 85%, 58 and 89%, 30 and 78%, respectively, when grown alone. However, when *M. indicus* was grown side by side with wheat, 50 and 100% plant extract of *D. muricata* decreased the shoot length, fresh weight and dry weight of *M. indicus* by 37 and 59%, 33 and 52%, 26 and 69%, respectively. On the other hand, synthetic herbicide completely eradicated the test weed at both test concentrations. Synthetic herbicide negatively affected the growth of wheat in comparison with extracts of *D. muricata*. As the aqueous extract of *D. muricata* exhibited potent toxicity towards test weed without harming the wheat crop, so the extracts of *D. muricata* can be used to explore novel herbicidal compounds. Spectroscopic analysis of *D. muricata* extract revealed the presence of β-caryophyllene, ferulic acid, d-pinitol, hexadecanoic acid, lupeol, quercetin, germacrene, 10-epi-γ-eudesmol, β-amyrone, stigmast-5-en-3-ol, oleate, hexatriacontane, and berberine, at higher concentrations. It was concluded that herbicidal activity of *D. muricata* was due to the presence of these allelochemicals.

**Keywords** Herbicide, Triticum aestivum, Digera muricata, Melilotus indicus

# A potentially sustainable weed control method using urease enzymes extracted from weeds

Md Mizanur Rahman[1], Isaac Ahenkorah[2], Bikash Devkota[3], Md Rajibul Karim[4], Simon Beecham[1], Christopher Chow[1] and Jack Desboilles[5]
[1] Professor, University of South Australia, UniSA STEM, SA 5000, Australia.
[2] Ph.D. in Civil Engineering, University of South Australia, UniSA STEM, SA 5000, Australia.
[3] Ph.D. candidate, University of South Australia, UniSA STEM, SA 5000, Australia.
[4] Senior Lecturer, University of South Australia, UniSA STEM, SA 5000, Australia.
[5] Research Fellow, University of South Australia, UniSA STEM, SA 5000, Australia.
(mizanur.rahman@unisa.edu.au)

**Summary** Weed management and control are essential for successful crop production. In recent years, there has been increased interest in the use of sustainable biological approaches for weed control due to their potential environmental and economic benefits. In this study, the enzyme-induced carbonate precipitation (EICP) approach was adopted to form a soil crust through calcium carbonate bonding using plant-sourced urease enzymes extracted from the weed paddy melon (*Cucumis myriocarpus* Naud.), urea and calcium chloride solution. The penetration and erosion resistance of the EICP-treated soil crust was then measured.

The results from this study show that the EICP-treated soil crust exhibited a significant surface hardening with a maximum penetration resistance of 1307 kPa and significantly high resistance to raindrops and wind erosion compared to untreated soil. The penetration and erosion resistance of the EICP-treated soil crust also increased with an increased number of treatment cycles. The outcome of this study shows that an EICP-approach, using crude enzymes extracted from weeds, can achieve a desirable crust penetration strength that may significantly reduce weed seedling emergence. The technique can also be developed as a potentially sustainable method for weed control for uncultivated land such as roadside shoulders and embankments.

**Keywords** weed control, paddy melon, EICP, urease enzyme.

## INTRODUCTION

Weeds are unwanted plants that grow outside their natural ecosystems where they may be of no positive economic importance (Oerke 2006). In many cases, the presence of weeds on farmlands affects the productivity of the land, crop development and yield. Conventional methods of weed control include manual removal and the use of chemical herbicides (Bajwa et al. 2018; Christoffoleti et al. 2007). However, the use of physical and/or chemical methods of weed control is often undesirable or insufficient in many situations. In recent times, the evolution of herbicide-resistant weed ecotypes across the world has further aggravated the situation (Bajwa et al. 2019). One of the most sustainable approaches for weed control can be by preventing weed seedling emergence through soil crusting. However, this approach has not been studied in the literature.

Nonetheless, the influence of naturally crusted soils on plant seedling emergence has been investigated in several studies (Anzooman et al. 2018; Joshi 1987; Laker and Nortjé 2019; Massingue 2002). The emergence of seedlings from crusted soils depends on the seedling emergence force and the soil crust strength (Anzooman et al. 2018; Sinha and Ghildyal 1979). Most importantly, the mechanical resistance of the soil crust often restricts seedling emergence. If the force exerted by a young seedling immediately after germination is less than the resistance of the soil crust, the seedling remains beneath the crust and does not emerge (Awadhwal and Thierstein 1985). Not only do soil crusts provide a mechanical impedance to seedling emergence, but they also obstruct soil moisture, temperature and gaseous exchange due to their low porosity, thereby limiting the supply of oxygen to germinating seeds. Most studies have reported a negative linear correlation between the percentage of seedling emergence and crust strength with a typical crust strength between 40-700 kPa required to fully hinder seedling emergence (Bennett et al. 1964; Joshi 1987; Massingue 2002; Parker Jr and Taylor 1965; Richards 1953). The variations in the reported threshold of crust strength required for fully hindering seedling emergence as reported in the literature are possibly due to the differences in crop seedlings and the crust strength testing method used in various studies.

Although most studies have investigated the influence of naturally crusted soils on the seedling emergence of crops, none of these has studied the potential of biologically induced crusted soils as a sustainable approach for controlling weed seedling emergence. Therefore, biocementation approaches, such as enzyme-induced carbonate precipitation

(EICP), can be a potentially sustainable means of preventing or reducing weed growth or emergence through natural ground solidification and/or hardening. EICP involves calcium carbonate ($CaCO_3$) precipitation via urea hydrolysis catalysed by plant-sourced urease enzymes (Ahenkorah et al. 2021b). The precipitated $CaCO_3$ forms bonds between the soil particles, which results in ground solidification. In this study, the penetration and erosion resistance of EICP-treated soil crusts was assessed as a potentially sustainable method of weed control for uncultivated lands such as roadside shoulders and embankments. A sustainable source of crude urease enzyme extracted from the weed paddy melon (*Cucumis myriocarpus* Naud.) was used in this study for the treatment of EICP-treated soil crusts.

## MATERIALS AND METHODS

**Soil type and sample preparation** Two soils were used in this study, namely Karoonda silty sand (KSS) and Adelaide industrial sand (AIS). These were both sourced from Adelaide, South Australia. Figures 1(a) and (b) show pictures of KSS and AIS, respectively. The soil samples were mixed with 10% moisture and prepared in round PVC containers (110 mm in diameter and 25 mm in height) to a relative density of ~30%.

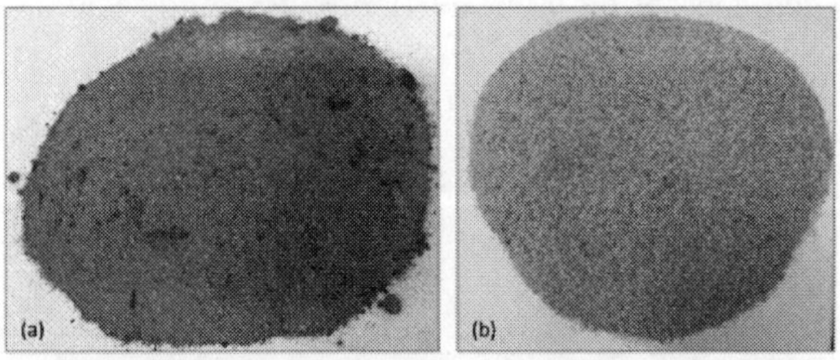


**Figure 1.** Pictures of different soils used in this study: (a) KSS; and (b) AIS.

**Preparation of EICP treatment solution** An EICP treatment solution consisting of a mixture of crude urease enzyme solution (extracted from paddy melon seeds) and (equimolar) 0.50 M of cementation solution (containing urea and $CaCl_2$) was prepared. To prepare crude urease enzyme extract, paddy melons were collected from areas surrounding Adelaide, South Australia. Figure 2 shows an image of paddy melon fruits and seeds. A 50 g amount of seeds obtained from the paddy melons were then soaked overnight in 200 mL of deionised water and the solution was homogenised in a blender for approximately 5 minutes. The enzyme-containing solution was then centrifuged twice at ~4400 rpm for 15 minutes and the supernatant was collected as crude urease enzyme extract (Ahenkorah et al. 2021a).

**Figure 2.** Paddy melon fruits and seeds as a source of plant urease enzyme.

**EICP soil treatment process** A new EICP treatment approach was developed in this study. Soil samples were treated with only one cycle of EICP treatment. The EICP treatment consisted of spraying 7.5 mL of crude urease enzyme extract from paddy melon seeds and 7.5 mL of cementation solution on top of the specimens from opposite directions at the same time and the sample was then cured at 30 °C for 40-48 hours. This approach allowed the percolation of the treatment solution to the desired depth and prevented concentrated precipitation at the surface or precipitation within the EICP solution before application. The volume of crude urease enzyme and cementation solution used was calculated based on the field capacity of the soil used and a target depth of cementation of ~5 mm. The strength and erosion resistance of the EICP-treated soil crusts were then determined.

**Penetration test** A handheld penetration test using a Mecmesin AFG500 force gauge fitted with a flat end circular probe of diameter 7-8 mm was used in this study to determine the strength of the EICP-treated soil crust. During the penetration tests, the crust strength at five different locations of the circular surface of the EICP-treated soils, i.e. top (T), bottom (B), left (L), right (R) and center (C) were measured.

**Erosion resistance test** To evaluate the erodibility of the EICP-treated soil crust, both wind and raindrop erosion resistance tests were conducted in this study. It should be noted that of the two soil types, the AIS was used to prepare samples used for these tests due to its loose and cohesionless nature, making it more susceptible to wind and rainfall erosion. For wind erosion tests, a 1 m long wind tunnel was developed and used in this study. A digital wind speed meter (anemometer) was used to measure the wind speed. Specimens were placed in the middle of the tunnel and were subjected to blowing wind for 1 hour at the set wind speed. The wind speed was adjusted to 10, 20, 30 and 40 km $h^{-1}$ in different tests. The mass of the specimen was measured before and after the test and mass loss was calculated per unit area per unit time. For raindrop erosion tests, deionised water fell from a constant height of 400 mm from the vertical centre of the specimen for 1 hour at various rates (3, 6, 9 and 12 mL $min^{-1}$). After the test, a vernier calliper was used to measure the erosion cavity diameter at the widest point as well as the erosion cavity depth at the deepest point.

## RESULTS AND DISCUSSION

**Formation of EICP-treated soil crust** By visual inspection, the EICP-treated soils formed a solid crust ~5 mm thick at the surface. The images of each soil before and after EICP treatment are shown in Figures 3(a) and (b) for KSS and 3(c) and (d) for AIS, respectively. Figures 3(e) and (f) show an image of the EICP-treated soil crusts.

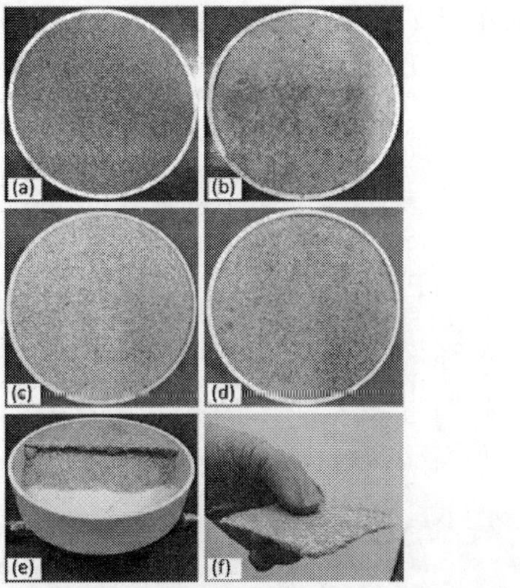


**Figure 3.** Soil specimen showing: (a) untreated KSS; (b) EICP-treated KSS; (c) untreated AIS; (d) EICP-treated AIS; and (e and f) EICP-treated soil crust.

**Strength of EICP-treated soil crust** Figure 4 shows the results of crust strength for KSS and AIS measured at five different locations for each sample. The results show that the strength of the crust at the centre of the sample was significantly higher, and indeed was almost twice that of other locations. For example, maximum crust strengths of 1307 kPa and 1050 kPa were achieved at the centre of KSS and AIS, respectively. This could possibly be due to the accumulation of EICP treatment solution at the centre of the sample during the treatment process, leading to high precipitated $CaCO_3$ bonding, resulting in higher strength.

By comparing the crust strength of the two soil types, the KSS soil showed slightly higher strength than the AIS, possibly due to the differences in chemical composition, particle size distribution and particle shape. Overall, both KSS and AIS showed a significantly high crust strength after just one cycle of EICP treatment. This indicates that an EICP-treated soil crust using urease enzyme from paddy melon seeds has enough strength to reduce weed emergence and therefore could be a sustainable approach for weed control, especially for uncultivated lands such as roadside shoulders and embankments.

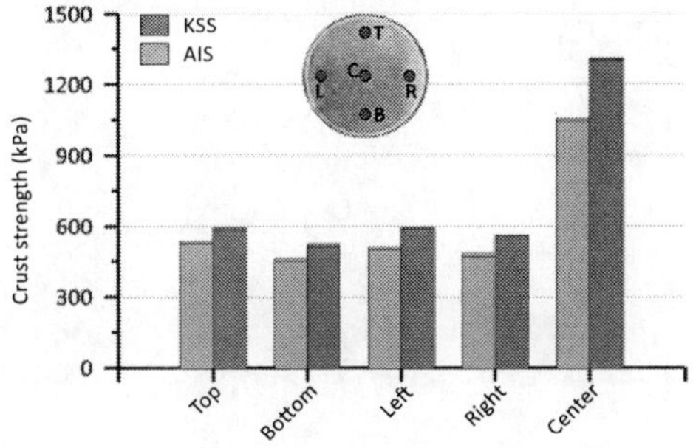


**Figure 4.** A plot of crust strength at different locations for EICP-treated KSS and AIS.

**Erosion resistance of EICP-treated soil crusts** Figure 5 shows the mass loss per unit area and unit time during the wind erosion resistance tests. As expected, the mass loss of the untreated AIS increased rapidly with increasing wind speed and reached nearly 42,000 g $m^{-2}$ $h^{-1}$ at a speed of 40 km $h^{-1}$. The EICP-treated AIS crust showed significantly higher resistance with almost 0 g $m^{-2}$ $h^{-1}$ mass loss up to a speed of 30 km $h^{-1}$. A relatively small increase in mass loss (~4.5 g $m^{-2}$ $h^{-1}$) was observed at a speed of 40 km $h^{-1}$. The results show that the EICP-treated AIS crust exhibited high resistance against wind erosion.

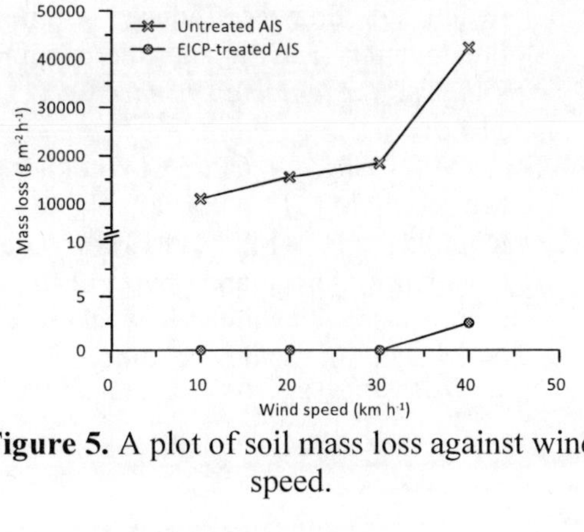


**Figure 5.** A plot of soil mass loss against wind speed.

Figure 6 shows the erosion cavity radius plotted against the flow rate of rain droplets for untreated and EICP-treated AIS crusts. Overall, the erosion cavity radius increased with an increased flow rate of rain droplets. The untreated AIS showed the largest cavity radius compared to the EICP-treated AIS crust. The high resistance to raindrop erosion exhibited by the EICP-treated AIS crust can be attributed to the presence of $CaCO_3$ bonding within the crust.

Concerning the crusts' erodibility, the EICP-treated AIS crust exhibited higher durability than the untreated AIS crust, and its erosion was less progressive under all conditions.

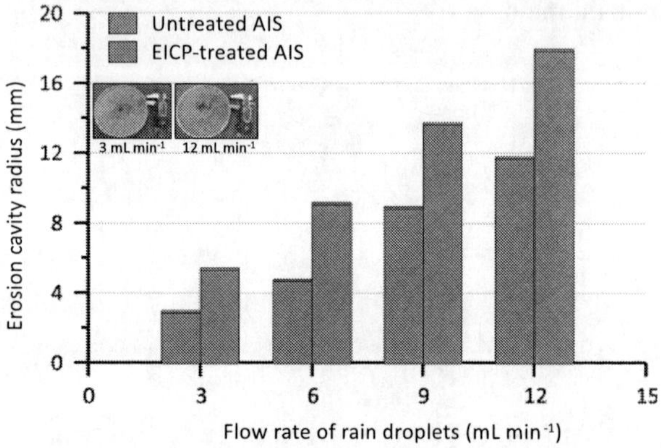


**Figure 6.** A plot of erosion cavity radius against flow rate of rain droplets.

ACKNOWLEDGMENTS

The authors would like to acknowledge technical team at UniSA STEM for their support and guidance.

REFERENCES

Ahenkorah, I., Rahman, M. M., Karim, M. R., and Beecham, S. (2021a). "Enzyme induced calcium carbonate precipitation and its engineering application: A systematic review and meta-analysis." *Construction and Building Materials*, 308(15), 125000.

Ahenkorah, I., Rahman, M. M., Karim, M. R., Beecham, S., and Saint, C. (2021b). "A review of Enzyme Induced Carbonate Precipitation (EICP): The role of enzyme kinetics." *Sustainable Chemistry*, 2(1), 92-114.

Anzooman, M., Dang, Y. P., Christopher, J., Mumford, M. H., Menzies, N. W., and Kopittke, P. M. (2018). "Greater emergence force and hypocotyl cross sectional area may improve wheat seedling emergence in sodic conditions." *Plant Science*, 277, 188-195.

Awadhwal, N., and Thierstein, G. (1985). "Soil crust and its impact on crop establishment: a review." *Soil Tillage Research*, 5(3), 289-302.

Bajwa, A. A., Khan, M. J., Bhowmik, P. C., Walsh, M., and Chauhan, B. S. (2019). "Sustainable weed management." *Innovations in Sustainable Agriculture*, Springer, 249-286.

Bajwa, A. A., Nguyen, T., Navie, S., O'Donnell, C., and Adkins, S. (2018). "Weed seed spread and its prevention: the role of roadside wash down." *Journal of Environmental Management*, 208, 8-14.

Bennett, O., Ashley, D., and Doss, B. (1964). "Methods of reducing soil crusting to increase cotton seedling emergence." *Agronomy Journal*, 56(2), 162-165.

Christoffoleti, P. J., Carvalho, S., Nicolai, M., Doohan, D., and VanGessel, M. (2007). "Prevention strategies in weed management." *Non-chemical weed management: principles, concepts technology*, CABI, Oxfordshire, 1-16.

Joshi, N. (1987). "Seedling emergence and yield of pearlmillet on naturally crusted arid soils in relation to sowing and cultural methods." *Soil Tillage Research*, 10(2), 103-112.

Laker, M. C., and Nortjé, G. P. (2019). "Review of existing knowledge on soil crusting in South Africa." *Advances in agronomy*, 155, 189-242.

Massingue, F. I. (2002). "The effect of crop residue cover and soil texture on crusting." University of the Free State.

Oerke, E.-C. (2006). "Crop losses to pests." *The Journal of Agricultural Science*, 144(1), 31-43.

Parker Jr, J. J., and Taylor, H. M. (1965). "Soil strength and seedling emergence relations. 1. Soil type, moisture tension, temperature, and planting depth effects." *Agronomy Journal*, 57(3), 289-291.

Richards, L. A. (1953). "Modulus of rupture as an index of crusting of soil." *Soil Science Society of America Journal*, 17(4), 321-323.

Sinha, A., and Ghildyal, B. (1979). "Emergence force of crop seedlings." *Plant Soils Foundations*, 51(1), 153-156.

# Supporting weed management including early invader weeds during bushfire recovery in Victoria after the summer of 2019/2020

Kate Blood[1], Bianca Gold[1]
[1]DELWP, Victoria, Australia
(kate.blood@delwp.vic.gov.au)

**Summary** The summer of late 2019 and early 2020 in Victoria was dominated by large-scale and intense bushfires in the north east and east Gippsland. Over 1.5 million hectares across the State were burnt. A short window after bushfire provides a great opportunity to treat emerging weed seedlings and regrowth before they have a chance to smother the ground preventing indigenous plant growth, and before they mature to seeding and reproduce on mass. The Weeds at the Early Stage of Invasion (WESI) Project was tasked with assisting bushfire recovery from the early months of 2020. We share a range of activities and support offered to biodiversity and public land managers, and private land holders to assist nature recover and reduce the burden of environmental weeds in the landscape. With restrictions on travel at the time, the project shifted to a series of webinars in November-December 2020 https://tinyurl.com/WeedsAfterFire.
We share our findings from running online webinars and hints for those thinking of doing the same https://www.environment.vic.gov.au/invasive-plants-and-animals/early-invaders
These webinars were funded by the Victorian Government's $22.5 million Bushfire Biodiversity Response and Recovery (BBRR) program. For more information on the BBRR program, visit www.wildlife.vic.gov.au/home/biodiversity-bushfire-response-and-recovery

**Keywords** Environmental weeds, early invader weeds, webinars, engagement

# Navigating Bushfire Recovery in the Adelaide Hills

Katrina Warner[1], Catherine Austin
[1]Hills And Fleurieu Landscape Board, Mount Barker, Australia
(katrina.warner@sa.gov.au)

**Summary** After witnessing the devastating effects of the 2019 Cudlee Creek bushfire on the landscape, the Hills and Fleurieu Board applied and was successful in receiving both federal and state funding to assist with recovery. The Cudlee Creek Bushfire Recovery project facilitates a targeted, tenure blind approach to pest plant control and supports landholders to implement activities that:

- Manage mass germination of weeds seeds in the soil seedbank that are advantaged by fire,
- Protect and enhance remaining stands of native vegetation, and
- Provide funding to local councils to undertake roadside pest plant control activities

This funding support has presented the Board with a unique opportunity to address weed control issues in a way which traditional programs have not allowed, and we have some great, success stories we would love to share. Broom Control Campaign - English Broom and Cape Broom produce seeds that can be viable for over 40 years and are notoriously fire-active with their seeds forced to germinate. Infestations explode in numbers after fire but there is a significant yet temporary reduction in the seed bank, presenting a time critical opportunity. We put the call out to landholders to sign up for free Broom control and had an excellent response. From September to November 2021 our contractors visited 63 properties and controlled a total of 195.6 Ha of English and Cape Broom prior to seed set. A fantastic result.

Gorse control at Birdwood – Fire prevention works
A Gorse infestation located to the North of Birdwood posed a major bushfire risk to the. The Gorse patch was smothering pasture and native vegetation above the Torrens River and provided a hiding spot for pest animals such as foxes and rabbits. We partnered with the National Parks and Wildlife Burning on Private Lands Program to coordinate a plan for the site which involves initial mulching, prescribed burning and then follow up spraying.

Biological Control Project - 'Nurseries' for Salvation Jane Crown Weevil
Salvation Jane has made a reappearance in the fire scar after almost being eradicated from the Hills landscape with successful biological control. We want to help kick start a weevil boom to work as another tool to help keep the weed under control. In the future, agents can be collected and spread beyond the release site to tackle other Jane infestations.

Bushfire Recovery Action Plans (BRAP)-The Bushfire Recovery Action Plan (BRAP) serves as a tool for landholders to guide their land management efforts over time, and may also serve as a contractual document committing funds for the provision of on-ground works.

# Unexpected outcomes on the road to recovery after the Black Summer Fires - and its mostly Australian natives!

Laura Williams[1], Jason Walter[2]
[1]PIRSA, Kingscote, Australia,
[2]Kangaroo Island Landscape Board, Kingscote, Australia
(laura.williams20@gmail.com)

**Summary** In the summer of 2019/20, Kangaroo Island experienced the most devastating bushfires in living memory. The fires burnt around half of the island, with 56 houses lost and many tens of thousands of animals killed. Kangaroo Island is world renowned for its unique and pristine natural environments and clean and green agricultural sectors. The island is free of many of the weeds, pests and diseases found in other parts of the country. Unfortunately the firefighting and recovery activities have exposed the island to many biosecurity threats. The introduction of weeds on firefighting units, army equipment and heavy machinery brought to, and moved around the island; weeds and pests in imported fodder brought over to feed stock and in grain and other food stuffs distributed for wildlife. The disturbance from the bushfires also makes these environments particularly susceptible to these biosecurity risks.
A number of projects have been developed to minimise these biosecurity risks to the island's fragile environmental and agriculture sectors. Some of the weed incursions have been as expected – woody weeds spreading along new fencelines dispersed by the fencing plant and declared weeds that have not been established previously on the island coming in with new stock. However there have been several unusual and unprecedented weed explosions. Declared weeds like Salvation Jane and Cape tulip have appeared in new areas. Fire adapted Western Australian natives have spread from gardens burnt in the fires. Even more concerning is the weed outbreak in the forestry sector which may prove to be one of the biggest challenges the island will face following the bushfires, but also has the opportunity for the community and many stakeholders to pull together and solve the challenge, thereby protecting the valuable and pristine environments Kangaroo Island is famous for.

**Keywords** Fire; unexpected; declared

# Beyond the boundary: implementing an area-wide weed management approach for mobile crop weeds

Rick Llewellyn[1], Tanja Morgan[2], Iva Quarisa[3], Rebecca Kirby[4], Jason Emms[5], Sonia Graham[7], Chris Preston[8], Christina Ratcliff[1], James Hereward[9], Tim Capon[10], Stuart Whitten[10], Marta Monjardino[1], Vicki Green[6]

[1]CSIRO, Adelaide, Australia,
[2]Mallee Sustainable Farming, Mildura, Australia,
[3]IREC, Griffith, Australia,
[4]Milmerran Landcare, Milmerran, Australia,
[5]GRDC, Adelaide, Australia,
[6]GRDC, Toowoomba, Australia,
[7]University of Wollongong, Wollongong, Australia,
[8]University of Adelaide, Adelaide, Australia,
[9]University of Queensland, Brisbane, Australia,
[10]CSIRO, Canberra, Australia
(rick.llewellyn@csiro.au)

**Summary** Several of the most costly weeds in Australian cropping regions are highly mobile via seed and pollen, with the potential to also spread herbicide resistance. It is well recognised that movement across farm boundaries occurs, often involving increasingly diverse neighbouring land uses such as where there are expanding areas of irrigated plantings alongside dryland grains, as well as adjoining roadsides. Although there are few existing examples in cropping regions, these factors point to increasing potential for area-wide approaches that can bring together multiple sectors to help reduce the impact of some key weed threats. We describe activities in three pilot regions (Darling Downs, Riverina - Murrumbidgee Irrigation Area and Sunraysia) where the potential for area-wide weed management is being evaluated and trialled in partnership with local stakeholders, including those from grains, cotton, horticulture, viticulture and local government. Weed, social, genetic, economic and spatial sciences have been brought together to investigate mobile weed issues prioritised by local growers. Backed by characterisation of local herbicide resistance status and mobility, trials led by regional partners include evaluation of summer weed control options to reduce seed set on grain and horticultural properties, channel bank vegetation for weed suppression, and roadside management options. Demand for greater cross-sector weed management extension activities has been demonstrated, along with improved opportunities to address roadside weeds. Local stakeholder expectations about the costs and benefits of implementing collaborative management of crop weeds highlight the need to clearly define management area scale and engagement demands. By understanding the benefits and key principles influencing success in the pilot regions, opportunities for development of localised approaches to reduce the cost of mobile weeds in other regions of Australia are being identified.

**Keywords** Resistance, mobile, area wide, management, participation

# Opportunities for area-wide management of cropping weeds: a survey of growers in Australia

Silja Schrader[1], Sonia Graham[1], Rebecca Campbell[1], Kaitlyn Height[1]
[1]University of Wollongong, Wollongong, Australia
(sonia.graham1@gmail.com)

**Summary** The mobility of weeds, use of biological controls and spread of herbicide resistance mean that weed management is a landscape-scale problem. Area-wide management (AWM) presents one way to address these challenges by bringing land managers, industry and government representatives together to collectively manage weeds on shared and private properties. Such an approach has been successfully used for other landscape-scale problems, such as managing some animal and insect pests. The aim of this study was to investigate the extent to which growers in three diverse cropping regions in Australia currently engage in area-wide management of weeds and their reasons for doing so. Survey responses of 604 land managers from the Riverina, NSW (n=218), Sunraysia, Victoria (n=200), and the Darling Downs, Queensland (n=186) were recorded between July and September 2021. Almost all (95%) growers agreed or strongly agreed that each land manager has a responsibility to the whole region to control weeds and 84% agreed or strongly agreed that effective control of weeds requires land managers to work together. Yet only 24% of growers currently work with others on weed management. Growers who are less likely to currently engage in AWM of weeds are those who are less concerned about herbicide resistant weeds spreading to neighbouring land and who are unlikely to share information with other land managers about weeds or attend meetings about local weed issues. To support greater uptake of AWM of weeds in the future will require increased awareness and education about the mobility of weeds, building of new networks among growers and other key stakeholders, and development of AWM activities that are accessible to all land managers regardless of time and financial constraints.

**Keywords** Community management, collective action, participation, grower cooperation, area-wide control, cooperative weed management

# A population genetics approach to evaluating weed movement and the role for area wide weed management

James Hereward[1], Jenna Malone[2], Christina Ratcliff[3], Iva Quarisa[4], Rebecca Kirby[5], Tanja Morgan[6] and Chris Preston[2]

[1]The University of Queensland, St Lucia, Brisbane, QLD 4068, Australia
[2] School of Agriculture, Food & Wine, University of Adelaide, PMB 1, Glen Osmond, SA 5064, Australia
[3] CSIRO Waite Campus, Locked Bag 2, Glen Osmond, SA 5064, Australia
[4] Irrigation Research & Extension Committee, 255 Research Station Rd, Griffith, NSW 2680, Australia
[5] Millmerran Landcare Group, 31 Campbell Street, Millmerran QLD 4357, Australia
[6] Mallee Sustainable Farming, 10189 Mildura Vic 3502, Australia
(j.hereward@uq.edu.au)

**Summary** This study used a population genetics approach to assess weed movement within and across each of three regions in Australia – the Darling Downs, the Riverina, and Sunraysia. Populations of three weed species (fleabane, feathertop Rhodes grass and annual ryegrass) were collected across varying land uses and gene flow was investigated using a genotyping by sequencing approach.

Annual ryegrass populations were very similar, with very little genetic differentiation across sites sampled in the Riverina region. This suggests high levels of mixing within the region. Feathertop Rhodes grass populations sampled across the Darling Downs were also genetically similar, indicating high levels of gene flow and movement in this weed species. While some evidence of long-distance dispersal between regions was found, fleabane populations revealed surprising evidence of genetic differences within the Riverina region, and between the Sunraysia and Riverina regions. The research suggests that herbicide resistance in annual ryegrass and feathertop Rhodes grass is moving widely across the regional landscape.

**Keywords** Population genetics, gene flow, weed, area wide management.

## INTRODUCTION

Understanding how far and how quickly a weed species can spread is important in planning an area wide weed management programme. Highly mobile weeds more rapidly become shared problems, especially as increasing numbers of weed species are becoming resistant to key herbicides such as glyphosate. The increase and spread of herbicide resistance are seen as key issues by growers in the regions investigated in this study (Height *et al.* 2022).

This study was designed to investigate the movement of key weeds in the three regions. Weed species were selected based on high potential mobility, the occurrence of herbicide resistance, and concerns of growers in these regions (Height *et al.* 2022). The three species selected were fleabane (*Conyza bonariensis*), annual ryegrass (*Lolium rigidum*), and feathertop Rhodes grass (*Chloris virgata*).

Fleabane has high fecundity and high dispersal capacity due to its wind (and potentially water) dispersed seeds (Wu 2007). Glyphosate resistance was first detected in Queensland in 2006 (Walker *et al.* 2011), but by 2018 all populations of fleabane screened for glyphosate resistance across Queensland were resistant (Jalaludin *et al.* 2019).

Feathertop Rhodes grass was always difficult to control with glyphosate and was considered tolerant of the herbicide, but widespread use of glyphosate has resulted in the evolution of several different target site resistance mutations in this species (Ngo *et al.* 2018) making it even harder to control.

In contrast to the other two species, annual ryegrass is a self-incompatible, outcrossing weed species. Pollen mediated gene flow of herbicide resistance genes has been measured up to the maximum tested distance of 3,000m in an experiment in Western Australia (Busi *et al.* 2008). The species is diploid, favouring the evolution of target site herbicide resistance, and outcrossing has led to the widespread occurrence of populations resistant to multiple modes of action (Matzrafi *et al.* 2021).

The goal of this study was to assess the movement of these three weed species by comparing the genetics of weeds sampled at the same sites across two seasons (2020; 2021). This paper reports the results of the genotyping of the populations collected in 2020 and assesses population structure within and between regions to infer patterns of gene flow and mobility.

## MATERIALS AND METHODS

In 2020, fleabane was collected in the Riverina and Sunraysia, feathertop Rhodes grass was collected in the Darling Downs, and annual ryegrass was sampled in the Riverina. At each site 32 individuals were sampled, and leaf material was placed directly into silica gel for DNA preservation.

DNA was extracted from all samples using a CTAB buffer and spin column extraction (Ridley *et al.* 2016). Genome-wide single nucleotide polymorphysim (snp) data were generated using a genotyping-by-sequencing method. The protocol and adaptor regime can be found online at http://www.jameshereward.org/GBS.html. We pooled 288 individuals per sequencing lane and sequenced the libraries with PE150 Illumina sequencing at Novogene (Beijing, China).

The sequence data were demultiplexed, assembled, and snps were called using STACKS (Catchen *et al.* 2013). A variant call format (vcf) file of the genotype data was output with any marker having a heterozygosity above 0.65 discarded. The genotypes were then filtered using vcftools (Danecek *et al.* 2011) to a minor allele count of three (one heterozygote and one homozygote), to conservatively remove singleton snps that are likely to be errors. We also set a minimum depth of five and kept only biallelic snps. We removed missing data in three steps. First, any marker missing more than 50% data was discarded to remove the markers most affected by missing data. Second, any individual that had missing data at more than 50% of the markers was discarded to remove the individuals that had bad quality genotyping. Finally, any marker missing more than 5% data was discarded to produce a final dataset with relatively little missing data (~3%).

For each species and each region, a principal component analysis (PCA) was conducted using the adegenet package in R (Jombart 2008). We then assessed geneflow and genetic clustering by performing a STRUCTURE analysis for each species in each region, this program uses a Bayesian algorithm to assign individuals to each of K hypothetical populations based on allele frequencies (Pritchard *et al.* 2000). For each species and each region, we performed 10 runs of the program using different starting seeds. For fleabane in the Riverina, we assumed K=3 populations, and for all the others K=2, the algorithm was run using the 'admixture' model and with 100,000 iterations of 'burn-in' followed by 1,000,000 iterations.

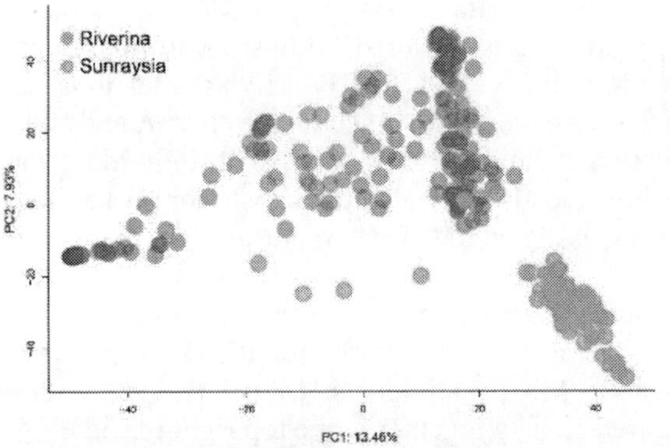


Figure 1. Genetic clustering (principal component analysis axes one and two) for all fleabane samples (Riverina and Sunraysia).

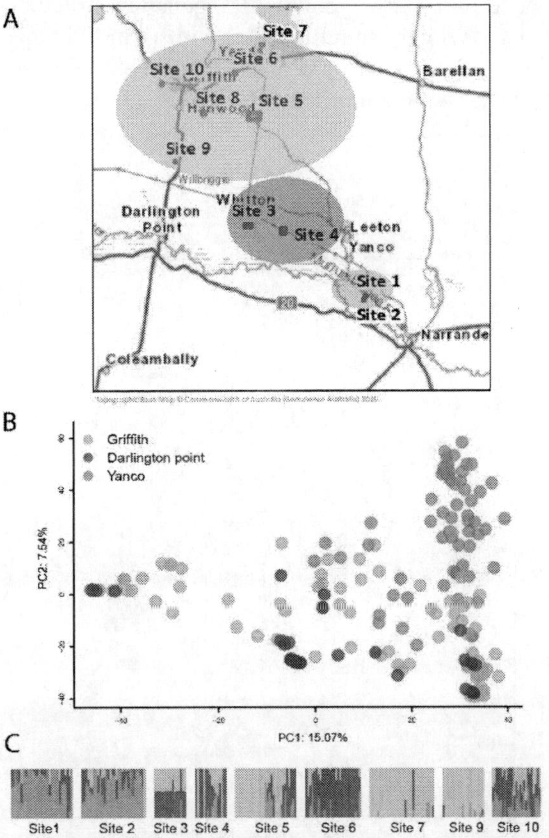


Figure 2. Genetic analysis of fleabane populations from the Riverina region, A; sampling locations, B; plot of principal components analysis of the genetic data showing axes one and two, C; plot showing the results of the STRUCTURE analysis.

## RESULTS

**Fleabane.** When the fleabane samples from the Riverina and Sunraysia were compared to each other, the samples from each region largely clustered independently (Fig. 1). Two individuals from Sunraysia placed within the Riverina cluster, and these likely represent movement of weeds between regions. The genotypic diversity recovered was quite high, indicating that fleabane populations do outcross under field conditions (although likely at very low frequencies).

In the Riverina region, individuals from around Yanco (sites 1 and 2, Fig. 2A) clustered together in the top right of the PCA plot (Fig. 2B). These sites were also assigned to a separate cluster from the other sites in the STRUCTURE analysis (Fig. 2C). Overall, the STRUCTURE plot indicated high levels of admixture as indicated by many individuals with a posterior probability of being assigned to more than one cluster. In the PCA (Fig. 2B), sites around Darlington were clustered with sites from around Griffith. In the STRUCTURE analysis, sites four six and ten were mostly assigned to one cluster, and sites five, seven and nine to another, with individuals at site three mostly having a 50/50 posterior probability of being assigned to these two clusters.

**Feathertop Rhodes grass.** Genetic and genotypic diversity was lower for feathertop Rhodes grass than fleabane. Most individuals from all sites were clustered closely together in the top right of the PCA plot (Fig. 3B) with only a few individuals being separated by principal components one and two. In the STRUCTURE analysis there was very little evidence for any geographic genetic structure associated with the sampling sites with all individuals having a high posterior probability of belonging to one genetic population (Fig. 3C).

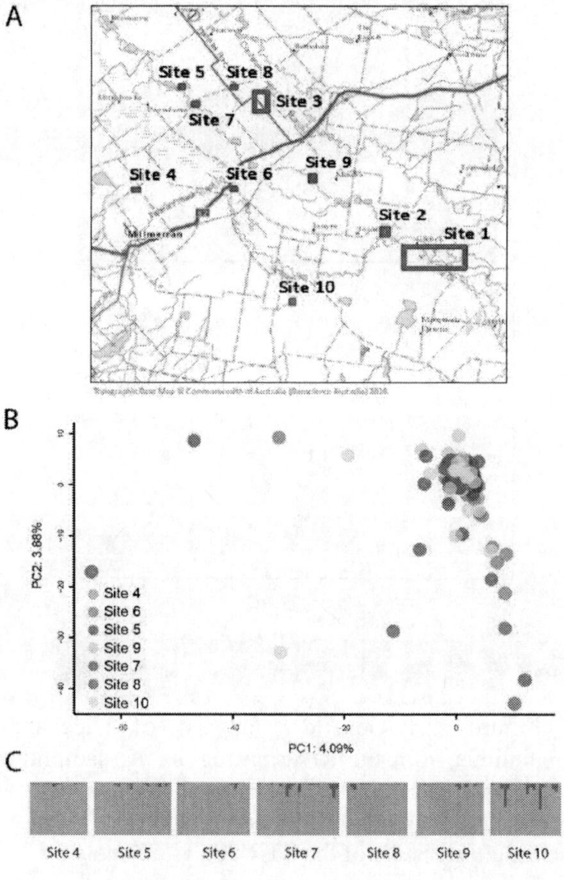


Figure 3. Genetic analysis of feathertop Rhodes grass, A; sampling locations B; plot of principal components analysis axes one and two, C; STRUCTURE analysis.

**Annual Ryegrass.** Samples were collected from ten sites during 2020 with the genetic analysis of three of them presented here. Site one (near Yanco), site two (near Griffith) and site three (near Darlington Point) were all clustered together in the PCA (Fig. 4B). The STRUCTURE analysis indicated that all three sites represented a single genetic population with every individual having around 10% posterior probability of being assigned to one population and 90% posterior probability of being assigned to the second one (Fig. 4C). There was very little evidence of any admixture in the structure analysis.

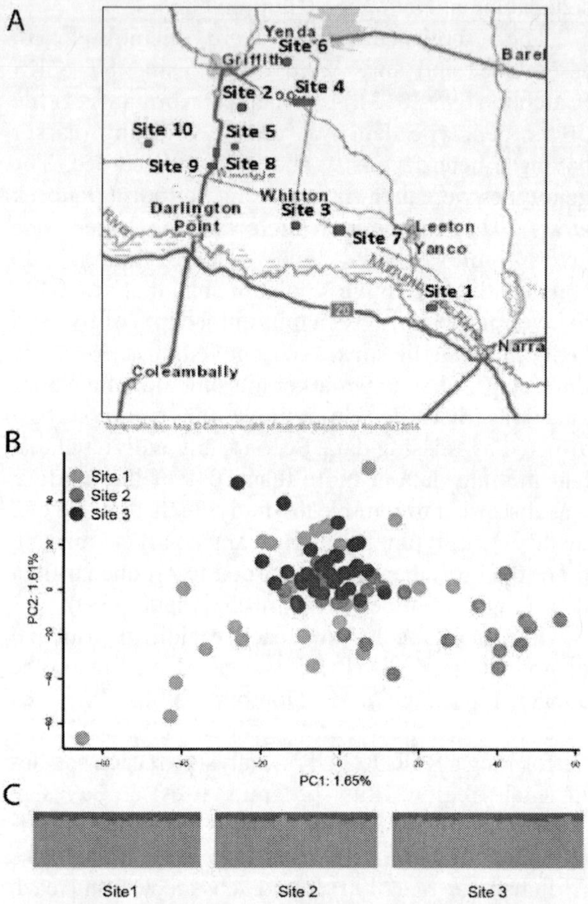


Figure 4. Genetic analysis of annual ryegrass populations from the Riverina region, A; sampling locations B; plot of principal components analysis axes one and two, C; STRUCTURE analysis.

## DISCUSSION

Despite its reputation for long distance dispersal by wind, we detected local genetic population structure in fleabane. While some evidence of long-distance dispersal was identified, substantial genetic differences were found between the Sunraysia and Riverina regions. Most notably, genetic differences were found within the Riverina region and, to a lesser extent, Sunraysia. Our results suggest that local dispersal may be a more important driver of fleabane population genetics than long-distance dispersal. The extremely high seed production of fleabane may mean that a few long-distance dispersal events tend not to contribute significantly to local population dynamics, the short life of the seed bank also leads to rapid population turnover in this species. The spread of glyphosate resistance across the whole of Queensland within 12 years of its first detection highlights the importance of weed movement in spreading herbicide resistance in this species (Walker *et al.* 2011, Jalaludin *et al.* 2019).

We found very little evidence for genetic structure in feathertop Rhodes grass populations within the Darling Downs region, and none for annual ryegrass within the Riverina. This indicates very high levels of movement of these two species within the regions investigated here. Our results for annual ryegrass are consistent with previous studies indicating very long-distance dispersal capability of resistance genes via pollen in this obligately cross-pollinating species (Busi *et al.* 2008). Previous population genetic analysis of feathertop Rhodes grass suggests that glyphosate resistance has evolved at least 12 times in this species, with one of the resistance alleles having been recorded at sites 700km apart from each other (Hereward unpublished data). Although this species is regarded as both cross and self-pollinating, our data indicated a close clustering of genotypes, suggesting that self-pollination has been predominant in feathertop Rhodes grass populations, as self-pollination tends to reduce genotypic diversity. Further research on the extent of cross or self-pollination in this species is warranted.

The high mobility of these three weed species within regions will lead to the rapid spread of herbicide resistance genes across the landscape (Preston *et al.* 2022), highlighting the importance of early detection and elimination of herbicide resistant populations. Co-ordinated efforts to control herbicide survivors and eliminate resistant populations across land uses would have area wide community benefits by reducing the spread of herbicide resistance, especially at the regional scale.


## ACKNOWLEDGMENTS

This project is funded through the Australian Government's Rural R&D for Profit program, GRDC, and CRDC, with additional support from CSIRO, AgriFutures, Universities of Queensland, Adelaide, and Wollongong, Mallee Sustainable Farming, Millmerran Landcare Group and IREC Inc.


## REFERENCES


Busi, R., Yu, Q., Barrett-Lennard, R. and Powles, S. (2008). Long distance pollen-mediated flow of herbicide resistance genes in *Lolium rigidum. Theoretical and Applied Genetics* 117(8), 1281.

Catchen, J., Hohenlohe, P., Bassham, S., Amores, A., and Cresko, W. (2013). Stacks: an analysis tool set for population genomics. Molecular Ecology, 22(11), 3124–3140.

Danecek, P., Auton, A., Abecasis, G., Albers, C., Banks, E., DePristo, *et al.* (2011). The variant call format and VCFtools. Bioinformatics, 27(15), 2156–2158.

Height, K., Graham, S., Campbell, R., Hawkes, G., Schrader, S., Blessington, L., and McKinnon, S. (2022). Opportunities to Manage Herbicide Resistance through Area-Wide Management: Lessons from Australian Cropping Regions. Sustainability: Science Practice and Policy, 14(3), 1793.

Jalaludin, A., Widderick, M., Broster, J., Chambers, A and Walsh, M. (2019). Herbicide resistance survey results of the Northern cropping region – GRDC update papers.

Jombart, T. (2008). adegenet: a R package for the multivariate analysis of genetic markers. Bioinformatics, 24(11), 1403–1405.

Matzrafi, M., Preston, C., and Brunharo, C. A. (2021). Review: evolutionary drivers of agricultural adaptation in Lolium spp. Pest Management Science, 77(5), 2209–2218.

Ngo, T. D., Krishnan, M., Boutsalis, P., Gill, G., and Preston, C. (2018). Target-site mutations conferring resistance to glyphosate in feathertop Rhodes grass (*Chloris virgata*) populations in Australia. Pest Management Science, 74(5), 1094–1100.

Preston C. *et al.* (2022). Glyphosate resistance in mobile weeds across land uses: implications for area wide management of weeds. Proceedings 22nd Australasian Weeds Conference.

Pritchard, J., Stephens, M., and Donnelly, P. (2000). Inference of population structure using multilocus genotype data. Genetics, 155(2), 945–959.

Ridley, A., Hereward, J., Daglish, G., Raghu, S., McCulloch, G., and Walter, G. (2016). Flight of *Rhyzopertha dominica* (Coleoptera: Bostrichidae)—a ppatio-temporal analysis with pheromone trapping and population genetics. Journal of Economic Entomology, 109(6), 2561–2571.

Walker, S., Bell, K., Robinson, G., and Widderick, M. (2011). Flaxleaf fleabane (*Conyza bonariensis*) populations have developed glyphosate resistance in north-east Australian cropping fields. Crop Protection, 30(3), 311–317.

Wu, H. (2007). The biology of Australian weeds 49. *Conyza bonariensis* (L.) Cronquist. Plant Protection Quarterly, 22(4), 122.

# Glyphosate resistance in mobile weeds across land uses: implications for area wide management of weeds

Christopher Preston[1], Fleur C Dolman[1], Tanja Morgan[2], Iva Quarisa[3], Rebecca Kirby[4] and Christina Ratcliff[5]
[1] School of Agriculture, Food & Wine, University of Adelaide, PMB 1, Glen Osmond, SA 5064, Australia
[2] Mallee Sustainable Farming, 7B Byrne Ct, Mildura VIC 3500, Australia
[3] Irrigation Research and Extension Committee, Lot 255 Research Station Road, Griffith NSW 2680, Australia
[4] Millmerran Landcare Group Inc, Campbell St, Millmerran QLD 4357, Australia
[5] CSIRO, Locked Bag 2, Glen Osmond SA 5064, Australia
(christopher.preston@adelaide.edu.au)

**Summary** Weeds that have the ability to rapidly disperse across the landscape can create ongoing problems for individual land managers. Efforts by land managers to reduce the risk of a weed problem can be negated by such mobile weeds moving across boundaries. When mobile weeds are easy to control using current practices, there is little incentive for more coordinated action among land managers. However, for more difficult to control weeds (or resistant weeds) there may be a case for collective action through an area wide weed approach. This project examined the distribution of glyphosate resistant weeds at a local scale in the Sunraysia, the Riverina and the Darling Downs. Based on local land manager consultation, the study focused on the three most concerning mobile weeds: *Conyza bonariensis* (flaxleaf fleabane) in the Sunraysia, flaxleaf fleabane and *Lolium rigidum* (annual ryegrass) in the Riverina and *Chloris virgata* (feathertop Rhodes grass) in the Darling Downs. Geo-referenced weed seed samples, plants in the case of annual ryegrass, were collected from a variety of land uses in each district in 2019-2020 and 2020-2021, including public land and roadsides, and the samples tested for resistance to glyphosate. Samples with survivors to glyphosate were considered to be resistant. High frequencies of glyphosate resistance were identified in all weeds and in all districts ranging from 5% of flaxleaf fleabane samples in the Sunraysia in 2021 to 81% of annual ryegrass samples in the Riverina in 2021. Spatial patterns of resistance status were analysed for each weed and region. Resistance occurred across the area sampled and was present on multiple land uses, including non-farm land. As land managers from multiple sectors already need to manage these resistant weeds, the incorporation by all land managers of practices that reduce seed production, and hence potential for spread of these weeds would be beneficial.

**Keywords** Area wide management, mobile weeds, glyphosate resistance.

## INTRODUCTION

Area wide pest management (AWM) is an approach to pest management where the pest is managed collectively across a geographic area, rather than treating individual fields for pests. AWM offers benefits through all populations of the pest within the geographic region being suppressed, rather than allow some populations to expand and risk infesting already treated or new areas (Kruger 2016). AWM has been used successfully for the management of many insect species, including *Helicoverpa armigera* Hübner and various fruit flies in Australia (Fitt 2000, Lloyd et al. 2010). AWM is most feasible where pest populations are mobile and there is an economic penalty from the pest on a large number of land holders (Lloyd et al. 2010, Kruger 2016).

There are few examples of AWM being used for weeds. Most AWM programs involving weeds are eradication programs where there are additional regulatory options for ensuring cooperation (Panetta 2014). Some of the challenges of employing AWM for weeds are the view that weeds are not very mobile, the perception that weeds are relatively easy to manage and land managers often managing a suite of weed species (Ervin and Frisvold 2016). In addition, area wide management is logistically complex (Kruger 2016).

Glyphosate is a herbicide widely used across many land uses in Australia. The intensive use of glyphosate has resulted in the evolution of resistance to this herbicide in 21 weed species in Australia (Heap 2022). Resistance occurs in situations where glyphosate is repeatedly used including grain production, horticulture, irrigation channels and roadsides (Preston 2010a). Mobile weed species with glyphosate resistance are likely to be challenging for land managers to control. Weeds such as these would be ideal to explore the benefits of AWM for weeds.

In this study, localised physical surveys in the Sunraysia, Riverina and Darling Downs of weed species were conducted to identify the proportion of weeds resistant to glyphosate and their location across land uses.

## MATERIALS AND METHODS

**Sample collection** Weed species sampled were identified following consultation with land managers in each of the regions to identify mobile weeds of concern for land managers to control. Surveys focused on *Conyza bonariensis* (L) Cronq. (flaxleaf fleabane) in the Sunraysia, flaxleaf fleabane and *Lolium rigidum* Gaud. (annual ryegrass) in the Riverina and *Chloris virgata* Sw. (feathertop Rhodes grass) in the Darling Downs.

During 2019-2020, project participants in each of the regions collected seed heads from individual, geo-referenced plants of flaxleaf fleabane and feathertop Rhodes grass and sent these to the University of Adelaide for testing. For annual ryegrass in the Riverina, 6 to 10 plants from a geo-referenced location were collected when well-tillered and sent to the University of Adelaide for testing. A second collection of weeds was made in 2020-2021 from the Sunraysia and Riverina only.

**Testing for resistance to glyphosate** Seed of flaxleaf fleabane and feathertop Rhodes grass were sown onto the surface of potting mix and watered. Once the seed had germinated, plants were transplanted into 0.55 L square pots (Masrac, SA) containing potting mix with 5 plants per pot, and grown outdoors at the Waite Campus, University of Adelaide (Boutsalis et al. 2012). At the 6 to 8-leaf stage for flaxleaf fleabane and 4 to 6-leaf stage for feathertop Rhodes grass, duplicate pots for each sample were treated with 1080 g a.e. $ha^{-1}$ glyphosate (WeedMaster® Argo®, 540 g $L^{-1}$ glyphosate present as the potassium and isopropylamine salts, Nufarm Australia) using a moving boom laboratory spray cabinet (Boutsalis et al. 2012). At 28 days after application survival was assessed. Samples with survivors were recorded as resistant and samples where all plants were killed were recorded as susceptible. Susceptible control populations were completely controlled by this rate of glyphosate.

Annual ryegrass resistance to glyphosate was assessed using the Quick Test method (Boutsalis 2001). Plants of annual ryegrass were divided into 3 pieces containing 2 to 4 tillers each. The plants were trimmed and potted into 0.55 L square pots containing potting mix. The three pieces of each plant were placed in the same position in three pots. A week after transplanting, when plants had produced new green leaf tissue, two pots of each population were treated with 540 g $ha^{-1}$ glyphosate as above. The final pot was an untreated control. A single pot of susceptible annual ryegrass plants was also treated with glyphosate. Plants were assessed for survival at 28 days after treatment. Samples with survivors were recorded as resistant and samples where all plants were killed were recorded as susceptible.

**Spatial pattern of resistance** Resistant and susceptible samples for each of the regions were plotted on maps using the geo-references. The land use was recorded at the time of collection. Further analysis of spatial patterns was supplemented by other freely available datasets like land cover, transport and hydrology networks.

## RESULTS

Not all samples collected could be tested for resistance to glyphosate. Some seed samples of flaxleaf fleabane and feathertop Rhodes grass failed to germinate. In addition, some annual ryegrass samples failed to grow once transplanted.

Resistance to glyphosate was identified in each of the weed species tested in 2020 (Table 1) and in 2021 (Table 2). There were differences in the percentage of glyphosate resistant samples within a region between years. Glyphosate resistance status by land use was analysed for fleabane in the Riverina and resistance was found across the range of land uses. This includes samples from within agricultural land (crop, orchard, vineyard) where 60% of samples were resistant and from the edge of roads, tracks and channels where 54% of samples were resistant.

Flaxleaf fleabane was sampled in Sunraysia and Riverina in both years. In 2020, the frequency of glyphosate resistance was high in both regions with 42% of samples tested from the Sunraysia resistant and 64% of tested samples from the Riverina resistant (Table 1). The frequency of resistance in this species was lower in both regions in 2021 (Table 2).

**Table 1.** Percentage of tested weed samples resistant to glyphosate in each of the weed species collected in Sunraysia, Riverina and Darling Downs during 2019-2020.

| Region | Weed species | Samples tested | Samples with resistance to glyphosate (%) |
|---|---|---|---|
| Sunraysia | Flaxleaf fleabane | 50 | 42 |
| Riverina | Flaxleaf fleabane | 64 | 64 |
| | Annual ryegrass | 18 | 67 |
| Darling Downs | Feathertop Rhodes grass | 36 | 50 |

**Table 2.** Percentage of tested weed samples resistant to glyphosate in each of the weed species collected in Sunraysia and Riverina in 2020-2021.

| Region | Weed species | Samples tested | Samples with resistance to glyphosate (%) |
|---|---|---|---|
| Sunraysia | Flaxleaf fleabane | 55 | 5 |
| Riverina | Flaxleaf fleabane | 57 | 37 |
| | Annual ryegrass | 16 | 81 |

Annual ryegrass was sampled in the Riverina in both years. The frequency of glyphosate resistance was high in both years of sampling, 67% of samples in 2020 (Table 1) and 81% in 2021 (Table 2). Feathertop Rhodes grass was sampled only in the Darling Downs in 2020 and 50% of the samples were resistant.

The spatial distribution of glyphosate resistant samples of each species was dispersed across the collection region and interspersed with susceptible samples. Figure 1 illustrates the distribution of flaxleaf fleabane in the Riverina in the two years. Resistant samples of each weed species were found across all land uses.

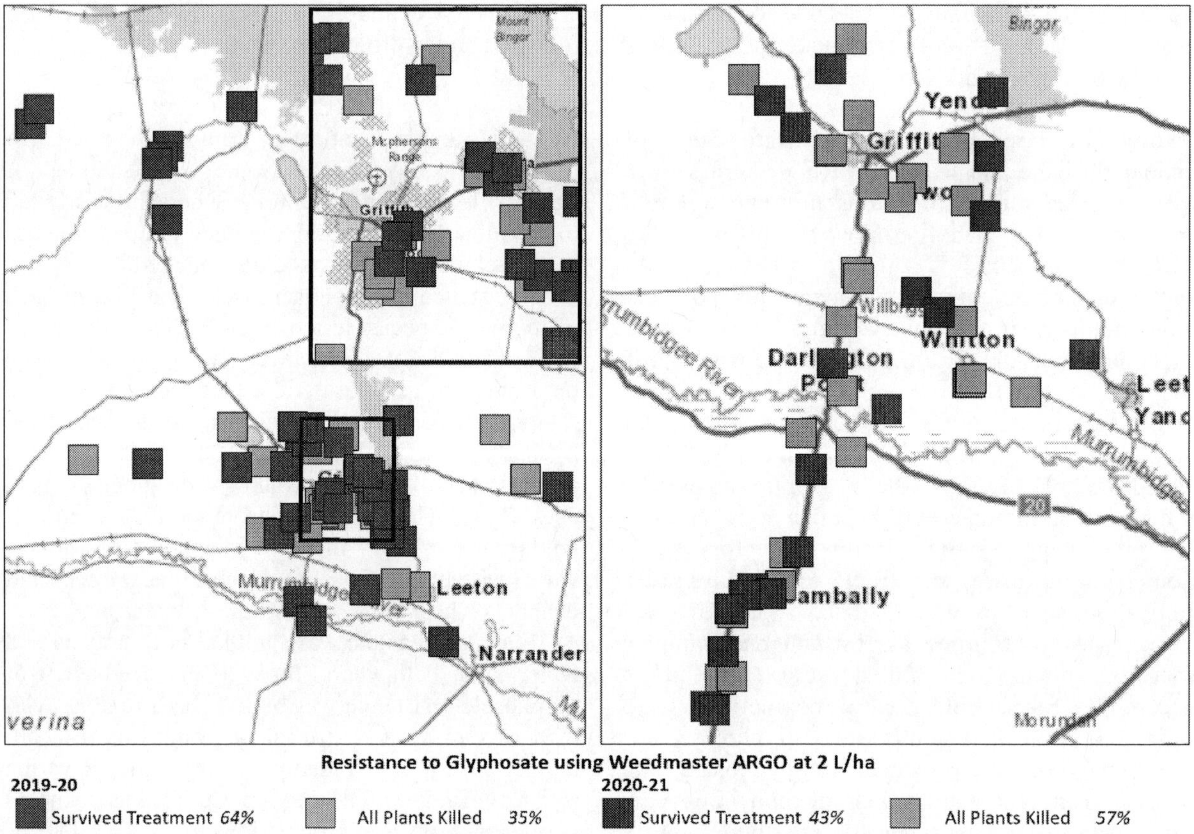


**Figure 1.** Distribution of glyphosate resistant samples of flaxleaf fleabane (dark symbols) and susceptible samples (light symbols) in the Riverina collected in 2019-2020 (Left) and 2020-2021 (Right).

## DISCUSSION

High frequencies of resistance to glyphosate in flaxleaf fleabane, annual ryegrass and feathertop Rhodes grass was identified in the localized surveys conducted here. Glyphosate resistant weed samples were dispersed across the landscape and occurred in multiple land uses. These are all relatively mobile weed species where glyphosate resistance had previously been identified in multiple land uses (Preston 2010b, Malone et al. 2012, Ngo et al. 2018, Aves et al. 2020).

There were variations between years in the frequency of glyphosate resistant samples identified, particularly for flaxleaf fleabane. In 2019-20, 42% of all flaxleaf fleabane samples collected in Sunraysia were resistant to glyphosate (Table 1), but only 5%

of samples collected in 2020-2021 (Table 2). While sampling was conducted across a localized area, the same geo-referenced locations were not sampled in both years (Figure 1). Sunraysia had a severe drought in 2019 resulting in fewer weeds being present and those were concentrated along water courses and road sides. Considerably more rainfall occurred in 2020 resulting in a more dispersed distribution of weeds. This suggests that some land uses may have higher likelihood of glyphosate resistant summer weed occurrence, which is being further investigated.

As glyphosate resistance in each of the three regions sampled occurred across multiple land uses, many land managers are already having to manage glyphosate resistant weeds. Land managers who fail to control these glyphosate resistant weeds on their land could have populations increase and spread to neighbouring land (Hereward et al. 2022). This creates an opportunity for collective management where all land managers could reduce seed set of the weeds and thereby reduce spread of glyphosate resistance. The challenge will be to encourage all land managers to participate and demonstrate the benefits that could arise from joint action (Height et al. 2022).

## ACKNOWLEDGMENTS

This project is supported by funding from the Australian Government Department of Agriculture, Water and the Environment as part of its Rural R&D for Profit program and involves Grains Research and Development Corporation, Cotton Research and Development Corporation, AgriFutures Australia, CSIRO, University of Queensland, University of Adelaide, University of Wollongong, Mallee Sustainable Farming, Millmerran Landcare Group and Irrigation Research & Extension Committee Inc.

## REFERENCES

Aves, C., Broster, J., Weston, L., Gill, G.S. and Preston, C. (2020). *Conyza bonariensis* (flax-leaf fleabane) resistant to both glyphosate and ALS inhibiting herbicides in north-eastern Victoria. *Crop and Pasture Science* 71, 864-71.

Boutsalis, P. (2001). Syngenta quick-test: a rapid whole-plant test for herbicide resistance. *Weed Technology* 15 257-263.

Boutsalis, P., Gill, G.S. and Preston, C. (2012). Incidence of herbicide resistance in rigid ryegrass (*Lolium rigidum*) across southeastern Australia. *Weed Technology* 26, 391-8.

Ervin, D.E. and Frisvold, G.B. (2016). Community-based approaches to herbicide-resistant weed management: lessons from science and practice. *Weed Science* 64, 609-26.

Fitt, G.P. (2000). An Australian approach to IPM in cotton: integrating new technologies to minimise insecticide dependence. *Crop Protection* 19, 793-800.

Heap, I. (2022). The International Herbicide-Resistant Weed Database. Available www.weedscience.org (accessed 30 March 2022).

Height, K., Graham, S., Campbell, R., Hawkes, G., Schrader, S., Blessington, L. and McKinnon, S. (2022). Opportunities to manage herbicide resistance through area-wide management: lessons from Australian cropping regions. Sustainability 14, 1793.

Hereward, J., Malone J.M., Ratcliff, C., Quarisi, I., Kirby, R. Morgan, T. and Preston, C. (2022) A population genetics approach to evaluating weed movement and the role for area wide weed management. Proceedings of the 22nd Australasian Weeds Conference.

Kruger, H. (2016). Designing local institutions for cooperative pest management to underpin market access: the case of industry-driven fruit fly area-wide management. *International Journal of the Commons* 10, 176–99.

Lloyd, A.C., Hamacek, E.L., Kopittke, R.A., Peek, T., Wyatt, P.M., Neale, C.J., Eelkema, M. and Gu, H. (2010). Area-wide management of fruit flies (Diptera: Tephritidae) in the Central Burnett district of Queensland, Australia. *Crop Protection* 29, 462-9.

Malone, J., Cook, A., Wu, H., Hashem, A. and Preston, C. (2012). *In* Management of glyphosate resistant weeds in non agricultural areas. In 18th Australasian Weeds Conference Proceedings, eds V. Eldershaw, pp. 184-6. (Weed Society of Victoria, Melbourne).

Ngo, T.D., Krishnan, M., Boutsalis, P., Gill, G. and Preston, C. (2018). Target-site mutations conferring resistance to glyphosate in feathertop Rhodes grass (*Chloris virgata*) populations in Australia. *Pest Management Science* 74, 1094-1100.

Panetta, F.D. (2014). Weed eradication feasibility: lessons of the 21st century. *Weed Research* 55, 226-38.

Preston, C. (2010a). Managing glyphosate resistant weeds in Australia. *In* 17th Australasian Weeds Conference Proceedings, eds S.M. Zydenbos, pp. 26-30. New Zealand Plant Protection Society, Christchurch, New Zealand.

Preston, C. (2010b). Glyphosate-resistant rigid ryegrass in Australia. *In* Glyphosate Resistance in Crops and Weeds: History, Development, and Management, ed V.K. Nandula, pp.233-47. John Wiley & Sons, Hoboken, NJ.

# Increasing cropping systems resilience to reduce the costs of new weed and resistance arrivals

Tim Capon[1], Marta Monjardino[1], Rick Llewellyn[1], Sonia Graham[2], Stuart Whitten[1]
[1]CSIRO, Australia,
[2]University of Wollongong, Australia
(tim.capon@csiro.au)

**Summary** The invasion of a new weed species or increasing herbicide resistance can be costly to rectify where they are not readily controlled by existing practices. Consultations with growers and advisors reveal concerns with mobile new weeds and the mobility of herbicide resistant weeds. Area-Wide Management (AWM) can help coordinate surveillance and communicate information about new and emerging weed problems to encourage improved preparedness and faster adjustment across growing regions with shared problems. Better control of new mobile weed problems at the farm-level could help prevent costly incursions at this scale and avoid further spread across cropping regions. Accordingly, we investigate whether Integrated Weed Management (IWM) can help to reduce costs and increase resilience to new weed problems, contributing to an AWM approach. We focus on summer fallow weeds in case studies for Western Australia, the South Australian/Victorian Mallee, and the Darling Downs in Queensland. We are collecting data by consulting with weed experts to better understand which additional summer fallow weeds would increase weed management costs. To complement this analysis, the Ryegrass Integrated Management (RIM) model was used to evaluate the costs of gaining glyphosate resistant annual ryegrass in winter cropping systems. Scenarios examined glyphosate applications and IWM practices including early seeding and harvest weed seed control. We found the cost of gaining herbicide resistance in this context was not necessarily extreme when more diverse weed control practices and competitive crops were in place. Overall, growers investing in a diverse weed management strategy are likely to achieve profitable management of existing weeds and reduce costs and risks from new resistance and weed introductions. Practices that are likely to reduce the seed set of new potentially mobile weeds can also improve AWM.

**Keywords** Integrated Weed Management (IWM), Ryegrass Integrated Management (RIM) model, herbicide resistance

# Developing practical tools to support biosecurity legislation

Matt Sheehan[1], S Potter[1]
[1]Wild Matters Pty. Ltd.,
(matt@wildmatters.com.au)

**Summary** Australia has seen significant biosecurity reforms in the past decade. New measures aim to streamline systems and offer a more flexible and consistent approach to invasive species management. With a key focus on risk-based decision making, governments are emphasising biosecurity as a shared responsibility. In Queensland the Biosecurity Act 2014 (the Act) includes several fundamental changes to previous legislation, including:

- The introduction of outcome-based legislation (through a General Biosecurity Obligation);
- The requirement to establish reasonable and practical management measures based on the level of risk for each situation;
- Changes to the way declared species are defined and categorised; and
- The introduction of the precautionary principle.

These changes represent a paradigm shift for those enforcing the Act, and have introduced uncertainty regarding appropriate planning, enforcement and administration processes. In recognition of this the Queensland Department of Agriculture and Fisheries, in partnership with the Local Government Association of Queensland, have invested in the development of simple and practical tools that assist local governments. The tools provide guidance and best practice approaches that can be used to develop local government biosecurity plans, a mandatory document under the Act. The presentation will present a brief overview of the Act, the engagement process to determine areas of greatest need and the development of a series of biosecurity planning and support tools for local governments. Detail will be provided on:

- How to identify and address the needs and varying capacities of local authorities to plan and prioritise weed and pest animal management;
- Instilling confidence in providing direction to meet a general biosecurity obligation; and
- Communicating the new approach to local communities. The state-wide roll-out of these tools seeks to improve collaboration and consistency across nearly eighty local government areas, enabling biosecurity officers to reduce risk posed by biosecurity matter in Queensland.

**Keywords** Biosecurity legislation, local government, planning, tools

# A new post-border weed biosecurity risk management system

Paul O. Downey[1], Simon Holloway[2] and Neville Plumb[2]
[1] Institute for Applied Ecology, University of Canberra, Bruce, ACT 2617, Australia
(paul.downey@canberra.edu.au)
[2] Queanbeyan–Palerang Regional Council, Bungendore, NSW 2621, Australia

**Summary** One of the key tools guiding weed management are risk assessment / management systems. Whilst a post-border weed risk management (WRM) system was established in Australia 15 years ago, very little further development and testing has occurred. More recently, the adoption of a biosecurity model for weed management predicated on the management response being proportional to the risk posed has been introduced. A review of the existing WRM system showed it was no longer fit-for-purpose as a standalone system, given the current biosecurity framework. Here we outline a new post-border weed biosecurity risk management system. The development of the new system is based on extensive research and testing, and multiple revisions over the past three years. The new system has also been successfully converted into a mirror draft version for pest animals and a freshwater aquatic weed version is in development. A short overview of the new system is outlined below.

**Keywords** Risk Assessment, biosecurity, prevention, eradication, containment, asset protection, species reduction, risk-response, general biosecurity duty.

## INTRODUCTION

The development of weed risk assessment systems in Australia started in the late 1980s and culminated with a pre-border Weed Risk Assessment (WRA) system for screening out future weed species prior to entry (Pheloung *et al.* 1999), and later a national post-border Weed Risk Management (WRM) system for weed species that have become established in Australia (Anon. 2006); also adopted to individual states (e.g. SA (Virtue 2008), NSW (Johnson 2009) and the NT (Setterfield *et al.* 2010)).

Whilst the WRA system has been tested widely and adopted internationally, the WRM system has not (see Downey *et al.* 2010a). In fact, there has been limited testing or development of the WRM over the past 15 years. Ironically recent testing/developments of the WRM system have been undertaken internationally (e.g. in Iran (Sohrabi *et al.* 2020) and Bhutan (Dorjee *et al.* 2021)).

Since the development of the WRM system Australia has adopted a biosecurity model for managing weeds and other invasive species (i.e. based on Beale *et al.* 2008). Despite this, the WRM system has not been revised, developed or evaluated to determine how it meets the biosecurity requirements, specifically determining and managing the risk posed by invasive species [biosecurity matter] to the economy, environment and community.

The lack of publicly available evaluations as to whether the WRM system is fit-for-purpose for biosecurity weed risk management, is a major shortcoming.

## REVIEW OF THE WRM SYSTEM

In 2019 a review of the WRM system was undertaken to determine how it could be modified to meet the requirements of the NSW *Biosecurity Act 2015*, specifically to account for the spatially variable nature of the risk posed by weed species (i.e. for most weed species its distribution and risk level is not uniform across the landscape). This review showed many issues with the WRM system including that: (a) the Feasibility of Control Component assessed generic control related questions, but then applied a specific management objective to the outcome (i.e. eradication) despite not asking any eradication-specific questions; (b) the system was not developed to account for the spatial variability of the risk and; (c) the scoring system led to outcomes which did not align to the risk for some species (Downey 2020). Further examination of 300+ completed WRM assessments using the NSW system showed that the system: (i) did not adequately handle new incursions when there was limited knowledge of the species; (ii) majority of assessments resulted in the ***manage weed*** outcome, which could indicate a problem; (iii) resulted in inconsistency between assessments of the same species; and (iv) some of the questions were problematic (Downey unpublished data).

Additionally, the WRM system does not readily align to more recent biosecurity legislation, for example, assessment of the impact is not specifically aligned to: (a) the economy; (b) the environment; or (c) the community; despite asking impact questions (i.e. the questions are not mutually exclusive on a sectoral basis). Lastly, the WRM system could be better aligned to the formal risk approach based on **likelihood** and **consequence**. This review showed that the WRM system was no longer fit-for-purpose as a standalone system for many species and that fixing the problems and making it so required a new system/approach rather than modifications to the existing WRM system.

In addition, to the issues raised in the review, other issues raised previously (i.e. Auld *et al.* 2012) were also considered during the development of the new system.

## A NEW POST-BORDER SYSTEM

The development of the new weed biosecurity risk management system started with three key aspects, being to create a: (a) spatially variable risk framework that enabled property and/or site level assessments to help determine the individual duty; (b) system that assessed the key management objectives of prevention, eradication, containment and asset protection; and (c) system that aligns to the biosecurity model.

A draft model was developed and Alpha-tested on a variety of weed species in 2020 (see Downey 2020). The draft version was Beta-tested in late 2021 with Weeds Officers and other stakeholders in south-east NSW, and a revised version developed (Downey 2022). The new biosecurity risk management system involves three components (Figure 1), which are briefly outlined below.

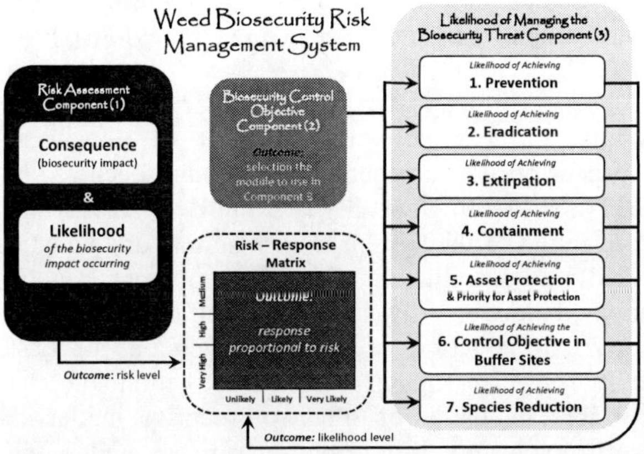


**Figure 1.** An overview of the three components of the new Weed Biosecurity Risk Management system.

**Component 1 – Weed Risk Assessment** Whilst loosely based on that of the existing WRM system, Component 1 has been totally revised to align to the biosecurity framework. For example, the assessment of impact has been developed to determine the impact level/severity to the: (a) economy; (b) environment; and (c) community; through standalone assessments (e.g. to provide an economic impact score). In addition, weed species which pose a current adverse effect [biosecurity impact] (e.g. a transformer species) are not assessed further to determine their invasiveness or potential distribution [likelihood] because the consequence has occurred and thus assessing the likelihood is moot. In fact, evaluation of the current WRM system revealed that for such species assessing their potential distribution could actually reduce their overall risk level if they were widespread or had a small or limited potential distribution; which is a major problem.

**Component 2 – Biosecurity Objective** This new component creates a system for determining the indicative biosecurity or control objective for a weed species spanning a range of spatial scales (i.e. property or region), based on the following nine biosecurity objectives:

1. Prevention;
2. Eradication;
3. Extirpation (local eradication);
4. Containment;
5. Asset Protection;
6. Buffer Sites (control aimed at supporting programs being managed on neighbouring lands for another objective (i.e. eradication);
7. Species Reduction (to reduce the overall threat at a landscape level, as opposed to a specific asset);
8. Inspect – monitoring and surveillance; and
9. Alternative Measures (for species which can't be managed through other objectives or for which control options are not available or practical).

These nine biosecurity objectives are an expansion of the four key management categories of the invasion curve (prevention, eradication, containment and asset-protection) (see EWWG 2007), to include a management objective for every weed species.

**Component 3 – Likelihood of Managing the Biosecurity Threat** The first seven objectives from Component 2 form the basis for individual modules in Component 3 (i.e. the Likelihood of Achieving Eradication). These seven modules were developed based on specific questions relating to the biosecurity objective derived from the literature. For example, the likelihood of achieving eradication is highly dependent on: (i) the number of known infestations; and (ii) the size of the area invaded (see Rejmánek and Pitcairn 2002; Panetta and Timmins 2004); both of which informed questions in the Eradication Module. The specific risk event associated with each objective is also defined. For example, the risk event associated with eradication is *the failure to destroy all infestations of a new weed species not previously known to occur in a region (new incursion).*

When assessing the likelihood of achieving asset protection, the assessment focus needed to be expanded to incorporate both the weed species and the asset to be protected; something that the current WRM system did not address, as highlighted by the outcome being *Protect Sites* and not assets. Furthermore, the importance of the asset being protected needed to be determined, to prioritize site management. This required a second part to the Asset Protection Module, to: (1) assess the likelihood of achieving asset protection through weed control, which is not dependent or determined in anyway based on the priority of the asset; and (2) rank the priority of the asset based on its relative value (i.e. control should be directed at high priority assets, where weed control can deliver protection to the asset (see Downey *et al.* 2010b)).

**Risk-Response Matrix** Each of the seven modules contain a specific risk-response matrix, based on the combination of the risk posed by the weed species and the likelihood [of achieving the control objective] levels. The cells of this risk-response matrix are tailored to the specific management action for the combination. For example, a *medium risk* and an *unlikely* [control] outcome combination contains a statement about revising the objective and/or undertaking risk reducing operations to determine if management could increase the likelihood level. **Note**: This is determined through a risk mitigation process in each of the modules.

The last two management objectives in Component 2 (Inspect and Alternative Measures), whilst not assessed through individual modules in Component 3, are assigned through the outcome of the risk-response matrix. For example, in the Prevention risk-response matrix one of the management outcomes assigned is inspect (surveillance), and in the Species Reduction risk-response matrix an unlikely outcome is investigate alternative control measures (e.g. biological control).

Lastly, the outcome of the risk-response matrix can provide an indicative statement about the likely general biosecurity duty associated with the risk level posed by the weed species and the likelihood of achieving a specific management outcome at a specific location.

**Technical manual and electronic scoresheet** The new weed biosecurity risk management system is underpinned by a several hundred page technical manual which provides justification for all aspects of the system and links to the literature that supports each question (see Downey 2022). For example, the Eradication module questions are based on published eradication studies or reviews of such studies (e.g. Panetta and Timmins 2004). Despite the extensive scale of the technical manual, the actual assessment process (i.e. questions, attributes and criteria) has been developed to be simple to use by a wide range of end users, as demonstrated by the successful Beta-testing stage. An accompanying electronic scoresheet has also been developed to help with assessments. Based on the outcomes and feedback from the Beta-testing stage, assessments for weed species known or familiar to the assessor are much quicker through the new system than the existing WRM system, with outcomes that more closely align to the on-ground reality.

## AN AQUATIC WEEDS VERSION

A freshwater aquatic weeds version of the risk analysis system is currently being developed and is scheduled to be completed by mid-2023.

## A MIRROR VERSION FOR PEST ANIMALS

A draft version of a pest animal mirror version of the weed biosecurity risk management system has been developed and alpha tested. Apart from the species-specific context differences (i.e. seed banks in weeds, and the mobility of pest animal species), the questions do not differ between weed and pest animal species versions of the biosecurity risk management system. The successful conversion of the system to pest animals means that both groups of invasive species can be assessed under a similar process; something that has not been achieved previously.

## FUTURE DIRECTION

Whilst the extensive development of the weed biosecurity risk management system has occurred in SE NSW over the past 3 years, the Beta-testing and conversion of the system to pest animal and aquatic weed versions illustrates that the approach can be adopted more broadly (i.e. to other regions). Also, the successful creation of a mirror version for pest animal species suggests that investigation into the possible inclusion of other biosecurity matter is worth exploring. Whilst the system has undergone extensive testing and revisions, further testing and use is needed to ensure broader adoption.

A key future development will be to transition the system from a technical manual and accompanying electronic scoresheet to an on-line system which is underpinned by a range of spatial layers, especially given the system is built to be spatially enabled. Furthermore, other datasets could also be integrated, for example information on assets (i.e. threatened species) to provide a more integrated assessment process. Lastly an online system would reduce assessment times by prefilling responses which do not change between assessments, as well as to enable access to the risk assessment process to all stakeholders. At present weed risk assessments are kept behind a restricted access departmental portal, which is a problem for the delivery of a shared model of biosecurity.

This biosecurity risk management system provides a significant development in the evolution of post border weed risk assessment systems, by addressing the shortcomings of the existing system, transitioning the approach to the biosecurity model for managing invasive species, integrating the invasion curve categories into the assessment system and accounting to the spatial variability of the risk.

ACKNOWLEDGMENTS

We thank Stephen Johnson for his inputs, advice and feedback on the development of the system. Megan Wyllie from the South East Local Land Services (NSW) (SE LLS) who provided advice on the development of the system. The Weeds Officers from the 11 LGAs in SE NSW and Mich Michelmore (NSW DPI) are also thanked for their inputs during the Beta-testing stage. The development of the system was funded by Queanbeyan-Palerang Regional Council & SE LLS.

REFERENCES

Anon. (2006). *National Post-Border Weed Risk Management Protocol*. HB 294. Standards Australia, Standards New Zealand, and Cooperative Research Centre for Australian Weed Management, Sydney.

Auld, B., Johnson, S., Ainsworth, N., Brunel, S., Csurhes, S., Downey, P., Ferdinands, K., *et al.* (2012). Further development of the National Weed Risk Management Protocol. Proceedings of the 18th Australasian Weeds Conference, ed. V. Eldershaw, pp. 317-319. (Victorian Weeds Society, Melbourne).

Beale, R., Fairbrother, J., Inglis, A. and Trebeck, D. (2008). *One biosecurity – a working partnership. The independent review of Australia's quarantine and biosecurity arrangements report to the Australian Government*. Canberra

Dorjee, Johnson, S.B., Buckmaster, A.J. and Downey, P.O. (2021). Developing a hybrid weed risk assessment system for countries with open and porous borders: insights from Bhutan. *Biological Invasions* **23**, 2945-2959.

Downey, P.O. (2020). *Draft - New feasibility of control component for use with the NSW Weed Risk Management system*. Unpublished report prepared for the Queanbeyan-Palerang Regional Council, Bungendore. 135pages.

Downey, P.O. (2022). *Likelihood of Managing the Biosecurity Threat Component of a Weed Risk Analysis System*. A Technical Manual prepared for Queanbeyan-Palerang Regional Council and South East Local Land Services. Bungendore.

Downey, P.O., Johnson, S.B., Virtue, J.G. and Williams, P.A. (2010a). Assessing risk across the spectrum of weed management. *CAB Reviews: Perspectives in Agriculture, Veterinary Science, Nutrition and Natural Resources* **5**, 038.

Downey, P.O., Williams, M.C., Whiffen, L.K., Auld, B.A., Hamilton, M.A., Burley, A.L. and Turner, P.J. (2010b). Managing alien plants for biodiversity outcomes – the need for triage. *Invasive Plant Science & Management* **3**, 1-11.

EWWG (Environmental Weeds Working Group) (2007). *Guidelines and procedures for managing the environmental impacts of weeds on public land in Victoria 2007*. Department of Sustainability and Environment, Melbourne.

Johnson, S. (2009). *NSW Weed Risk Management System - Instruction book*. Industry and Investment New South Wales, Orange.

Panetta, F.D. and Timmins, S.M. (2004). Evaluating the feasibility of eradication for terrestrial weed incursions. *Plant Protection Quarterly* **19**, 5-11.

Pheloung, P.C., Williams, P.A. and Halloy, S.R. (1999). A weed risk assessment model for use as a biosecurity tool evaluating plant introductions. *Journal Environment Management* **57**, 239-251.

Rejmánek, M. and Pitcairn, M.J. (2002). When is eradication of exotic pest plants a realistic goal? *Turning the Tide: the eradication of invasive species*, eds C.R. Veitch, C.R. and M.N. Clout pp. 249-253, IUCN SSC Occasional Paper no. 27, IUCN, Gland.

Setterfield, S., Ferdinands, K., Rossiter-Rachor, N. and Douglas, M. (2010). The Northern Territory's weed risk management system: an application of the national post-border WRM Protocol. *Plant Protection Quarterly* **25**, 86-89.

Sohrabi, S., Downey, P.O., Gherekhloo, J. and Hassanpour-bourkheili, S. (2020). Testing the Australian Post-Border Weed Risk Management (WRM) system for invasive plants in Iran. *Journal of Nature Conservation* **53**, 125780.

Virtue, J.G. (2008). *SA Weed Risk Management Guide*: February 2008. Department of Water Land and Biodiversity Conservation, Adelaide.

# Understanding the system matters – Stepping back for general surveillance

Jenifer Ticehurst[1], Heleen Kruger[1], Catherine McInerney[2] and Melinda Laidlaw[3]
[1] Department of Agriculture, Water and the Environment, GPO Box 858, Canberra ACT 2601, Australia
[2] Agriculture Victoria, 70A Camp Road, Anglesea, Victoria 3230, Australia
[3] Department of Environment and Science, 55 Priors Pocket Rd, Moggill, Queensland 4070, Australia
(Jen.Ticehurst@awe.gov.au)

**Summary** General surveillance, also known as passive surveillance, is a process whereby people from all walks of life monitor and report weeds, pests and diseases. General surveillance programs are complex systems, and actions in one part of the program can create unintentional consequences elsewhere. Social scientists applied systems thinking to explore what works well for nine general surveillance case study programs from across Australia and New Zealand. Two of the case studies focus on weeds, namely the Weed Spotters programs in Victoria and Queensland. These well-established and long-running initiatives, based on the same 'Weed Spotters' framework, adapted and evolved to best meet their goals and local contexts. This paper highlights some of the similarities and differences in these weed-based programs, and how systems thinking provides valuable insight.

**Keywords** Systems thinking, general surveillance, weed spotters

## INTRODUCTION

As our climate changes and international travel and trade increase, the risk to Australia's biosecurity also increases, including the introduction, establishment and spread of weeds. Surveillance is a vital component of weed management because it supports early detection; helps understand weed spread; and informs prioritisation and management. General surveillance engages people from all walks of life in the monitoring and reporting of pests, weeds and diseases. It has elements of opportunism that enable broad surveillance coverage and/or more cost-effective surveillance than targeted, active surveillance. It also fits well within the current paradigm of shared biosecurity responsibility between government, industry and the community.

**Background to Weed Spotters programs** There have been several different weed detection systems in Australia including the Weed Alert program in Victoria from the late 1990s to 2002 (Morton 2007), which made the first call to community members to become 'Weed Spotters'. Weed Spotters programs are general surveillance programs designed to support the monitoring and reporting of targeted weed species for a particular region. Drawing from the Weed Alert program and others, Morton (2007) proposed a conceptual framework for a weed detection network to guide a national program of surveillance. It identifies five key program components: 1) establish community and professional detection and surveillance networks to enhance surveillance and reporting potential; 2) provide the capacity for rapid and accurate identification of reports; 3) have notification systems in place to act if a new plant is confirmed; 4) provide the means for a rapid risk assessment if a new plant is confirmed; and 5) have an information management system(s) to support the storage and use of personal and plant-based data collection.

Like most general surveillance programs, Weed Spotters programs are complex systems involving various functions (including monitoring, species identification, data management and use, and supporting weed spotters). The complexity stems from the interactions between the elements or parts thereof.

Systems thinking describes an approach to consider how a group of interdependent components interact through time to achieve a purpose (Arnold and Wade, 2015). Systems thinking facilitates management of complex problems with principles. For example, seemingly inconsequential actions in one part of the system may create unexpected and undermining consequences elsewhere. Feedback loops can also occur, and their impacts may be delayed. People from throughout a system will often view it from a different perspective. Systems are also inherently dynamic and change through time.

In this research, systems thinking was used to explore two Weed Spotters case studies, Weed Spotters Network Queensland (WSNQ) and Weed Spotters Victoria (WSV). The research aimed to inform what is needed to make weed general surveillance programs sustainable, practical and effective.

## MATERIALS AND METHODS

**Research framework** We adjusted the Agricultural Innovation Systems structural framework (Wieczorek and Hekkert 2012) to guide our research. Based on this framework, the structural components of innovations systems used are (i) actors and their interactions, (ii) the institutions (rules) that influence their behaviour, and (iii) physical, financial and knowledge infrastructure. We add a biophysical component to include considerations related to invasive species and their environment that shape general surveillance programs. Through the resulting framework (shown in Kruger et al 2022; Figure 1) we use systems thinking to consider the components of both Weed Spotters programs and the interactions between them to (i) identify the process and system around data flow; (ii) develop a timeline of program change and development; and (iii) enquire about what works and doesn't work in meeting program goals.

**Data collection** In July and August 2020 research began on the WSNQ and WSV, respectively. For each case study, we reviewed relevant literature and webpages provided by the program representatives. For each program semi-structured interviews were carried out with up to 11 people from throughout the system including weed spotters, weed spotters coordinators, experts who identify submissions, data managers and users, funders and policy makers, and government officers who respond to high-risk detections. Interview topics were tailored to the interviewee's role in the program, including what works and doesn't work from their perspective. Interview transcripts were analysed with NVIVO software. A focus group with another six to eight people from throughout the system reviewed a summary of the interview findings. The main findings were discussed with program coordinators and key program staff for clarification and feedback. An online survey link was emailed to weed spotters in both programs in November 2020 to capture motivations and barriers to participating, and what they think works and does not work. Seventy-two surveys were completed for the WSNQ and 83 for WSV. A more detailed description of the data collection methods is available in Kruger et al. (2022), including the interview and survey questions and detailed survey results.

## RESULTS

**The Weed Spotters framework** Data flow diagrams for WSNQ and WSV can be found on the project website (WSNQ and WSV). The diagrams show how both programs conform to the Weed Spotters framework. As such, (1) weed spotters provide the surveillance network, (2) the Queensland Herbarium and Agriculture Victoria identify submissions, (3) Biosecurity Queensland and Agriculture Victoria Biosecurity Officers act if a detection is made, 4) if a new plant is confirmed, detections are assessed by Biosecurity Queensland and the High Risk Invasive Plants team (HRIP) (WSNQ and WSV respectively) for their potential risk, and 5) they have information systems to support data storage.

**Comparison of the Weed Spotters programs** Table 1 shows various similarities and differences between the two Weed Spotters programs structured around the four components of the framework. Figure 1 shows key points in evolution and change for each program.

## DISCUSSION

Systems thinking provides valuable insight into the effective functioning of the Weed Spotters programs. Integrating knowledge, keeping weed spotters engaged and evolving the program through time is important for it to remain successful.

**Knowledge integration benefits program development** Sharing knowledge and the experiences of people from throughout a program is beneficial because changes or weaknesses within one component are likely to have implications elsewhere. This avoids making wrong assumptions about other parts or people in the program, produces nuanced knowledge about design to make the program more effective and practical, and helps people to feel more connected. For example, the pilot study that initiated the WSNQ involved a small number of Natural Resource Management Regions with an interest in weeds and the Queensland Herbarium to co-design the initial program. It allowed the herbarium to make the needed adjustments, e.g. introducing new protocols relating to dealing with weeds, hygiene and notifications. As part of this design, regional coordinators were enlisted to provide a friendly local face to notifiers, and to apply regional knowledge and context to the state wide information provided by Biosecurity Queensland and Queensland Department of Environment and Science (DES). As such they provide an important knowledge broking connection between the weed spotters and the program coordinator. Knowledge is also shared throughout the program in the newsletters, which provide regional information on program activities and detections, as well as through informal meetings between the coordinator and regional coordinators in conjunction with other weed related meetings.

WSV also use a newsletter to keep weed spotters informed, including about what is being found across the state and potential new threats. Their half-day in-person training course is also very important for sharing knowledge about what the target species are, and how they can be identified, and for weed spotters to interact directly with Weed Spotters staff, providing feedback on the effectiveness of the training.

**Table 1.** Comparison of Weed Spotters characteristics against the research framework.

| | Topic | WSNQ | WSV |
|---|---|---|---|
| Actors and their relationships | **Weed spotter motivations** | Protect the environment and ecosystems services | |
| | **Weed spotter barriers** | Lack of time<br>Limited motivation/interest in target species | |
| | **Engagement and outreach** | Targeted to people with skill, motivation and ability to make accurate, timely and complete (i.e. quality) reports<br>Newsletters containing interesting finds and background information are emailed to weed spotters directly | |
| | | 10 newsletters/year report regional activity<br>Handbook<br>Android app<br>In 2019/20, used Facebook to advertise for members | 3 newsletters/year report state activity<br>Annual calendar<br>Weed ID cards<br>Have not used social media for communication |
| | **Regionally based positions** | Volunteer regional coordinators know local context, triage some reports & are 'trusted friendly faces' | No longer have regionally based positions. Reports are triaged by the state coordinator |
| | **Identification** | State government | |
| | | Queensland herbarium (DES) | Experts within Agriculture Victoria<br>State/national herbarium (if required) |
| | **Funding bodies** | State government | |
| | | Biosecurity Queensland (Qld Dept for Agriculture and Fisheries)<br>Qld herbarium (DES)<br>Local governments | Agriculture Victoria |
| Infrastructure | **Reporting tools** | Email, Android app or specimen submission | Website, hotline or dedicated email |
| | **Training** | Face-to-face, regionally based training to add local context and respond to local interests<br>Online training under development | Face-to-face training coordinated and run at state level to ensure consistency<br>Online training also available<br>Live specimens available for viewing at regional centers |
| | **Data users** | Biosecurity Queensland<br>Atlas of Living Australia<br>Community groups<br>Local governments | Agriculture Victoria High Risk Invasive Plant team |
| | | Australian Virtual Herbarium | |
| | **Resourcing** | Legislative requirements support ongoing funding | |
| Institutions | **Key external institutions** | Qld *Biosecurity Act 2014*<br>DES Strategic Plan 2019–23<br>Qld Biosecurity Strategy 2018-2023 | *Catchment and Land Protection Act 1994* |
| | **Focus species** | 240 species in scope – includes prohibited and restricted plants | 8-12 specific State Prohibited Weeds |
| | **Specimen submissions** | Specimen requested if photos indicate a species of concern | Prefer plants remain *in situ* to reduce risk of spread |
| Weeds and their env. | **Surveillance spread** | Weed spotters focus on monitoring regions they are most knowledgeable about | Address monitoring gaps by targeting regions that have few weed spotters |

**Figure 1:** Timeline showing key milestones and developmental change for WSNQ and WSV

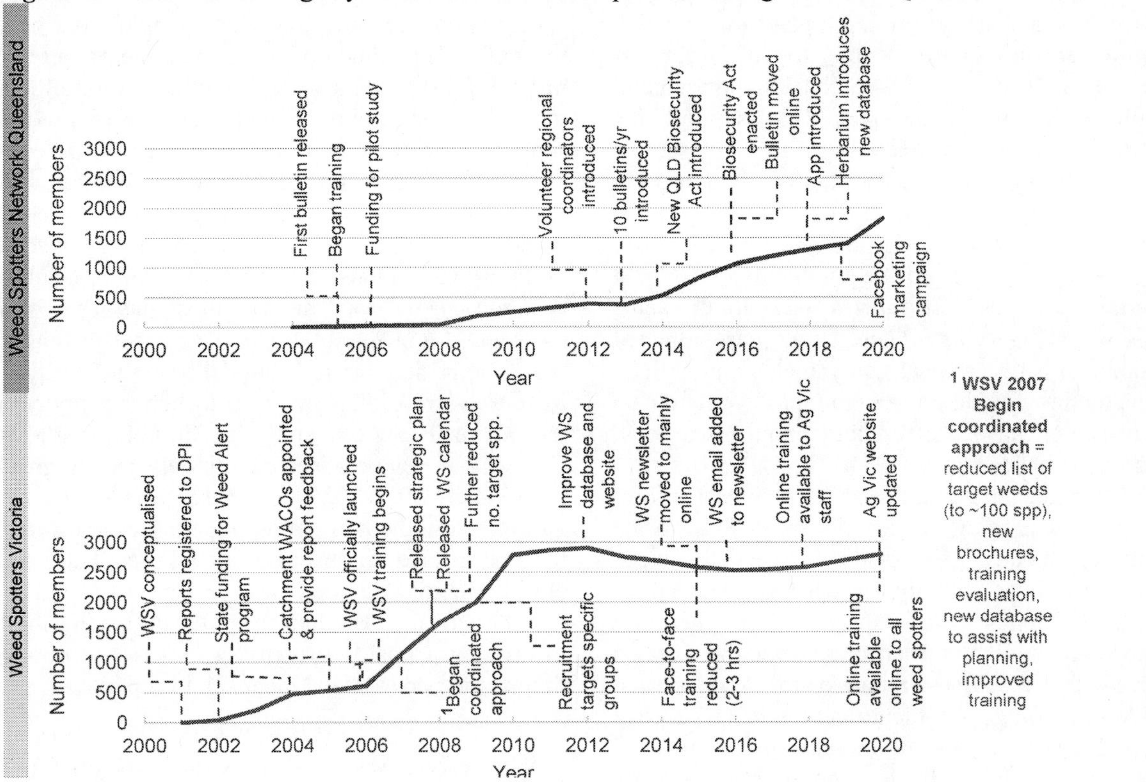


**Keeping weed spotters engaged** To limit transaction costs for training new weed spotters and identifying inaccurate reports, both programs target people with motivation, opportunity and skill to make quality reports (including various field staff to who work outside, e.g. council weed officers, as well as gardener groups in WSNQ). Quality reports refer to reports that are timely, accurate and complete. Both programs make participation simple and easy knowing that weed spotters are time poor. For example, WSV provide multiple reporting options, such as a hotline, a dedicated email address and its web form, to suit people's personal preferences. WSNQ introduced a reporting app in 2018 to minimise the need for more laborious specimen submissions. To minimise costs, the app was developed for Android devices only, by Masters students from the University of Queensland. Developing and maintaining an app for WSV is deemed too costly to service a few reports for a small number of target species, especially when three reporting options are available and do not require additional cost or maintenance.

Keeping people motivated can be challenging when target species aren't of interest or seldom present. To help, both programs focus on delivering a positive reporting experience by providing personal feedback to all reports, including the species' identity and updates on follow-up outcomes. For example, WSNQ provides management information for out-of-scope weeds or connects weed spotters with local government officers to address their concerns. Furthermore, providing a win-win situation will secure people's ongoing engagement. For example, the WSNQ provides local governments with valuable services such as weed identification training and easy access to Qld Herbarium staff. In return, the Qld Herbarium receives more specimens for their collection and Biosecurity Queensland have a greater spread of surveillance effort.

**Adapting and evolving through time** Systems are dynamic and change through time. Both programs have conducted considerable monitoring and evaluation since they began to identify and address challenges and utilise opportunities to remain effective and relevant. Methods include conducting Weed Spotters surveys, dedicated research (e.g. WSV program review and resultant strategic plan) and collating and analysing notification data. Monitoring and evaluation can identify important system behaviours such as leverage points, most limiting factors and feedback loops.

*Leverage points* are areas in the system where a small shift can deliver considerable beneficial change in other points(s) or the whole program. For example, the Queensland *Biosecurity Act 2014* supports the

WSNQ because it sets out information sharing requirements for reporting of notifiable species: This facilitates data sharing between organisations which in turn, enables more people to look out for incursions. The Act's emphasis on shared responsibility encourages people to support the program. Notifiable species under the Act set the scope of the program, and as the need to prevent related incursions is set in legislation, it encourages departmental investment in the program. The Act also sets weed related requirements for local governments, which encourages local governments to use WSNQ services. The resulting strengthened relationship between local governments and WSNQ means the program has access to more eyes and ears and can refer out-of-scope public enquiries to local government officers.

WSV reduced the number of target species from *all* newly emerging weeds to only 8-12 State Prohibited Weeds. This made it easier for weed spotters to focus intently on species assessed as posing the biggest risk to the state, and therefore increase the likelihood of accurate detections. It also reduced the resources required to develop and maintain training and identification material (e.g. Weed ID cards) and made it easier to target weed spotters who are most likely to be in high risk areas for the target weeds. Clearly defining which species are in scope also makes it easier for the program coordinator to triage incoming reports.

The *most limiting factor* is the variable that is most important to the system to bring about change. Once identified, the most limiting factor can be managed to improve program effectiveness. The most limiting factor may be time constraints for weed spotters to make reports (discussed above), or could be the capacity of staff to identify incoming submissions. In the latter case, interviewees in both programs emphasised the importance of receiving quality, rather than many, reports. Accurate and complete reports minimise the need for lab/herbarium staff to wade through species that are out of scope or to follow-up with notifiers for missing information. Notifier training, providing well considered reporting tools, triaging reports (such as via the WSNQ regional coordinators or WSV coordinator) and providing feedback on every report to assist in learning and increasing awareness, help to maintain report quality and avoid overloading the system.

*Feedback loops* occur when changes in one component flows through the system creating effects back on the same component. For example, there are feedback implications if the quality of incoming notifications is too low. WSV initially conducted widespread recruitment of anyone interested in being involved in reporting weeds. This grew weed spotters' numbers considerably, but most target species notifications came from people working in the field. Thus, much of the training costs were not justifiable. If weed spotters' numbers were allowed to increase unchecked, it could have caused out-of-scope notifications to flood the system. This could slow the triage and identification process, and confirmation of state prohibited weed detections could be delayed. The program team moved to targeting individuals well placed to find and submit accurate notifications, such as those having existing knowledge and interest, being involved in outdoor work activities or residing at certain locations. Resources were then freed up to better support the most effective weed spotters. Similarly, although QWSN ran a generalised recruitment campaign via Facebook in 2019-2020, targeted at people interested in the environment or gardening, they were aware of the risk in creating an influx of reports. Thus, they closely monitored reports to ensure they did not overwhelm lab/herbarium staff to maintain the ability to provide a positive reporting experience through timely and personalised feedback to reporters.


## ACKNOWLEDGMENTS

This research was funded by the Department of Agriculture, Water and the Environment. The authors thank all case study representatives for allowing their programs to participate in this research. Our gratitude goes to all interview, focus group and survey participants for their time and input. Thanks also to the project steering committee for their time, input and guidance into this research.


## REFERENCES


Arnold, R.D. and Wade, J.P. (2015). A definition of systems thinking: A systems thinking approach, *Procedia Computer Science* 44, 669-78 (DOI 10.1016/j.procs.2015.03.050).

Kruger, H., Ticehurst, J. and Van der Meer Simo, A. (2022). Making general surveillance programs work – Lessons from nine case studies, ABARES, Research Report 22.10, Canberra, April, CC BY 4.0, DOI: https://doi.org/10.25814/j4wh-1639

Morton, J. (2007). Building a national, community-based model for preventing new weed incursions, Final phase 4 report, ID 46963, (CRC for Australian Weed Management).

Wieczorek, A.J. and Hekkert, M.P. (2012). Systemic instruments for systemic innovation problems: A framework for policy makers and innovation scholars, *Science and Public Policy* 39, 74-87.

# A tool to assess knowledge and risk level of Exotic Perennial Grass invasion in NSW native communities

Julia Rayment[1,2]
[1]University of Wollongong, Wollongong, Australia,
[2]Department of Planning and Environment, Parramatta, Australia
(jrayment@uow.edu.au)

**Summary** Exotic Perennial Grasses, or EPGs, have negative impacts and pose significant risks to native species and ecosystems. Native grasslands and open woodland communities are particularly susceptible to invasion, with the impact of EPGs more frequently framed in an agricultural context. A risk assessment tool specific to functionally similar EPGs was developed to rank EPGs and identify knowledge gaps. In conjunction with field surveys of 139 sites across nine grassy communities we identified levels of invasion and tested the usefulness of the risk assessment tool. Five widespread invaders were particularly established in all regions and communities. Invasion by pasture grasses was the most significant threat to grassy communities. Species with higher risk rankings were recorded in more sites although a few grasses were more invasive than their ranking predicted. Our study found higher levels of invasion were associated with species that were ranked more invasive, with evidence this ranking could be used to influence management priorities in native communities. Our findings indicate that management of exotic perennial grasses in grassy communities should be undertaken at the community level although there are some species that are important invaders in the whole landscape. The risk assessment tool has the potential to become an important tool for aid in management.

**Keywords** Exotic Perennial Grasses, management, invasion, threatened communities

# Determining new Weeds of National Significance

John G. Virtue[1,2], Matthew R. Sheehan[1] and Shauna Potter[1]
[1] Wild Matters Pty Ltd, 11 Templeton Street, Castlemaine Victoria 3450, Australia
[2] JG Virtue Biosecurity Services Pty Ltd, 142 Range Rd, Rockleigh South Australia 5254, Australia
(weedpriorities@wildmatters.com.au)

**Summary** The National Established Weed Priorities Framework proposes the determination of new Weeds of National Significance (WoNS). Since WoNS were last determined in 2012 there have been important domestic and international developments in pest prioritisation. A review of these has identified improvements so that future determination of new WoNS is contemporary and more participatory for industry and community stakeholders. A proposed nomination, assessment and selection process is described.

**Keywords** WoNS, weed risk assessment, pest prioritisation, national significance.

## INTRODUCTION

The draft National Established Weed Priorities (NEWP) Framework (Wild Matters 2022) aims to guide the prioritisation and management of established weeds and associated weed issues across Australia. Building on the successes of the >20-year Weeds of National Significance (WoNS) initiative, the NEWP Framework has been designed to build better long-term partnerships and collaborations between governments, industry and community organisations in tackling shared weed priorities.

WoNS is a core component of the NEWP Framework. For each of the current 32 WoNS, advances have been made in knowledge, information, tools and strategic actions to better reduce their spread and impacts. One of the greatest legacies of the WoNS initiative is the enduring national network of partnerships that continue to manage WoNS from local and regional control programs to national RD&E collaborations.

There have been two selection rounds for WoNS; the original twenty in 1999 (Thorp and Lynch 2000) and an additional twelve in 2012 (Hennecke 2012). The methodology to select WoNS should evolve with advances in pest risk assessment, biosecurity policy and stakeholder expectations. Taking account of these, this paper summarises considerations for the nomination, assessment and selection of new WoNS.

## METHODS

A literature review of national and international pest risk prioritisation was undertaken to inform improvements to the WoNS selection assessment processes. This focused on national standards for post-border weed risk management (WRM) and risk management more generally (anon. 2006, ISO 2018).

The main biosecurity policy guidance for determining new WoNS is the National Framework for the Management of Established Pests and Diseases of National Significance (EPDNS; NBC 2016). It sets three overarching criteria in determining nationally-significant species threats: national impact; feasibility of management intervention; and benefits from taking a nationally coordinated approach.

A process to determine new WoNS is proposed in the draft NEWP Framework. This process was progressively refined through workshops, meetings and formal feedback from a large cross-section of stakeholders in established weed management across Australia.

## RESULTS

General requirements of a contemporary weed prioritisation model include the following considerations (Heikkila 2011, Leung *et al.* 2012, Vanderhoeven *et al.* 2017, Bartz and Kowarik 2019, Osunkoya *et al.* 2019a and Vila *et al.* 2019, plus additional references below):

**Explicitly addresses uncertainty** To reduce misinterpretation, questions must be clearly written and unambiguous. Lack of information, data variability, conflicting evidence and subjective judgement needs to be explicitly considered when designing scoring approaches, expert elicitation methods and/or statistical measures of confidence.

**Systematic and structured** The prioritisation model should have a logical, scientific basis and be validated for accuracy. The determination of risk should align with the standard formula of likelihood × consequence (which is equivalent to weed spread × impacts). The National Post-Border WRM Protocol (anon. 2006) gives standard decision criteria for determining overall rankings for weed risk and feasibility of control, which are independent considerations to be compared in determining pest priorities (Canessa *et al.* 2021). A robust species ranking model needs sufficient, defensible questions to confidently distinguish species. Questions that poorly differentiate species or questions that are correlated with others should be avoided. Definitions for multiple choices within questions should, where possible, be quantitative and scaled geometrically (Evans *et al.* 2019) or exponentially (Blackburn *et al.* 2014, Ireland *et al.* 2020) to help distinguish species.

**Stakeholder involvement in weighting criteria** Whilst questions in a prioritisation model must have a scientific basis and align with standards, their relative importance (weightings) also needs to explicitly consider human values, including economic, cultural, social and environmental factors. Techniques to select stakeholders and survey their values have been applied to weighting impacts of weeds (e.g. Hurley *et al.* 2010, Kumschick *et al.* 2012).

**Transparent and inclusive** Trust in the results of risk assessment and prioritisation comes from understanding the model and how its components are scored and combined mathematically. Individual species scoring must be visible and documented, with opportunity for peer review (experts and stakeholders). An expert elicitation approach to scoring, with the structured use of groups of people to assess species through rounds of review and consensus building, provides an inclusive, robust process (Booy et al. 2017, Hemming *et al.* 2017, Osunkoya *et al.* 2019b, Evans *et al.* 2019).

**Accesses best available information** Ideally, species assessments would be completed based solely on published literature. However, even for widespread weeds, there are likely to be gaps in the literature, yet a wealth of personal observations and experience with experts who have studied or managed weeds. Species should not be disadvantaged in a prioritisation process by a lack of documented information. Thus there is the need to compile available relevant literature and personal observations and experience to inform a structured expert elicitation process (see above).

Where national spatial datasets are available these should be used to create maps of risk and feasibility of control, to give a more informed and accurate prioritisation (Kriticos *et al.* 2018). Potential distribution mapping under future climate scenarios, is needed to inform risk and future impact (Roger *et al.* 2015).

**Broadly applicable to any weed** The prioritisation model should allow assessment of weed risk and impact in any land use, ecosystem, climate and region. Questions need to be generic so that the model can be applied to all vascular plant lifeforms, including aquatic herbs, grasses, geophytes and woody plants.

## DISCUSSION

A proposed approach for determining new WoNS, based on the above technical and stakeholder engagement considerations and EPDNS policy requirements, is outlined in Figure 1.

The selection process for WoNS needs to be transparent, inclusive of all stakeholder sectors, fair, logical, defensible and systematic. These requirements will be met through a multi-stage nomination and assessment process. The process must handle uncertainty and identify and manage any potential conflicts of interest. This includes ensuring independence between those people who design and implement the assessment process and those nominating weed species.

**Nominations** Any industry, community or government stakeholder would be able to nominate weeds to be assessed for WoNS consideration. Groups of closely related weeds could also be nominated as a WoNS under the banner of a single species, where they are similar in life-form and management requirements, as per some current WoNS (e.g. opuntioid cacti and Asparagus weeds).

Through an initial, confidential expression of interest, organisations with mutual interests in nominating a species would be 'joined-up'. This facilitated partnering to do joint nominations will enable efficiencies and resource sharing in completing a template of required information to support the nomination. The template would include screening questions to filter out candidates that would not meet EPDNS requirements.

**Assessing impact** This equates with weed risk and it is proposed that the determination of new WoNS should evolve the weed risk ranking model used in 2012 (Hennecke 2012) as the starting point. Impacts questions should seek to align with EPDNS and definitions in international pest impact standards. The model will assess weeds' current and potential impacts on economic, environmental and social assets across Australia, taking account of regional differences and uniqueness and climate change.

Scoring individual weeds in the updated model will consider both high quality published information and the expert opinions of a panel of scientists and weed control practitioners. Uncertainty will be considered using a structured elicitation process that ranks impacts whilst also recording levels of confidence in scoring.

**Figure 1.** Proposed process for determination of new WoNS (as at May 2022).

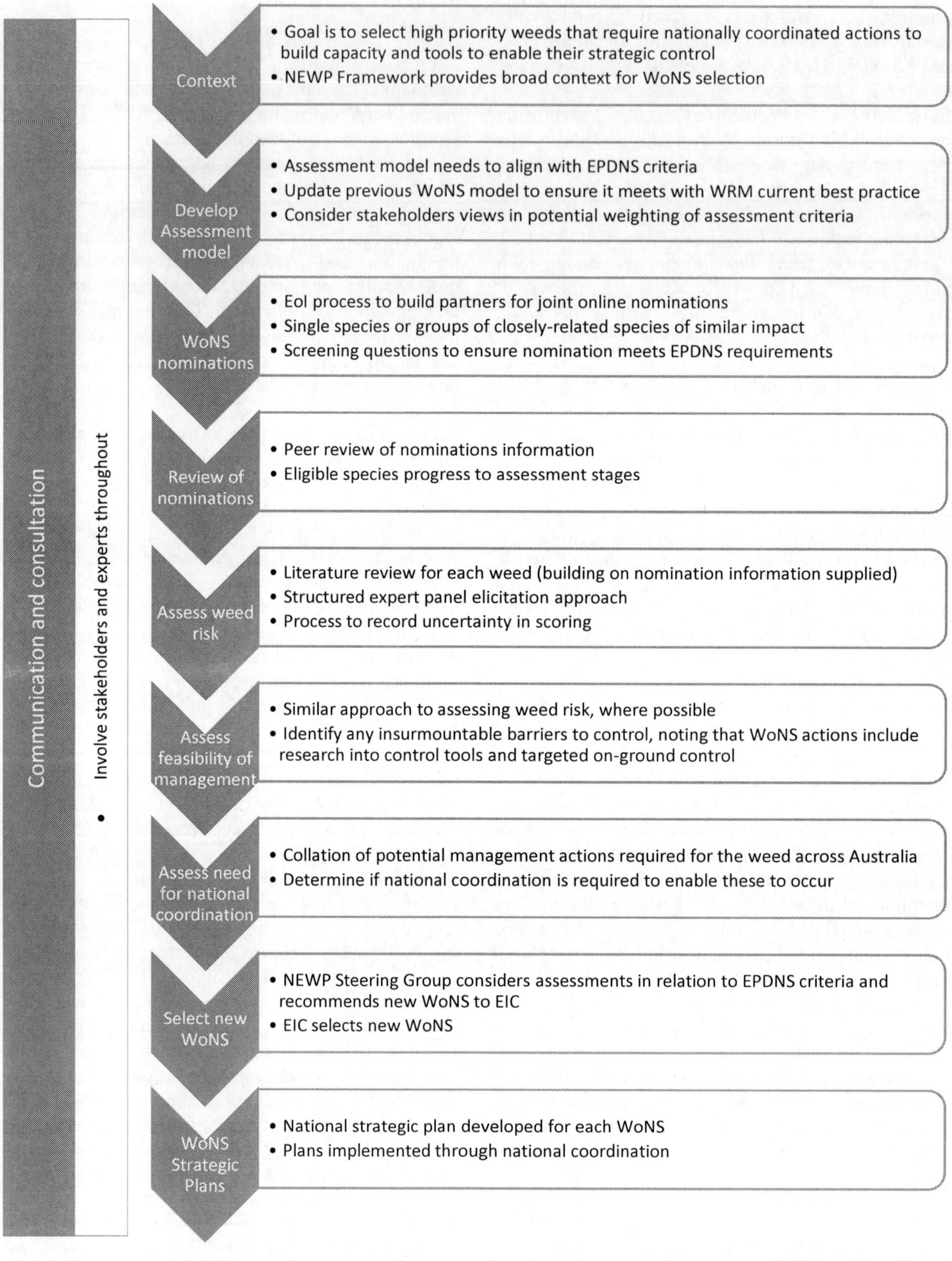

**Feasibility of management intervention** The EPDNS list four factors to be considered in assessing feasibility of management intervention; technical feasibility of implementing a management approach, potential role of regulatory mechanisms, cost-effectiveness of the proposed approach and level of socio-political support (NBC 2016). There is no existing weed ranking system for feasibility of management intervention that integrates all of these factors. For WoNS, such interventions could include on-ground control or containment programs, new control techniques, research, extension, regulation, coordination and/or spread prevention.

Assessment of feasibility of management intervention for WoNS candidates would require a combination of technical and policy analysis. The intent would be to determine whether substantial progress could potentially be made to better manage the national impacts of a WoNS candidate.

**Benefits from national coordination** During the 2012 WoNS determination, potential management actions of national benefit were identified by government, which in turn informed national strategic plans. These actions were collated for each candidate WoNS under broad action categories of prevention of spread, asset protection and increased management capacity. This approach would be improved by seeking input from community and industry organisations to determine specific, cost-effective actions. These actions would focus on those requiring coordination of cross-jurisdictional/cross-sectoral partnerships to bring about measurable, long-term benefits in addressing a WoNS' spread and impacts.

**Selection of WoNS** The process for determination of new WoNS would be overseen by a national NEWP Steering Group made up of representatives of government, industry and community stakeholders. The Steering Group would recommend proposed new WoNS to the Environment and Invasives Committee (EIC) for its approval.

## ACKNOWLEDGMENTS

This work has been funded by the Australian Government Department of Agriculture, Water and the Environment. Oversight has been provided by the EIC Weeds Working Group. Thanks to Hillary Cherry for comments on the paper.

## REFERENCES

anon. (2006). 'HB 294:2006 National Post-Border Weed Risk Management Protocol' (Standards Australia, Sydney).

Bartz, R. and Kowarik, I. (2019). Assessing the environmental impacts of invasive alien plants: a review of assessment approaches. *NeoBiota* 43, 69–99.

Blackburn, T. M., Essl, F., Evans, T., Hulme, P. E., Jeschke, J. M., Kühn, I., Kumschick, S., Marková, Z., Mrugała, A., Nentwig, W., Pergl, J., Pyšek, P., Rabitsch, W., Ricciardi, A., Richardson, D. M., Sendek, A., Vilà, M., Wilson, J. R. U., Winter, M., Genovesi, P. and Bacher, S. (2014). A Unified Classification of Alien Species Based on the Magnitude of their Environmental Impacts. *PLoS Biology* 12(5), 1–11.

Booy, O., Mill, A. C., Roy, H. E., Hiley, A., Moore, N., Robertson, P., Baker, S., Brazier, M., Bue, M., Bullock, R., Campbell, S., Eyre, D., Foster, J., Hatton-Ellis, M., Long, J., Macadam, C., Morrison-Bell, C., Mumford, J., Newman, J., Parrott, D., Payne, R., Renals, T., Rodgers, E., Spencer, M, Stebbing, P., Sutton-Croft, M., Walker, K.J., Ward, A., Whittaker, S. and Wyn, G. (2017). Risk management to prioritise the eradication of new and emerging invasive non-native species. *Biological Invasions* 19(8), 2401–17.

Canessa, S., Trask, A.E. and Ewen, J.G. (2021). Mind the gap (between assessing risks and prioritizing management). *NeoBiota* 68, pp. 1–4.

Evans, J., Parsons, S., Kancans, R., Ng, K., Downey, R. and Raphael, B. (2019). 'Prioritisation methodology for exotic environmental pests and diseases: scientific rationale' (Australian Bureau of Agricultural and Resource Economics and Sciences, Canberra).

Heikkilä, J. (2011). A review of pest prioritisation schemes of pathogens, pests and weeds: principles and practices. *Agricultural and Food Science* 20, 15–28.

Hemming, V., Burgman, M. A., Hanea, A. M., McBride, M. F. and Wintle, B. C. (2018). A practical guide to structured expert elicitation using the IDEA protocol. *Methods in Ecology and Evolution* 9(1), 169–80.

Hennecke, B.R. (2012). Assessing new Weeds of National Significance candidates. Proceedings of the 18th Australasian Weeds Conference, ed V Eldershaw, pp. 191–4. (Weed Society of Victoria Inc., Melbourne).

Hurley, M. V., Lowell, K. E., Cook, D. C., Liu, S., Siddique, A.-B. and Diggle, A. (2010). Prioritizing biosecurity risks using a participatory decision-making tool. *Human and Ecological Risk Assessment* 16(6), 1379–94.

Ireland, K. B., van Klinken, R., Cook, D. C., Logan, D., Jamieson, L., Tyson, J. L., Hulme, P. E., Worner, S., Brockerhoff, E. G., Fletcher, J. D.,

Rodoni, B., Christopher, M., Ludowici, V. A., Bulman, L., Teulon, D., Crampton, K. A., Hodda, M. and Paini, D. (2020). Plant Pest Impact Metric System (PPIMS): Framework and guidelines for a common set of metrics to classify and prioritise plant pests. *Crop Protection* 128, 105003.

ISO (2018). 'ISO 31000:2018 Risk Management' (International Standards Organisation,).

Kriticos, D.J., Beautrais, J.R. and Dodd, M.B. (2018). WRASP: A spatial strategic weed risk analysis tool reveals important subnational variations in weed risks. *Weed Research* 58(6), 398–412.

Kumschick, S., Bacher, S., Dawson, W., Heikkilä, J., Sendek, A., Pluess, T., Robinson, T. and Kühn, I. (2012). A conceptual framework for prioritization of invasive alien species for management according to their impact. *NeoBiota* 15, 69–100.

Leung, B., Roura-Pascual, N., Bacher, S., Heikkilä, J., Brotons, L., Burgman, M. A., Dehnen-Schmutz, K., Essl, F., Hulme, P. E., Richardson, D. M., Sol, D. and Vila, M. (2012). TEASIng apart alien species risk assessments: a framework for best practices. *Ecology Letters* 15, 1475–93.

NBC (2016). 'National framework for the management of established pests and diseases of national significance' (National Biosecurity Committee, Department of Agriculture and Water Resources, Canberra).

Osunkoya, O.O., Froese, J.G. and Nicol, S. (2019a). Management feasibility of established invasive plant species in Queensland, Australia: A stakeholders' perspective. *Journal of Environmental Management* 246, 484–95.

Osunkoya, O. O., Froese, J. G., Nicol, S., Perrett, C., Moore, K., Callander, J. and Campbell, S. (2019b). 'A risk-based inventory of invasive plant species of Queensland, Australia: Regional, ecological and floristic insights. *Austral Ecology*, 44(7), 1123–38.

Roger, E., Duursma, D. E., Downey, P. O., Gallagher, R. V., Hughes, L., Steel, J., Johnson, S. B. and Leishman, M. R. (2015). A tool to assess potential for alien plant establishment and expansion under climate change. *Journal of Environmental Management* 159, 121–7.

Thorp, J. R. and Lynch, R. (2000). 'The determination of Weeds of National Significance'. (Commonwealth of Australia and National Weeds Strategy Executive Committee, Canberra).

Vanderhoeven, S., Branquart, E., Casaer, J., D'hondt, B., Hulme, P. E., Shwartz, A., Strubbe, D., Turbé, A., Verreycken, H. and Adriaens, T. (2017). Beyond protocols: improving the reliability of expert-based risk analysis underpinning invasive species policies. *Biological Invasions* 19(9), 2507–17.

Vilà, M., Gallardo, B., Preda, C., García-Berthou, E., Essl, F., Kenis, M., Roy, H. E. and González-Moreno, P. (2019). A review of impact assessment protocols of non-native plants. *Biological Invasions* 21(3), 709–23.

Wild Matters (2022). 'National Established Weed Priorities: Draft Framework March 2022' (Wild Matters Pty Ltd, Castelmaine).

# Herbicide use in sugarcane and Great Barrier Reef: Can growers reduce their pesticide losses via runoff

Emilie Fillols[1], A Davis[2], S Lewis[2]
[1]Sugar Research Australia Ltd, Meringa, Australia,
[2]Tropwater, James Cook University, Townsville, Australia
(efillols@sugarresearch.com.au)

**Summary** The Great Barrier Reef (GBR) ecosystems have been recognised at risk from the exposure of agricultural pesticide runoff, particularly herbicides from sugarcane. To mitigate this risk, the Sugar industry is required to transit towards an 'alternative' herbicide suite, and to move away from diuron. This shift to new weed management strategies involving regulatory product changes in recent years, has become increasingly complicated for sugarcane farmers to adapt to their specific farming system. This paper quantifies and compares the efficacy and environmental risk profile of a range of established, emerging, and recently registered pre-emergent herbicides across three rainfed field trials in green cane trash blanketed ratoons in the Wet Tropics region of North Queensland, a major contributing region to annual herbicide loads to the GBR coastal environment. Herbicide efficacy trials were implemented as randomised complete block with three replicates and adjacent untreated controls and were monitored fortnightly for six months after pre-emergent herbicide application at 300L/ha water rate. Losses of the tested pre-emergent herbicides in runoff were monitored using replicated rainfall simulations, delivering 80mm of simulated rain, 48h after herbicide application. Imazapic (95g/ha) + hexazinone (475g/ha) was found as efficient as the now restricted diuron (1872g/ha) + hexazinone (528g/ha), while other tested active ingredients like imazapic (96g/ha), isoxaflutole (150g/ha) and amicarbazone (980g/ha) were effective only on some weed species and would require to be apply in a mixture if a wider weed spectrum is targeted. All tested herbicides were found in runoff water at levels aligned with their application label rate, and all tested alternatives were likely more environmentally friendly than diuron on basis of available ecotoxicity data. Herbicides such as isoxaflutole and imazapic had minimal environmental runoff footprints (14 to 250 times less risk) when compared with diuron. Results demonstrated that alternatives to some of the more environmentally problematic herbicides are available, but considerable challenges still face canegrowers in making cost-effective decisions on sustainable herbicide selection.

**Keywords** Great Barrier Reef, herbicides, runoff, sugarcane, diuron

# Management of Navua sedge (*Cyperus aromaticus*): a role of competition using two pasture species

Boyang Shi [1], Mogomotsi Moilwa [2], Olusegun Osunkoya [1], Kunjithapatham Dhileepan [1] and Steve Adkins [2]
[1] Biosecurity Queensland, Department of Agriculture and Fisheries, Ecosciences Precinct, Dutton Park 4102, Queensland, Australia
[2] School of Agriculture and Food Science, The University of Queensland, Gatton, Queensland, Australia
(Boyang.Shi@daf.qld.gov.au)

**Summary** Navua sedge (*Cyperus aromaticus* (Ridley) Mattf. & Kük.), is a grass-like perennial weed and is native to equatorial Africa and islands in the Indian Ocean, off the coast of southeast Africa. It was accidentally introduced to Northern Queensland, Australia where it has invaded rangelands and cropping [e.g. sugarcane (*Saccharum officinarum* L.)] areas. Currently, the weed is causing a significant loss to agricultural productivity by smothering pasture, crops, and native plants due to its fast growth rate as well as the absence of sustainable weed control methods. Therefore, it is essential to find a weed control method which can suppress Navua sedge growth to prevent further spread and to control the weed's invasiveness. The hypothesis that vigorously growing grasses can outcompete and suppress the growth of this invasive weed was tested.

A replacement series competition study using five density ratios, at two soil moisture levels of 50 and 100% of field capacity, was established to determine the intraspecific and interspecific competition effect of two fodder grass; Rhodes grass (*Chloris gayana* Kunth) and humidicola (*Brachiaria humidicola* (Rendle) Schweick). on the growth of Navua sedge. At all density ratios and for all three species, there were minimal growth differences in response to soil moisture level. Navua sedge managed to produce more tillers and leaves under all moisture levels as well as at all intraspecific and interspecific competition levels than Rhodes grass and humidicola. However, Rhodes grass produced 4.2 times more biomass in mixtures with Navua sedge than when with humidicola. The results indicate that Rhodes grass had a greater competitive ability than Navua sedge or humidicola, and hence a good candidate to compete with Navua sedge in the field.

**Keywords** Navua sedge, competition, pasture species, invasive, management.

## INTRODUCTION

Navua sedge, *Cyperus aromaticus* (Ridley) Mattf. & Kük. (Cyperaceae), is having a massive impact on the farming communities in Northern Queensland, Australia (Shi et al. 2021). The weed was first detected in Cairns, Northern Queensland, Australia in 1979, where it is now the dominant weed in sugarcane crops and pasture lands. Navua sedge is causing significant losses by strongly out-competing and smothering many tropical pasture species (Vitelli et al. 2010; Vogler et al. 2015). Currently, the control of Navua sedge is restricted to physical, mechanical and chemical control methods, which are not sustainable as they bring only a temporary solution and could also cause potential adverse impacts to the environment. Moreover, the only registered herbicide, Sempra® (halosulfuron-methyl) is not effective in controlling the subterranean rhizomes of Navua sedge (Vogler et al. 2015; Vitelli et al. 2010). A biological control project was initiated in 2017 to prospect for natural enemies of Navua sedge in equatorial Africa and several fungal pathogens were identified. Host specificity testing of promising smut fungus (*Cintractia kyllingae* J. Kruse and R.G. Shivas) and rust (*Uredo kyllingae-erecta* J.M. Yen) are in progress at CABI-UK (Dhileepan et al. 2022).

These observations have brought attention to the search for additional approaches that can assist in the integrated weed management of Navua sedge by reducing seed production and depletion of the soil seedbank. The use of vigorous and competitive fodder plant to suppress the growth of Navua sedge can complement other management approaches. Rhodes grass (*Chloris gayana*) and humidicola (*Brachiaria humidicola*) were selected for a competition experiment with Navua sedge as they are two widely used pasture species with higher growth performance and fodder characteristics in the north Queensland wet tropics. The aim is to explore the competitive effect of these two fodder grasses on the growth performance of Navua sedge using a replacement ratio approach.

## MATERIALS AND METHODS

**Test species and seedling production** Seeds of humidicola (*Brachiaria humidicola*) and Rhodes grass (*Chloris gayana*) were obtained from Heritage Seeds Pty. Ltd. Seed of Navua sedge was collected from South Johnstone, northern Queensland. Seedlings were raised in seedling trays (350 × 295 mm) filled with a sterilized potting mix consisting composted bark (0–12 mm), coco peat, organic minerals, and hydroPlex wetter to boost water holding capacity to up to 50% (Centenary Landscaping (2018). Each species was germinated into 10 trays placed inside a growth cabinet under illuminated conditions with a 14/10-hour photoperiod and a matching 27/22 ± 5°C (day/night) thermoperiod. Irrigation of seedlings was conducted twice a day. Fertiliser foliar spray was applied on a weekly basis in the last 2 weeks of seedling production stage.

**Replacement series** Approximately equal size and healthy seedlings of ca. 7 days old were transplanted into 25 cm diameter plastic pots filled with black Vertosol soil (60% clay, pH 7.3) obtained from a field at The University of Queensland Research Station, Gatton, Australia. Transplanting was conducted following a replacement series of five ratios of 4:0, 3:1, 2:2, 1:3 and 0:4 of each species in combination with Navua sedge. In this way, a population of four seedlings per pot was maintained, whereas the proportion of each species in each pot were varied.

In each pot, the seedlings were equally spaced 5 cm apart to allow unbiassed competition. Each replacement ratio had five replicates, resulting 25 pots and 100 seedlings per each species per ratio proportion. The 50 pots were further imposed with two moisture levels of 100% and 50% water field capacity. All the pots were labelled according to the water field capacity level, density ratio, type of species and replicate number. Pots were maintained on a bench in a glasshouse under 27/22 + 5°C (day/night) thermoperiod with 60% humidity. Randomisation of the pots were conducted on a weekly basis to allow equal experience of temperature, light and humidity among the pots and seedlings. The pots and their treatments were also randomly distributed within each bench. All transplanted seedling mortalities in the first 7 days of the trial were replaced with seedlings of the same size. The trial ran for 8 weeks with daily monitoring to maintain the water field capacity.

Various plant growth parameters, including plant height, number of tillers, number of inflorescences and biomass, were measured at the termination stage.

**Data analysis** The relative yield (RY) was calculated using the following equations which describes the relative biomass of each species in interspecific competition as a percentage of its intraspecific biomass under the same growing conditions (Prince et al. 2018): $RY_x = X_{inter}/X_{intra}$, whereby $RY_x$ is the relative yield of a plant x, $X_{inter}$ is the biomass of x growing in mixtures and $X_{intra}$ is the biomass of x growing alone (monoculture). A three-way analysis of variance in SPSS (IBM version 25) was used to perform the statistical analysis, with species (three levels), moisture condition (two levels) and competition type (two levels) as the main factors. The 4:0 or 0:4 ratios are measured of intraspecific competition, while 3:1, 2:2 and 1:3 ratios measure intensity of interspecific competition. The comparisons of means between group effects were performed by using the Tukey's test and a p-value smaller than 0.05 was considered significantly different.

## RESULTS

**Overall observations** Main effects of species, competition type and moisture levels were significant. For all treatments, there was no significant interaction effect of moisture and competition for the height, but with marginal effect on number of tillers, the number of Navua sedge inflorescences and biomass (Table 1).

**Plant height** Rhodes grass excelled in plant height attainment. There was significant difference in plant height for all species with Navua sedge being shortest (Figure 1, Table 1).

Figure 1. Plant height of Navua sedge (mean ± SE) and two pasture species (humidicola and Rhodes grass) under two moisture capacity levels (50 and 100%).

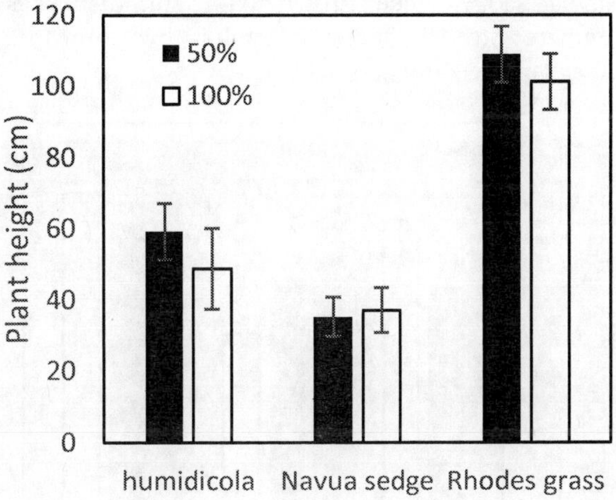


**Tiller production** At all replacement ratios, Navua sedge produced greatest number of tillers, followed by humidicola and Rhodes grass (Figure 2, Table 1).

Table 1. Summary tests of ANOVA for the effects of interspecific or intraspecific competition for Navua sedge, humidicola and Rhodes grass under intra/interspecific competition and with two soil moisture capacity levels. Note that the number of inflorescences is only recorded for Navua sedge.
* $P \leq 0.05$; ** $P \leq 0.02$; *** $P \leq 0.00$

| Factor | df | *F* ratio and probability | | | |
|---|---|---|---|---|---|
| | | Tiller number | Plant height | Inflorescences number | Biomass gained |
| Species (S) | 2 | 239 *** | 1390 *** | 556 *** | 112 *** |
| Competition (C) | 1 | 0.78 | 2.38 | 13.8 *** | 12.35 *** |
| Moisture (M) | 1 | 4.18 * | 0.48 | 1.13 | 4.68 * |
| S × C | 2 | 3.72 * | 3.47 * | 12.5 *** | 30.04 *** |
| S × M | 2 | 0.48 | 37.85 *** | 0.68 | 5.47 ** |
| C × M | 1 | 0.5 | 0.5 | 0.27 | 2.89 |
| S × C × M | 2 | 0.18 | 1.63 | 1.85 | 1.59 |

Figure 2. Number of tillers (mean ± SE) produced by Navua sedge and two pasture species (humidicola and Rhodes grass) under two moisture capacity levels (50 and 100%).

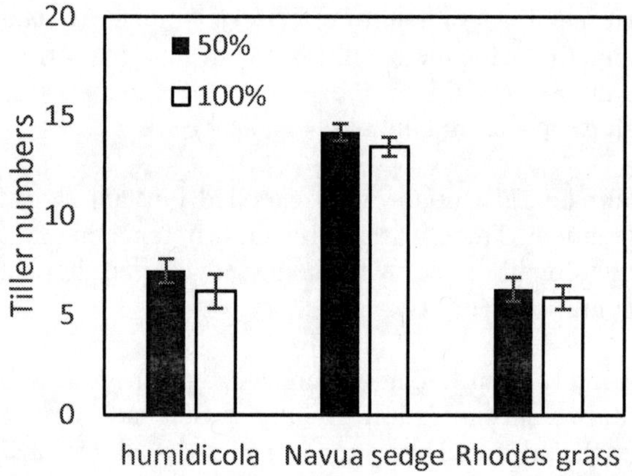


Figure 3. Inflorescence numbers (mean ± SE) of Navua sedge when grown with humidicola as compared to Rhodes grass under two moisture capacity levels (50 and 100%).

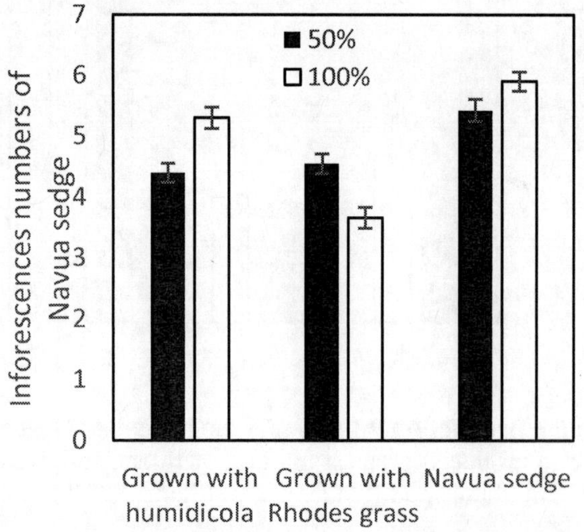


**Inflorescence production** During the time of the study, only Navua sedge produced inflorescences due to short duration of the study. More inflorescences of Navua sedge were produced in the monoculture (intraspecific competition) than when mixed with other species (interspecific competition). Additionally, similar number of inflorescences of Navua sedge were produced when grown with humidicola or with Rhodes grass (Figure 3).

**Biomass production** Rhodes grass produced more biomass than both Navua sedge and humidicola under both high and low moisture capacity. Interestingly, Rhodes grass produced more biomass under interspecific combinations than when it was growing in the monoculture (Figure 4 and 5).

Figure 4. Relative yield (mean ± SE) of total biomass for humidicola, Navua sedge and Rhodes grass under two moisture capacity levels (50 and 100%).

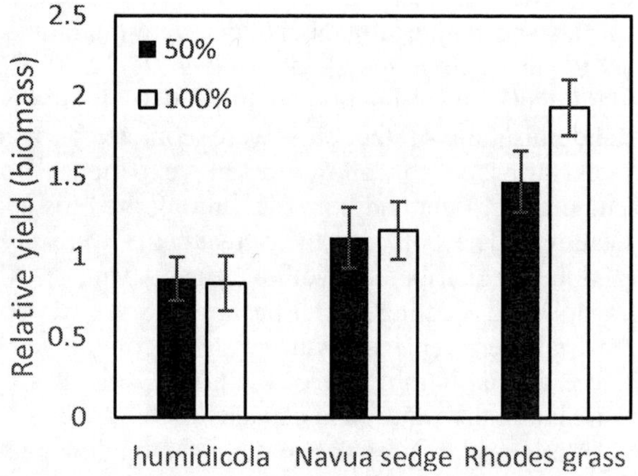


DISCUSSION

A mixture plant growth competition study should examine the plant responses to competition as well as the components of the plant interactions (Radosevich 1987). A replacement series method was selected to find the competitive effect of humidicola and Rhodes grass on the growth of Navua sedge. The approach was chosen in this study

Figure 5. Total biomass of humidicola (white), Rhodes grass (black) and Navua sedge (grey) under interspecific and intraspecific competition.

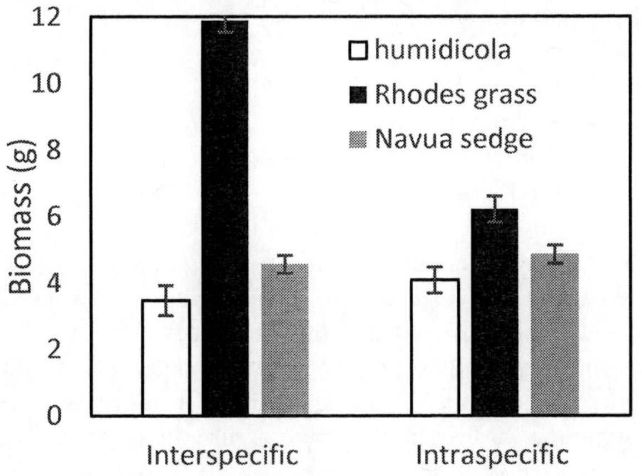


because it has the ability to avoid criticism when compared to additive method which inadequately accounts for the influence of density and species proportion on the outcome of competition (Radosevich 1987).

Height is a competitive advantage character where taller growing plants can shade the shorter species and inhibit them from accessing light energy resource (Falster and Westoby 2003). Light is a significant resource for plant competition which can result in a reduction of reproduction and growth rate. Both humidicola and Rhodes grass were able to grow taller than Navua sedge which in turn could reduce the biomass of Navua sedge by shading/inhibiting access to sunlight. The initiation and development of tillers in a plant is a basic unit of production which is also correlated to the vegetative period of a species and it affects the plant height as well as the structure of a plant (Anwar et al. 2012). Seavers and Wright (1999) stated that the tillering capacity of a plant is a critical element that can influence the competitiveness of a species to establish in an area, indicating that large tiller number production of Navua sedge will help its invasiveness. The duration of the trial was too short to reach the maturity stage of reproduction of all species, hence the competition experiment may not be robust enough to give a conclusive outcome of winners and losers. Further field trials to validate the results are needed. Nonetheless, the growth behavior of Rhodes grass indicates increasing yield under interspecific competition compared to growth in intraspecific competition. Thus Rhodes grass has a comparatively stronger ability to compete with Navua sedge (Szymura et al. 2018), and maybe used as a candidate grass species in pasture field infested with Navua sedge.


ACKNOWLEDGMENTS

We sincerely thank the Queensland Department of Agriculture and Fisheries for partly funding this project and the University of Queensland for providing access to facilities. We also thank Stephen Setter and Melissa Setter for the help with seed collection of Navua sedge. This is dedicated to the memory of Ms Mogomotsi Moilwa who sadly passed away this year.


REFERENCES


Anwar, M., Akmal, M., Shah, A., Asim, M. and Gohar, R. (2012). Growth and yield comparison of perennial grasses as rainfed fodder production. *Pakistan Journal of Botany* 44, 547-552.

Dhileepan, K., Musili, P.M., Ntandu, J.E., Chukwuma, E., Kurose, D., Seier, M.K., Ellison, C.A. and Shivas, R.G. (2022). Fungal pathogens of Navua sedge (*Cyperus aromaticus*) in equatorial Africa as prospective weed biological control agents. *Biocontrol Science and Technology* 32, 114-120.

Falster, D.S. and Westoby, M. (2003). Plant height and evolutionary games. *Trends in Ecology & Evolution* 18, 337-343.

Prince, C.M., MacDonald, G.E., Ferrell, J.A., Sellers, B.A. and Wang, J. (2018). Impact of soil pH on Cogongrass (*Imperata cylindrica*) and Bahiagrass (*Paspalum notatum*) competition. *Weed Technology* 32, 336-341.

Radosevich, S.R. (1987). Methods to study interactions among crops and weeds. *Weed Technology* 1, 190-198.

Seavers, G. and Wright, K. (1999). Crop canopy development and structure influence weed suppression. *Weed Research* 39, 319-328.

Shi, B., Osunkoya, O.O., Chadha, A., Florentine, S.K. and Dhileepan, K. (2021). Biology, Ecology and Management of the Invasive Navua Sedge (*Cyperus aromaticus*)—A Global Review. *Plants* 10, 1851. doi:https://doi.org/10.3390/plants10091851

Szymura, M., Szymura, T., Wolski, K. and Świerszcz, S. (2018). Can native grass species outcompete invasive goldenrods? Results of a replacement series experiment. *Weed Research* 58, 304-317.

Vitelli, J.S., Madigan, B.A. and van Haaren, P.E. (2010). Control techniques and management strategies for the problematic Navua sedge (Cyperus aromaticus). *Invasive Plant Science and Management* 3, 315-326.

Vogler, W.D., Carlos, E.H., Setter, S.D., Roden, L. and Setter, M.J. (2015). Halosulfuron-methyl: A selective herbicide option for the control of the invasive *Cyperus aromaticus* (Ridley) Mattf. and Kukenth (Navua sedge). *Plant Protection Quarterly* 30, 61-66.

# Precise Robotic Weed Spot-Spraying for Improved Environmental and Economic Outcomes in the Sugarcane Industry

Mostafa Rahimi Azghadi[1], Alex Olsen[2], Jake Wood[2], Brendan Calvert[1], Bronson Philippa[1], Terry Granshaw[3], Emilie Fillols[3]
[1]James Cook University, Townsville, Douglas, Australia,
[2]AutoWeed Pty Ltd, Townsville, Australia,
[3]Sugar Research Australia, Burdekin, Australia
(mostafa.rahimiazghadi@jcu.edu.au)

**Summary** This paper presents a novel robotic spot spraying solution for weed management in sugarcane that exploits the latest advances in deep learning, machine vision, and robotics to reduce herbicide usage on sugarcane farms and potentially deliver significant economic and environmental impact. Most herbicides can be lost by runoff and have been detected in Great Barrier Reef (GBR) ecosystems at concentrations high enough to affect organisms. The AutoWeed smart spot spraying system utilises deep learning to detect and spray grass and broadleaf weed species within any target crop or pasture environment using their image features as the basis for detection. As part of a project funded by the partnership between the Australian Government's Reef Trust and the Great Barrier Reef Foundation, James Cook University, AutoWeed, and Sugar Research Australia are conducting trials of the novel AutoWeed spot spraying technology on sugarcane farms in the Burdekin region. Field trials compare the performance of the new spot spray technology against existing broadcast spraying practices by measuring the efficacy to control the weeds, and water quality improvements in runoff. So far, the average results across 15 hectares of field trials show that AutoWeed spraying of nutgrass in sugarcane is 95% as effective as broadcast spraying and reduces herbicide usage by 35%, proportionally to the weed coverage. For specific trial strips with lower weed pressure, spot spraying reduced herbicide usage by up to 62%. Irrigation inducted runoff, three days after spraying, also showed that spot spraying reduced the mean concentration of ametryn and trifloxysulfuron in runoff by 49% and 60% respectively compared to broadcast spraying. These promising early results reveal the capability of this technology to reduce herbicide usage on sugarcane farms without impacting weed control and potentially providing sustained water quality benefits in the GBR catchment.

**Keywords** Precision robotic, spot spraying, deep learning, artificial intelligence, sugarcane

# Biological control of African Lovegrass (*Eragrostis curvula*): native-range surveys in Africa

Guy Sutton[1], Liam Yell[1], Clarke van Steenderen[1], Andrew McConnachie[2], Kerinne Harvey[3], Iain Paterson[1]
[1]Centre for Biological Control, Rhodes University, Makhanda, South Africa,
[2]Weed Research Unit, Invasive Species Biosecurity, NSW Department of Primary Industries Biosecurity and Food Safety, Orange Agricultural Research Institute, 1447 Forest Road, Orange, Australia,
[3]Weed Research Unit, Invasive Species Biosecurity, NSW Department of Primary Industries Biosecurity and Food Safety, Grafton Primary Industries Institute, PMB 2 Grafton, Australia
(guysutton41@gmail.com)

**Summary** African lovegrass, *Eragrostis curvula* (Schrad.) Nees (Poaceae), is a problematic alien invasive grass in Australia, particularly in New South Wales (NSW), where it reduces the grazing capacity of invaded pastures and increases fuel loads during the fire season. As such, African lovegrass is currently in the process of being nominated as suitable for weed biological control research in Australia. In the interim, surveys for potential biological control agents are being performed across its native range in southern Africa. We surveyed the insect communities on a range of sympatric *Eragrostis* spp. and closely-related genera to determine the field host-range of any potential biological control agents. Two stem-boring wasps (*Tetramesa* spp.; Eurytomidae) were recorded on *E. curvula* and prioritised as potential agents. Subsequent field surveys, however, yielded several morphologically indistinguishable *Tetramesa* spp. on a number of non-target grasses. We used DNA barcoding to delineate different *Tetramesa* spp. and assess their field host-range. No-choice oviposition trials were performed under greenhouse conditions to evaluate the host-range of both *Tetramesa* spp. Here, we discuss the progress of this project to date, highlighting the prospects and challenges for the biological control of African lovegrass in Australia.

**Keywords** Grass; Poaceae; host-specificity; field surveys; no-choice tests; *Tetramesa*

# *Listronotus frontalis* (Curculionidae: Coleoptera): host-specificity testing

Umar Lubanga[1], Raelene M. Kwong[1]
[1]Invertebrate & Weed Sciences, Agriculture Victoria Research Division, Department of Jobs, Precincts and Regions, AgriBio Centre, Bundoora, Australia
(umar.lubanga@agriculture.vic.gov.au)

**Summary** *Sagittaria platyphylla* (Alismataceae) is an emergent aquatic herb native to north America that has become a serious weed of shallow ephemeral or permanent water bodies, in natural and ruderal habitats. This weed is a serious invader of irrigation channels and drains in south-eastern Australia, where it forms dense monocultures that impede water flow, increase the risk of flooding and damage irrigation infrastructure. In natural waterways, extensive infestations threaten native biodiversity and potentially impede the movement of native fish. *Listronotus frontalis* was identified as a promising biological control agent for *S. platyphylla* alongside two other weevils (*Listronotus appendiculatus* and *Listronotus sordidus*). Since little was known about this weevil, host-specificity testing was preceded by pre-host specificity studies of the weevils' basic biology, ecology and behaviour. Results showed that larval development was negatively impacted by high water levels and plant phenology (reduced development on young plants compared to old plants). Laboratory host-specificity testing showed that non-Alismataceae species tested are not at risk off-target attack. Conversely, all of the exotic congeneric *Sagittaria* spp. tested may be at risk since *L. frontalis* larvae completed development on all of these species. Among native Alismataceae, three species *Alisama plantago-aquatica*, *Hydrocleys nymphoides* and *Damasonium minus* supported complete larval development and are therefore predicted to be at risk of off-target attack. *Caldesia oligococca* could not be effectively tested because this species only grows well under submerged conditions that are unfavourable for oviposition and larval development. Native species (*H. nymphoides* and *Caldesia oligococca*) that grow under submerged conditions may not be at risk, however, the safety of *A. plantago-aquatica* and *D. minus* that grow alongside *S. platyphylla* in the field cannot be guaranteed based on these results. Consequently, an application for release of *L. frontalis* will not be submitted until further studies ascertain the safety of these species. This work is part of the AgriFutures Biocontrol of Weeds project, funded by the Australian Government Department of Agriculture, Fisheries and Forestry as part of its Rural R&D for Profit program.

**Keywords** Alismataceae, Biocontrol, Curculionidae, *Listronotus frontalis*

# The blackberry cane-boring sawfly – what does its DNA tell us about its host specificity?

Raelene Kwong[1], Mark Blacket[1], Lea Rako[1], Vincent Lesieur[2], Thierry Thomann[2]
[1]Agriculture Victoria, Department of Jobs, Precincts and Regions, Bundoora, Australia,
[2]CSIRO Health and Biosecurity - European Laboratory, Montferrier sur Lez, France
(rae.kwong@agriculture.vic.gov.au)

**Summary** European blackberry taxa (*Rubus fruticosus* L. agg.) are a major threat to natural and agricultural ecosystems in Australia. Biocontrol using the leaf rust fungus, *Phragmidium violaceum* has achieved some advances in the suppression of susceptible *Rubus* genotypes, but the rust is not effective in low rainfall or moisture-stressed habitats. New agents for blackberry are still required and research should concentrate on natural enemies attacking blackberry crowns and primocanes (first year canes) because they are likely to have greater impacts on infestation characteristics. Larvae of the cane-boring sawfly, *Phylloecus faunus* (= *Hartigia albomaculata*) tunnel within the primocanes causing them to weaken and break thereby reducing daughter plant production. Initial host specificity testing conducted during the 1970s indicated that some *Rosacea* species might be at risk of attack, although it was suspected that the lab-based trials might have overestimated the true host range of this insect. Using DNA barcoding to rapidly identify larval specimens, we conducted a field survey in Mediterranean Europe to further our understanding of the field host range of *P. faunus* as a first step to assessing its potential as a candidate agent for the biological control of European blackberry in Australia. All specimens of *P. faunus* were collected exclusively from *Rubus fruticosus* and no evidence of the sawfly was found in *Rosa canina* plants growing nearby. Instead, a different sawfly species, *Cladardis elongatula* was found within *R. canina* canes. This study gathered supporting evidence that the ecological host range of the blackberry sawfly might be more restricted than initial studies suggest. Further investigation to assess its safety for introduction into Australia should be considered.

**Keywords** European blackberry, DNA barcoding, *Phylloecus faunus*

# A phylogeographic role in host-specificity testing – a case study of *Acacia auriculiformis* herbivores

Muhammad Nawaz[1,2], Dean R. Brookes[2,3], Graham A. McCulloch[2,4], Gimme H. Walter[2]
[1]Weed Research Unit, Invasive Species Biosecurity, New South Wales Department of Primary Industries, Orange, Australia,
[2]School of Biological Sciences, The University of Queensland, St. Lucia, Australia,
[3]CSIRO Health and Biosecurity, Brisbane, Australia,
[4]Department of Zoology, University of Otago, New Zealand
(muhammad.nawaz@dpi.nsw.gov.au)

**Summary** *Acacia auriculiformis* (Mimosaceae), also known as earleaf acacia, is a native Australian tree that has now become a category 1 invasive weed in Florida, USA. This research focused on identifying and prioritizing potential biocontrol agents against this weed. Field surveys were conducted (2016-2019) in its native range in Australia i.e. Far North Queensland and the Northern Territory (FNQ and NT, respectively). Over 1,000 specimens, from various insect groups, were collected from *A. auriculiformis* and related species (the latter distributed mainly in southern Queensland), and identified by COI DNA barcoding. Two insect groups were identified as highly damaging to the target weed: the leaf-tying caterpillars (mainly belonging to the cosmopterigid genus Macrobathra) and chrysomelid beetles identified as *Calomela intemerata*. Phylogenetic relationships within Macrobathra species were reconstructed using Bayesian inference. Seven moth lineages were identified from the 102 specimens sequenced across both FNQ and the NT. Haplotype networks were also constructed for the different lineages involved. The molecular analyses identified a deep genetic disjunction within many species across the Gulf of Carpentaria, a well-known biogeographic barrier. These disjunctions match the spatial genetic disjunction previously found in *A. auriculiformis*. Some of the lineages, mainly those ones sourced from GenBank and BOLD, were also collected beyond the native range of the target weed. Our findings suggest that the phylogenomic and phylogeographic approaches are helpful in addressing some of the key questions regarding the field host range of a potential weed biocontrol agent at the native range survey stage of a program.

# Biological control of *Cabomba caroliniana*: biology and host range of the cabomba weevil *Hydrotimetes*

Nagalingam Kumaran[1], Tim Vance[1], David Comben[2], Quinton Dell[1], MI Oleiro[3], C Mengoni Goñalons[3], Guillermo Cabrera Walsh[3], S Raghu[1]
[1]CSIRO, Dutton park, Australia,
[2]Queensland Department of Agriculture and Fisheries, Dutton park, Australia,
[3]Fundación para el Estudio de Especies Invasivas, Hurlingham, Argentina
(kumaran.nagalingam@csiro.au)

**Summary** *Cabomba caroliniana* Gray is a submerged aquatic weed, invasive in the waterways of Australia and several other countries. In Australia, *C. caroliniana* is a Weed of National Significance, and its detrimental effects include choking of waterways, reducing the water holding capacity of dams supplying drinking water and affecting native flora. During preliminary surveys and host specificity tests in the native range (Argentina and Paraguay), the aquatic weevil, *Hydrotimetes natans* Kolbe has been identified as a potential biological control agent to control *C. caroliniana*. We imported *H. natans* from the native range to Australia and studied its biology and host specificity. From biology studies, we found that eggs were laid on submerged leaves of *C. caroliniana* semi-embedded in a small divot and hatched in 7.65 ± 0.86 days. Larvae developed tunnelling through the leaves (early instar) or stems (late instar) and pupated outside the stem near the base of petioles after 25 to 27 days of development. Pupae developed into adult in 14.3 ± 2.7 days. The full lifecycle, from oviposition through to adult eclosion, took 46.5 ± 4.4 days. Host specificity trials were setup with *Brasenia*, *Nymphaea* and *Trithuria* species selected based on the centrifugal phylogenetic method, and data on oviposition, larval development, pupation and lifecycle completion were recorded. We found no evidence of oviposition and development of H. *natans* on any of the *Nymphaea* or *Trithuria* plant species tested. While *Brasenia schreberi* supported partial development (which was significantly lower than that on *C. caroliniana*), it did not support multiple generations of *H. natans*. In this talk, we discuss these results in light of risks of *H. natans* to native and other non-target species in Australia, and its potential to be part of the integrated weed management of *C. caroliniana*.

**Keywords** Biocontrol, cabomba, *brasenia*, WoNS, invasion, weeds, weed management, host specificity, erirhininae

# Herbicide resistance status of barley grass (*Hordeum glaucum* Steud.) populations in low rainfall zones of southern and Western Australia

Gurjeet Gill[1], Daniel Petersen[1] and Benjamin Fleet[1]
[1]School of Agriculture, Food and Wine, University of Adelaide, Glen Osmond SA 5064
(gurjeet.gill@adelaide.edu.au)

**Summary** Barley grass populations from the random survey of 2018 (n=143) showed no resistance to glyphosate or paraquat. However, resistance to FOP (group 1) herbicide quizalofop was present in 4% of the populations tested (n=7). Resistance to imidazolinone (IMI) herbicide Intervix® was only detected in two populations of barley grass from the Eyre Peninsula. These populations also exhibited cross-resistance to sulfonylurea (SU) herbicide mesosulfuron. There were large regional differences in the level of resistance detected. Barley grass populations from New South Wales (n=42) and Victorian Mallee (n=3) showed no resistance to any of the four herbicide groups used in resistance screening. In contrast, resistance to the SU herbicide mesosulfuron was identified in 16% of the populations from South Australia (SA) and Western Australia (WA).

Butroxydim had greater efficacy on quizalofop resistant populations of barley grass than clethodim. In the short term, it may be possible to improve weed control of clethodim resistant populations in the field by adding butroxydim to the mixture or by using it on its own.

Targeted sampling of barley grass in SA and Victoria in 2019 and 2020 showed higher levels of resistance to group 1 herbicides than in the random survey of 2018. Resistance to knockdown herbicides glyphosate and paraquat was also confirmed in some of the samples collected in 2020.

**Keywords** Barley grass, herbicide resistance.

## INTRODUCTION

Over the past 10 years, many growers in southern Australia have reported an increase in barley grass (*Hordeum glaucum* Steud.) infestation in cereal crops. There are several possible explanations for this increasing incidence of barley grass in these farming systems. Adoption of early sowing (sometimes dry sowing) has increased in this region, which could allow more plants to escape pre-sowing weed control. Another possibility is that barley populations have developed adaptive mechanisms to escape pre-sowing weed control practices used in crop production. Fleet and Gill (2012) showed that barley populations collected from cropping fields in the Eyre Peninsula and the Mid North regions of SA had a much longer seed dormancy than those from non-crop habitats.

Prior to this survey, the extent of herbicide resistance in barley grass was unclear. In a previous survey of barley grass in the Upper North and Eyre Peninsula in 2012 by Shergill et al. (2015a), group 1 resistance was detected in 15% of the populations. Grain growers in low rainfall regions are still seeking information on the current status of herbicide resistance in barley grass. The aim of this random survey was to determine herbicide resistance status of barley grass in low rainfall regions of NSW, VIC, SA and WA.

## MATERIALS AND METHODS

Barley grass populations (n= 143) were collected from farms in NSW, VIC, SA and WA in late spring and summer of 2018. To avoid any bias, fields for sampling were selected randomly on the basis of presence of barley grass without any consideration of previous control failures or management history. Additional targeted surveys of barley grass were undertaken in 2019 in Eyre Peninsula and Mid North of SA and in Eyre Peninsula and Victorian Mallee in 2020.

Barley grass samples to be used in resistance testing were stored at ambient conditions at the Roseworthy campus of the University of Adelaide. In April of each year, barley grass seeds were sown into potting mix (cocoa peat) in seedling trays and irrigated if needed. At the one leaf stage, barley grass seedling were carefully uprooted and transplanted into pots (10 plants $pot^{-1}$) for resistance screening.

**Herbicide resistance screening** Barley grass seedlings were sprayed with the label rates of group 1 (quizalofop as Leopard®), 2 (mesosulfuron as Atlantis® and imazamox + imazapyr as Intervix®), 9 (glyphosate as Weedmaster® DST®) and 22 (paraquat as Para-Ken®) herbicides. Adjuvants recommended by the manufacturers were added to the spray solution of all herbicides. Herbicide treatments were applied in a spray chamber (De Vries Manufacturing, Hollandale, United States), which was calibrated to deliver 100 L $ha^{-1}$ through a single TeeJet® 8002E (TeeJet Technologies, Illinois, United States) flat-fan nozzle at a speed of 3.6 km $h^{-1}$. Herbicide susceptible barley grass population collected from Yaninee in 2006 was used as the susceptible control. This population has been used in previous studies of herbicide resistance at the University of Adelaide. Plants were assessed for survival 4 weeks after the herbicide treatment and individuals with new shoot growth were counted as survivors.

## RESULTS

**Random survey 2018** All populations of barley grass collected in NSW and VIC were susceptible to the four herbicide groups used in resistance screening (Figure 1). However, some populations from SA and WA showed resistance to group 1 and 2 herbicides. Resistance to the SU herbicide mesosulfuron was identified in 16.1% of the populations tested. The level of growth inhibition of barley grass plants differed considerably between mesosulfuron resistant populations. Some of the populations showed 100% survival and no reduction in plant growth when sprayed with mesosulfuron, whereas others showed high survival but >50% reduction in barley grass height and biomass. It is quite likely that the mechanisms of resistance present in these two types of populations are different. The presence of resistance to the imidazolinone herbicide imazamox + imazapyr (Intervix®) was relatively low (1.4%).

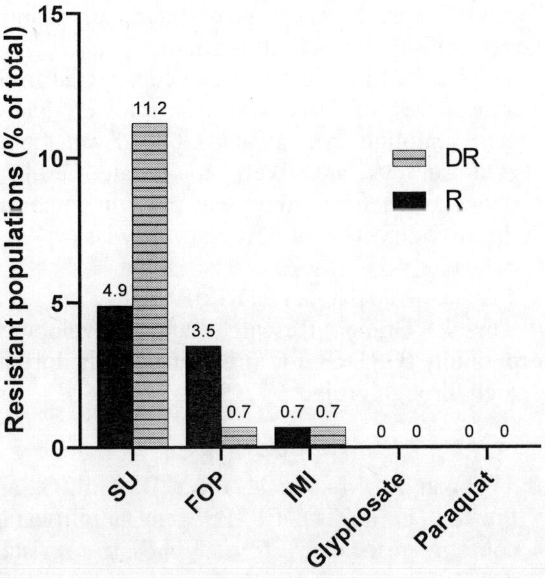


Figure 1. Detection of resistance to different herbicide groups in the random survey of barley grass (n=143). R = resistant (>20% survival) and DR = developing resistance (6-19% survival).

Resistance to quizalofop was detected in 4.2% of the barley populations tested. Four of these populations came from the upper Eyre Peninsula in SA and 2 from WA. There was no resistance detected to glyphosate or paraquat in barley grass samples in this survey.

**Cross-resistance to group 1 herbicides** Five barley grass populations confirmed to be resistant to quizalofop were also resistant to haloxyfop (Verdict®) (data not shown). Four of the five FOP resistant populations also showed complete (95-100%) survival when sprayed with clethodim (Clethodim®). Butroxydim (Factor®) provided much greater control of barley grass than quizalofop, haloxyfop and clethodim. At the higher rate of butroxydim (Factor® 180 g/ha), there was a complete kill of all barley grass plants even in resistant populations that had 100% survival when sprayed with clethodim.

**Targeted surveys 2019 and 2020** In the targeted survey of 2019, there was a high level of resistance to quizalofop in barley grass populations from Eyre Peninsula and mid to upper north regions of SA. Out of 32 barley grass populations investigated, 50% were classified as resistant and 19% were developing resistance (Table 1). The frequency of resistance to clethodim (44%) was slightly lower than to quizalofop (69%) but still a cause for concern. The level of resistance to quizalofop in the targeted survey of 2019 was much greater than in the random survey of 2018 (4.1% V 69%). On a positive note, none of the populations collected in 2019 were resistant to glyphosate (Weedmaster® DST® @ 760 mL $ha^{-1}$) or paraquat (Para-Ken® @ 1200 mL $ha^{-1}$).

Resistance to imidazolinone herbicide Intervix® was very low in 2019 with only 1 population classified as resistant and 1 developing resistance. This low frequency of imidazolinone resistance is consistent with the results from resistance screening of randomly collected samples from previous year. However, it is important to note that this imidazolinone resistant population from Eyre Peninsula showed no adverse response to Intervix® at 375 or 750 mL $ha^{-1}$. Interestingly this imidazolinone resistant population was not resistant to the FOP or DIM herbicides, which indicates direct selection through the use of group 2 herbicides. This result also highlights the importance of resistance testing when planning weed management strategies.

**Table 1.** Frequency (%) of resistance detected to different herbicides in targeted surveys. R = resistant >20% survival; DR = developing resistance 6-19% survival.

| HERBICIDE (GROUP) | 2019 | | 2020 | |
|---|---|---|---|---|
| | R | DR | R | DR |
| Fop (1) | 50.0 | 19.0 | 37.0 | 0.0 |
| Dim (1) | 38.0 | 6.0 | 25.9 | 3.7 |
| Imidazolinone (2) | 3.1 | 3.1 | 0.0 | 0.0 |
| Glyphosate (9) | 0.0 | 0.0 | 7.4 | 0.0 |
| Paraquat (22) | 0.0 | 0.0 | 11.1 | 3.7 |

Barley grass populations collected from Eyre Peninsula (n=12)) and Victorian Mallee (n=15) in 2020 were tested for herbicide resistance status in the winter of 2021. Resistance to quizalofop and clethodim (group 1) was confirmed in samples from both regions but the level of resistance was higher in the samples from EP. There was no resistance detected to group 2 herbicide Intervix® in 2020 populations. Some of the populations from the Victorian Mallee were resistant to glyphosate (13%) and paraquat (27%) but none of the samples possessed resistance to both of these herbicides. There was no resistance detected to glyphosate or paraquat in barley grass samples from Eyre Peninsula.

## DISCUSSION

Resistance to group 1 and 2 herbicides was confirmed in all 3 years of testing. Based on previous experience with other weed species, the highest level of resistance was expected to group 2 herbicides. However, this was not the case with resistance to group 1 herbicides being much more frequent than to group 2 herbicides.

Survivors of FOP herbicide quizalofop (group 1) were vigorous and usually showed no inhibition in growth. As these herbicides have been extensively used for in-crop and pasture weed control, herbicide resistance was expected. There is no doubt, presence of resistance to group 1 herbicides in the southern and western region will complicate management of barley grass in break crops and pastures. In previous research at the University of Adelaide, sequencing of the CT domain of the ACCase gene from barley grass populations confirmed the presence of previously known mutations Ile1781Leu and Gly2096Ala (Shergill *et al.* 2015b). These mutations can provide effective levels of resistance to quizalofop but variable level of resistance to the DIM herbicides.

It was interesting to observe much greater efficacy of butroxydim on quizalofop resistant populations. This unexpected greater sensitivity of group 1 resistant barley grass to butroxydim may prove beneficial for weed control in the short term. However, use of integrated weed management practices would be needed to delay the onset of butroxydim resistance.

Among group 2 herbicides, there was much greater resistance to mesosulfuron than imazamox + imazapyr (Figure 1). The level of resistance detected to imidazolinone (IMI) herbicides was relatively low (~3%). Therefore, growers can still use Clearfield® crops and imidazolinone herbicides with confidence but efforts should be made to diversify crop rotations and herbicide use as well as integration of non-chemical weed control tactics.

These studies also confirmed presence of resistance to glyphosate and paraquat in barley grass populations collected in the Victorian Mallee. Samples with paraquat resistance came from paddocks with extensive use of paraquat in lucerne, which is consistent with previous reports of paraquat resistance in barley grass in Australia (Powles 1986). Recently resistance to glyphosate was reported in barley grass from non-crop habitat in South Australia (Adu-Yeboah *et al.* 2020). In our survey, glyphosate resistance was identified in two barley grass populations collected from cropping paddocks in Victorian Mallee. Therefore, growers need to be vigilant so that resistance to these important knockdown herbicides can be detected early prior to a large build-up in weed infestations.

The level of herbicide resistance detected in the random survey of 2018 was much lower than in targeted sampling in 2019 and 2020. Even though targeted surveys are likely to inflate herbicide resistance frequencies, they can play an important role in early detection of new resistance issues.

## ACKNOWLEDGMENTS

We thank Grains Research and Development Corporation (GRDC) for providing funds for this research through project 9176981.

## REFERENCES

Adu-Yeboah, P., Malone, J., Fleet, B., Gill, G., and Preston, C. (2020). EPSPS gene amplification confers resistance to glyphosate resistant populations of *Hordeum glaucum* Steud. (northern barley grass) in South Australia. *Pest Management Science* 76, 1214-21.

Fleet, B. and Gill, G. (2012). Seed dormancy and seedling recruitment in smooth barley (*Hordeum murinum* ssp. *glaucum*) populations in southern Australia. *Weed Science* 60, 394-400.

Powles, S.B. (1986). Appearance of a biotype of the weed, *Hordeum glaucum* Steud., resistant to the herbicide paraquat. *Weed Research* 26, 167-72.

Shergill, L.S. Fleet, B., Preston, C. and Gill, G. (2015a). Incidence of herbicide resistance, seedling emergence, and seed persistence of smooth barley (*Hordeum glaucum*) in South Australia. *Weed Technology* 29, 782-92.

Shergill, L. S., Malone, J., Boutsalis, P., Preston, C., and Gill, G. (2015b). Target-site point mutations conferring resistance to ACCase-inhibiting herbicides in smooth barley (*Hordeum glaucum*) and hare barley (*Hordeum leporinum*). *Weed Science* 63, 408-15.

# Identification of field resistance to HPPD-inhibiting herbicides in wild radish

Roberto Busi[1], Danica Goggin[1], Hugh Beckie[1]
[1]Australian Herbicide Resistance Initiative, Crawley, Australia
(roberto.busi@uwa.edu.au)

**Summary** Reliance on 4-hydroxyphenylpyruvate dioxygenase (HPPD)-inhibiting herbicides for the control of multiple-resistant wild radish (*Raphanus raphanistrum*) populations has been common practice for the last decade in Australian wheat crops. Such an overreliance on HPPD herbicides has increased the risk for resistance evolution and resulted in reduced weed control in wild radish-infested crops. RESULTS Two wild radish populations (86-2020 and 91-2020) identified as putative resistant in an initial large-scale screening were characterized and confirmed to be 5- to 8-fold (comparison of LD50 values) resistant to the HPPD inhibitor pyrasulfotole, when plants were treated at the four-leaf stage, than the susceptible control population. Consistently, the two pyrasulfotole-resistant populations exhibited up to 4-fold resistance to the pre-formulated synergistic herbicide mixture pyrasulfutole + bromoxynil and up to 9- and 11-fold cross-resistance to mesotrione and topramezone, respectively. Results were confirmed by a small-plot trial conducted in the original field suspected of resistance where the mixtures pyrasulfotole + bromoxynil or topramezone + bromoxynil applied post-emergence delivered a significantly lower control of wild radish (79-87%) than mesotrione pre-emergence (>99%).
Conclusion
The first case of field resistance to HPPD herbicides in wild radish urges a turn in weed control practices. The mitigation of herbicide resistance in continuous no-till cropping requires constant optimization of the herbicide technology via the alternation and mixtures of multiple (old and novel) modes of action, use of pre-emergence herbicides and delivery to most sensitive plant stages. This must be integrated by proactive adoption of non-chemical weed control options, weed seed destruction and crop rotation.

**Keywords** Herbicide resistance

# Resistance surveys and commercial testing services – similarities and differences for wild oats, barley grass and brome grass across south eastern Australia

John Broster[1], Peter Boutsalis, Chris Preston, Gurjeet Gill
[1]Gulbali Institute, Charles Sturt University, Wagga Wagga, Australia
(jbroster@csu.edu.au)

**Summary** Wild oats (*Avena* spp.), barley grass (*Hordeum* spp.) and brome grass (*Bromus* spp.) are important weeds of cropping across south eastern Australia with herbicide resistance detected across this region in several herbicides to all three species. For this reason these species are all collected as part of the GRDC sponsored random surveys for herbicide resistance. Additionally, they are also provided to commercial resistance testing services for screening from locations where resistance is suspected. In this paper we compare the incidence of herbicide resistance for these species between random weed surveys and commercial testing services for samples collected from South Australia, Victoria, New South Wales and Tasmania across the 2015-2019 cropping seasons. Over 1000 wild oat, 300 barley grass and 400 brome grass samples have been collected from these two sources. For all three species a higher percentage of the testing service samples are resistant to Group A 'fop' (testing services - wild oats 61%, barley grass 55%, brome grass 20%; surveys - wild oats 29%, barley grass 2%, brome grass 1%) and 'dim' (testing services - wild oats 8%, barley grass 20%, brome grass 31%; surveys - wild oats 1%, barley grass 1%, brome grass 0%) herbicides compared with random survey samples. However for the Group B 'SU' herbicides, resistance was more common in the random survey samples than the testing services (testing services - wild oats 6%, barley grass 0%, brome grass 12%; surveys - wild oats 4%, barley grass 17%, brome grass 23%). Variability in resistance levels between states and regions for samples from both sources resulted in increases and/or decreases in these differences. Potential reason for these differences will be discussed.

**Keywords** Herbicide resistance, wild oats, brome grass, barley grass

# Dinitroaniline herbicide resistance evolution in *Lolium rigidum*

Qin Yu[1], Heping Han[1], Stephen Powles[1], Hugh Beckie[1]
[1]AHRI, University of Western Australia, Crawley, Australia
(qin.yu@uwa.edu.au)

**Summary** Dinitroaniline herbicides have been used for pre-emergence weed control for decades. Trifluralin is widely used in Australia as one of the most important pre-emergence herbicides to control annual ryegrass (*Lolium rigidum*) populations. However, periodic herbicide resistance surveys in southern Australia indicate that trifluralin resistance evolution is on the increase but is relative slow compared to resistance to major post-emergence herbicides. Like resistance to other herbicide modes of action, both target-site and non-target-site resistance mechanisms have been identified in some trifluralin-resistant ryegrass populations. A recent resistance mechanism survey with more than 20 field-evolved resistant populations from Western Australia suggests target-site resistance to trifluralin is more prevalent than non-target-site resistance. Based on this survey, novel resistance mutations of I235M, L238L, R243S and V252M have also been identified and functionally characterized in transgenic rice, in addition to previously identified resistance mutations of V202F, T239I, R243M and R243K in ryegrass. Among them, the V202F mutation exhibiting a fitness advantage is the most frequent one and resistance with this mutation is inherited as a recessive trait. Other resistance mutations occur at low frequencies, particularly the R243M mutation causing plant helical growth that is nearly lethal. Based on our work, relatively slow evolution of high-level dinitroaniline herbicide resistance in ryegrass is likely associated with recessive target-site resistance due to the existence of multiple target tubulin isoforms, and low frequencies of multiple resistance alleles possibly due to associated fitness costs. Other contributing factors slowing resistance include plants surviving the pre-emergence herbicide treatment early in the season controlled by the application of post-emergence herbicides as well as by non-chemical control measures, and use of mixtures of pre-emergence herbicides of different modes of action that significantly improves weed control efficacy. Together, these factors help delay and mitigate trifluralin resistance evolution.

**Keywords** Dinitroaniline herbicides, trifluralin, ryegrass, tubulin mutations, genetic transformation, fitness advantage/cost

# Strategic tillage for soil amelioration — how does it change the weed management strategy?

Catherine Borger[1], Sultan Mia[1], George Mwenda[1], Sarah Collins[2], Stephen Davies[3], Gausul Azam[1]

[1]Department Of Primary Industries And Regional Development, Northam, Australia
[2]Department Of Primary Industries And Regional Development, South Perth, Australia
[3]Department Of Primary Industries And Regional Development, Geraldton, Australia
(catherine.borger@dpird.wa.gov.au)

**Summary** Soil amelioration via strategic tillage is occasionally necessary for no-tillage systems to alleviate soil constraints. Amelioration may include deep ripping in compact soil, soil mixing to incorporate surface soil amendments like lime or a full soil inversion to bury the surface soil due to water repellence, herbicide-resistant weed seeds etc. Deep ripping, mixing and full inversion were applied at two field sites at Yerecoin and Darkan WA in 2019. The sites were used to investigate the impact of tillage practices on weed seed burial, emergence, and growth in the following three years (2019 to 2021). Full inversion buried 88% to 89% of annual ryegrass and great brome seed to a depth of 10-20 cm. Ripping and spading left 31% to 91% of seed in the top 0-10 cm of soil, with broad variation between sites (i.e., soil type). Of the seeds that were buried, most were at 10-20 cm. Even though tillage depth was 30-40 cm, very few weed seeds were buried below 20 cm. Soil inversion reduced weed density and annual ryegrass remained at a density of <1 plant m-2 for the three years after amelioration. Great brome density was reduced compared to the no-tillage control but recovered more successfully than annual ryegrass in the three years after amelioration, reaching 11 plants m-2 at Yerecoin and 147 plants at Darkan. This was possibly because great brome seed can have high emergence from a depth of 5-15 cm whereas annual ryegrass seeds have low emergence at 5 cm. Deep ripping and spading had no consistent impact on weed density. A comprehensive weed management plan is required following soil amelioration, to control existing weeds or maintain the benefit of soil inversion for large-seeded species like great brome that may reinfest the system.

**Keywords** soil renovation, mouldboard, integrated weed management, seed bank, Lolium rigidum, Bromus diandrus

# The role of crop competition in managing early emerging summer weeds in wheat

Asad Shabbir[1], Michael Widderick[2], Greg Harvey[2], Linda Heuke[1], Michael Walsh
[1]School of Life and Environmental Sciences, University of Sydney, Sydney, Australia,
[2]Agri-Science Queensland Department of Agriculture and Fisheries, Australia
(asad.shabbir@sydney.edu.au)

**Summary** In the northern grain region of Australia, it is increasingly common for 'summer' weed species to emerge in early spring within winter grown crops. This emergence typically coincides with the loss of residual herbicide activity, at a stage when advanced crop growth and canopy closure make seedlings difficult to target. However, these weeds can potentially be suppressed by enhancing winter crop competition effects on late-season emerging weed seedlings. The aim of this study then was to determine if reduced row spacing and increased wheat plant density prevented the establishment of summer weeds late in the winter growing season. Field trials at Narrabri, NSW and Hermitage, Qld measured the impact of wheat row spacing (25 cm and 50 cm) and plant density (75, 100 and 125 plants m-2) on the growth and seed production of common sowthistle, fleabane, awnless barnyard grass and feathertop Rhodes grass plants emerging at wheat pre-booting stage. Weeds biomass was significantly reduced to 83% when row spacing was reduced from 50 to 25cm. Increasing wheat plant density from 75 to 125 plants m-2 reduced the biomass of weeds by 49%. These results indicate that early emerging summer weeds can be controlled in wheat by maximizing crop competition through manipulating agronomic practices.

**Keywords** Row spacing, crop density, summer weeds, wheat, northern grains region

# Effect of combinations of sowing time, seed rate and herbicides on ryegrass management in faba beans

Ben Fleet[1], Gurjeet Gill[1]

[1]School of Agriculture, Food and Wine, The University of Adelaide, Roseworthy Campus, Roseworthy, Australia

(benjamin.fleet@adelaide.edu.au)

**Summary** Legume crops tend to be weak competitors with weeds and weed seedbanks can build up after the legume phase. Delaying crop sowing can reduce weed seedbank before seeding, but later sown crops can be less competitive against weeds. Seed rate has already been found to influence competitive alibility of many crops with weeds. A field trial was conducted at Roseworthy (SA) in 2019 to investigate factorial combinations of sowing time, seed rate and herbicides on the management of annual ryegrass (ARG) in faba beans. A three week delay in seeding faba beans (7 May to 31 May) did not reduce ARG plant density. However herbicide treatments that included post-emergent (POST) applications of clethodim and butroxydim had much greater efficacy when they were applied later in the 2nd time of sowing treatments. The same trend was evident in ARG seed set when crop seeding was delayed. ARG seed production was strongly influenced by faba bean seed rate (P<0.001). The high faba bean seed rate had 43% less ARG seed set compared to the low faba bean seed rate. Faba bean grain yield was significantly influenced by crop seed rate (P<0.001), with the high seed rate yielding 14% and 30% higher than the medium and low seed rates, respectively. Herbicide treatment had a significant effect on faba bean grain yield. When POST clethodim was applied after the simazine + trifluralin IBS (1.55 t/ha), faba bean grain yield improved by 43% to 2.21 t/ha. This field study has shown that at high plant density, faba beans can provide a significant suppression of ARG. However, faba beans were very intolerant to weeds as shown by the large yield losses. Higher seed rates, are important to both increase suppression of ARG by faba beans and to maintain grain yield.

**Keywords** Annual ryegrass, faba beans, weed management, seed rate, sowing time, herbicide

# Parkinsonia biological control: Establishment, spread and impact of UU1 and UU2 across northern Australia

Michelle Rafter[1], Kelli Pukallus[2], Wenting Su[3], Gimme Walter[3], Andrew White[1]
[1]CSIRO, Brisbane, Australia,
[2]QDAF, Charters Towers, Australia,
[3]UQ, Brisbane, Australia
(michelle.rafter@csiro.au)

**Summary** Parkinsonia (*Parkinsonia aculeata*) is a major weed of rangeland in northern Australia costing between $2-$300/ha/y to control depending on the density of infestations. Reducing control costs and improving pasture productivity can therefore assist in improving the profitability of rangeland production systems. Having a landscape-scale self-perpetuating form of control like biological control in these systems may aid in the integrated management of parkinsonia. This was the basis for the research pipeline of projects funded by Meat & Livestock Australia (B.NBP.0366; B.NBP.0620; B.WEE.0134) to identify candidate biological control agents and test their safety, mass rearing and release of agents that feed on parkinsonia but not on other plants. Currently the program has released over > 1 million *Eueupithecia cisplatensis* (UU1) and 337,638 *Eueupithecia vollonoides* (UU2) both of which are leaf defoliating moths. Field assessments have been conducted to determine the establishment, spread and impact of UU1 and UU2 on parkinsiona across northern Australia since 2020. An online survey of land managers has also been conducted to better define the management expectations of biological program on parkinsonia. Management objectives and results of the laboratory and field assessments of UU1 and UU2 effectiveness in the field and future directions for parkinsonia biological control will be discussed.

**Keywords** *Parkinsonia aculeata*, biocontrol, stakeholder engagement, field release, establishment, spread

# Pheromone trapping for monitoring the establishment and spread of *Eueupithecia cisplatensis* and *E. vollonoides*, biological control agents for *Parkinsonia aculeata*.

Cody-Ellen Murray[1,2], Michelle Rafter[1,2], Gimme Walter[1]
[1]The University of Queensland, St. Lucia, Australia,
[2]CSIRO, Dutton Park, Australia
(c.murray6@uq.edu.au)

**Summary** The accurate evaluation of the establishment and spread of a biological control agent is integral to monitoring the efficacy of weed management programs. Traditional monitoring techniques, such as in-person active sampling, while effective, are often labour intensive and expensive to conduct routinely. We are developing a pheromone trapping system that can be used to monitor agent establishment and spread, both spatially and temporally, for the biological control agents of *Parkinsonia aculeata*, *Eueupithecia cisplatensis* and *E. vollonoides*. Delta traps baited with live virgin females or lures made up of pheromone gland extracts were tested as monitoring tools for the presence of Eueupithecia in Parkinsonia infestations. The attractiveness of the pheromones across species was also tested, and trapping data revealed males were responsive to the pheromone profile of heterospecific females, although the rate of catch relative to that of conspecific females has yet to be quantified. Gas chromatography was used to determine the major chemical compounds (and their relative concentrations) in the pheromone profile of each species, and both laboratory and field assays were conducted to determine which compounds are behaviorally active. The most active compounds have now been prioritized as candidates for the development of synthetic lures. Trap data from North Queensland demonstrates that pheromone lures are a viable option for monitoring Eueupithecia populations in Parkinsonia infestations and that the development of a single synthetic lure for both species should be possible because the males are attracted to the pheromones of both species. Alternatively, combinations of the pheromones may be more effective.

**Keywords** Biological control, Chemical ecology, pheromones, lure monitoring

# First release and establishment of the biological control agent *Cecidochares connexa* for the management of *Chromolaena odorata* (L.) R.M. King & H. Rob (chromolaena) in Australia

Kelli Pukallus[1], Ainsley Kronk[1], Michelle Franklin[2]
[1] Biosecurity Queensland, Queensland Department of Agriculture and Fisheries, Tropical Weeds Research Centre, Charters Towers, Queensland, Australia.
[2] Weed Management Branch, Department of Environment Parks and Water Security, Northern Territory Government.
(kelli.pukallus@daf.qld.gov.au; michelle.franklin@nt.gov.au)

**Summary** *Chromolaena odorata* (L.) R.M. King & H. Rob (chromolaena or Siam weed) is a scrambling invasive shrub native to tropical America that significantly impacts terrestrial systems in Africa, SE Asia and, more recently, northern Australia. After its northern Queensland detection in 1994, *C. odorata* was part of a national cost-share eradication program until 2012 when eradication was deemed unfeasible. Anticipating the risk of *C. odorata* to the Australian environment, host testing commenced on the stem-galling fly *Cecidochares connexa* (Macquart), which was approved for release in 2018. This is Australia's first biological control agent for *Chromolaena odorata*.

**Keywords**: *Chromolaena odorata,* Siam weed, biological control agent, Australia's first, Queensland, Northern Territory.

## INTRODUCTION

*Chromolaena odorata* (L.) R.M. King & H. Rob (chromolaena) was first detected in Australia, near Bingil Bay, Queensland (QLD) in 1994 (Waterhouse 1994) and subsequently in July 2019 in the Northern Territory (NT) (NT Govt. 2020).

Chromolaena is a fast-growing multi-stemmed perennial shrub of the Asteraceae family. Plants grow two to three metres unsupported, and up to 10 metres when supported by other vegetation (Zachariades *et al.* 2009). Chromolaena can form dense thickets that can prevent the movement of livestock in grazing lands, affect agricultural crops and plantations, permanently alter ecosystems and contribute to hotter fires which destroy native vegetation (Zachariades *et al.* 2009). Chromolaena can be toxic to cattle and stock (QDAF 2020).

Chromolaena reproduces vegetatively and via seeds which are produced following the peak flowering period of May to August in Australia. Seed dispersal occurs through wind, attachment to machinery, animals or clothing and the movement of plant material along watercourses (QDAF 2020).

Chromolaena was a target for national cost-share eradication in Australia until 2012, when eradication was no longer deemed feasible. Subsequently, chromolaena was identified as a target for biological control and the stem-galling fly *Cecidochares connexa* was selected as the most feasible agent. Host testing commenced at Ecosciences Precinct in Brisbane, QLD, by the QLD Department of Agriculture and Fisheries (QDAF) in 2012. In 2018 release of *C. connexa* in Australia was approved by the Australian Government Department of Agriculture and Water Resources. The overall likelihood of off-target effects and potential consequences associated with the release of *C. connexa* was determined as being negligible (Australian Government Department of Agriculture and Water Resources 2018).

*C. connexa* was first released as a biocontrol agent in Indonesia in 1995 (McFadyen *et al.* 2003). It has since been released or detected in 11 countries in Africa, the Americas, Asia, and Oceania (Winston *et al.* 2014, Day *et al.* 2016). Mass-rearing commenced in 2019 at QDAF's Tropical Weeds Research Centre (TWRC) in Charters Towers, QLD. Shortly after, a collaborative breeding program was set up between QDAF and the NT Department of Environment, Parks and Water Security (NTDEPWS) Weed Management Branch. Releases began in both QLD and the NT in November 2019. This paper reports on the mass-rearing and release program of *C. connexa* in Australia.

## AGENT BIOLOGY

*C. connexa* is a small stem-galling fly (Diptera: Tephritidae) native to central America (McFadyen *et al.* 2003). Adults are between three and five millimetres long and females can be distinguished from males by the presence of an ovipositor at the end of the abdomen (Figure 1). Females use their ovipositor to deposit eggs into the stem tips (growing points) of chromolaena plants. Once these eggs hatch, the larvae feed on plant material inside the stem. Over the next 30-45 days, galls form around the larvae and these galls act as nutrient sinks, limiting the plant's ability to flower and produce seed, reducing its reproductive potential.

One to ten adults can emerge from the gall depending on the size of the gall and number of eggs laid. Emergence is through tunnels created prior to pupation ending with an emergence 'window' (Figure 2). The lifecycle from egg to adult takes approximately 50-80 days depending on climatic conditions. During hotter and wetter conditions, the lifecycle duration decreases and increases during cooler conditions.

**Figure 1**. Female *C. connexa* (left) have a black abdomen and ovipositor, males (right) have a brown abdomen and no ovipositor.

**Figure 2.** Multiple emergence 'windows' on glasshouse plant gall (left). Female *C. connexa* recently emerged from an emergence hole formed in a mature gall within an established release site (right).

## REARING PROCEDURE

At the primary breeding facility in QLD, TWRC, adult flies were collected following their emergence from galled stems in holding cages. Six to eight adult flies of each sex were placed in rearing cages containing three 200mm potted multi-stemmed chromolaena plants (Figure 3). These rearing cages were kept in temperature-controlled glasshouses where temperatures ranged from 27°C day to 22°C night and with natural lighting.

**Figure 3.** *Chromolaena odorata* plants in a rearing cage at TWRC.

In the NT similar female:male and plant ratios were used, however the rearing cages were housed in an outdoor shaded area exposed to natural temperatures throughout the year ranging from 38°C day to 15°C night.

Three weeks after cage set-up date, the plants were removed from the cages and cages were washed and sterilised with chlorine solution. Each plant was labelled, and the plants continued to grow and develop galls over the following six to eight weeks. Leaves were removed periodically to assist with pest control and improve gall detection.

Once emergence windows appeared, galls were collected by cutting stems approximately 10-15cm below the gall. All remaining green leaf material was removed, and any pests hosed off. Galled stems were placed in water filled glass jars within galled stem holding cages. Most flies emerged during the first few weeks following collection. Stems were kept watered for approximately one month until they turned brown and dried off. Flies emerged from the dried stems in the months following, but in smaller numbers. Several holding cages were used concurrently, with different collection dates, ensuring a consistent emergence of flies.

## RELEASES

Two methods were used for releasing *C. connexa*: galled stems or adults. Releases in QLD started with galled stems and then moved to adult releases as the main method. Adults were easier to send and release, and oviposition started immediately once released. Releases in the NT began with adults sent over from QLD and then moved into a combined adult/gall release system once the local colony became sustainable.

Suitable release sites need to have actively growing plants, with new shoots and limited flowers or seeds, in full or partial sun. Initial sites were targeted based on eco-climatic suitability modelling for *C. connexa* in QLD (Day *et al.* 2016). Releases were conducted on at least a 20m x 20m area containing a minimum of 20 larger chromolaena plants, spread relatively evenly across the site. Sites adjoining another chromolaena infestation were preferred for establishment and to increase spread potential. In the NT, a single suitable release site in the core of the infestation was used in the first year. The release site was also chosen as it was likely to be accessible during monsoonal rains.

Releases have taken place on private property, national parks and reserves, Defence land, local government and state land, in roadside verges, forestry plantations, quarries, riparian areas, open paddocks, gullies, hillsides, and rocky outcrops. Stems or flies were overnight couriered to clients for release or released directly by QDAF or NTDEPWS staff.

Releases from November 2019 to March 2022 (28 months), totalled 27,534 flies and 3,357 galls in QLD and 1,319 flies and 2,982 galls in the NT (Table 1). The total number of release sites for Australia is 114. This encompasses seven Local Government Areas in QLD and the Western Top End region in NT (Figure 4).

**Table 1.** Release data of *C. connexa* from November 2019 to March 2022 in Australia, showing release numbers and locations within each state.

| Local Government Area or location | Number of release sites | Number of adults/galls released | Number of sites with galls present/ monitored |
|---|---|---|---|
| **QLD** | | | |
| Burdekin | 5 | 1,864/0 | 4/5 |
| Cassowary Coast | 43 | 7,814/ 2,510 | 20/30 |
| Charters Towers | 10 | 2,127/ 110 | 6/6 |
| Douglas | 1 | 0/101 | 1/1 |
| Hinchinbrook | 9 | 994/636 | 4/4 |
| Tablelands | 5 | 1,754/0 | 2/3 |
| Townsville City | 39 | 12,981/0 | 23/28 |
| **Total** | **112** | **27,534/ 3,357** | **60/77** |
| **NT** | | | |
| Western Top End | 2 | 1,319/ 2,982 | 2/2 |
| **Total** | **2** | **1,319/ 2,982** | **2/2** |

**Galled stems**

In QLD, bundles of 40 to 50 stems were gathered and bound with a rubber band towards the base of the stems. Ends were trimmed to a uniform length and inserted through the lid of biodegradable coffee cup. A concentrated mixture of water crystals forming a slurry was placed in the bottom 1/3rd of the coffee cup. Once positioned in the field at the base of chromolaena plants, 5-6m apart, the coffee cups were filled with water. Refilling was sometimes required at weekly intervals. In the NT, bundles of 15 to 30 stems were placed in water-filled 100ml solid plastic containers with sponge around the top. Up to 16 containers were placed in an eight litre plastic bucket with drainage holes drilled in the base. The buckets were hung approximately one metre above ground level and in a shady location within the chromolaena site. The buckets were collected during later monitoring visits.

The flies emerged gradually over the following few weeks. The number of galled stems per release site ranged from 31 - 405 in QLD and 56 – 344 in NT.

**Figure 4.** Release locations in Northern Australia of *C. connexa*, shown by yellow dots.

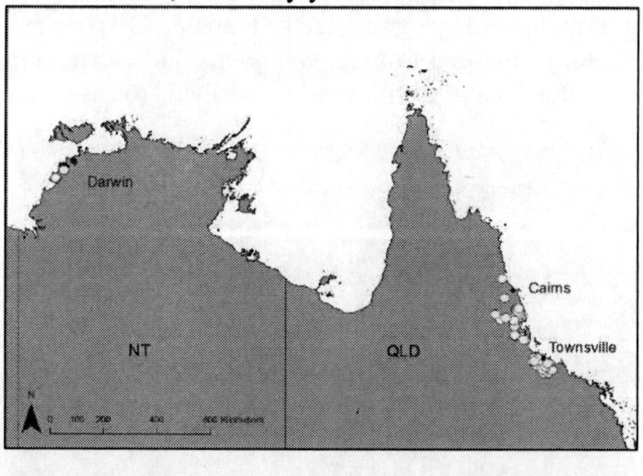


**Adults**

Flies were collected from glasshouse galled stem holding cages over four to five days and placed into 250ml round plastic containers. Large holes in the lids and gauze allowed air into the containers and prevented flies escaping. Approximately 20 females and 20 males were placed in each container, along with pieces of moistened paper towel. Flies were released directly onto plants in the field within seven days of collection and females oviposited straight away. One container was released every 3 to 5 metres. The total number of flies released per site ranged from 28 – 1,223. In the NT similar ratios and spacing were used, however the first four releases included the addition of a fly-screen cage placed over a single large plant at the release location. This was for the first few weeks after release and restricting all flies onto that plant during oviposition.

The number of releases per site ranged from one to six, except for one NT site that had 18 releases, using either adults, galled stems or a combination of both. The release numbers varied due to site suitability, site access, rearing colony production and establishment detection.

## MONITORING

Monitoring of sites commenced one month after adult releases and two months after gall releases. Follow-up monitoring for signs of galls continued over the following months. Establishment is declared at a field site when it contains galls at different stages of maturity over several months. Of the 79 monitored release sites, 62 (78.48%) met these criteria, which is considered a high level of establishment over the initial release period (Table 1).

Gall numbers have fluctuated at sites but persisted throughout the seasons. Plants at sites with average annual rainfall greater than 1,200mm, have longer periods of active growth and have produced more galls throughout the year. Typically, the wet season in northern Australia runs from November to April during which time most of the annual rainfall can occur. In QLD, 87% of releases were conducted during the wet season to capitalize on active growth periods of chromolaena. Outside of these months, plants display leaf drop and stem dieback due to the long period of low rainfall. Whereas in the NT, access to sites is restricted by heavy rainfall during the wet season, making it necessary to release more in the dry season when road conditions are more favourable.

During host specificity testing in Australia, *C. connexa* developed on *Praxelis clematidae* (Day *et.al* 2016). To date, no galls have been detected in wild field populations of *P. clematidae*, even growing within *C. connexa* established chromolaena sites.

Continued releases and monitoring are required for further evidence of *Cecidochares connexa* establishment, spread and impact on *Chromolaena odorata* within Australia.

## ACKNOWLEDGMENTS

The authors would like to thank landholders, QLD Local Governments, Department of Environment & Sciences, Biosecurity QLD, Defence Australia, NR&M groups, Townsville Correctional Farm, Tully Canegrowers, Bush Heritage, Department of Transport & Main Roads and NQ Plantations for their staff assistance with releases and involvement in the project and landholders who assisted with access to sites, traditional owners of the Delissaville / Wagait / Larrakia Aboriginal Land Trust, NT Department of Environment, Parks and Water Security staff, and the QLD and NT Governments for funding.

## REFERENCES

Australian Government Department of Agriculture and Water Resources (2018). Final risk analysis report for the release of *Cecidochares connexa* for the biological control of *Chromolaena odorata.*

Day, M.D., Riding, N. & Senaratne, K.A.D.W. (2016) The host specificity and climatic suitability of the gall fly Cecidochares connexa (Diptera: Tephritidae), a potential biological control agent for Chromolaena odorata (Asteraceae) in Australia. *Biocontrol Science and Technology* 26(5), 691-706.

McFadyen, R.E.C., Desmier de Chenon, R. & Sipayung, A. (2003). Biology and host specificity of the chromolaena stem gall fly, Cecidochares connexa (Macquart) (Diptera: Tephritidae), *Australian Journal of Entomology* 42, 294–297.

NT Govt., Northern Territory Government (2020). Siam Weed in the Northern Territory, https://nt.gov.au/__data/assets/pdf_file/0009/979722/siam-weed-brochure.pdf (accessed 3 May 2022).

QDAF, State of Queensland, Department of Agriculture and Fisheries (2020). Siam weed, *Chromolaena odorata and Chromolaena squalida* Fact Sheet. https://www.daf.qld.gov.au/__data/assets/pdf_file/0015/50028/siam-weed.pdf, (Accessed 1 May 2022).

Zachariades, C., Day, M., Muniappan, R. and Reddy. G.V.P. (2009). Chromolaena odorata (L.) King and Robinson (Asteraceae). *In* 'Biological Control of tropical weeds using arthropods' eds R. Muniappan, G.V.P. Reddy and A. Raman. pp. 130-62. (Cambridge University Press, New York).

Waterhouse, B. (1994). Discovery of Chromolaena odorata in northern Australia. Chromolaena odorata Newsletter, 9, 1–2.

Winston, R.L., Schwarzländer, M., Hinz, H.L., Day, M.D., Cock, M.J.W. and Julien, M.H. (2014). Biological Control of Weeds: A World Catalogue of Agents and Their Target Weeds, 5th edn. (USDA Forest Service, Morgantown, West Virginia).

# A thorny tale: *Cylindropuntia pallida* (Hudson pear) biocontrol in New South Wales, Australia

Andrew McConnachie[1], P Jones[2], A Fletcher[3], M Savage[3], A Patterson[1], R Holtkamp[4], L Snow[2], T Taylor[2], J Skewes[5], P Dawson[6], C Bergin[7], K Harvey[1], P Turner[1], R Shilpakar[8], M Nawaz[1]
[1]New South Wales Department of Primary Industries, Orange, Australia,
[2]Department of Agriculture and Fisheries, Biosecurity Queensland, Brisbane, Australia,
[3]Castlereagh Macquarie Count Council, Australia,
[4]Horizon Ecological Consulting, Australia,
[5]Northern Slopes Landcare Association, Australia,
[6]North West Local Land Services, Australia,
[7]North West Local Land Services, Australia,
[8]Greater Sydney Local Land Services, Australia
(andrew.mcconnachie@dpi.nsw.gov.au)

**Summary** *Cylindropuntia* spp. (Cactaceae) are weeds of arid and semi-arid regions of mainland Australia, with eight species currently recorded as naturalised. All of these are recorded in the north west of New South Wales (NSW), however, *Cylindropuntia pallida* (Hudson pear) is considered the most problematic in this region, with the weed currently thought to occupy ca. 100,000 hectares. (Modelling has shown that it has the potential to spread to 600,000 hectares in NSW and 112 million hectares across Australia in the next two to three decades if left unchecked.) Hudson pear reduces the viability of agricultural enterprises, land values and severely impacts native fauna and flora. A biocontrol program was initiated in Australia in 1925 for the control of *Cylindropuntia imbricata*, with the introduction of a cochineal, *Dactylopius tomentosus* (Dactylopiidae). More recently, exploratory work in the southern United States of America and Mexico yielded 22 lineages of *D. tomentosus*. Of these, six lineages were identified (through a systematic and quantifiable process) as having the greatest impact on each of their eight *Cylindropuntia* spp. targets. The *D. tomentosus* lineage 'californica var. parkeri', was earmarked to tackle the core of the Hudson pear infestation in NSW, Australia. To understand the potential dispersal and impact of the cochineal post-release, two long-term field monitoring sites were established in 2017 and are currently sampled every three months. To enhance the biocontrol effort in the core Hudson pear infestation, a decision was made to invest in a dedicated cochineal mass-rearing facility which would be able to produce large numbers of cochineal-infested cladodes as part of an augmented approach. Here we discuss the progress of the biocontrol programme for Hudson pear and its prospects for the future.

**Keywords** Mass-rearing, impact, lineage

# Sensitivity of lupin (*Lupinus angustifolius*) and other plant species to Overwatch® Herbicide (bixlozone) when applied at rates simulating spray drift exposure

Marco Montagna[1], Geoff Robertson[1], Peter Boutsalis[2], Sam Kleemann[2]
[1]FMC Australasia Ltd, North Ryde, Australia,
[2]Plant Science Consulting Pty Ltd, Prospect, Australia
(marco.montagna@fmc.com)

**Summary** Overwatch® Herbicide (400 g/L bixlozone) is a new herbicide developed in Australia by FMC for the pre-emergence control or suppression of a wide range of grass and broadleaf weeds. Overwatch® Herbicide was granted first registration in 2020 at an application rate of 1.25 L/ha (500 g a.i./ha) rate in wheat (*Triticum aestivum*), barley (*Hordeum vulgare*), canola (*Brassica napus*). FMC investigated the sensitivity of certain crop and weed species to bixlozone following in-field reports of off-target movement. Six small plot replicated field trials were installed in lupin (*Lupinus angustifolius*) crops across Western Australia and New South Wales where Overwatch® Herbicide was applied at 1.25 to 250 mL/ha (0.5 to 100 g a.i./ha). Overwatch® Herbicide exposure was also assessed in an outdoor controlled environmental trial on lupin, lentil (*Lens culinaris*), chickpea (*Cicer arietinum*), serradella (*Ornithopus sativus*), canola (*B. napus*), oat (*Avena sativa*), wheat, and milk thistle (*Sonchus oleraceus*). In addition, the effect on lupin, of 2.5 to 12.5 ml/ha (1 to 5 g a.i./ha) Overwatch® Herbicide mixed with glyphosate (2.5 to 12.5 g a.i./ha), paraquat (1 to 5 g a.i./ha) alone or plus trifluralin (2.88 to 14.4 g a.i./ha) was also investigated. The field trials demonstrated the high sensitivity of lupin to bixlozone with symptoms of bleaching being visible from 2.5 mL/ha Overwatch® Herbicide (1 g a.i./ha); however, the crops recovered with no significant yield penalties up to 50 mL/ha Overwatch® (20 g a.i./ha) ($p<0.05$). Serradella was the most sensitive tested species to bixlozone, whereas wheat showed a high level of tolerance. Bixlozone tank-mixes with glyphosate or paraquat did not increase the level of crop damage apart for minor necrosis of some lupin leaves treated with the paraquat mixes.

**Keywords** Bixlozone, lupin, off-target, bleaching

# Weed seed bank mitigation using cover crops in maize (*Zea mays*)

Deborah Hackell[1], Trevor James[1], Ben Wynne-Jones[1], Steve Payne[2] and Sam McDougall[2]
[1]AgResearch, Ruakura Research Centre, Private Bag 3123, Hamilton 3240, New Zealand
[2] Foundation for Arable Research, PO Box 23133, Hornby, Christchurch 8441, New Zealand
(deborah.l.hackell@agresearch.co.nz)

**Summary** Winter cover crops used in maize (*Zea mays* L.) grain or silage production systems can mitigate weed seed banks over time which ultimately allows for fewer herbicide applications. A winter cover crop trial in maize, using four different cover crops (or left fallow) under five herbicide regimes, was conducted over a 4-year period in Waikato, NZ. Cover crops used were gland clover (*Trifolium glanduliferum* Boiss.), faba bean (*Vicia faba* L.), oats (*Avena sativa* L.) and Italian ryegrass (*Lolium multiflorum* Lam.).

Results show the ryegrass winter cover crop was most effective in reducing the weed seed bank, but under the oats cover crop the weed seed bank was less diverse. Herbicide treatments significantly reduced the weed seed bank further, but a single post-emergence treatment was not significantly different from the more complex combination of a pre- and post-emergence herbicide treatment or the double post-emergence herbicide treatments. A winter cover crop can mitigate weed pressure from the seed bank in a repeat maize cropping system when combined with a single post-emergence herbicide treatment.

**Keywords** Maize, winter cover crop, herbicide, integrated weed management.

## INTRODUCTION

The seedbank has a critical role in the population dynamics of species that reproduce by seeds. In cropping, most weeds will be annuals with the seed being the longest-lived part of the plant's lifecycle. With a large part of these plants' evolutionary strategy being to produce a seed, mitigating weed seed banks needs to be at the forefront of weed management in our agricultural systems (Merfield, 2019).

With the advent of herbicides, agricultural production has increased, but the heavy reliance on the use of this one tool for weed management has increased the evolutionary pressure towards herbicide resistance. Herbicide resistant weeds are now a global issue and an increasingly emerging one in New Zealand with 13 weed taxa showing herbicide resistance in 2020 (Buddenhagen et al, 2020) and a series of recent publications increasing that number to 19 species (Buddenhagen, 2022, pers. commun.). This developing herbicide resistance issue means growers need to manage this tool's longevity. Growers will require more diverse integrated weed management options to maintain and enhance production. The use of a cover crop is one weed management tool that can be used to mitigate the weed seed bank return and allow for fewer herbicide applications over time.

Cover crop terminology and their application, however, can be quite diverse and widely applied. In this paper a cover crop is a single species winter crop, used as a dead mulch (terminated before crop planting) in spring planted maize. The cover crop therefore acts as a short-term "soil blanket" between seasons and in early maize growth, minimizing the open, light filled areas where weeds grow and replenish the weed seed bank. In New Zealand, maize grain systems are typically left fallow over winter, while maize silage systems may be planted in a winter cover crop, it would usually be harvested for fodder prior to spring cultivation and maize planting (James et al, 2007). Both pre- and post-emergence herbicides are also often applied to reduce weed impacts in the maize crop. This paper investigates, in a maize grain system, the effectiveness of four single species winter cover crop treatments and five herbicide treatment combinations. We focus on whether cover crops can mitigate the weed seed bank and potentially allow for fewer herbicide treatments needing to be applied in the following maize crop.

## MATERIALS AND METHODS

A trial site was established in June 2016 at the Foundation for Arable Research Northern Crop Research site (NCRS) at Tamahere, Hamilton, NZ. The soil was Horotiu silt loam. The average annual precipitation is 1202 mm with a mean air temperature of 13.7°C (NIWA 2021). The trial location and layout remained consistent over the trial period to establish a multi-season crop effect within a spring planted no-till maize grain system where maize stover was left in situ. The trial set up was a randomised strip split plot design with four replicates. Each of the five cover crop treatments were sown horizontally in 6 m wide by 30 m long plots. Winter cover crops for each year were sown between 22 May to 11 June. The vertical split comprised five herbicide regimes across the different cover crops. Each of the 100 plots (cover crop plus herbicide) were 6 x 6 m. In spring (between 27 October and 9 November), maize was planted no-till with a John Deere planter so that each plot contained eight rows of maize. Plots were managed similarly for each season for 4 years with nutrients applied as required.

**Treatments** Following several weeks fallow after the previous maize harvest, the winter crops were sown with a John Deere 750A box drill down the strips. The cover crop treatments were gland clover (*Trifolium glanduliferum* Boiss.) cv Prima; faba bean (*Vicia faba* L.) cv Ben; oats (*Avena sativa* L.) cv Milton; and Italian ryegrass (*Lolium multiflorum* Lam.) cv Tama, with an unplanted winter fallow used as a comparative control. Sowing rates were 6.6, 300, 100, 25 and 0 kg ha$^{-1}$ respectively. Winter cover crops were terminated in late spring by first rolling crosswise with a tractor-mounted crimp roller set at minimal pressure then sprayed the following day with glyphosate (Weedmaster TS540 at 3 or 4 L ha$^{-1}$) + organosilicone adjuvant (Pulse Penetrant 0.1% v/v) using a quad bike mounted sprayer.

The herbicide treatments for the maize crop were: no herbicide control; a pre-emergence application of acetochlor + saflufenacil; a pre-emergence of acetochlor + saflufenacil followed by topramezone + atrazine applied 5-6 weeks after maize emergence; a single post-emergence application of topramezone + atrazine applied 3-4 weeks after maize emergence; and two post-emergence applications firstly mesotrione + atrazine and then nicosulfuron applied 2-3 and 5-6 weeks after maize emergence respectively. All herbicides were applied at label recommended rates and with recommended adjuvants. They were applied to the appropriate plots using a $CO_2$ powered backpack sprayer with a water rate of 200 L ha$^{-1}$ at 160 kPa, through a 3 m boom holding four Teejet® 11003 air induction nozzles, spaced at 75 cm to spray inter-row.

**Assessments** Cover crop biomass (DM ha$^{-1}$) was sampled before termination in late spring. A 1 m$^2$ quadrat was placed in a representative area within each plot and foliage was cut to 10 mm above ground. Fresh weight was recorded and then a $\geq$300 g sub-sample dried at 80°C for 48 h to determine dry matter (DM) and biomass ha$^{-1}$. Cover crop residues were assessed by a visual percent ground cover assessment four days after cover crop termination.

Maize plants were also harvested for silage yield in March each year from a 2 m representative strip in the third maize row of each plot. All plants were cut 10 cm above ground, their fresh weight recorded and then plants mulched, and a 200 g sub sample dried at 80°C for 48 hours to determine DM.

Grain yield was measured in early May. Plants were harvested by hand from three 2 m long sections of central rows in each plot, avoiding rows harvested for silage. Plants in these lengths were counted to determine plant populations and yield ha$^{-1}$. The cobs were shelled, and the moisture content measured with a Dickey-John GAC2100 grain moisture meter.

After maize harvest in the fourth season of the trial, six 75 mm diameter soil cores to a depth of 75 mm (total average weight of 1273 g) were collected randomly from each plot. In the laboratory, they were broken up and thoroughly mixed, and then a subsample of 500 g soil from each was washed in a mesh bag (0.25 mm mesh) in a modified agitator washing machine. Samples were left to dry and then seed separation was achieved by manually removing seed while viewed under a binocular light microscope (Nikon SMZ745). Seed from each of the 100 samples were identified, sorted and abundance recorded.

## RESULTS

After four consecutive years using winter cover crops in a maize grain system, there was a reduction in the weed seed bank compared with the fallow and no-herbicide management regimes. Overall, 20,710 weed seeds were counted from the 100 plots. Weed seeds recorded under fallow, clover, ryegrass, oats and faba bean cover crops totaled 7,454, 3,128, 1,745, 5,100 and 3,283 respectively. Weed seed bank mitigation was greatest under the ryegrass winter cover crop with 77% fewer weed seeds in the soil compared with fallow (Figure 1).

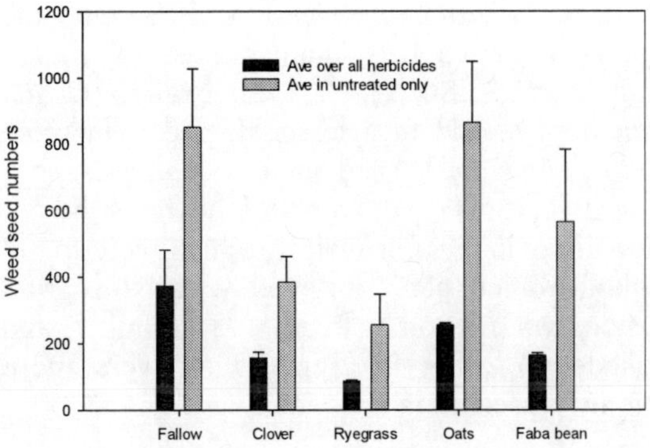


Figure 1. Comparison of average weed seed bank (per 500 g soil sample) under different cover crops after 4 years with pooled herbicide treatments and in the no herbicide control. Error bars are the SEM for each data set.

As expected, significantly fewer weed seeds were present in the herbicide treatments compared with no herbicide, both across all cover crops and in fallow only (Figure 2). When comparing the different herbicide treatments, results show no significant difference between the three herbicide treatments which used a post-emergence application either with or without cover crops (Figure 2).

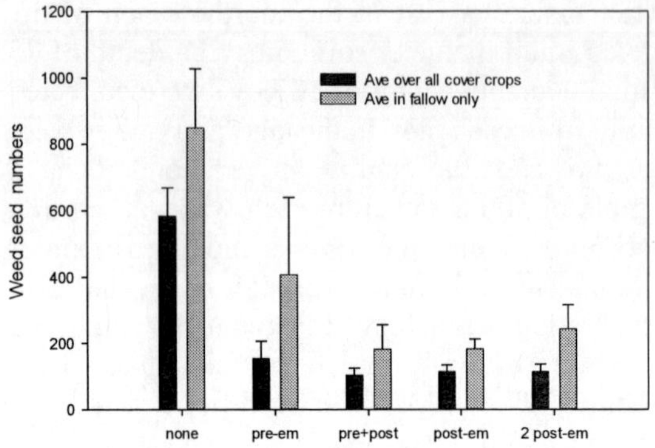


Figure 2. Comparison of average weed seed bank (per 500 g soil) under different herbicide treatments with pooled cover crops and under the no cover crop fallow. Error bars are the SEM for each data set.

Weed seed bank diversity was similar across the different cover crops, except for oats which had half the species variety compared to fallow despite oats having greater seed abundance (255 mean seed number 500 g$^{-1}$ ± 5.6 SEM) than the other cover crops (Table 1). In the weed seed bank, the most common weed seeds found were annuals: amaranths (*Amaranthus* L. spp), fathen (*Chenopodium album* L.) and chickweed (*Stellaria media* (L.) Vill.). Other common weed seeds found across most plots were bitter cress (*Cardamine hirsuta* L.), purslane (*Portulaca oleracea* L.), summer grass (*Digitaria sanguinalis* (L.) Scop.), horned oxalis (*Oxalis corniculata* L.), and to a lesser degree twin cress (*Lepidium didymium* L.) and mallow (*Malva* L. spp.). The less diverse oat cover crop plots however did not contain any twin cress or mallow which were found throughout other plots, or yellow bristle grass (*Setaria pumila* (Poir.) Roem. & Schult.) and nightshade (*Solanum* L.) spp. which were more commonly present in fallow plots.

Table 1. Weed seed bank diversity after four years under different cover crop and herbicide treatments.

| Treatment | Species | Treatment | Species |
|---|---|---|---|
| Fallow | 30 | No Herbicide | 21 |
| Clover | 28 | Pre-em only | 26 |
| Ryegrass | 22 | Pre- + post-em | 31 |
| Oats | 15 | 1 post-em | 33 |
| Faba bean | 27 | 2 post-em | 23 |

Comparing absolute seed numbers in these herbicide and cover crop regimes, the ryegrass cover crop combined with both a pre- and post-emergence herbicide treatment was the most effective at reducing the weed seed bank. The fallow/no herbicide plots had seven times more weed seeds than this combination of treatments. However, when assessing the herbicide treatments on weed seed bank abundance, there were no significant differences between the various herbicide combinations.

Cover crop residues are presented in Table 2 for the 4 years of the trial. Overall oats and ryegrass provided the best residue coverage in the first 3 years. However, stover build up over time impeded the cover crop drill and establishment of the small-seeded cover crops, particularly clover and ryegrass. In 2019 the plots to be planted in clover and ryegrass were cultivated post maize harvest to reduce the maize residue.

Table 2. Cover crop biomass and residue coverage at termination

| Cover crop | 2016 | 2017 | 2018 | 2019 |
|---|---|---|---|---|
| | Biomass (DM t ha$^{-1}$) | | | |
| Ryegrass | 4.6 | 2.3 | 1.4 | 8.4[1] |
| Oats | 6.7 | 4.9 | 7.0 | 5.7 |
| Faba bean | 4.0 | 3.3 | 5.7 | 6.2 |
| Clover | 1.2 | 0.3 | 0.03 | 4.2[1] |
| LSD | 1.5 | 0.9 | 1.5 | 1.2 |
| | % Residue cover | | | |
| Fallow | 14.5 | 0.9 | 10.3 | 24.1 |
| Ryegrass | 79.9 | 39.1 | 46.8 | 98.3 |
| Oats | 78.7 | 42.8 | 87.8 | 86.5 |
| Faba bean | 43.2 | 14.8 | 45.2 | 86.9 |
| Clover | 35.1 | 7.8 | 8.5 | 99.1 |
| LSD | 14.4 | 13.0 | 12.1 | 10.5 |

[1]cultivation undertaken prior to sowing.

Grain and silage yield under the different cover crops over 4 years are presented in Table 3. All years had good growing climate conditions except 2019 which experienced a drought over summer.

Table 3. Silage and grain yield over 4 years

| Cover crop | 2017 | 2018 | 2019 | 2020 |
|---|---|---|---|---|
| | Silage yield (t DM ha$^{-1}$) | | | |
| Fallow | 21.5 | 19.9 | 17.3 | 11.9 |
| Ryegrass | 17.6 | 19.1 | 18.4 | 13.5 |
| Oats | 19.2 | 21.0 | 17.5 | 13.9 |
| Faba bean | 23.2 | 21.2 | 18.5 | 14.2 |
| Clover | 24.7 | 20.6 | 18.8 | 14.0 |
| LSD | 2.4 | 2.2 | 1.6 | 1.8 |

| Grain yield (t DM ha$^{-1}$) | | | | |
|---|---|---|---|---|
| Fallow | 11.2 | 13.5 | 6.7 | - |
| Ryegrass | 12.5 | 14.3 | 7.9 | - |
| Oats | 12.8 | 15.4 | 7.6 | - |
| Faba bean | 9.4 | 13.3 | 7.4 | - |
| Clover | 8.3 | 12.4 | 8.1 | - |
| LSD | 1.7 | 1.0 | 0.8 | - |

- Covid lockdown prevented data collection.

## DISCUSSION

Weed seed bank mitigation is important in improving farm resilience to weed pressures and herbicide resistance. One of the benefits of using a winter cover crop in maize evident in this trial was they reduced weed seed bank abundance sufficiently so that a single post emergence herbicide treatment was adequate for effective weed control.

Cover crops, once established, provide a physical barrier or blanket against new weed establishment. Adequate competition is best achieved by dense cover crop residues.

In New Zealand, current practice with ryegrass cover crops is to use it as a cash or fodder crop, leaving no residue thus removing its ability to restrict weed emergence and prevent further weed seed set. Additionally, the crop residue is cultivated in preparation for maize planting bringing weed seeds to the surface to germinate. No-till is not widely practiced in New Zealand maize cropping. Work is continuing to investigate how much ryegrass residue would be required to reap the benefit of reducing the weed seed bank while returning some fodder to the farmer. Residue coverage is just one factor influencing the weed seed bank. Oats and ryegrass cover crops in this trial provided the best residue coverage in the first three years but both showed quite different effects on the weed seed bank. Mean weed seed bank abundance was higher for the oats cover crop but weed species were less diverse. The differences in diversity of weed species indicates there may also be an allelopathic benefit.

Both oats and ryegrass have both been shown to have low level allelochemicals from root leachates (Bezuidenhout et al, 2012). Bertoldi et al. (2009) demonstrated oats released allelochemicals into the soil during its decomposition and this inhibited both seed germination and reduced growth of new seedlings. Allelochemicals released from oats have also been shown by Bertoldi (2009) to be weed species specific. Oats plots compared with fallow did not show weed seeds from twin cress, mallow, nightshades, or yellow bristle grass.

Although ryegrass was shown to be the most effective in weed seed bank mitigation, its impact on yield suggests more research is required, particularly on nutrient requirements, before recommending a winter cover crop. Ryegrass decomposes more rapidly than oats thus utilising available soil N at a time when the establishing maize plants most require it. This may also be partly or wholly overcome by combining ryegrass with a legume. However, there is some indication from this trial that the initial costs from reduced maize yield caused by ryegrass cover crops may be offset over time by the weed seed bank mitigation, resulting in fewer herbicide applications and improved yield resilience under dry growing conditions (as shown in 2019).

Seed bank studies require longer term funding for greater understanding of seed bank dynamics, and it seems clear that the use of a winter cover crop needs an ongoing commitment by growers and researchers to quantify all the benefits. However, the use of an Italian ryegrass winter cover crop in a maize production system with a single post-emergence herbicide treatment would mitigate weed seed bank abundance over time.

## ACKNOWLEDGMENTS

This study was funded by Ministry of Business, Innovation and Employment (MBIE) with support from Foundation for Arable Research (FAR) in the Managing Herbicide Resistance programme (C10X1806).

## REFERENCES

Bertoldi C., Leo M.D., Braca, A. and Ercoli, L. (2009) Bioassay-guided isolation of allelochemicals from *Avena sativa* L.: allelopathic potential of flavone C-glycosides. *Chemoecology* 19,169-179.

Bezuidenhout S.R., Reinhardt C.F. and Whitwell M.I. (2012). Cover crops of oats, stooling rye and three annual ryegrass cultivars influence maize and *Cyperus esculentus* growth. *Weed Research* 52, 153-160.

Buddenhagen, C.E, Gunnarsson, M., Rolston, P., Chynoweth, R.J., Bourdot, G. and James, T.K. (2020) Costs and risks associated with surveying the extent of herbicide resistance in New Zealand. *New Zealand Journal of Agricultural Research*. 63, 430-448.

James T., Rahman, A. and Trolove, M. (2007) Optimising time of planting and herbicide application for control of problem weeds in maize. *New Zealand Plant Protection* 60, 183-188.

Merfield, C. N. (2019). Integrated weed management in arable crop systems. Report written for the Foundation of Arable Research (FAR).

# Gene technologies in weed management: what we need to know?

Nagalingam Kumaran[1], Anupma Choudhary[1], Mathieu Legros[2], Andy Sheppard[2], Luke Barrett[2], Donald Gardiner[3], S Raghu[1]
[1]CSIRO, Dutton park, Australia,
[2]CSIRO, Canberra, Australia,
[3]CSIRO, Brisbane, Australia
(kumaran.nagalingam@csiro.au)

**Summary** Contemporary gene technologies such as 'gene silencing' and 'gene drive' hold enormous potential to develop novel tools for weed management, and approaches to develop genetically-based tools are gaining significant momentum. These technologies can expose a variety of pathways for development of options for sustainable weed management. For instance, gene silencing can switch-off genes mediating adaptation (e.g. growth, herbicide resistance), and gene drive can be used to spread modified traits and to engineer wild populations with reduced fitness. Developing and applying these technologies are expected to be inherently complex, however, as their application is constrained by several methodological, technological, regulatory, ecological (e.g. genome editing, delivery, resistance, reproductive biology) and ethical challenges. In this talk, we highlight these challenges and discuss strategies to accelerate the development of gene-tech based tools for weed control.

**Keywords** Gene drive, gene silencing, RNAi, CRISPR, gene-drive

# Voraxor® Herbicide: An Alternative to Paraquat in Fallow Double-Knock - Managing Glyphosate and Paraquat Resistant Weeds.

Russell Ison, Ian Francis, Georgia Readett, Melissa Brown
BASF Australia Ltd, Level 12, 28 Freshwater Place, Southbank, Victoria 3006, Australia
(russell.ison@basf.com)

**Summary** Glyphosate resistant weeds pose a significant risk, threatening the sustainability of conservation farming systems in Australia. To counter glyphosate resistance, the 'double-knock' herbicide technique has been developed for fallow control of problematic weeds. The double-knock strategy refers to the sequential herbicide approach, in which different herbicide groups are applied with at least a one-week interval between the application of the first and second sprays. Predominantly glyphosate (Group 9) is followed by paraquat (Group 22).

A combination of field and pot trials were conducted to test the hypothesis that the Group 14 herbicide Voraxor® is a viable alternative to paraquat in the fallow double-knock system. Results from these trials indicated that Voraxor at 37.5 g ai ha$^{-1}$ is equivalent or superior to paraquat at 1320 g ai ha$^{-1}$ for control of several problematic fallow weeds when applied using the double-knock method following glyphosate at 855 g ai ha$^{-1}$.

**Keywords** Voraxor®, fallow, weeds, knockdown, double-knock, residual control, resistance management.

## INTRODUCTION

Worldwide, thirty-eight weed species have now evolved resistance to glyphosate, distributed across 37 countries and in 34 different crops and six non-crop situations (Heap *et al.* 2018).

Since 2007, 76 populations of *Echinochloa colona*, 57 populations of *Conyza bonariensis*, 10 populations of *Chloris truncata* R.Br. and 3 populations *Urochloa panicoides* P. Beauv. have been identified as glyphosate-resistant (group 9) in Australia (Preston 2010 & Cook *et al.* 2008 - 2018). There are also 7 recorded cases of paraquat (group 22) resistance in *Conyza spp.* in Australia (Chauhan *et al.* 2018).

The double-knock method relies on the sequential herbicide application to control any survivors following the initial glyphosate application. The technique seeks to optimise glyphosate efficacy and acts as a means of combating increasing glyphosate resistance.

A limitation of the current double-knock strategy is the reliance on a single herbicide group – group 22. This trial work sought to evaluate the addition of an alternate mode of action (MOA), Group 14, into the double-knock strategy as the use of a different herbicide MOA can delay the evolution of herbicide resistance (Beckie & Reboud, 2009).

Voraxor® herbicide contains a combination of trifludimoxazin plus saflufenacil (Tirexor® and Kixor®) - two potent protoporphyrinogen IX oxidase (PPO or Protox) inhibiting herbicides that provide complimentary activity. This combination has demonstrated high levels of knockdown and residual activity against grass and broadleaf weeds in comparison to other Group 14 chemistry (Witschel *et al.* 2018 & Armel e*t al.* 2020). Also, notably against Group 14 resistant broadleaf weeds (Porri *et al.* 2022).

This paper outlines the results from a series of field and pot experiments that examined an alternate use pattern for Voraxor® as a paraquat replacement in the double-knock weed control system.

## MATERIALS AND METHODS

**Herbicide evaluation** Experiments were performed on several weed species (Table 1) that are known to be tolerant or resistant to glyphosate at the BASF CropSolutions Research Farm, Loomberah, NSW, Australia (-31.1814, 151.063), which is located near Tamworth.

**Field trials** Between 2019 - 2022, 18 field trials took over place several soil types, ranging in timing from September to February. The field trial sites were managed to encourage populations of weeds that are known to be tolerant to glyphosate as a solo application.

**Table 1.** Weeds evaluated

| Common Name | Species |
|---|---|
| Flaxleaf fleabane | *C. bonariensis* |
| Feathertop Rhodes grass | *C. virgata* |
| Windmill grass | *C. truncata* |
| Barnyard grass | *E. colona* |
| Liverseed grass | *U. panicoides* |

Trial weed populations ranged between 58 and 89 plants per m². Site management ensured an even growth stage of each weed species was present at the time of application. Individual trials were conducted using a complete randomised block design of four replications, plot size was 20 m². Data for this paper was drawn from the relevant treatments that pertain to this hypothesis.

**Pot trials** Between 2019 and 2022, five randomised pot experiments tested both Group 9 resistant and susceptible biotypes of *C. virgata* and *E. colona.* The resistant biotypes were collected from Northern NSW and Southern QLD. The biotypes are known to be Group 9 resistant though resistance testing and screening. Seeds where pre-germinated and then four replicates of each biotype planted into commercially available potting mix.

**Application** Occurred when weed species were at the 4-6 leaf stage. Pot and field experiments were each subjected to the double-knock sequential herbicide strategy, see Table 2. Herbicides were applied at a 7-day interval, using a hand boom operating at 3 BAR, utilising AIXR110015 nozzles, applying a coarse droplet in 100 L ha$^{-1}$ total volume.

**Experimental design – Treatments** Data was extracted from the relevant treatments of the 23 trials then, the percentage weed control from both double-knock treatments and a single application of glyphosate compared relative to the unsprayed control.

**Table 2.** Trial treatments

| | | |
|---|---|---|
| glyphosate 570 g/L | Single | 855.0 g ai ha$^{-1}$ |
| glyphosate 570 g/L *fb* paraquat 360 g/L | Sequential | 855.0 g ai ha$^{-1}$ *fb* 1320.0 g ai ha$^{-1}$ |
| glyphosate 570 g/L *fb* Voraxor® 375 g/L | Sequential | 855.0 g ai ha$^{-1}$ *fb* 37.5 g ai ha$^{-1}$ (+ 1% v/v MSO) |

*fb* = followed by
MSO -Methylated seed oil

**Assessment** Visual assessment of herbicide efficacy using a scale of 0-100% relative to the untreated control. Results presented are at 14 days after the sequential application occurred.

Statistical analysis conducted using a one-way analysis of variance (ANOVA) at the 95% confidence level utilising ARM software, Revision 2020.2 (GDM Solutions). Where significant treatment effects occurred (p=0.05) a LSD / Tukey's HSD mean separation test was conducted to determine treatment differences.

## RESULTS

The subset data from field experiments involving the species listed in Table 1 demonstrated that applying the current industry standard, for difficult to control grasses - glyphosate followed by paraquat, provided between 83.2 – 97.7 % control of the weed species listed in table 1. The *C. truncata* data is a notable exception, where the industry standard, whilst a significant improvement in comparison with a solo glyphosate application, resulted in suppression only (84.3% control).

In the same series of subset data from field trials, results indicated that applying glyphosate followed by Voraxor® + MSO provided 94.6 – 98.3 % control of all grass species listed in Table 1 when applied at the 4-6 leaf stage.

The pot trial data which evaluated control of Group 9 susceptible and resistant species demonstrated the validity of the double-knock methodology with a significant increase in the control of these glyphosate resistant grasses.

Pot trial data demonstrated applying glyphosate followed by Voraxor® + MSO provided 94.6 – 94.8 % control of glyphosate resistant grasses. The level of group 9 resistant grass weed control of glyphosate followed by Voraxor® + MSO was statistically equivalent to the current industry standard of glyphosate followed by paraquat. (83.2 – 96.0 %).

**Table 3.** Effect of herbicide treatments on percentage control of glyphosate-susceptible and glyphosate resistant weed biotypes at 14 days after application (relative to the untreated control, UTC = 0%).

| Treatment | % Weed control | | | | | | |
|---|---|---|---|---|---|---|---|
| | *E. colona* | | *C .virgata* | | *C. truncata* | *C. bonariensis* | *U. panicoides* |
| | Susceptible | Resistant | Susceptible | Resistant | Susceptible | Susceptible | Susceptible |
| glyphosate | 91.1 b | 13.8 b | 60.6 b | 3.5 b | 19.8 b | 3.3 c | 91.2 a |
| glyphosate *fb* paraquat | 97.7 a | 83.2 a | 93.9 a | 96.0 a | 84.3 a | 45.8 b | 95.6 a |
| glyphosate *fb* Voraxor® + MSO | 98.3 a | 94.8 a | 97.5 a | 94.6 a | 98.6 a | 97.9 a | 98.2 a |

Compendium of data collected from the following: *C. bonariensis* – 18 field trials, *C. virgata* – 9 field & 5 pot trials, *C. truncata* - 18 field trials, *E. colona* – 18 field & 5 pot trials, *U. panicoides* – 18 field trials

The current industry standard for *C. bonariensis* control in fallows is glyphosate + 2,4-D followed by paraquat. The use of 2,4-D can lead to off-target vapour drift and damage of highly susceptible crops such as tomatoes, cotton, sunflowers, soybeans, and grapes. (https://www.dpi.nsw.gov.au/biosecurity/weeds/weed-control/herbicides/spray-drift)

This trial work also sought to evaluate the potential of removing 2,4-D in situations where *C. bonariensis* is a target species.

Glyphosate followed by paraquat has been identified as an effective control tactic for *Conyza spp*. (Werth *et al*. 2010; Widderick *et al*. 2014).

The subset data extracted from field trials indicated that glyphosate followed by Voraxor® + MSO provided statistically superior control of C. bonariensis compared to glyphosate followed by paraquat (97.9 compared to 45.8 % control).

The combination of glyphosate followed by Voraxor® + MSO provided a statistically significant increase in control of *C. bonariensis* when compared to glyphosate followed by paraquat. (Table 3).

## DISCUSSION

These experiments indicate Voraxor® 375 g/L when applied 7 days after an initial application of glyphosate, at a rate of 37.5 g ai $ha^{-1}$ + 1% v/v of MSO, is a viable alternative to paraquat 360 g/L at a rate of 1320.0 g ai $ha^{-1}$ when applied as the sequential partner in a double-knock scenario, for the control of glyphosate tolerant and resistant weed species.

These experiments validate this new use pattern and suggest Voraxor® is a robust option for weed control and as a herbicide resistance management tool in cropping fallows targeting the weeds listed in Table 1. The use of Voraxor® as the partner to glyphosate in double-knock can also negate the risk to highly susceptible crops in areas where this use pattern can be adopted to control *C. bonariensis.*

## ACKNOWLEDGMENTS

Tony Cook. Department of Primary Industries, Tamworth Agricultural Institute, 4 Marsden Park Road, Calala, NSW 2340, Australia, for the provision of glyphosate resistant weed seed.

## REFERENCES

Australian Government. (2012). Impact of weeds. Available from URL: www.weeds.gov.au (Accessed April 2022).

Armel, G, Francis, I, Nielson, R, Witschel, M & Liebl, R. (2020) Trifludimoxazin: A new PPO inhibitor that controls PPO resistant weed biotypes. 21st Australasian Weeds Conference.

Beckie, H. J. & Reboud, X. (2009). Selecting for weed resistance: Herbicide rotation and mixture. Weed Technology, 23, 363–370.

Cook, T. (2015) Reducing Herbicide Spray Drift. Available from URL: https://www.dpi.nsw.gov.au/biosecurity/weeds/weed-control/herbicides/spray-drift

Diggle, A.J., Neve, P.B. and Smith, F.P. (2003). Herbicides used in combination can reduce the probability of herbicide resistance in finite weed populations. Weed Research 43, 371-82.

Heap, I. (2022). International survey of herbicide resistant weeds. Available from URL: http://www.weedscience.org/summary/country.aspx?countryid=1 (cited 9 May 2022).

Heap, I., & O Duke, S. (2018). Overview of glyphosate-resistant weeds worldwide. Pest Management Science 74, 1040-1049.

Hereward, J.P.; Werth, J.A.; Thornby, D.F.; Keenan, M.; Chauhan, B.S.; Walter, G.H. Gene expression in response to glyphosate treatment in fleabane (Conyza bonariensis)—Glyphosate death response and candidate resistance genes. Pest Manag. Sci. 2018, 74, 2346–2355.

Keating, B.A., Carberry, P.S., Hammer, G.L., Probert, M.E., Robertson, M.J., Holzworth, D.P., Huth, N.I., Hargreaves, J.N.G., Meinke, H., Hochman, Z., McLean, G., Verburg, K., Snow, V.O., Dimes, J.P., Silburn, D.M., Wang, E., Brown, S.D., Bristow, K.L., Asseng, S., Chapman, S.C., McCown, R.L., Freebairn, D.M. and Smith, C.J. (2003). An overview of APSIM, a model designed for farming systems simulation. European Journal of Agronomy 18, 267-88.

McGillion, T. and Storrie, A. (eds) (2006). Integrated weed management in Australian cropping systems – a training resource for farm advisors. CRC for Australian Weed Management, Adelaide, South Australia.

Osten, V.A., Walker, S.R., Storrie, A., Widderick, M., Moylan, P., Robinson, G.R. and Galea, K. (2007). Survey of weed flora and management relative to cropping practices in the north-eastern grain region of Australia, Australian Journal of Experimental Agriculture 47, 57-70.

Preston, C. (2014). Australian Glyphosate Resistance Register. Available from URL:

http://www.glyphosateresistance.org.au (cited 19 May 2014, Accessed April 2021).

Thornby, D., Walker, S. & Whish, J. (2006). Simulating weed persistence and herbicide resistance in the northern grain region using a validated crop growth model with extensions for seedbank dynamics and mating. Proceedings of the 15th Australian Weeds Conference, eds C. Preston, J.H. Watts and N.D. Crossman, pp. 499-502. (Weed Management Society of South Australia, Adelaide).

Porri, A., Matheus, M., Noguera, Betz, M., Sälinger, D., Brändle, F., Bowe, S., Lerchl, J., Meyer, J. & Knapp, M. (2022). Can double PPO mutations exist in the same allele and are such mutants functional? View ORCID Profile Nilda Roma-Burgos.

Witschel, M., Aponte, R., Armel, G,. Bowerman, P., Mietzner, T., Newton, T., Porri, A., Simon, A. & Seitz, T. (2018). Tirexor®—design of a new resistance breaking protoporphyrinogen IX oxidase inhibitor. BASF SE, Global Research Herbicides, Ludwigshafen, Germany.

Walker, S., Bell, K., Robinson, G. & Widderick, M. (2011). Flaxleaf fleabane (*Conyza bonariensis*) biotypes have developed glyphosate resistance in north-east Australian cropping fields. Crop Protection, 30(3), 311–317.

Werth, J.A., Walker, S.R., Boucher, L.R. & Robinson, G.R., (2010). Applying the double knock technique to control Flaxleaf fleabane (*Conyza bonariensis*). Weed Biology and Management 10, 1e8.

# Transforming Wild and Weedy Australia: Outcomes of the 2020 Fenner Conference on the Environment

Sonia Graham[1], N Gill[1]
[1]University of Wollongong, Wollongong, Australia
(sonia.graham1@gmail.com)

**Summary** Over the last two decades there have been repeated calls for greater collaboration among policy makers, practitioners and researchers involved in weed management. Simultaneously, there have been major changes in the ways weeds are being managed, now forming part of biosecurity management, as well as significant environmental change, such as that incurred through the extreme bushfire season. In response to the calls for collaboration and changing policy, environmental and agricultural landscapes, thirty expert researchers, policy makers and practitioners met in February 2020 in Charlotte Pass, Kosciuszko National Park, to answer the question: What are the highest priority actions we can commit to, working together across boundaries and disciplines, that will deliver the greatest contribution to radically improved weed management research, policy and practice in Australia? This interactive presentation will provide an overview of the innovative workshop format and the top five priority actions that were identified and committed to by the workshop participants. It will then invite further discussion and expressions of interest for making these high priorities a reality across and beyond Australia.

**Keywords** Transformation; collective action; decision-making; networks

# Prevention is best: Protecting Australia from future environmental weed threats

Simon Brooks[1], Hillary Cherry[2], Katherina Ng[3], Barbara Waterhouse[4] and Paul Champion.[5]
[1] Department of Agriculture and Fisheries, PO Box 976, Charters Towers 4820, QLD
[2] Department of Planning and Environment, 12 Darcy St, Parramatta 2150, NSW
[3] Australian Bureau of Agricultural and Resource Economics and Sciences, GPO Box 858, Canberra, 2601
[4]Department of Agriculture, Fisheries and Forestry, Cairns Regional Office, GPO Box 858 Canberra, 2601
[5] National Institute of Water and Atmospheric Research Ltd., Hamilton 3251, New Zealand
(Hillary.Cherry@environment.nsw.gov.au)

**Summary** The national priority list of exotic environmental pests, weeds and diseases was released in November 2020. An implementation plan to identify and prioritise risk reduction actions was finalised in 2022. The list, developed collaboratively with over 100 experts, informs activities to prevent the entry, establishment and spread of exotic species that can negatively impact Australia's environment and/or social amenity. As examples of pests known to cause severe negative impacts, the future threats of four of the weeds on the list are detailed here to support preventative action.

**Keywords** prevention, environmental biosecurity, Mikania vine, Manchurian wildrice, spiked pepper, mouse-ear hawkweed.

## WHAT IS THE LIST?

In 2017, the independent review into the capacity of the national biosecurity system and its underpinning intergovernmental agreement delivered its final report, *Priorities for Australia's biosecurity system* (Craik *et al.* 2017). Recommendation 11 of the report was for a national priority list for exotic environmental pests and diseases to be developed in partnership with system participants. The Australian Bureau of Agricultural and Resource Economics and Sciences (ABARES), and the then Department of Agriculture, led the development of this national priority list (ABARES 2021).

The resulting list, also known as the exotic environmental pest list or the EEPL, identifies pests, weeds, and diseases that are not present in Australia (or if present, under eradication) that pose the greatest threat to Australia's environment and social amenity. It is distinct from other national lists primarily focused on agricultural risks as it aims to strengthen the national biosecurity system by improving focus and awareness on environmental biosecurity. The EEPL is not an exhaustive list and does not aim to limit jurisdictions or industries from pre- or post-border actions targeting other pests or diseases. Rather, the list highlights the diverse array of environmental biosecurity risks that Australia faces from pests and diseases.

## WHAT'S ON THE LIST?

The list contains 168 exotic species across eight thematic groups (weeds and freshwater algae, vertebrates, marine pests, native animal diseases, aquatic animal diseases, plant pathogens, terrestrial invertebrates, and freshwater invertebrates), and includes 19 weed species; see ABARES (2021) for full list. Of these, four species were assessed as higher risk weeds that pose the greatest risk to Australia's environmental biosecurity: Manchurian wildrice (*Zizania latifolia* (Griseb.) Turcz. ex Stapf, Poaceae); Mikania vine (*Mikania micrantha* Kunth, Asteraceae); Mouse-ear hawkweed (*Pilosella officinarum* Vaill., Asteraceae); and Spiked pepper (*Piper aduncum* L., Piperaceae).

## HOW WERE WEEDS SELECTED?

Cross-sectoral collaboration was a key component of EEPL development, with ABARES working with over 100 experts across all taxonomic groups from governments, research institutions, museums, and other organisations. Experts participated in workshops to facilitate joint decisions on the purpose and methodology to determine the priority list. Experts also shortlisted candidate species and took part in a structured modified-Delphi expert elicitation process with a semi-quantitative assessment (ABARES 2021).

A total of 20 species in the weeds and freshwater algae group (which consist of 19 weeds) were shortlisted for further assessment. To be included in the short list, a species must: have demonstrated negative impacts on environment and/or social amenity; be exotic to Australia (i.e., not currently known to be present in Australia or, if present, subject to nationally agreed eradication); have at least one known or potential pathway of entry to Australia; have the potential to establish and spread in Australia; and have the potential for nationally important negative impacts on Australia's environment or social amenity.

The semi-quantitative assessment allowed for risk scoring and ranking within the thematic groups to determine the higher risk EEPL species based on the likelihood of entry, establishment and spread, and the environmental and social amenity impacts. The four above-mentioned weed species were assessed as having a higher overall risk ranking for Australia.

FOUR EEPL WEEDS

**Mikania vine** *Mikania micrantha* is a rampant, smothering vine, native to Central and South America (Waterhouse 2003). It has invaded a range of tropical and sub-tropical areas, to become one of the most serious weeds across tropical Asia, the Indian sub-continent, and Pacific regions (Day *et al.* 2016). Seeds of mikania vine are light and readily dispersed by wind, but can also be dispersed by animals, water, human activity, and vehicles (S. Brooks, unpublished data). Abundant seasonal seed production forms a persistent soil seed bank, and it is readily capable of vegetative propagation from each stem node (Brooks and Jeffery 2018).

Mikania vine is a vigorous, perennial vine that is capable of climbing, entangling, and choking trees, shrubs, and fences (Day *et al*. 2016). It has invaded a wide variety of agricultural and environmental land uses across an exceedingly wide variety of damp habitats including wetlands. It is recognized as a serious environmental weed across many Pacific Islands and tropical to subtropical Asia, stretching north into Nepal and southern China (Day *et al.* 2016 and references there-in). Mikania vine is also a serious pest of pastures, plantations, and orchard crops. Day *et al.* (2016) provides a summary of 23 crops severely impacted by Mikania vine. The small seed could enter Australia as a contaminant of agricultural goods (Waterhouse 2003).

Mikania vine was listed on the first Northern Australia Quarantine Strategy (NAQS) weed list by Michael (1989). In Australia, Mikania vine was first discovered near Mission Beach in north Queensland in 1998 and was included in the nationally cost-shared 'National Tropical Weeds Eradication Program' when it commenced in late 2003 (Waterhouse 2003). While present on mainland Australia, it is the subject of an eradication program. Mikania vine was included on the EEPL due to the risks of further introductions leading to wide-spread impacts over wet coastal areas of northern and eastern Australia.

**Mouse-ear hawkweed** *Pilosella officinarum* [syn. *Hieracium pilosella* L.], is a perennial daisy, native to Eurasia. It spreads via aggressive stolons, forming new rosettes and creating monocultures that can exclude all other species (French and Watts 2020). It has invaded areas in Asia, North and South America, and New Zealand, where it causes severe impacts (CABI 2022). Currently, two small incursions are under formal eradication in the Australian Alps (New South Wales and Victoria; Hamilton *et al.* 2015).

Mouse-ear hawkweed can have a major impact on native plant communities and associated biodiversity by altering soil properties, nutrient cycling, and overall community structure (Espie 2001). It grows vigorously (French and Watts 2020), reproduces asexually, and its seeds are spread by wind. It can outcompete native plants by secreting chemicals into the soil that prevent the germination and growth of other plants (McIntosh *et al.*1995). Seeds of mouse-ear hawkweed are very small and may be unintentionally introduced into Australia on clothing, shoes, or outdoor equipment. Hygiene is critical, especially given the extensive infestations in New Zealand and the likelihood of recreational travel between those areas and Australian alpine regions.

Alpine regions are extremely vulnerable to Mouse-ear hawkweed invasion, as it can rapidly displace native vegetation, including inter-tussock vegetation in alpine environments (Espie 2001). This may reduce their aesthetic value and could cause the loss of rare and threatened plants and animals that depend on these alpine communities.

The impact of Mouse-ear hawkweed is severe in New Zealand, where over six million hectares are invaded, with significant impacts to conservation and production values (Espie 2001). Mouse-ear hawkweed is also a serious risk to agricultural productivity, as it is unpalatable to stock. Invasions are estimated to reduce the value of agricultural production by up to NZ$4.4 million annually (Grundy 1989) in New Zealand, with further social impacts to farmers and land managers. Mouse-ear hawkweed could establish in a broad range of habitats across large areas of southeastern Australia (Weed Futures 2019), including New South Wales, Victoria, Tasmania, and South Australia. A preventative approach is warranted given the risks mouse-ear hawkweed poses to a variety of conservation and production environments.

**Spiked pepper** *Piper aduncum* is a slender-stemmed tree native in Central and South America (Mexico to Argentina), where it occurs from sea level to *c.* 2000 m ASL (POWO 2022). It is invasive in deforested areas of the Amazon Basin within its native range, and Malaysia, Indonesia, the Philippines, Papua New Guinea, Solomon Islands, Vanuatu, Fiji, Hawaii and Florida where it is an introduced species (Hartemink 2010, Padmanaba and Shiel 2014, Waterhouse pers. obs.) The only Australian records are from Christmas Island where it has been known since 1987 (AVH 2022, Hartemink 2010, Waterhouse 2003). Recognition of the threat posed by spiked pepper to northern Australia led to its inclusion on the second and subsequent editions of the NAQS weed list (Waterhouse and Mitchell 1998) and recently the EEPL. Its introduction to Australia is prohibited (BICON 2022) and it is classified as a prohibited species under Queensland and Northern Territory biosecurity legislation.

Spiked pepper is a pioneer species which readily establishes, persists, and spreads in natural and human-induced disturbances such as tree fall gaps, landslips, riparian zones, roadsides, recently logged forests, mine sites, and gardens (Hartemink 2010, Padmanaba and Shiel 2014). Growing to 8 m tall and often multi-stemmed, it forms impenetrable stands (B. Waterhouse pers obs). Its invasiveness results from a rapid growth rate to reproductive maturity, high biomass accumulation, prolific and continuous seed production, domination of the soil seed bank and ability to resprout vegetatively from stem fragments or after damage (Hartemink 2006). The tiny seeds are dispersed long distances by birds, bats and as contaminants on footwear, vehicles, and forestry/mining equipment (Hartemink 2010, Padmanaba and Shiel 2014). It is cultivated as an ornamental shrub and a 'living fence' in Papua New Guinea (Hartemink 2006, Waterhouse, pers. obs) and has ethnobotanical and medicinal uses, raising the potential for seeds to be traded over the internet (Hartemink 2010). Spiked pepper could become a serious environmental weed in tropical and subtropical forests from far north Queensland through to north-eastern NSW, as well as adjacent agricultural and pastoral lands. These forests are highly fragmented and experience increasingly frequent seasonal disturbances from tropical cyclones, damaging storms, floods, and bushfires. Continued surveillance for this species is essential due to its potential to reach northern Australia associated with nomadic fruit bats and birds moving from neighbouring landmasses, or via ports and airports as an accidental contaminant with travelers and equipment from infested areas.

**Manchurian wildrice** *Zizania latifolia* is a giant perennial rhizomatous aquatic and wetland grass up to 4 m tall. Plants form dense, tall infestations on waterbody margins, riverbanks, floodplains, tidal flats, and flood-prone pasture and cropping land. Manchurian wildrice is native to eastern Asia where it is a common emergent aquatic species (Hung *et al.* 2008). Manchurian wildrice is reported as naturalised in relatively few countries, with New Zealand being the first reported country outside of Asia (Champion 2020). It is also naturalised in five European countries and Hawaii but there are few reported sites in any of these countries. There are also records of cultivated food plant specimens from mainland North America (Champion 2020).

Manchurian wildrice forms monocultures within both its native and introduced range. Dense stands are a major barrier to water body use, regeneration of native species, particularly where the plant is perennial with no winter die-back. Manchurian wildrice displaces short-stature vegetation and envelops taller individual indigenous plants that are unable to produce progeny within the dense sward (Champion 2020). Based on impacts seen in New Zealand, it could cause serious impacts to both freshwater and estuarine native plant and animal communities including internationally important Ramsar wetlands, protected freshwater and coastal wetlands, and lagoons across most of Australia.Manchurian wildrice contributes to riverbank slumping and bank failure, causing flooding of adjacent pastures and providing further habitat for Manchurian wild rice. Although short young foliage is palatable to cattle, tall, dense plants become unpalatable (Champion 2020).

Introduction to New Zealand was via mudbricks contaminated with viable plant material and it was erroneously spread as an erosion/flood control plant within New Zealand (Champion 2020). The threat posed by Manchurian wildrice has been recognised in New Zealand since its introduction around 1900 (Arnold 1937). However, a concerted effort to manage the plant nationally did not occur until 2008, when it was classified as a Notifiable Organism and one of 13 National Interest Pest Responses (Champion and Hofstra 2010), with an operational plan aimed to eradicate all populations of this species from the four regions where it occurs. Manchurian wildrice was ranked as New Zealand's third worst aquatic weed (Champion and Clayton 2000), behind *Phragmites australis* (Cav.) Trin.ex Steud. and *Hydrilla verticillata* (L.f.) Royle, both native to Australia but also subject to national eradication programmes in New Zealand.

## OVERVIEW

Australia's National Priority List of Exotic Environmental Pests, Weeds and Diseases (EEPL) provides an opportunity to raise awareness and target prevention activities at organisms that pose the greatest environmental biosecurity risks. For weeds, the four species described here are critical targets for prevention activities and for considering actions, such as eradication, should prevention fail. Implementing activities to ensure all weeds on the list do not establish and impact Australia in future is recommended. An implementation plan is being developed to coordinate mitigation actions for EEPL species.

## ACKNOWLEDGMENTS

All participants and experts involved in the project are thanked for their valuable input to the design of the prioritisation methodology, the shortlisting of species and the assessment of species.

REFERENCES


ABARES (2021). The National Priority List of Exotic Environmental Pests, Weeds and Diseases: Information Paper (V2.0), ABARES report to client prepared for the Chief Environmental Biosecurity Officer, Department of Agriculture, Water and the Environment, Canberra, ACT. CC BY 4.0.

Arnold, E.H. (1937). Manchurian rice grass (*Zizania latifolia*): its occurrence and distribution in the Northern Wairoa District. *The New Zealand Journal of Agriculture* 55, 3, 129–33.

Australasian Virtual Herbarium (AVH) (2022). Council of Heads of Australasian Herbaria https://avh.chah.org.au, accessed 01 May 2022

Australian Biosecurity Import Conditions Database (BICON), (2022). Department of Agriculture, Water and the Environment BICON - Import Conditions (agriculture.gov.au), accessed 01 May 2022.

Brooks, S. and Jeffery, M. (2018). Progress in the eradication of *Mikania micrantha* from Australia. Proceedings of the 21st Australasian Weeds Conference, eds S Johnson, L Weston, H Wu and B Auld. pp. 350–3. (The Weed Society of New South Wales, Sydney).

CABI (2022). *Pilosella officinarum* (mouse-ear hawkweed). In, Invasive Species Compendium. Wallingford, UK: CAB International. www.cabi.org/isc. accessed 01 May 2022.

Champion, P.D. (2020). Management of Manchurian Wild Rice (*Zizania latifolia*): Information review. NIWA Client Report 2020208HN, Hamilton.

Champion, P.D. and Clayton, J.S. (2000). Border control for potential aquatic weeds. Stage 1 Weed risk model. *Science for Conservation,* 141. Department of Conservation, Wellington.

Champion, P.D. and Hofstra, D.E. (2010). Manchurian wild rice (*Zizania latifolia*) control. Proceedings of the 17th Australasian Weeds Conference, Christchurch. 318–20.

Craik, W., Palmer, D. and Sheldrake, R. (2017). Priorities for Australia's biosecurity system, An independent review of the capacity of the national biosecurity system and its underpinning Intergovernmental Agreement, Canberra.

Day, M.D., Clements, D.R., Gile, C., Senaratne, K.A.D., Shen, S., Weston, L.A. and Zhang, F. (2016). Biology and impacts of Pacific Islands invasive species. 13. *Mikania micrantha* Kunth. (Asteraceae). *Pacific Science* 70, 3, 257–85.

Espie, P. (2001). Hieracium in New Zealand: ecology and management, AgResearch Ltd, Mosgiel, New Zealand). 66 pp.

French, K. and Watts, E. (2020) Differences in vegetative growth of two invasive hawkweeds at temperatures simulating invaded habitats at two altitudes. *Scientific Reports* 10, 2180.

Grundy T.P. (1989). An economic evaluation of biological control of Hieracium. Research Report - Agribusiness & Economics Research Unit, Lincoln College Canterbury, New Zealand, No. 202: 41pp.

Hamilton, M.A., Cherry, H. and Turner, P.J. (2015). Hawkweed eradication from NSW: Could this be the first? *Plant Protection Quarterly* 30, 110–5.

Hartemink, A.E. (2006). *Invasion of Piper aduncum in the shifting cultivation systems of PNG.* CIP-Gegevens Koninklijke Bibliotheek, Den Haag.

Hartemink, A.E. (2010). The invasive shrub *Piper aduncum* in Papua New Guinea: A Review. *Journal of Tropical Forest Science* 22, 2, 202–13.

Hung, L.Q., Asaeda T., Kallibala, M. and Mnaya, B.J. (2008). Effects of nutrient concentration and litter cover on quantitative shoot parameters and belowground biomass of *Zizania latifolia* (L). *Chemistry and Ecology* 24, 5, 357–65.

McIntosh, P.D., Loeseke, M., and Bechler, K. (1995). Soil changes under mouse-ear hawkweed (*Hieracium pilosella*). *New Zealand Journal of Ecology* 19 1, 29–34.

Michael, P.W. (1989). *Review Paper on Weeds of Concern to Northern Australia.* Report to the Bureau of Rural Resources.

Padmanaba, M. and Shiel, D. (2014). Spread of the invasive alien species *Piper aduncum* via logging roads in Borneo. *Tropical Conservation Science* 35–44.

Plants of the World Online, (2022). (POWO) Facilitated by the Royal Botanic Gardens, Kew; http://www.plantsoftheworldonline.org/ (retrieved 4 May 2022)

Waterhouse, B.M. (2003). Know your enemy: Recent records of potentially serious weeds in northern Australia, Papua New Guinea and Papua (Indonesia). *Telopea* 10, 1, 477–85.

Waterhouse, B.M. and Mitchell, A.A. (1998). *Northern Australia Quarantine Strategy Weeds Target List*. Second edition. Australian Quarantine and Inspection Service Miscellaneous Publication 6/98.

Weed Futures (2019). Macquarie University; http://www.weedfutures.net/species.php?id=2067 http://www.weedfutures.net/species.php?id=2067(retrieved 20 June 2022).

# Abating the threat of exotic vines and scramblers in Threatened Ecological Communities

Adam Bernich[1]
[1]School of Earth, Atmospheric and Life Sciences, University of Wollongong, Wollongong, Australia
(abernich@uow.edu.au)

**Summary** The invasion and establishment of exotic vines and scramblers (EVS) are listed as a key threatening process to biodiversity in NSW. However, little is known of the ecology, life history, or potential threat of many EVS species. Knowledge of which species are the largest threat to biodiversity, particularly for threatened ecological communities, will allow for the prioritisation of species for removal, and thus resources can be spent efficiently. We present work involving field surveys across 12 Threatened Ecological Communities (TEC's) in NSW to determine which EVS species pose the largest threat to native biodiversity. Initial results have shown that the prevalence and impact of EVS species varies amongst the different communities suggesting different prioritisation is needed in different TECs. Strategies to establish and invade communities differs between species, with some species having wide distributions, though relatively low abundance in sites, whereas other species have only been recorded in a small number of sites, though in extremely high densities. This highlights the need to understand the ecology of EVS species in order to control them effectively.

**Keywords** Prioritisation, exotic vines and scramblers, transformer species, management

# Adopting standard methods for assessing biodiversity impacts of invasive alien species

Jens G Froese[1], Ben Gooden[2], Rocio Ponce-Reyes[1], Alana L. Burley[3], David A. Nipperess[3], Kristen Williams[2]
[1]CSIRO, Dutton Park, Australia,
[2]CSIRO, Canberra, Australia,
[3]Department of Planning, Industry and Environment, Parramatta, Australia
(jens.froese@csiro.au)

**Summary** Invasive alien species (IAS) are considered the number one threat to vulnerable native species in Australia. In order to invest limited management resources wisely and demonstrate the effectiveness of threat mitigation, measuring the biodiversity impacts of IAS is of vital importance. However, IAS impact is a complex and highly context-dependent issue. Assessments have mostly relied on the use of surrogate measures or assumptions about potential risks from evidence elsewhere. Direct estimates of impact are rare and poorly generalisable. In response to this monitoring need, various impact assessment protocols have been developed that aim to synthesize evidence from a range of sources in a systematic, repeatable and comparable form. Among these, the IUCN Environmental Impact Classification for Alien Taxa (EICAT) standard for measuring the magnitude of IAS impacts is gaining widespread attention by researchers and policy makers alike. We adapted the EICAT method to conduct an IAS impact assessment for the purposes of the New South Wales (NSW) Biodiversity Indicator Program. In this pilot study we assessed the current evidence of impact caused by 22 significant weed and pest animal species on 97 listed threatened species and ecological communities in NSW using a structured expert elicitation protocol. Results showed that many weeds currently cause significant biodiversity impacts, including local population extinctions of native species. We suggest that standardized impact assessment methods such as EICAT provide a consistent yet flexible framework for comparing IAS impacts across biological groups and spatial or temporal scales. When integrated with measures of uncertainty, they provide a valuable way for monitoring, evaluating and reporting impacts to support outcome-oriented IAS management in Australia.

**Keywords** Weeds, pest animals, invasive species, impact assessment, magnitude of impact, structured expert elicitation, monitoring, evaluation, reporting

# Host-specificity testing for the tradescantia leaf beetle (*Neolema ogloblini*)

Umar Lubanga[1], Greg K. Lefoe[1], David McLaren[1], Tony M. Dugdale[1], Raelene M. Kwong[1]
[1]Invertebrate & Weed Sciences, Agriculture Victoria Research Division, Department of Jobs, Precincts and Regions, AgriBio Centre, Bundoora, Australia
(umar.lubanga@agriculture.vic.gov.au)

**Summary** *Tradescantia fluminensis* was declared a Target for Biological Control in December 2015 by the Invasive Plants and Animals Committee (IPAC). New Zealand researchers identified three leaf beetles and the smut-like fungus *Kordyana brasiliensis* from the native range (south-eastern Brazil) as potential biocontrol agents of the weed. Host-specificity studies of all four agents were completed, and agents approved for release in New Zealand. To date, Australian host-range studies have mainly focussed on *K. brasiliensis* and the tradescantia leaf beetle, *Neolema ogloblini*. Host-specificity testing of *K. brasiliensis* was completed by CSIRO and the agent was approved for release in December 2018. Host-specificity testing of the *N. ogloblini* was carried out by Agriculture Victoria. Standard choice and no-choice tests showed *N. ogloblini* to be highly host specific to *T. fluminensis*. Although larvae completed development on seven of the twenty-five plants tested under no-choice conditions, oviposition occurred on only five of these species and none of the eggs - hatched to develop into adults. Minor damage (unlikely to impact the fitness of affected plants) was recorded on some non-target species in starvation trials highlighting the possibility of spill-over damage for affected non-target species that occur in close proximity to *T. fluminensis*. None of the tested plants sustained *N. ogloblini* populations over successive generations, suggesting reduced risk of consistent adult feeding damage in the field. Lastly, assessment of risk of off-target attack in the field using Paynter et al. (2015)'s scoring system showed that it is highly unlikely that any of the non-target species tested will be attacked in the field. An application for release of *N. ogloblini* will be submitted to the Department of Agriculture, Fisheries and Forestry and if approved for release N. *ogloblini* will complement *K. brasiliensis* in drier environments that are not favourable for fungal growth.

**Keywords** Biological control, Commelinaceae, Host-specificity testing, Wandering trad

# Towards the Development of Herbicides Targeting Lysine Biosynthesis

Emily Mackie[1], Andrew Barrow[1], Marie-Claire Giel[1], Mark Hulett[1], Anthony Gendall[2], Santosh Panjikar[3], Tatiana Soares da Costa[1,2]
[1]La Trobe Institute for Molecular Science, Melbourne, Australia,
[2]AgriBio Centre for Agri-biosciences, Melbourne, Australia, [3]Australian Synchrotron, ANSTO, Melbourne, Australia
(17694478@students.latrobe.edu.au)

**Summary** Herbicide resistance is one of the biggest threats to our natural environment and agricultural industry. The rapid emergence of herbicide-resistant weeds has been driven by repeated application of the same herbicides, combined with a lack of new herbicides entering the market in the past 30 years. Therefore, we urgently require herbicides with new modes of action to tackle the herbicide resistance crisis we are facing. Although many commercial herbicides inhibit the biosynthesis of amino acids in plants, the lysine biosynthesis pathway is yet to be exploited as a herbicide target. Our goal is therefore to validate lysine biosynthesis enzymes as herbicide targets and develop inhibitors of these enzymes as novel herbicide leads. To achieve this, we are characterising the structure and function of plant lysine biosynthesis enzymes using enzyme kinetic assays, circular dichroism spectroscopy, analytical ultracentrifugation and X-ray crystallography. We are developing small molecule inhibitors of several of these enzymes, with screening being conducted in vitro using biochemical assays, as well as in planta against *Arabidopsis thaliana*. The herbicidal efficacy of lead inhibitors against weeds is being validated against the most problematic weed in Australia, *Lolium rigidum*. The relationship between lead inhibitor mechanisms and physiological outcomes is being probed using X ray crystallography, toxicity assays, and systems biology. This work has the potential to provide novel herbicide modes of action to help combat the global herbicide resistance crisis.

**Keywords** Herbicide-resistance, lysine, amino acid biosynthesis, herbicide, enzyme, ryegrass

# Progress towards the eradication of *Limnocharis flava* from Australia

Simon Brooks[1], Kimberley Erbacher[2] and Jacob Maher[3]
[1] Tropical Weeds Research Centre, Biosecurity Queensland, Department of Agriculture and Fisheries, PO Box 976 Charters Towers, Queensland 4820, Australia
[2]Biosecurity Queensland, Department of Agriculture and Fisheries, PO Box 20, South Johnstone Queensland, 4859 Australia.
[3]School of Biological Sciences, The University of Adelaide, North Terrace, Adelaide, South Australia, 5005 Australia.
(kimberley.erbacher@daf.qld.gov.au )

**Summary** *Limnocharis flava* (Limnocharis) is an anchored aquatic weed preferring shallow silty tropical habitats. *Limnocharis flava* was discovered near Kuranda in north Queensland in 2001 and included in the nationally cost-shared National Tropical Weeds Eradication Program when it commenced in late 2003. Small and cultivated *L. flava* loci have been discovered between Cape Tribulation and Townsville, particularly around the greater Cairns area. In 2021, *L. flava* was discovered in the Northern Territory as well as in in southeast Queensland. The cultivation of this invasive plant as a garden ornamental or as a vegetable, increases the risk of populations established beyond the current known extent.

**Keywords** tropics, declaring eradication, aquatic weed, edible

## INTRODUCTION

*Limnocharis flava* (L.) Buchenau. (Limnocharis) is an invasive anchored aquatic plant native to Central and South America. It has become naturalized in shallow water courses in tropical regions across Asia, where it is so widespread it is utilized as a local edible vegetable (Weber and Brooks 2013).

In Australia, *L. flava* is declared under state and federal legislation. The Biosecurity Act 2015 provides legislative measures to prevent the entry/reintroduction of *L. flava* into Australia. *Limnocharis flava* is declared under legislation in Queensland, New South Wales, Northern Territory and Western Australia. It is also Restricted Biosecurity Matter under the Queensland Biosecurity Act 2014, and it is illegal to distribute, move, keep and / or not report *L. flava*. Officers authorized under the Queensland Act can issue a Penalty Infringement Notice (PIN) or on the spot fines for keeping *L. flava*.

*Limnocharis flava* is a species targeted for eradication under the National Tropical Weed Eradication Program (NTWEP). The following information discusses changes to the NTWEP reporting procedures and to the national risk profile. There is also an update on overall progress towards eradication of *L. flava*, last published by Brooks *et al.* (2008b).

## ERADICATION DATA PROCESSING

The NTWEP reports eradication progress data from discrete locations (loci). These are categorised into either 'contained water features' (garden ponds, water features through which water doesn't readily flow) and 'uncontained habitat' such as creeks, dams and drainage lines.

Field officers search areas which have previously recorded *L. flava* plants, including areas 200m downstream from loci within unconfined water systems. Active loci are revisited at monthly intervals throughout the year, which provides one or more opportunities to detect seedlings which take at least 46 days to produce immature fruit (Brooks *et al.* 2008a). Annual extended downstream surveys of 1km (or until salt water is reached) and 500m upstream are undertaken at loci within uncontained water systems.

Field records including, the date of discovery, precise location, number of plants and the reproductive status are stored in BORIS (Biosecurity Online Resources & Information Systems). This is the Biosecurity Queensland portal that houses all the NTWEP data records including compliance records. Presence or absence is derived from field records for every known unique site identification number (waypoint). Sites are added if plants are detected more than 30 m away from a known location.

From 2010, eradication progress reporting adopted a grid-based system of fixed one hectare 'management areas' (100m x 100m). Previously a system based on loci of a range of sizes was used (Brooks *et al.* 2008b). All data prior to 2010 was re-analysed using the 'management area' scale, which allows spatially consistent annual reporting.

Every six months, point records are summarized to allocate a 'control phase' status where plants are present, or 'monitoring phase' status where plants are absent for each management area. Management areas only enter a monitoring phase when absence data is recorded in the last 2 x 6-month periods; progression is via evidence of absence. The time that management areas have been in the monitoring phase is categorised to an annual value of 'years in monitoring phase'. If plants are recorded in a management area which is in the monitoring phase, it relapses to a control phase for at least a 12-month period. The number of years of monitoring (prior to a relapse) is tallied to determine 'monitoring relapse' frequency data. The allocation to control and years in monitoring status is updated every 6 months, with this dataset updated to the end December 2021.

The NTWEP also uses the 'time since last reproduction' as a measure of eradication progress (Brooks and Jeffery 2018). In cases where no seed production has been observed, the discovery date is used to calculate the time since last seed production. The time since last seed production (or discovery) accrues annually unless there is a seed production event (reproductive escape), causing the management area to suffer a 'reproductive relapse'. The last reproduction data is determined at the end of each financial year from a single (discovery or reproductive relapse) date for each management area. The last detection and last reproduction or discovery data have the same sample size (Figures 1 and 2) but are calculated differently. The following information contains examples of data reported annually to cost-sharing partners.

## ERADICATION PROGRESS

**Discovery and delimitation** There were 81 management areas as of December 2021. Twenty-one management areas are in (contained) domestic ponds or water features. The remaining 60 management areas are in dams, creeks and drainage channels through which water can flow.

Discoveries in 2021 include two new management areas in the Northern Territory (Table 1), these were cultivated specimens and are currently being treated as 'contained' loci. Viable seed has been found beside tubs at one location (S. Brooks, unpublished data) and it may be reclassified as unconfined if further plants are found. The Northern Territory government (Department of Environment, Parks and Water Security) is preparing their national response plan (N. Weston, pers comm 2022), but the two locations are included in this data set, until their response plan is endorsed.

Since 2016-17, six new loci have been in contained features, with some deliberately cultivated for sale as an edible vegetable. One of these discoveries was traced to south-east Queensland from a new locus in Townsville. Another was a cultivated plant at a residence near Cape Tribulation. Combined with the two new loci in the Northern Territory, these discoveries are beyond the extent of the previous known locations, between Townsville and Port Douglas in north Queensland.

**Table 1.** Number and type of *Limnocharis flava* loci in Australia

| Locus Type | QLD | NT | Total management areas |
|---|---|---|---|
| Contained (ponds or water features) | 19 | 2 | 21 |
| Unconfined habitat | 22 | | 60 |

The last new unconfined *L. flava* locus was discovered in May 2017. The unconfined loci occupy between one and eight management areas. They are small and discrete and appear limited by salty water or rocky substrates. To date, field surveys have revealed limited downstream spread, even after intense tropical rainfall events. However, further downstream dispersal is possible in shallow freshwater systems that are not under tidal influence. No new loci have been found by specific ground or water search activities.

Across both locus types, the establishment of new *L. flava* loci appears to be by the deliberate or accidental cultivation of plants. Using the categories of Brooks and Galway (2008), detection has resulted from awareness amongst weed allied professionals (43.2%), information from the public via engagement activities (40.9%) and tracing forward and backward information (15.9%). Whilst detection by tracing was more common at the start of the NTWEP, three of the last six confined loci were discovered using tracing information obtained from other occurrences.

All the current unconfined loci are small and occur across a similar geographic extent. The detection of new contained loci over a much broader geographic area has been traced to the deliberate cultivation of this species as a vegetable in ethnic communities, which poses a significant risk to eradication of *L. flava* in Australia. Discoveries reflect the deliberate or unintentional cultivation of this plant, including being grown as a vegetable. This risk is managed principally through maintaining and expanding the surveillance and control efforts at current loci; advertising campaigns, including targeted social media; TV advertisements; community engagement activities and compliance.

Social media groups and websites were targeted by web scrapers to detect online domestic trade of *L. flava,* following methods described in Stringham *et al.* (2021). Over 380,000 online advertisements and social media posts were searched across four websites. Thirty-five search terms in three languages were used to locate *L. flava*. This search found no evidence of *L. flava* being sold or traded through four e-commerce 'surface' websites. It appears that the greater risk is through 'closed groups' on the internet.

**Plant absence and extirpation** Of the 81 management areas, 73% had progressed to a monitoring stage (plant absence for more than a year) by December 2021 (Figure 1). Several of the confined habitats have been removed or permanently capped, which means they are not monitored regularly and are considered eradicated although they are still progressing along 'years in monitoring axis' in Figure 1. Most control phase management areas are either new contained features or six active loci along the wet tropical coast of north Queensland between Tully and Cairns. These loci include management areas that are still in the control phase despite more than 10 years after the last mature plant was controlled (Figure 2). The field crew data and ongoing field and glasshouse research (S. Brooks unpublished data) shows *L. flava* forms a persistent soil seed bank, particularly in constantly wet habitats. After 10 to 13 years, small numbers of seedlings continue to emerge in management areas that had high pre-discovery seed input. The *L. flava* seed bank is much more persistent than the evidence available to Brooks *et al*. (2008b) suggested.

**Figure 1**. Years in monitoring phase of *Limnocharis flava* management areas (n=81), categorized by locus type.

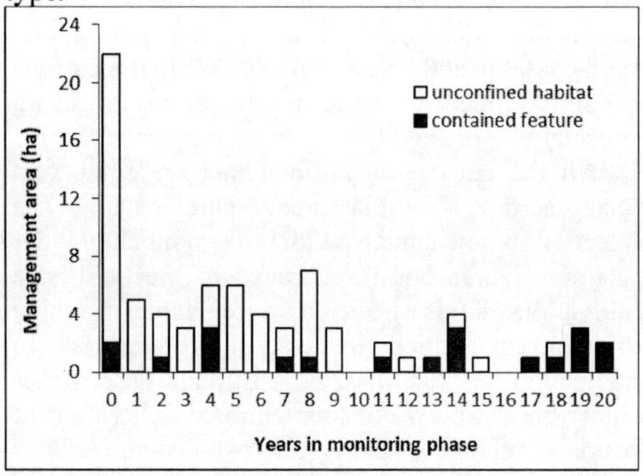


Fifty-eight of 81 management areas are in 1 to 20 years of monitoring (Figure 1). There have been 42 monitoring to control phase relapse events recorded, and some management areas have relapsed 2 or 3 times. Most of these events (39 of 42) have been after 4 years or less in monitoring phase. The remaining 3 events occurred between 7 and 8.5 years, with two of these at a contained loci in central Cairns. Multiple eradication criteria, including an absence of 9 to 10 years will be developed to cover all field situations. Given the spread of management areas in Figure 1 and with 92% of relapses from 1 to 4 years of monitoring, the provisionally eradicated category could be the fifth or sixth year of the monitoring phase for *L. flava*.

**Figure 2**. Years since last reproduction or discovery of *Limnocharis flava* management areas (n=81), categorized by current control or monitoring status.

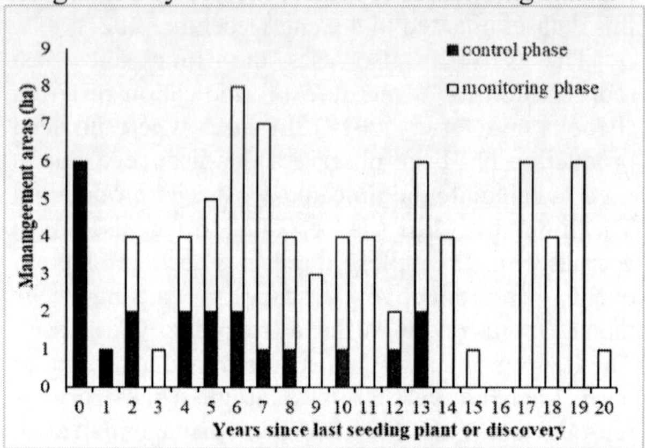


The information presented in Figure 2 is primarily driven by the discovery data, as plants are likely to have produced seed prior to discovery and the rate of reproductive relapses is low. There has never been a reproductive relapse recorded in a contained locus and the overall occurrence of reproductive relapses at unconfined loci is low (0, 1 or 2 events a year since 2003-2004). Many of these events involve the production of intact fruit which may not have spilt seed from the follicles (Weber and Brooks 2008). Throughout the year, monthly survey and control activities are preventing all but an occasional potential seed production event in known management areas. Whilst the seed is persistent, the ongoing recruitment at more active loci reflects seed production prior to discovery more-so than reproductive relapses.

As an example of possible eradication criteria, there are 17 management areas currently in 11+ years monitoring (Figure 1) which largely overlap with the 17 management areas with 16+ years since last reproduction or discovery (Figure 2). Where these data points overlap, all management areas form part of the same locus and if they have a consistent coverage of visits over time and space, then proposed eradication criteria could be met. Ultimately decisions about declaring loci eradicated are not seen as solely a combination of data points, but also involve local field manager input as to how confident they are to reduce the visit frequency to zero.

## OPPORTUNITIES AND CHALLENGES

The net progression of management areas transitioning to a monitoring phase is outpacing the discovery of new areas, ensuring that progress towards eradication accrues gradually. The longevity of the soil seed bank and detection of all small loci and cultivated specimens are the main issues confronting the eradication program.

The overall progress towards the eradication of this serious tropical aquatic weed is prompting discussion about the application of program resources to areas with a continuous history of plant absence. Although management areas within loci show different stages of last reproduction and years in monitoring, the loci are small and discrete. These decisions will be informed by using a range of criteria including time since last mature plant, years in monitoring, status of neighbouring management areas and local field manager confidence as to the frequency, duration, and extent of loci management. Although, the field activities targeting *L. flava* consume approximately 4% of the total recorded field effort. So, a lower frequency of contained loci visits or declaring loci eradicated, will have a small impact on the NTWEP budget. Even with new discoveries, the Queensland-based resource use was approximately 120 field workdays a year over the last decade.

Continuing the theme of constant improvement (Jeffery and Brooks 2016), the NTWEP is encouraging photographic evidence of all reproductive plants and herbarium specimens from all new loci. The Program is also investigating the capacity to deliver quick DNA results for confirmation of identification in the cases where desiccated or juvenile specimens are found at new loci. The program may explore environmental DNA samples from adjacent water bodies, to assist with determining proof of freedom as an additional criterion to declaring eradication.

The NTWEP is addressing the challenge of detecting all *L. flava* plants being cultivated in Australia through a) the utilization of web scraping technology, b) encouraging the reporting of suspected plants through policy and legislation (illegal not to report), c) extension and education through targeted social media campaigns and compliance activities. The NTWEP recognizes increased risk apparent in the sourcing of *L. flava* online across state and potentially international borders, and the potential for cultivation within and beyond areas currently considered suitable habitat. Each jurisdiction should be aware of the trade of this plant as a vegetable amongst ethnic communities to avert further cultivation and potential naturalization events.

## ACKNOWLEDGMENTS

We thank Christopher Collins, Louis Elliott and Nigel Weston from the Department of Environment, Parks and Water Security Northern Territory Government for additional site information in the Northern Territory. We are grateful to all Biosecurity Officers, and Local Government Pest Management Officers for undertaking inspections and the reporting of suspected *L. flava* loci.

## REFERENCES

Brooks, S.J. and Galway, K.E. (2008). Processes leading to the detection of tropical weed infestations during an eradication program. Proceedings of the 16th Australian Weeds Conference, eds R.D. van Klinken, V.A. Osten, F.D. Panetta and J.C. Scanlan. pp. 424-6. (Queensland Weeds Society, Brisbane).

Brooks, S.J, Weber, J.M, Setter, S.D, and Akacich, B.A. (2008a). Seed production and maturation of *Limnocharis flava* (L.) Buchenau in the field and glasshouse. Proceedings of the 16th Australian Weeds Conference, eds R.D van Klinken, V.A. Osten, F.D. Panetta and J.C. Scanlan. pp. 180-2. (Queensland Weeds Society, Brisbane).

Brooks, S.J, Panetta, D, and Galway, K.E. (2008b). Progress towards the eradication of Mikania vine *(Mikania micrantha)* and Limnocharis *(Limnocharis flava)* in Northern Australia. *Invasive Plant Science and Management* 1, 296-303.

Jeffery, M. and Brooks, S.J. (2016). Eradication in the tropics: constantly changing and adapting. Proceedings of the 20th Australasian Weeds Conference, eds R. Randall, S. Lloyd and C. Borger. pp. 23-7. (Weeds Society of Western Australia, Perth).

Stringham, O.C., Toomes, A., Kanishka, A.M., Mitchell, L., Heinrich, S., Ross, J.V. and Cassey, P. (2021). A guide to using the internet to monitor and quantify the wildlife trade. *Conservation Biology*, 35, 4, 1130-9.

Weber, J.M. and Brooks, S.J. (2013). The biology of Australian Weeds 62. *Limnocharis flava* (L.) Buchenau. *Plant Protection Quarterly* 28, 4, 101-13.

# Age and size of flowering *Mikania micrantha* plants raised in a controlled environment

Simon J. Brooks and Kirsty L. Gough.
Tropical Weeds Research Centre, Department of Agriculture and Fisheries, PO Box 976, Charters Towers, Queensland 4820, Australia
(simon.brooks@daf.qld.gov.au)

**Summary** *Mikania micrantha* is a rampant vine and target of a national cost-shared weed eradication program on mainland, Australia. A series of trials were conducted in a quarantine glasshouse to inform the surveillance activities of this serious weed. The trials aimed to determine the time taken for *M. micrantha* to grow from seed or fresh cuttings to flowering in a glasshouse. They also investigated seasonal patterns of flowering behaviour and documented plant sizes at first flowering. Distinct seasonality is reflected in the trial results and field records of mature plants from control teams. *Mikania micrantha* grows within and on the margins of tropical forests. The minimum size of plants when they first flowered is considered in relation to detecting vines in this complex tropical environment.

**Keywords** Mikania, eradication, maturity, tropics.

## INTRODUCTION

*Mikania micrantha* Kunth (Mikania vine) is a rampant, smothering tropical weed, readily capable of vegetative dispersal and seed dispersal by wind, water, or machinery. It is one of the most serious weeds across tropical Asia, the Indian sub-continent, and Pacific regions (Day *et al.* 2016). *Mikania micrantha* was first discovered near Mission Beach in north Queensland and included in the nationally cost-shared 'National Tropical Weeds Eradication Program' (NTWEP). An update on the progress towards the eradication of *M. micrantha* from mainland Australia was presented by Brooks and Jeffery (2018b). *Mikania micrantha* was also discovered near Forrest Beach, Ingham, Bingil Bay and Speewah, but current active infestations are near Mission Beach and Bingil Bay.

Weed eradication programs conduct surveys to prevent the occurrence of mature plants and dispersal events, and to deplete the soil seed bank (Brooks and Setter 2014a). The rate and time at which eradication teams revisit infestations is driven by the need to detect seedlings in the juvenile life phase. Field research on mature eradication target species can be impractical (Brooks and Setter 2014a). So, trials are often limited to quarantined environments where ideal growth conditions provide information on the minimum time to, and size of, mature plants. The first flowering behaviour of *M. micrantha* was investigated by raising plants from seed and cuttings at regular intervals over three years.

## GLASSHOUSE TRIAL METHODS

The pot trials were on plants raised under ideal conditions inside a quarantine glasshouse at Charters Towers. Temperatures are a minimum 20 °C night and 25 °C day, although with evaporative cooling the day temperature may go up to 35 deg C. *Mikania micrantha* readily propagates vegetatively, with plants readily establishing from double node cuttings (Macanawi *et al.* 2011).

**Treatment timing** In trial one, six pots were established per month for 12 months between the 1/8/2012 and 1/7/2013; three pots were from cuttings and three from seed. This is the only trial that raised seedlings from seed. In trial two, six pots of cuttings were established at two- or four-week intervals between 20/10/2014 and 30/1/2015. In trial three, six pots were established from cuttings fortnightly from the 10/11/2015 until the 2/2/2016.

**Establishment from seed** In trial one, every 30-31 days for a year, 5 seeds were germinated in each of 10 peat jiffy pots in a Thermoline® incubator for 28 days. Seeds were randomly taken from a pool of 2000+ seeds collected near Mission Beach in 2011. Seedlings were transplanted into individual tube pots (1-part vermiculite +1-part sand +2-parts peat) standing in shallow water containers in the glasshouse for 16 days. The largest 9 seedlings were distributed between 3 x 29cm pots each month. Seedlings were thinned after a further 30 days to keep only the largest seedling per pot.

**From cuttings** Twelve source plants were cultivated from the same seed source as is used for the 'seed raised' plants above. Stem cuttings (20) with 2 nodes were taken at the same time as the seeds (trial 1 above) were germinated. Cuttings were planted in constantly wet tube pots containing the same mix as the seed tube pots and grown for 44 days. Then 3 cuttings from these pots per block per month were planted in the larger 29 cm diameter pots. Cuttings were thinned to keep the largest plant after

30 days. This process was repeated for trials 2 and 3 at the treatment timing mentioned above.

**Growing conditions** Pots and soil were steam treated prior to the trial commencing. The 29 cm pots were filled with local garden soil mix (2-parts organic garden soil + 2-parts bulk clay soil + 1-part sand), with a layer of weed mat at the base. Five grams of a slow-release fertilizer was added to each pot on day 74 when the thinning was complete. All pots in trials one and two were standing in large troughs of approximately 5 cm deep tap water. Pots were added to the water pool 48 hours before the seedlings and cuttings were planted. Pots used in trial 3 were regularly watered for 2 minutes every 6 hours with 2 irrigation drippers flowing at 6 L per hour, which commenced 48 hours before the cuttings were added. Pots were spaced to mitigate any neighbour shade effects from the oldest plants, and vines were trained up 3m bamboo poles to limit interference between pots and to assist with growth measures.

**Measurements** Plant leader length was recorded when seedlings were transferred to pots and at thinning. Weekly checks of flowering bud presence or absence were conducted. Monthly data of maximum plant height (longest leader length), stem diameter at ground level and number of distinct leaders were conducted but the interim growth data is not shown. Height data was not collected once the leaders reached 4 m tall and became inter-twined at the top of the glasshouse. Length of flowering leader (4 m or less), date of first flowers and basal diameter at flowering were recorded. Plants were destructively harvested four weeks after flowering was observed. There was no attempt made to record the time of seed viability as plants are insect pollinated (Day *et al.* 2016). Additional growth data, wet and dry biomass (stem, leaf and flower) at harvesting and flower production data was collected but is not reported here. Soil and all biomass samples were disposed of by deep burial on site after baking at 80+°C for 48+ hours.

## GLASSHOUSE TRIAL RESULTS

**Trial 1** All the plants that flowered were from the first six treatments established between the 1/8/12 and 31/12/12 at a minimum of 130 days (Table 1). Flowering was observed between 29/4/13 and 21/6/13. Sixteen flowering plants had reached 4 m tall and the roof of the glasshouse, the remaining 4 were at least 3 m tall. The flowering plants tended to have larger diameters than the non-flowering plants in the same cohort. However, diameter was related to age and no threshold size for flowering was evident in the data.

There were fewer flowering plants from seed source, and they tended to be smaller but still at least 3 m tall. The plants from cuttings were taller when thinned at day 74, but the seed raised plants had similar heights when flowering (Table 1). The sample of flowering plants was less than expected in trial 1 and statistical comparisons of seed and cutting plant sources were not conducted.

Plants from treatment 7 established on 30/1/13, were between 92 and 156 cm tall with diameters less than 0.36 cm by late April 2013. Plants from treatment 8 (established 28/2/13,) were under 1 m tall and not thinned when most of the earlier plants started flowering in mid-May. Plants established in treatments 9 to 12 (1/4/13, 1/5/13, 31/5/13 and 1/7/13) did not flower until April 2014.

Plants that did not flower between late-April and June 2013 continued to grow and commenced flowering in April 2014, when the vines had become intertwined and per plant data was not collected at harvest time.

**Table 1.** Establishment dates and first flowering times and sizes of *Mikania micrantha* raised in trial 1 and flowering in 2013. Maximum number of plants flowering is 3.

| N | Establish date | Plant source | Number flowering | First flowering date | Days to flowering | Height range (cm) | Mean stem diameter (cm) |
|---|---|---|---|---|---|---|---|
| 1 | 1/8/2012 | cuttings | 1 | 17/5/13 | 289 | 400 | 1.05 |
| | | seed | 1 | 10/5/13 | 282 | 400 | 1.29 |
| 2 | 31/8/12 | cuttings | 3 | 17/5/13 | 259-280 | 400 | 1.15 |
| | | seed | 2 | 17/5/13 | 259 | 350-400 | 1.04 |
| 3 | 2/10/2012 | cuttings | 3 | 10/5/13 | 220-227 | 350-400 | 1.10 |
| | | seed | 1 | 10/5/13 | 220 | 400 | 0.85 |
| 4 | 31/10/2012 | cuttings | 3 | 29/4/13 | 180-198 | 400 | 0.78 |
| | | seed | 1 | 21/6/13 | 233 | 400 | 0.84 |
| 5 | 30/11/2012 | cutting | 1 | 21/6/13 | 203 | 400 | 1.08 |
| | | seed | 1 | 17/5/13 | 168 | 400 | 0.71 |
| 6 | 31/12/2012 | cuttings | 2 | 10/5/13 | 130-165 | 300-400 | 0.85 |
| | | seed | 1 | 17/5/13 | 137 | 300 | 0.64 |

**Trial 2** Growth and establishment in trial 2 was poorer than the other trials and led to the decision to use drippers in trial 3. One of six plants established on 20/10/2014 flowered after 203 days on the 11/5/15. Four of 6 plants established on the 19/11/14 flowered 159 to 176 days later starting on the 27/4/14. One of the 6 plants established on 19/12/14 flowered after 146 days on 14/5/15. None of the plants established on the 2/1/15, 16/1/15 or 30/1/15 flowered. Heights at flowering ranged between 167 and 350 cm. Plants that survived and did not flower were harvested at an age of 300 days.

**Trial 3** Growth and flowering behaviour was more consistent in this trial (Table 2), and all plants survived to flowering or destruction. Across the first 5 fortnightly establishment dates the plants flowered at an earlier age and smaller size (Table 2). The minimum was for plants established on the 5/1/16 which all flowered after 112 days on the 26/4/16 and less than 3.2 m tall. Subsequent treatments established on the 19/1/16 (6 plants) and 2/2/16 (2 plants) flowered late May and early June and had larger heights and diameters than treatment 5 (5/1/16). There were 14 cuttings struck in 2016 that flowered in the same year. Plants that did not flower in April to June 2016 were destructively harvested at an age of 300 days.

**Table 2.** Establishment dates and first flowering times and sizes of *Mikania micrantha* raised in trial 3 from cuttings and flowering in 2016. Maximum number of flowering plants is 6.

| N | Establish date | Number flowering | First flowering date | Days to flowering | Height range (cm) | Mean diameter (cm) |
|---|---|---|---|---|---|---|
| 1 | 10/11/15 | 3 | 2/5/16 | 174-211 | 400 | 1.13 |
| 2 | 24/11/15 | 5 | 18/4/16 | 146-161 | 400 | 0.87 |
| 3 | 8/12/15 | 6 | 18/4/16 | 132-143 | 320-400 | 0.74 |
| 4 | 22/12/15 | 6 | 18/4/16 | 118-126 | 236-400 | 0.63 |
| 5 | 5/1/16 | 6 | 26/4/16 | 112 | 150-320 | 0.52 |
| 6 | 19/1/16 | 6 | 30/5/16 | 132-146 | 250-380 | 0.62 |
| 7 | 2/2/16 | 2 | 4/6/16 | 123-129 | 200-380 | 0.66 |

FIELD INFORMATION

**Field data** Presence data collected by the NTWEP (Brooks and Jeffery 2018b) in the survey and control on *M. micrantha* was summarized to record the frequency of seedling, flowering and seeding events. There was no separation of the initial control records for new infestations (which often contain mature plants) and revisit records for known infestations (predominantly seedlings). Presence records (n=1774) between 2003 and 2021 were classified into reproductive status and week of the year (1 to 52). Each point and date may have more than one plant type present.

The field data clearly shows the flowering plants being observed from late April (week 17 onwards) (Figure 1), with most flowering observations in May and June, leading to the observation of seeding plants from week 27 onwards. The retention of seed on plants may account for the prolonged time of seed observations. Occasional later season flowering behaviour is observed in the field; however, no glasshouse plants commenced flowering after July in each calendar year.

**Figure 1.** Count of field presence records with seedlings, flowering or seeding plants, aggregated from all *M. micrantha* infestations between 2003 to 2021 and categorized by week of the year.

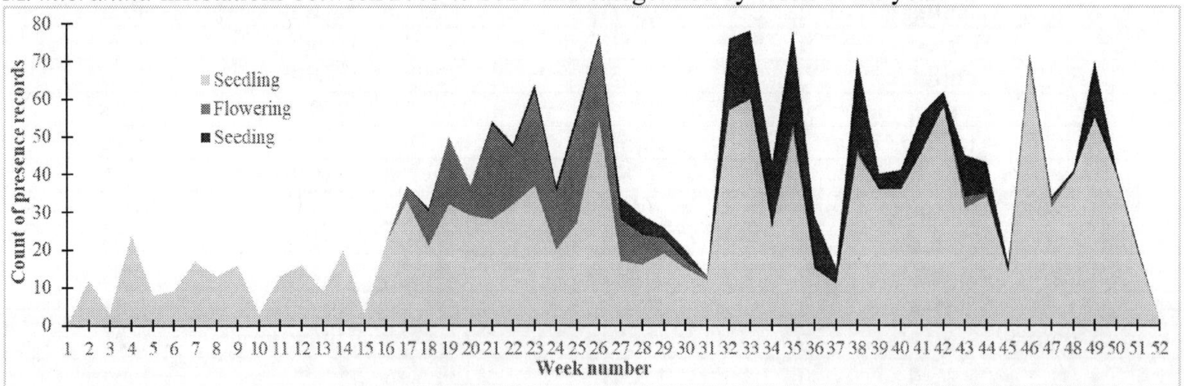

**Vegetation types** To provide context to the glasshouse trial data, information on the types of vegetation present in the *M. micrantha* search area in the Mission Beach area (~281 ha - K. Erbacher pers comm.) was extracted from the Regional Ecosystem database (Queensland Herbarium 2021). Regional ecosystems are classified by a numerical code of bioregion, land zone and vegetation community. The biodiversity classification of 'endangered' is also noted. The predominant remnant vegetation types in the Mission Beach search area include 'Mesophyll to notophyll vine forest' (7.12.1a), 'open areas in vine forests, dominated by sprawling vines, from cyclone damaged forest' (7.12.40a), 'Simple mesophyll vine forest' (7.8.1d, endangered), 'Mesophyll vine forest' (7.3.10a, endangered) and '*Melaleuca leucadendra* L. (L) open forest and woodland (7.3.25a). These descriptions show that the Mission Beach search area contains forest types typified by native vines. The area includes a variety of complex tropical forest types including cyclone disturbed forest (Brooks and Jeffery 2018a). *Mikania micrantha* can also grow in cleared 'non remnant' vegetation, such as grazing land. The *M. micrantha* patches treated by Brooks and Setter (2014b) at Mission Beach were growing horizontally amongst tropical grasses.

## DISCUSSION

A NTWEP field team, mostly dedicated to *M. micrantha* control, are searching for juvenile green vines amongst tropical rainforests with native vines and woodland vegetation. Most plants flowering in the glasshouse trials were several metres tall and at least 1.5 m when growing with support. Whilst the diameters are slender, the plants are likely to be of a reasonable size when first flowering.

Under field conditions it could be expected that in moist well-lit habitats a subset of plants could establish and reproduce in the same calendar year. Across three trials, initial flowering was observed between the 18th of April and the 6th of July. This is consistent with the field data and demonstrates that first flowering behaviour is seasonally driven. Flowering behaviour was determined by time of the year rather than plant size.

Most plants commenced flowering in May or June at ages between 112 and 280 days, depending on their time of establishment. The latest establishment date for a plant flowering in the same year was the 2nd of February when flower buds recorded 112 days later. Under controlled conditions, there were cohorts where not all plants flowered, and a range of plant sizes at which some did.

The field data and information from the glasshouse trials was broadly consistent. February to May is a particularly important time to find and control *M. micrantha* prior to seed production. The timing of surveys is more critical than the interval between surveys. Ideally surveys would provide two opportunities to detect plants before they produce seed. However, in some years, late tropical wet season rainfall can limit access to coastal infestations until April or May (K. Erbacher pers comm.). In these years only a single search and control visit would be possible prior to potential seed production.


## ACKNOWLEDGMENTS

We acknowledge the assistance of Judy Clarke in conducting the glasshouse trials. Comments on the paper were received from Kim Erbacher, Clare Warren and Wayne Vogler. Field data is courtesy of the diligent and hard-working NTWEP crews and Hinchinbrook Shire Council.


## REFERENCES


Brooks, S.J. and Jeffery, M. (2018a). The effects of cyclones on a tropical weed eradication program. Proceedings of the 21st Australasian Weeds Conference, eds S Johnson, L Weston, H Wu and B Auld. pp. 119-23. (The Weed Society of New South Wales, Sydney).

Brooks, S.J. and Jeffery, M. (2018b). Progress in the eradication of *Mikania micrantha* from Australia. Proceedings of the 21st Australasian Weeds Conference, eds S Johnson, L Weston, H Wu and B Auld. pp. 350-3. (The Weed Society of New South Wales, Sydney).

Brooks, S.J. and Setter, S.D. (2014a). Issues and solutions for researching weed eradication target species. Proceedings of the 19th Australasian Weeds Conference, ed. M. Baker, pp. 255-8. (Tasmanian Weed Society, Hobart).

Brooks, S.J and Setter, S.D. (2014b). Increased options for controlling mikania vine (*Mikania micrantha)* with foliar herbicides. Proceedings of the 19th Australasian Weeds Conference ed. M. Baker. pp. 409-12. (Tasmanian Weed Society, Hobart).

Day, M.D., Clements, D.R., Gile, C., Senaratne, K.A.D., Shen, S., Weston, L.A. and Zhang, F. (2016). Biology and impacts of Pacific Islands invasive species. 13. *Mikania micrantha* Kunth (Asteraceae). *Pacific Science* 70 3, 257-85.

Macanawai, A. R., Day, M.D., Tumanaeng-Diete, T. and Adkins, S.W. (2011). Vegetative growth and development of mikania micrantha in taro and cassava production in Viti Levu, Fiji. Proc of 23rd Asian-Pacific Weed Science Society Conference. pp 322-9.

Queensland Herbarium (2021) Regional Ecosystem Description Database (REDD). Version 12.1 (December 2021) (DES: Brisbane). Accessed 26/4/2022.

# Enhancement of the Colonisation Process of a Bioherbicide in Chinese Elm (*Celtis sinensis*) through Co-Treatment

Ciara O'Brien[1], Brooke Johnstone[1], Victor Galea[1]
[1]The University of Queensland, Gatton, Australia
(ciara.obrien@uqconnect.edu.au)

**Summary** The dominant paradigm for woody weed management is currently the application of liquid or granular synthetic herbicides. Despite being highly efficacious, there is increasing evidence of collateral damage to non-target native vegetation through herbicidal drift, runoff or leaching thereby promoting significant developments in the field of biological control. However, the commercialisation of bioherbicides has had a chequered history often resulting in non-establishment or reduced control of the target species. To increase host susceptibility, a preliminary pilot trial investigated the effectiveness of a sub-lethal dose of glyphosate (0.05 g/capsule a.i.) on the colonisation of an encapsulated bioherbicide (*Macrophomina phaseolina*, *Lasiodiplodia pseudotheobromae*, *Neoscytalidium novaehollandiae*) in juvenile Chinese elm [*Celtis sinensis* Pers.] plants located near Grandchester, south-east Queensland. This study demonstrated enhanced fungal colonisation under co-treatment by serving as a source of systemic physiological distress. The interaction was further explored by assessing three proprietary inoculums sourced from distressed *Celtis sinensis* plants located at Kholo, south-east Queensland. A replicated trial (n = 72) was established near Laidley, south-east Queensland, following a randomised complete block design (three blocks) whereby the eight treatments (three inoculums, three co-treatments, two controls) were assigned to a total of three plants per block. The treated plants were destructively sampled at four-week intervals and their internal stem lesions were measured. The addition of a sub-lethal dose of glyphosate (0.0497 g/capsule a.i.) resulted in a significantly ($p < 0.05$) higher degree of inoculum colonisation. This was evidenced by increased internal stem lesion lengths relative to the independent inoculum treatments (i.e., without glyphosate). Other symptoms of physiological distress were also recorded, such as discolouration or splitting of the outer bark tissue, sap seepage and canopy browning or mortality. The full extent of this stress-fungus interaction could not be discerned given the greater than expected rate of colonisation, and limited scope (trial size and duration) of these trials. Further research under field conditions is needed.

**Keywords** Chinese elm, *Celtis sinensis*, woody weed, weed management, biocontrol, bioherbicide, stem implantation

# Invasive grasses management program – piloting a different approach to an old problem

Cath Walsh[1], Shauna Potter[2]
[1]Centre for Invasive Species Solutions, Canberra, Australia,
[2]Wild Matters Pty Ltd, Castlemaine, Australia
(shauna@wildmatters.com.au)

**Summary** Worldwide, grasses are desirable species due to their value as pasture, crops, and as amenity and soil stabilisation plantings. These productive and competitive traits, coupled with a history of introduction spanning over 200 years, have resulted in exotic grasses becoming some of Australia's most problematic weeds. Although many introduced species were selected for their palatability, a sub-set of species that are fast-growing, with high biomass, high seed production and variable palatability have become particularly problematic.

Invasive grasses invade and impact both pastures and conservation areas, often forming dense monocultures that transform both natural and productive environments. The challenge of managing exotic perennial grasses is complex. Contributing factors to this are:

- Invasive grasses are difficult to identify (and can be confused with native species)
- The Influence of climatic events such as drought, fire and flood on management success
- Outdated best practice management information
- A lack of coordinated research, development and engagement (RD&E) effort.

Managing landscape scale infestations is extremely challenging so focussing on preventing establishment and strategies to live with infestations is also necessary.

The National Invasive Grasses Management Program will be the first initiative to emerge from the Centre for Invasive Species Solution Centre's 10 Year Weeds RD&E Plan 2020-30. The initial phase of the Program is being led by the NSW Government's Department of Primary Industries, with funding from the NSW and Australian Governments as well as Meat and Livestock Australia. The program will invest in invasive grass activities, with an emphasis on engagement and promotion of holistic management approaches to deliver benefits broader than just weed management. It will pilot an engagement approach focussed around a series of action-oriented sites with land managers testing science against their experiences and knowledge and adapting, as necessary, to local conditions. The initial focus is on seven Proof and Demonstration Sites across three states and around 40 Adaptation Sites across all states and the ACT.

**Keywords** Invasive grass management, research development and engagement

# Fund a landowner's weed control and you've weeded for a season, teach a landholder long-term weed control strategies and you've weeded for a lifetime

Shannon Robertson[1]
[1]Primary Industries and Regions, Urrbrae, Australia
(Shannon.robertson@sa.gov.au)

**Summary** Invasive unpalatable grasses reduce carrying capacity of pasture and selective grazing results in them dominating, further compounding their impact. African lovegrass (*Eragrostis curvula*), Coolatai grass (*Hyparrhenia hirta*), Chilean needlegrass (*Nassella neesiana*), Texas needlegrass (*Nassella leucotricha*) and fountain grass (*Cenchrus setaceus*) are all unpalatable and impact primary production. These weeds are being managed by landholders throughout regions of South Australia, but many continue to spread to previously uninfested areas. A South Australia wide project is being undertaken to improve community led management of unpalatable grasses. Funded through the Federation Funding Agreement, Enhancing National Pest Animal and Weed Management (Department of Agriculture, Water and the Environment) and in partnership with Landscapes SA, the project works closely groups of neighbouring landholders. With assistance from Landscape Board staff, landholders were contacted and encouraged to take part in the project. Participating landholders receive training in best practice incursion response and integrated weed control. Assistance with long term planning was provided using the Early Intervention Handbook and associated resources. Some landholders weren't aware of the presence of these weeds in the landscape, or the threat posed, mainly in areas where they aren't widespread. There was also an underutilisation of some herbicides registered for these weeds. On farm practices that help prevent spread were often overlooked.

It has been established that if invasive species management tools are not used appropriately by landholders there can be adverse outcomes (Taggart et al. 2022). This project has demonstrated that this is also the case for community led weed control. Investing in coordinated landholder capacity building reduces the threat of weed spread thereby reducing weed management costs in the long term. Landholder training and coordination as conducted by in this project is crucial for optimal management of weeds in SA.

**Keywords** Coordination, community, landholder, landscape scale, capacity

# Comparison of adjuvants with soil-binding properties to reduce herbicides runoff losses in sugarcane

A Davis[2], Emilie Fillols[1]
[1]Sugar Research Australia Ltd, Meringa, Australia,
[2]Tropwater, James Cook University, Townsville, Australia
(efillols@sugarresearch.com.au)

**Summary** Elevated herbicide concentrations coming from sugarcane cropping land may disturb sensitive marine ecosystems already affected by other pressures such as climate change. To mitigate its impact and maintain its productivity, the Sugar industry has been looking at innovative options to reduce the movement of herbicides off site. Three different oil-based adjuvants (Grounded® applied at 3L/ha, Atpolan® soil Maxx applied at 0.4L/ha and Ad-HereTM applied at 1L/ha), a terpene-based adjuvant (Flextend® applied at 1.2L/ha) and a polyol-based adjuvant (Watermaxx®2 applied at 9.35L/ha), have been tested on tilled plant cane and untilled bare soil or trash blanketed ratoon for their potential to reduce runoff losses as well as improving the weed control efficacy of four registered pre-emergent herbicides applied at full label rate: imazapic (96g/ha), hexazinone (975g/ha), isoxaflutole (150g/ha) and amicarbazone(700g/ha). Herbicide efficacy trials were implemented as randomised complete blocks with three replicates and adjacent untreated controls. Herbicide loss in runoff was monitored using replicated rainfall simulations, delivering 80mm of simulated rain, 48h or three weeks after herbicide application. None of the tested adjuvants significantly increased herbicide efficacy on weeds in the efficacy trials. On green cane trash blanket, all adjuvants did not affect or significantly increased the runoff of the tested herbicides, except the Watermaxx®2 which non-significantly reduced the runoff by up to 21% when the rain occurred 48h after herbicide application. On tilled bare soil (plant cane), Grounded®, Atpolan® soil Maxx and Watermaxx®2 significantly reduced herbicide runoff losses when rain occurred 48h after application, with Grounded® achieving the best performance (17 to 40% herbicide runoff losses reduction) across all tested herbicides. On untilled bare soil (ratoon), Grounded® added to the spray tank did not decrease herbicide loss via runoff. This study identified the adjuvant Grounded® could reduce the environmental impact of pre-emergent herbicides in tilled plant cane.

**Keywords** Great Barrier Reef, herbicides, runoff, sugarcane, adjuvant

# riskmapr: a web tool for mapping weed risk to support operational decisions

Jens G Froese[1], Moya Calvert[4], Simon Brooks[4], Alan Pearse[2], Grant Hamilton[3]
[1]CSIRO Health & Biosecurity, Dutton Park, Australia,
[2]University of Wollongong, Wollongong, Australia,
[3]Queensland University of Technology, Brisbane, Australia,
[4]QLD Department of Agriculture and Fisheries
( jens.froese@csiro.au)

**Summary** New alien plant species are constantly introduced across borders or landscapes, and some are likely to become problematic invasive weeds from experiences elsewhere. Initially, weed population growth and spread is typically slow. This 'invasion lag' often presents the only window of opportunity where containment and eradication can be achieved, and long-term negative impacts avoided. We developed a weed risk mapping tool that allows land managers to identify priority areas for monitoring and management relatively quickly and with limited data. The tool considers both habitat suitability (where weeds are likely to grow well and reproduce) and susceptibility (where weed propagules are likely to arrive from known source plants). 'Risk factors' that reflect the environmental conditions a weed needs to grow, reproduce and spread (like rainfall, land use, soil type or road networks) are identified and linked to spatial data. A collection of open-source web apps called 'riskmapr' allow land managers and researchers to plug in their own spatial data and generate weed risk maps. These maps may be used to direct on-ground resources, along with other operational considerations. Risk models can also help structure our understanding of the factors and processes driving weed invasions.

Here, we showcase an application of riskmapr to support surveillance planning for *Miconia calvescens* in the National Tropical Weeds Eradication Program. Previously, surveillance effort was evenly allocated to forested areas within concentric buffer zones surrounding known mature infestations. Using riskmapr and spatial analysis to gain a more nuanced understanding of invasion risk allowed the operations team to focus on-ground effort. At one site, high risk areas subject to frequent surveillance were reduced by 24% (from 613ha within buffer zones to 467ha using riskmapr outputs); medium risk areas were reduced by 27% from 833ha to 612ha; but less frequently surveyed low risk areas increased from 236ha to 672ha.

**Keywords** Invasion lag, risk model, risk map, risk factors, suitability, susceptibility, monitoring, surveillance

# Influence of broadacre crop rotational sequence on the weed seedbank in the Riverina

K M Shamsul Haque[1,2], Saliya Gurusinghe[1], Graeme Heath[1], Paul A. Weston[1], Agasthya Thotaga[1], William B. Brown[1], Leslie A. Weston[1]

[1] Graham Centre for Agricultural Innovation (NSW Department of Primary Industries and Charles Sturt University), Wagga Wagga, Australia,

[2]School of Agricultural and Wine Science, Charles Sturt University, Wagga Wagga, Australia (khaque@csu.edu.au)

**Summary** Incorporation of effective crop rotations for supplemental weed control has been shown to reduce the growth and establishment of annual weeds and deplete the weed seed bank in broadacre cropping systems. Long term rotational trials were established in 2014-2018 to quantify the impact of strategic management practices upon weed infestation, with a focus on key winter annual and summer fallow weeds, in the moderate and low rainfall zones of Wagga Wagga and Condobolin, NSW, respectively. Rotations assessed both grain and pasture crops to determine if long-term weed management was facilitated through successful manipulation of the weed seedbank over five growing seasons. Seedbank dynamics were assessed under glasshouse conditions by recording continuous weed seedling emergence in field soil collected yearly from each rotational treatment over a 5 year period. Total weed seed density was successfully depleted in all rotations receiving average and above-average rainfall from 2014 to 2017. However, limited rainfall in 2017-2018 negatively impacted rotational crop biomass and canopy closure, and a dramatic and significant rebound in weed seedbank numbers was observed in subsequent seedbank assessments. Several rotations were particularly effective in suppressing weed seedbank numbers over time, suggesting crop selection is an important consideration. The most successful rotations included a diverse selection of cereals and/or pulses (i.e. 5 separate species), while the least successful rotations included a lucerne monoculture, a lucerne/grass pasture and rotations with multiple years of wheat or field pea.

**Keywords** Weed seedbank, broadacre crop, rotation, Riverina

# Cover crops in maize (*Zea mays*) enable reduced herbicide use

Trevor James[1], Michael Trolove[1], Sam McDougall[2], Steve Payne[2]
[1]AgResearch, Ruakura Research Centre, Hamilton, New Zealand,
[2]Foundation for Arable Research, Innovation Park, Templeton, New Zealand
(trevor.james@agresearch.co.nz)

**Summary** A 4-year study investigated weed control in maize following four different cover crops compared to winter fallow. Trial site was at Tamahere near Hamilton NZ on a sandy loam soil. The four winter cover crops; gland clover (*Trifolium glanduliferum*) cv Prima, faba bean (*Vicia faba*) cv Ben, oats (*Avena sativa*) cv Milton, and ryegrass (*Lolium multiflorum*) cv Tama were first planted on 2 June 2016 then on 11 June 2017, 29 May 2018 and 22 May 2019 on the same plots. Each years the cover crops were terminated 3-4 days prior to direct drilling maize by spraying with glyphosate (2.16 kg ai/ha + Pulse 0.1%). Five herbicide regimes were evaluated on each of the cover crops and fallow each year, viz. no herbicide, pre-emergence only, pre- and post-emergence, a single post-emergence and two post-emergence herbicide applications. Maize production for both silage and grain were measured each year. Establishment of all cover crops was excellent in the first year, but the small seeded gland clover failed in years 2 and 3 due to a build-up of maize residue and these plots were cultivated in year 4 to enable establishment. Ryegrass failed to establish in year 3 and plots were cultivated in year 4. For maize silage, Gland clover, faba beans and oats all shared the highest yield on different years while for grain, the legumes always performed best. Yields in the untreated fallow plots dropped each year. For faba beans and oats, maize yields were never significantly improved by any of the herbicide regimes compared to untreated whereas in all but the first year, yields in the fallow plots were. Gland clover and ryegrass gave variable results, but these reflect the establishment difficulties. Overall, post-emergence weed control tended to result in higher yields than regimes involving a pre-emergence application.

**Keywords** Cover crops, maize, weed control

# Genetic mutation in a putative Aux/IAA gene in common sowthistle proposed as the basis for 2,4-D resistance

Mahima Krishnan[1], Tijana Petrovic[1], Alicia Merriam[1], Geetha Velappan[1], Christopher Preston[1]
[1]The University of Adelaide, Adelaide, Australia
(mahima.krishnan@adelaide.edu.au)

**Summary** Seven common sowthistle (*Sonchus oleraceus*) populations with resistance to Group I herbicides, including 2,4-D, were identified in South Australia. This has been due to an increased reliance on Group I herbicides as a result of the increasing failure of glyphosate and ALS herbicides to control common sowthistle. Study into the mode of inheritance suggests the resistance is caused by a single dominant gene. The Aux/IAA gene family, the likely target of 2,4-D and similar auxin-mimic herbicides according to recent findings (LeClere et. al. 2018), was genotyped in these populations. A small deletion, on either side of a highly conserved degron region in a putative Aux/IAA gene, IAA20, was observed in resistant populations but, not in susceptible ones and is the probable cause of 2,4-D resistance.

**Keywords** 2,4-D resistance, mutation, Aux/IAA, IAA20, common sowthistle, *Sonchus oleraceus*

# The ACT Experience with Mobile Device Mapping of Invasive Plants

Michael Tweedie[1], Harley Baker[1], Kirsten Tasker[1], Ian Lenon[1]
[1]ACT Parks and Conservation Service | Resilient Landscapes | EPSDD, Canberra, Australia
(michael.tweedie@act.gov.au)

**Summary** Invasive plant control on Australian Capital Territory (ACT) public land is mapped on shared feature layers using the ArcGIS Field Maps app, and formerly the ArcGIS Collector app. The app syncs to ArcGIS Online, and can work off-line. The ACT Government was an early adopter of this Esri mobile device mapping system, having commenced use of the original Collector app in 2014-15. There have been many benefits: prioritisation of control work, locating follow-up control sites, reporting early invaders, adaptive management, and securing increased budgets.

**Keywords** Mapping, mobile devices, management thresholds, apps

# How effective is Satusteam as a tool for weed control and restoring biodiversity?

Jeremy Winer[1]
[1]Weedtechnics
(jeremy@weedtechnics.com)

**Summary** In this presentation Jeremy Winer will refer to case studies and outline the challenges of chemical weed management that have led to decision makers adopting non-toxic methods. Pilot programs which have used Satusteam© include Mother of Millions, Tiger Pear, Harrissa cactus, Gorse and Blackberry. An unexpected result in some areas was regeneration from the native seed bank. Case studies include experiences from urban landscapes, indigenous rangers, Landcare and Coastcare as well as specialist contractors in organic weed control. In Cities and regional areas, farms, verges, natural margins, parks and recreation areas are part of the local ecosystem, which when nurtured, to build soil organic matter represent an opportunity to sequester atmospheric carbon and build biodiversity. Does Satusteam assist in this process? Delegates will gain an understanding of the range of effective non-toxic methods available and the differences in thermal methods including 'Steam' and Satusteam©. When are the applications most suitable and where other methods are going to deliver better outcomes? Attendees will walk away with knowledge of the 7 steps necessary to implement and document the efficacy of their thermal weed control program and have access to online support materials.

**Keywords** organic, steam, herbicide, resistant, weeds, Satusteam, Weedtechnics, Winer, biodiversity, indigenous, MOM, Tiger Pear, Harrissa, cactus, Gorse, Landcare, Coastcare, carbon

# Against the Odds - Policy Challenges for managing buffel grass invasion in non-pastoral arid lands

Jeffery Foulkes[1]

[1]Alinytjara Wilurara Natural Resource Management Board, Australia

(adam.wood3@sa.gov.au)

**Summary** The Alinytjara Wilurara (AW) Landscape Board region covers the north-west of South Australia covering more than 280,000km2, stretching from the NT and WA borders south to the Great Australian Bight. The land is mostly dedicated to conservation and traditional Aboriginal use and occupation. This includes Aṉangu Pitjantjatjara Yankunytjatjara (APY) Lands, Maralinga Tjarutja Lands, Yalata Lands, and co-managed parks and reserves. Buffel grass (*Cenchrus ciliaris* and *C. pennisetiformis*), or mamu tjanpi/tjanpi kura (Pitjantjatjara: devil grass/bad grass) is an introduced perennial tussock grass that has emerged as a significant threat to the culture and safety of remote communities in the region It was introduced for dust suppression in central Australia in the 1960's. It has since colonised large areas of the APY Lands. It outcompetes native grasses and shrubs; threatens woodlands, communities and infrastructure with destructive high fuel load fires. It establishes a dense monoculture, unsuitable as habitat and unpalatable to wildlife. Heavy infestations prevent traditional hunting, foraging and cultural activities. It's now recognised as one of the worst transformer weeds in Australia's arid rangelands. In some parts of northern Australia it is considered a reliable fodder grass and not a "weed". The negative and positive aspects of Buffel grass compromises definitive policy developments and there isn't a National strategy for sustainable management of Buffel grass. South Australia is the only state to have it declared under the Landscape South Australia Act 2019 and a state-wide Buffel Grass Strategic Plan. Control and eradication is one of the highest management priorities for AW Region. This is mirrored in the Healthy Country Plans and Indigenous Protected Area Plans of the remote aboriginal communities in the AW region. The AW Buffel Grass Operational Strategy 2018-2023 was developed to guide strategic management approaches. A key program is to eradicate where possible and otherwise control it in southern AW region by 2025 as its distribution is currently limited.

**Keywords** Buffel grass, *Cenchrus ciliaris*, Environmental Weed, Alinytjara Wilurara, traditional hunting, cultural activities

# Estimating tropical weed seed longevity with a laboratory test

Simon J. Brooks, Dannielle A. Brazier and Clare Warren.
Tropical Weeds Research Centre, Department of Agriculture and Fisheries, PO Box 976, Charters Towers, Queensland 4820, Australia
(simon.brooks@daf.qld.gov.au)

**Summary** Longevity of weed seeds in the soil drives the cost and duration of weed control activities. Traditional methods for estimating weed seed longevity, such as repeated field soil sampling and buried packet trials can take many years and substantial resources to complete. A laboratory process, a Controlled Ageing Test (CAT) exposes seeds to an 'ageing environment' of 45 °C temperatures and 60% humidity. Data from this test is used to sort species into relatively transient, short lived or long-lived categories of weed seed longevity. This paper reports on examples from a series of trials that seek to correlate the data from CAT batches with longevity data from buried packet trials.

**Keywords** controlled ageing test, soil seedbank, weed seed persistence, tropics.

## INTRODUCTION

In the absence of seed input, the longevity of weed seed in the soil will drive the length of weed control activities. This length influences the cost of activities and follow-up seedling control should be considered in landholder weed management plans. There are several sources of information that can be used to determine the likely persistence of weed seeds in the soil. The five main sources are buried packet trials, soil samples collected within infestations, controlled age testing, field control records and native range studies; each method has its own pros and cons (Brooks and Setter 2012). The burying and retrieval of seed packet trials has, for many years been the main and sometimes only formal source of local information on weed seed persistence. This remains a common, standard, and robust way of estimating weed seed longevity (reference section has 8 examples). Buried packet trials require seed, land and technical resources and take 1 to 15 years to complete. For newly identified weed species there may be little known about the seed longevity, limited seed available for long-term study and lack of a climatically suitable long-term site. So initial decisions on management, including eradication are made in the absence of seed longevity information.

Hay *et al.* (2006) and Probert *et al.* (2009) use a procedure to estimate the longevity of seed lots held in storage. Where seeds are exposed to an ageing environment and removed after 1 to 125 days and germinated under standard incubator conditions. The germination data is used to determine the number of days in the ageing environment when germination drops below 50% of an initial, unaged value (day 0), this value in days is called $P_{50}$. A study by Long *et al.* (2008) found a broad correlation between field trials of weed seed longevity and $P_{50}$ values from the Controlled Ageing Test (CAT). The CAT has also been used to rank seed of species in the same plant families by relative persistence (Probert *et al.* 2009).

Selected data from a series of controlled ageing tests is reported and compared to the results of local buried seed packet trials. Some weed species without local burial trials are included as their results are applicable to local management efforts.

## MATERIALS AND METHODS

Between 2018 and 2021, a series of CATs were conducted in seven batches containing seed lots of 1 to 13 species. A subset of the test results is reported in this paper (Table 1).

**Seed** Seed was collected at the date and location in Table 1, then carefully separated from fruit or pods and stored at laboratory room temperature until used in the CAT batch or batches. Unless otherwise noted, seed was sorted into 24 lots of 50 for testing. *Ziziphus mauritiana* Lam. was tested using 48 lots of 30 intact endocarps (kernels). Half the kernels were kept 'intact', with the seeds kept within the kernels for the hydration and ageing phases. Seeds were then removed from the intact kernels for germination at each retrieval time. Seeds were removed (cracked) from the other 24 kernels before exposure to the hydration and ageing phases. Seed totals for each lot of 24 kernels was recorded. *Cascabela thevetia* (L.) Lippold was sorted into 22 lots of 20 kernels and were kept in the kernels for the hydration and ageing phases. At each retrieval time seeds were removed from 2 x 20 kernels and counted prior to germination.

**Experimental conditions** Seed lots were subjected to a 'hydration' phase then an 'ageing' phase following a protocol of Hay *et al.* (2006). Each CAT batch used two replicate IP67 electrical boxes (labelled A and B). Seed lots were placed in individual open glass vials, plastic jars or centrifuge tubes, half in each of the sealed boxes. For the hydration phase, lots were exposed to a 47% relative humidity lithium chloride solution (320 g/L $H_2O$) in a dark 20°C Thermoline® incubator for 14 days.

For the ageing phase, the temperature was increased to 45°C with 60% relative humidity (lithium chloride at 370 g/L $H_2O$). Seeds remained in the dark ageing environment for 2 to 203 days and removed at each retrieval interval and germinated. Test conditions were checked with Onset® Hobo® temperature and humidity loggers and the lithium chloride solution was adjusted as necessary. Unless otherwise stated a seed container was removed from each box in the ageing environment on days 0, 2, 7, 14, 21, 28, 35, 42, 56, 77, 98, 126. This was the standard retrieval schedule. In batch 4, the final two retrievals for *Tecoma stans* L. Kunth were retrieved at days 91 and 98 due to no germination in the previous retrieval. In batch 6, there was no day 126 retrieval. The retrievals for batch 7 were on days 0, 7, 14, 28, 42, 56, 77, 98, 119, 147, 175 and 203. Batch 2 started on the 31/1/2019, batch 3 on the 3/7/2019, batch 4 on the 21/1/2020, batch 5 on the 3/6/2020, batch 6 on the 18/1/2021 and batch 7 on the 3/3/2021.

**Germination** A reference sample of each seed lot was removed from boxes A and B and germinated prior to the ageing phase (day 0). Each retrieved seed lot was placed in a 90mm petri dish, on top of moistened filter paper and an inverted watch glass. All petri dishes were kept moist with distilled water and germinated in a Thermoline® incubator running at 30/20 °C 12hr diurnal cycle, except for the *Senecio madagascarensis* Poir. which was germinated at 23/17 °C 12hr diurnal cycle reflecting its cooler distribution. Germinated seeds (identified by radicle emergence) were counted and removed periodically. Ungerminated seed of the Fabaceae species was scarified after approximately 28 days. Scarification was conducted by either, submergence in 98.08% solution of sulfuric acid for 25 minutes or knicking the outer seed coat with secateurs.

Table 1. Species, collection details, CAT batch, $P_{50}$ value and category. (+) indicates the regression line did not drop below the 50% in the duration of the test. * P50 category as defined by Long *et al.* (2008).

| Weed species | Collection location, month and year | CAT batch | $P_{50}$ value (day) | $P_{50}$ category* |
|---|---|---|---|---|
| *Acaciella angustissima* | Calcium 2019 | 3 | 111 | Long-lived |
| *Andropogon gayanus* | Mareeba 2018 | 2 | 16 | Transient |
| *Cascabela thevetia* | South Townsville 08/2020 | 6 | 25 | Short-lived |
| *Calotropis procera* | Upper Burdekin River 12/2019 | 4 | 109 | Long-lived |
| *Calotropis procera (repeat)* | Upper Burdekin River 12/2019 | 5 | 89 | Long-lived |
| *Cryptostegia grandiflora* | Charters Towers 12/2019 | 4 | 125 | Long-lived |
| *Cyperus aromaticus* | Innisfail 2018 | 3 | 82 | Long-lived |
| *Leucaena leucocephala* | Charters Towers 06/2018 | 2 | 69 | Long-lived |
| *Leucaena leucocephala (repeat)* | Charters Towers 06/2018 | 7 | 203+ | Long-lived |
| *Parkinsonia aculeata* | Upper Burdekin River 12/2020 | 7 | 127 | Long-lived |
| *Senecio madagascarensis* | Herberton 07/2018 | 4 | 19 | Transient |
| *Senna alata* | Townsville 07/2019 | 3 | 126+ | Long-lived |
| *Senna alata (repeat)* | Townsville 07/2019 | 7 | 190 | Long-lived |
| *Stevia ovata* | Ravenshoe 10/2018 | 2 | 16 | Transient |
| *Tecoma stans* | Charters Towers 07/2019 | 4 | 30 | Short-lived |
| *Ziziphus mauritiana (cracked)* | Charters Towers 11/2019 | 4c | 40 | Short-lived |
| *Ziziphus mauritiana (intact)* | Charters Towers 11/2019 | 4i | 20 | Short-lived |

**Data analysis** The total germination from the retrieval at day 0 was used a reference value. Subsequent germination was calculated as a proportion of the day 0 germination per box. Proportion data from the A and B boxes was used to create a negative logistic regression curve (equation 2 in Long *et al.* 2008) in Genstat® 19th, 21st edition VSNi®. The $P_{50}$ value in Table 1 is nearest whole day determined from the regression line. Where seed lots were repeat tested, data from 4 boxes and 2 batches was used to create the regression line, this combined value is mentioned below. The $P_{50}$ categories (Table 1) classified the seeds into longevity categories as defined by Long *et al.* (2008), with transient seed banks less than 1 year ($P_{50}<20$), to be short-lived seed banks of one to three years ($20<P_{50}<50$) and long-lived seed banks over three years ($P_{50}>50$).

## COMPARISON WITH BURIAL TRIALS

**Apocynaceae** The seed of *Cryptostegia grandiflora (*Roxb.) R. Br. in Table 1 indicated a long-lived seed based on a fall in the germination on the last (day 126) of retrievals. This is different to the field trial reported by Bebawi *et al.* (2003), who reported 0%

viable after 3 years, which would fit in the short-lived category. They also recorded a drop below 50 % viability between 11 and 20 years of cool, dry storage. The CAT can be applied to seeds held in cold storage (Probert *et al.* 2009) and generally less persistent seeds are correlated with a drop below 50% viable in under 20 years of storage. The other species in this family tested was *C. thevetia*. There was more consistency between the CAT data in Table 1 (short-lived) and the study of Bebawi *et al.* (2016) who found a small proportion of seed viable after 12 months burial and none after two years.

**Asclepiadaceae** There are large differences between the CAT data reported for *Calotropis procera* Aiton (W.T Aiton) in Long *et al.* (2008) ($P_{50}$ = 28.7) and Probert *et al.* (2009) ($P_{50}$ = 18.5 days) and the data in Table 1. After storage, the value for batch 5 was lower than 4 (Table 1). However, in both boxes in both batches the germination data from the first nine retrievals (day 56) was above 90% and was classified as long-lived. Bebawi *et al.* (2015) described the results of the buried packet trials as fitting in the short-lived category (0% after 18 months) and consistent with the findings of Long *et al.* (2008). Despite running the same seed source of *C. procera* in two CAT batches our results were inconsistent with the other published studies.

**Asteraceae** The germination of *S. madagascarensis* seed dropped quickly in the CAT (Table 1) and it could be classified as transient, although close to the short-lived category. Through the extrapolation of data, *S. madagarscarensis* seed was suspected to be longer lived (3-5 years, maybe more, Sindel, (2009)), though longer-term field data sources have not been identified. A short seed life would aid the management of the isolated northern incursion. The second Asteraceae, *Stevia ovata* Willd. also showed a transient seed bank (Table 1). In buried packet trials in the wet tropics this species was exhausted in 18 months and after 36 months in the dry tropics (Bebawi *et al.* 2018a). The CAT data appears to be an under-estimate of the field seed longevity, as *S. ovata* burial trials show a short-lived seed life.

**Bignoniaceae** An unpublished field trial near Moura in central Queensland found *T. stans* had short lived to transient seed bank when buried (W Vogler pers comm). The results from the CAT indicate a short seed life, which may be a slight over-estimate of buried seed longevity.

**Cyperaceae** Viable seed of *Cyperus aromaticus* (Ridley) Mattf. & Kük was found in packets that had been buried for 15 years in soil at South Johnstone (M Setter and J Vitelli unpubl. data). The CAT data in Table 1 supports the field test results that this species forms a long-lived seed bank, although the CAT data may be an underestimate of buried seed persistence in the wet tropics.

**Fabaceae** Probert *et al.* (2009) reported CAT results that showed species in this family form long-lived seed banks. So, CAT batch 7 (Fabaceae seed lots) was aged for up to 203 days and all seed lots were found to fit the persistent category (Table 1). Long *et al.* (2006) reported a $P_{50}$ value of 122 for *Parkinsonia aculeata* L. seed, which was scarified prior to ageing. This is like the value in Table 1, which was obtained for seed scarified after ageing. Field trials of buried *P. aculeata* found persistent seed banks (4+ years) were formed in wetter habitats (van Klinken *et al.* 2008). Buried packet trials and the CAT data are consistent in indicating long-lived seed persistence beyond 3 years. Data from a *Leucaena leucocephala* (Lam.) de Wit. seed longevity trial was summarized in Campbell *et al.* (2019). They reported viable buried seed was still present (>4%) after eight years. This indicates a long-lived seed bank, which is also reflected in the data of Probert *et al.* (2009) ($P_{50}$ = 75.2) and Table 1 (batch 2). The CAT data for batch 2 was influenced by values in both boxes below 50% at days 75 and 126. In CAT batch 7, there was a more consistent decline to 60% at day 203, although the combined line plateaued at 62%. All trials show *L. leucocephala* forms a long-lived seed bank which may extend well beyond three years.

There was little data available on the weeds *Senna alata* (L.) Roxb. and *Acaciella angustissima* (Mill.) Brit. & Rose. *Senna alata* has been noted as an emerging weed in north Queensland and showed no decline below $P_{50}$ in batch 3. Seed from the same collection was used in batch 7 and this only approached $P_{50}$ after 190 days (the combined line plotted at 196 days). The predicted seed life for *S. alata* may be well beyond 3 years and it would be expected to form a long-lived seed bank. *Acaciella angustissima* is being controlled at several north Queensland infestations and the data in Table 1 indicates this species also has a long-lived seed.

**Poaceae** Buried trials show *Andropogon gayanus* Kunth. forms a very short-lived seed bank with viability dropping to below 10% after 6 months, 1% after 12 months and 0% after 30 months (Bebawi *et al.* 2018b). The results in Table 1 fit into the transient category. It is possible some buried seed (10-20 cm) would be short lived, but both data sources indicate low overall viability beyond 12 months.

**Rhamnaceae** The study by Bebawi *et al.* (2016) found seed from buried 'intact' *Z. mauritiana* kernels was exhausted quickly (6-18 months) and surface kernels reached 0% viable in 24 months. The intact treatment in Table 1 reached $P_{50}$ in 20 days, which is consistent with a short-lived seed bank and the field trials. The $P_{50}$ for the seed removed from the kernels was 40 days, although this is higher than the intact kernels, it is still classified as short-lived.

## DISCUSSION

Much of the CAT data from most of the tropical weed species tested was broadly consistent with the published data from buried packet trials. The greatest differences occurred when the CAT data on *C. procera* and *C. grandiflora* was compared to buried packet trials and other sources. There were also inconsistencies in the CAT results for *L. leucocephala*. The CAT remains a useful tool to broadly categorize seed longevity where time, seed and field sites are in short supply, such as with new incursions. However, until the source of some of these differences is better understood it could be misleading to use the CAT data as a sole source of longevity information.

There can be useful, predictive trends in relative seed persistence amongst some plant families and within short lived genera (Probert et al. 2009). Ultimately the CAT data applies to the source population and a single value will not capture variation between populations (Long *et al.* 2008). The overall mean seed longevity data from buried packet trials often reflects the combination of depth, ground cover and soil factors, which are not factors covered by the laboratory test.

The three tropical species that were classified as transient based on the CAT data in Table 1, have been found to be short-lived in field trials. The transient range described by Long *et al.* (2008), may better reflect later field data if the $P_{50}$ was less than 16. More testing, ranking and correlation analysis may provide the scope to refine the zero-to-three-year categories into more defined or overlapping zones. Further analysis may also help to categorize $P_{50}$ values for very long-lived species that are found to be viable after five or ten years of field burial.

This series of experiments is continuing to compare the results from the CAT with field longevity trials and refine the predictability of this shorter laboratory-based test. This could be a useful and efficient tool to inform weed control programs where little seed longevity information is available. However, the CAT may not be a consistent predictor for all tropical seed lots and is best interpreted in conjunction with other field or trial data.

## ACKNOWLEDGMENTS

We would like to acknowledge assistance from Tony Salisbury, Kelli Pukallus, Kirsty Gough, Stephen and Melissa Setter and Mario Di Carlo with seed collection. Also, Melissa Setter and Wayne Vogler for providing comments on the paper.

## REFERENCES

Bebawi F.F., Campbell S.D., Mayer R.J. (2016). Seed bank persistence and germination of chinee apple (*Ziziphus mauritiana* Lam.). *The Rangeland Journal* 38, 17-25.

Bebawi, F.F. Campbell S.D. and Lindsay A.M (2003). Effects of burial and age on viability of rubber vine (*Cryptostegia grandiflora*) seeds *Plant Protection Quarterly* 18, 4, 147-51.

Bebawi, F.F., Campbell, S.D., Mayer, R.J., Setter, M.J. and Setter, S.D. (2018a). Effects of temperature and burial on seed germination and persistence of the restricted invasive *Stevia ovata* in northern Queensland. *Australian Journal of Botany* 66, 388-97.

Bebawi, F.F., Campbell S.D. and Mayer R.J. (2018b). Gamba grass (*Andropogon gayanus* Kunth.) seed persistence and germination temperature tolerance. *The Rangeland Journal* 40, 463-72.

Bebawi, F. F., Campbell, S. D. and Mayer, R. J. (2015) Seed bank longevity and age to reproductive maturity of *Calotropis procera* (Aiton) W.T. Aiton in the dry tropics of northern Queensland. *The Rangeland Journal* 37, 239-7.

Bebawi, F.F., Campbell, S.D. and Mayer, R.J. (2017). Seed ecology of Captain Cook tree [*Cascabela thevetia* (L.) Lippold] - germination and longevity, *Rangeland Journal* 39, 4, 307-16.

Brooks, S.J. and Setter, S.D. (2012). Soil seed bank longevity information for weed eradication target species. *Pak. J. Weed Sci. Res.* 18, 73-83.

Campbell, S.D., Vogler, W.D., Brazier, D.A., Vitelli, J.S and Brooks, S.J. (2019). Weed leucaena and its significance implications control. *Tropical Grasslands-Forrajes Tropicales*. 7, 4, 280-9.

Long, R.L., Panetta, F.D., Steadman, K.J., Probert R., Bekker, R.M., Brooks, S.J. and Adkins, S.W. (2008). Seed persistence in the field may be predicted by laboratory-controlled ageing. *Weed Science* 56, 523-8.

Hay, F. R., J., Klin, and R. J. Probert. (2006). Can a post-harvest ripening treatment extend the longevity of Rhododendron L. seeds? *Sci. Hortic. (Amst.)* 111, 80-3.

Probert, R.J., Daws, M.I. and Hay, F.R. (2009). Ecological correlates of ex situ seed longevity: a comparative study on 195 species. *Annals of Botany* 104, 57. 57-69.

Sindel, B.M. (2009). Fireweed in Australia. Directions for future research. Report for the Bega Valley Fireweed Association. University of New England, Armidale. 65p.

Van Klinken, R.D., Lukitsch, B. and Cook, C. (2008) Interaction Between Seed Dormancy-release Mechanism, Environment and Seed Bank Strategy for a Widely Distributed Perennial Legume, *Parkinsonia aculeata* (Caesalpinaceae), *Annals of Botany*, 102, 2, 255-64.

# Bixlozone: A new Isoxazolidinone herbicide for a wide range of major crops

Marco Montagna[1], Geoff Robertson[1]
[1]FMC Australasia Pty Ltd, North Ryde, Australia
(geoffrey.robertson@fmc.com)

**Summary** Bixlozone, trademarked Isoflex™ active, is a new herbicide from the isoxazolidinone family discovered and developed by FMC's research and development organization. It provides a new and unique selective residual weed control solution in a wide range of crops including, cereals, corn (*Zea mays*), legumes, rapeseed (*Brassica napus*), rice (*Oryza sativa*) and sugarcane (*Saccharum officinarum*), and will offer a new mode of action herbicide solution for many of these crops and crop rotations. Isoflex™ active provides both systemic and contact activity, with residual control and can be applied pre-emergence, early post-emergence or incorporated by sowing, across a wide range of agronomic environments. It controls major problematic grass weeds including annual ryegrass (*Lolium rigidum*) and several key broadleaf weeds by inhibiting 1-deoxy-D-xylulose 5-phosphate synthase resulting in the disruption of plastid isoprenoid biosynthesis. Isoflex™ active is classified as a Group 13 herbicide mode of action by the Herbicide Resistance Action Committee and will offer a new tool for resistance management, especially annual ryegrass (*L. rigidum*). Trials conducted in Australia between 2015 and 2020 demonstrate that pre-emergent control of annual ryegrass (*L. rigidum*) from Isoflex™ active is comparable to leading industry standards. It will also be an ideal complementary mixing partner for other pre-emergent herbicides as it can extend the utility of existing molecules by expanding the weed spectrum and increasing weed efficacy. It will also be safe to a wide variety of rotational crops seeded after initial crop planting. Isoflex™ active received first global registration in 2020 in Australia with subsequent launches planned in Asia Pacific, Latin America, and Europe. In Australia, Isoflex™ active is registered for pre-emergence application at the rate of at 500 g a.i. x ha-1 in wheat (*Triticum aestivum*), barley (*Hordeum vulgare*), canola (*B. napus*), field pea (*Pisum sativum*) and faba bean (*Vicia faba*) under the tradename Overwatch® Herbicide.

**Keywords** Bixlozone, isoxazolidinone, ryegrass, Overwatch, isoflex

# Survival of tropical weed species propagules after immersion in fresh, brackish and salt water

Stephen Setter[1], Melissa Setter[1] and Wayne Vogler[2].

[1] Biosecurity Queensland, Department of Agriculture and Fisheries, Centre for Wet Tropics Agriculture, South Johnstone, Queensland, Australia

[2] Biosecurity Queensland, Department of Agriculture and Fisheries, Tropical Weeds Research Centre, Charters Towers, Queensland, Australia

(stephen.setter@daf.qld.gov.au)

**Summary** Neem (*Azadiractha indica* A.Juss.), Navua sedge (*Cyperus aromaticus* (Ridl.) Mattf. & Kük.), Hymenachne (*Hymenachne amplexicaulis* (Rudge) Nees), Bellyache bush (*Jatropha gossypifolia* L.) and Leuceana (*Leucaena leucocephala* ssp. *glabrata* (Rose) Zarate) are serious weeds in tropical north Queensland. Water movement is a significant dispersal vector for seeds of each of these species, particularly along water courses, but little is known about seed survival following immersion and movement in water. To improve knowledge of the effect of water immersion on seed viability seeds from each species were placed in fish tanks containing fresh, brackish and salt water for set periods of time before testing their viability.

Seeds of four of the species tested remained viable for the maximum test period of 98 days (14 weeks). Only Neem lost all viability after seven days in salt water and 14 days in the fresh and brackish water treatments. After 98 days in salt water, Leucaena seed retained almost half its original viability, whereas Bellyache bush and Navua sedge seed viability was reduced to approximately a third. Notably, Hymenachne seed viability remained relatively unaffected by immersion in all water types.

Although seeds can be transported in water, a suitable habitat is still required for germination and plant establishment and survival. By determining the length of time seed can remain viable after immersion in water, and whether immersion promotes, delays or inhibits viability gives an insight into the role water plays in the dispersal of these introduced weed species.

**Keywords** water dispersal, hydrochory, seed immersion, *Azadiractha indica, Cyperus aromaticus, Hymenachne amplexicualis, Jatropha gossypifolia, Leucaena leucocephala* ssp. *glabrata*

## INTRODUCTION

"During the spring of last year, it occurred to me that it would be worthwhile, in relation to the distribution of plants to test how long seeds could endure immersion in sea water, and yet retain their vitality. As far as I knew, this had not been tried by botanists, who would have been far more capable of doing it than myself." Darwin (1857).

Determining the seed longevity of a weed species is a fundamental requirement for success when planning control activities (Panetta 2004, Brooks and Setter 2012). Relevant studies detailing the seed longevity of weed species in soil within Tropical North Queensland, such as for Bellyache bush exist (Bebawi *et al.* 2012) but information on their longevity in water is lacking.

Water dispersal of plant propagules (hydrochory) plays a major role in the expansion and establishment of populations of many weed species (Setter *et al.* 2008). Brooks *et al.* (2017) demonstrated this by observing a strong correlation between Siam weed (*Chromolaena odorata* (L.) R.M.King & H.Rob.) plants establishment and its spread along watercourses.

Populations of some species such as Hymenachne are closely associated with wetlands, drains, creeks, and rivers (Houston 2002), while other species such as Pond apple (*Annona glabra* L.) successfully disperse via ocean currents (Mason *et al.* 2008). Water is also one of several dispersal vectors for tree species such as Neem and as a result many infestations in north Queensland are located in creeks and river systems (Setter *et al.* 2009).

Many free floating or anchored aquatic species have developed specialist mechanisms for dispersal via water. Hygrophilla (*Hygrophilla costata* Nees) disperses successfully via stem fragments (Setter *et al.* 2017) while the anchored aquatic weed Limnocharis (*Limnocharis flava* (L.) Buchenau) produces fruit which readily dehisce and float, dropping seeds as they are transported by flowing water (Weber and Brooks 2013). Others, such as Navua sedge, grow and reproduce in areas of high rainfall that are frequently inundated with freshwater, thus facilitating dispersal (Vogler *et al.* 2015).

Within many creek/river systems there exists a natural transition from freshwater through to brackish and saltwater. Within the freshwater environs ideal conditions exist for localized seed dispersal and seedling establishment. As the water becomes more saline, either periodically or more permanently, conditions for successful recruitment decline. This is the case with Pond apple which can produce fruit in a freshwater location, deposit fruit/seed into a water body for dispersal then have the seeds transported via marine currents onto a suitable beachhead for seedling establishment (Mason *et al.* 2008).

Information gained through seed longevity studies in water can be used in conjunction with known water flow rates and directions to estimate the amount of seed dispersal from an infestation, as well as likely 'destinations' for deposition of viable seed (Mason *et al.* 2008).

In this study we test the effect of immersion of seed in water on the viability of seed from five species known to have water as a dispersal vector in tropical Queensland.

## MATERIALS AND METHODS

**Treatments:** Sixteen fish tanks were filled with one of three water treatments: sea water, creek water, 50/50 sea and creek water mix (brackish). Water samples were collected from the ocean and a mountain fed creek in the Wet Tropics. An additional four tanks were not filled with water but used as an 'air' comparison with fewer retrieval times as it was anticipated that little decline in seed viability would occur over the 14-week trial period. Samples of 50 seeds were randomly selected from a field-collected bulk seed pool of each species and sown into 130 μm nylon sieve mesh bags then placed into the tanks. Retrieval times were: 2, 4, 7, 14, 28, 42, 70 and 98 days for each water treatments. Bags were randomly selected for each species from each tank at each retrieval time.

**Water Quality:** The water/air tanks were housed in a temperature-controlled environment set at 20°C with abundant external ambient light. The water remained at a constant pH 10 and salinity was measured at 0, 15 and 34 ppm for the fresh, brackish, and saltwater treatments respectively. Aeration was provided to mimic natural water conditions and assist with water /seed interaction over time.

**Germination and Viability testing:** At each retrieval time, seeds were placed in a Petri dish on top of a distilled water moistened Whatman no.5 filter paper, over an inverted watch glass which was on top of another sheet of filter paper. The filter papers were kept moistened during incubation at 20°C with 12 hrs of darkness alternating with 30°C and 12 hrs of light. Records of germination (radicle emergence of 2 mm or more) was made every 3–7 days for the duration of the trial and germinates removed. Monitoring concluded when no further germination was recorded for two consecutive weeks. Bellyache bush seed was checked for viability using the tetrazolium method (Moore 1985). Due to the small size of the seeds of the other test species, any that failed to germinate were subjected to tests of rigidity with forceps to assess viability (Borza *et.al.* 2007).

Tetrazolium testing was used to determine the viability of large ungerminated seeds. Seeds were placed into Petri dishes containing a tetrazolium chloride solution. The dishes were then wrapped in foil and incubated at 25°C for 24 hrs. Viable seeds were those that showed red staining on the embryo and those that failed to stain were deemed non-viable. (Mao *et al.* 2019).

**Statistical Analysis:** Mean seed viability and the standard error of the mean (SE) for each species in each treatment and retrieval time was calculated using Genstat (VSN International 2017) and presented graphically.

## RESULTS

Neem seed lost all viability after seven days in salt water and 14 days in the fresh and brackish water treatments. Seeds of Hymenachne, Navua sedge, Bellyache bush and Leuceana remained viable for the maximum test period of 98 days (14 weeks). After 98 days in salt water, Leucaena seed retained almost half its original viability, whereas Bellyache bush and Navua sedge seed viability was reduced to about a third (Figure 1). It is notable that Hymenachne seed viability remained relatively unaffected by immersion in any of the water types.

## DISCUSSION

This research demonstrated that seeds of all species except Neem could survive for at least 98 days in fresh, brackish and saline water (Figure 1). Seeds in the tanks either germinated in the water, died, or entered/maintained dormancy that was broken upon retrieval and viability testing, which changed conditions. High levels of salt generally either inhibit seed germination by producing low osmotic potential preventing water uptake or the salt is toxic to seeds and can kill a significant portion of immersed seeds (Guia *et al.* 2010, Vincente *et al.* 2020).

The fate of the seeds within the tanks was determined by factors such as the ability of the seed coat to remain intact, thus preventing the initiation of germination or degradation of the seed and/or tolerance to the toxic effects of salt on the embryo within the seed. For each species the effect of salt water on seed viability was generally not significant compared to that of brackish or fresh water following 98 days of immersion (Figure 1). Irrespective of whether the tested species are aquatic specialists or

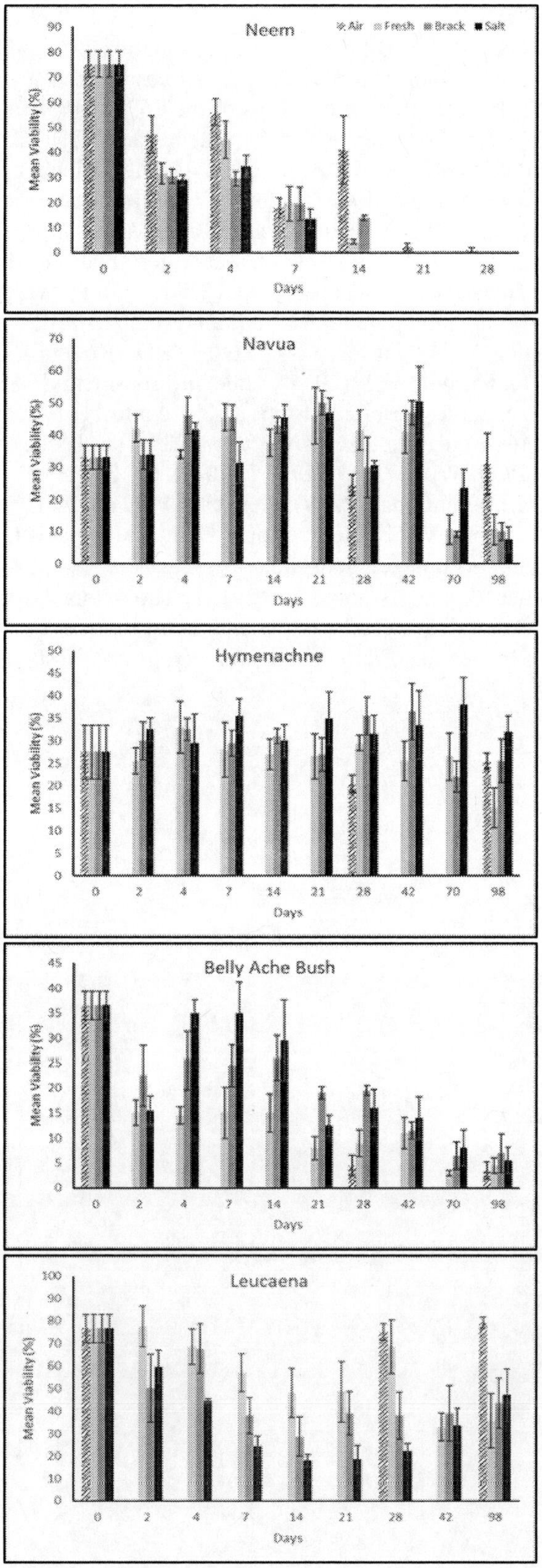


**Figure 1:** Effect of water treatments on seed viability of Neem, Navua sedge, Hymenachne, Belly Ache Bush and Leucaena. Vertical lines indicate ± 1 SE.

not they retained a significant level of viability for a considerable time. This is similar to the observations of Brooks *et al.* (2017) for Siam weed seeds after 18 weeks of immersion in similar water treatments to those used in this study

This survival of seed of each species would allow both short and long-distance water dispersal, dependent on other factors such as current flows and buoyancy of seed. Even if seeds were only buoyant for short periods, suspended in the water column or located on the stream bed they could be moved downstream by the flow of water or as bed load within the stream.

Together with viability, other factors will determine how long and how far seed may be distributed by water. For example, De Jager *et al.* (2019) found that large seeds generally disperse longer distances than smaller seeds in lowland streams, and that stream vegetation also greatly influenced dispersal distance. The results of Mason *et al.* (2008) showed how modelling the water dispersal of seeds and mapping the locations of deposition allowed for targeted management programs.

Although the duration of seed buoyancy was not quantified in this study, buoyancy along with seed survival when immersed in water is an important aspect when considering seed dispersal by water and developing effective management programs.

ACKNOWLEDGMENTS

Thanks to Michael Graham, Katie Patane and Kirsty Gough for assistance in data collection.

REFERENCES

Beames, L., Pickard, I. and Hamaguchi, N. (2017). Interim Neem (*Azadirachta indica*) Management Plan for the Kimberly, WA. Environs, Broome.

Bebawi, F.F., Campbell, S.D. and Mayer, R.J. (2012). Persistence of Bellyache bush (*Jatropha gossypiifolia* L.) soil seed banks. *The Rangeland Journal* 34, 429–438.

Borza, J., Westerman, P. and Liebman, M. (2007). Comparing estimates of seed viability in three Foxtail (*Setaria*) species using the imbibed seed crush test with and without additional tetrazolium testing. *Weed Technology* 21(2), 518–522.

Brooks, S.J. and Setter, S.D. (2012). Soil seed bank longevity information for weed eradication target species. *Pakistan Journal of Weed Science Research* 18, 73–83.

Brooks, S.J., Setter, S.D. and Gough, K.L. (2017). Siam Weed Dispersal Mechanisms. In:

Proceedings of the 14th Queensland Weed Symposium, Port Douglas, Queensland.

Campbell, S.D., Vogler, W., Brazier, D.A., Vitelli, J.S. and Brooks, S.J. (2019). Weed leucaena and its significance, implications and control. Tropical Grasslands. *Forrajes Tropicales* 7(4), 280–289.

Darwin, C.R. (1857). On the action of seawater on the germination of seeds. *Journal of the Proceedings of the Linnean Society of London (Botany)* 1, 130–140.

de Jager, M., Kaphingst, B., Janse, E.L., Buisman, R., Rinzema, S.G.T., Soons, M.B. (2019). Seed size regulates plant dispersal distances in flowing water. Journal of Ecology 107, 307–317.

Guja, L.K., Merritt D.J. and Dixon K.W. (2010). Buoyancy, salt tolerance and germination of coastal seeds: implications for oceanic hydrochorous dispersal. *Functional Plant Biology* 37, 1175–1186.

Houston, W.A. and Duivenvoorden, L.J. (2002). Replacement of littoral native vegetation with the ponded pasture grass *Hymenachne amplexicaulis*: effects on plant, macroinvertebrate, and fish biodiversity of backwaters in the Fitzroy River, Central Queensland, Australia. *Marine and Freshwater Research* 53, 1235–1244.

Mason, L.B., Setter, M.J., Setter, S.D., Hardy, T. and Graham, M.F. (2008). Ocean dispersal modelling for propagules of pond apple (*Annona glabra* L.). In: Proceedings of the 16th Australian Weeds Conference, Cairns, Queensland. 519–521.

Mao, R., Nguyen, T.L.T., Osunkoya, O.O. and Adkins, S.W. (2019). Spread pathways of the invasive weed *Parthenium hysterophorus* L.: The potential for water dispersal. *Austral Ecology* 44(7), 1111–1122.

Moore RP (1985) 'Handbook on tetrazolium testing.' (International Seed Testing Association: Zurich, Switzerland)

Panetta, F.D. (2004). Seed banks: The bane of the weed eradicator. Proceedings of the 14th Australian Weeds Conference, Wagga Wagga, NSW, Australia, 6–9 September 2004. 523–526.

Setter, M.J., Setter, S.D. and Styman, D.T. (2017). Survival and buoyancy of *Hygrophila costata* stem fragments in salt, brackish and fresh water. In: Proceedings of the 14th Queensland Weed Symposium, Port Douglas, Queensland

Setter, S.D and Patane, K.A. (2009). The spread of neem (*Azadirachta indica*) within a tropical riparian system. In: Proceedings of the 10th Queensland Weed Symposium, Yeppoon, Queensland.

Setter, S.D., Setter, M.J., Graham, M.F. and Vitelli, J.S. (2008). Buoyancy and germination of pond apple (*Annona glabra* L.) propagules in fresh and salt water. In: Proceedings of the 16th Australian Weeds Conference, Cairns, Queensland. 140–42.

Vicente, M.J., Martínez-Díaz, E., Martínez-Sánchez, J.J., Franco, J.A., Bañón, S. and Conesa, E. (2020). Effect of light, temperature, and salinity and drought stresses on seed germination of *Hypericum ericoides*, a wild plant with ornamental potential. *Scientia Horticulturae* 270.

Vogler, W.D., Carlos, E.H., Setter, S.D., Roden, L. and Setter, M.J. (2015). Halosulfuron-methyl: A selective herbicide option for control of the invasive *Cyperus aromaticus* (Navua sedge). *Plant Protection Quarterly* 30(2), 61–66.

VSN International (2017). Genstat *for Windows* 19th Edition. VSN International, Hemel Hempstead, UK. Web page: Genstat.co.uk

Weber, J. and Brooks, S. (2013). The biology of Australian weeds 62. *Limnocharis flava* (L.) Buchenau. *Plant Protection Quarterly* 28(4), 101–113.

# Effect of cultivar and seed weight on triticale competitiveness with annual ryegrass, *Lolium rigidum*

Husam Khalaf[1], Brian Sindel[1], Paul Kristiansen[1], Craig Birchall[1], Robin Jessop[1]
[1]University of New England, Armidale, Australia
(bsindel@une.edu.au)

**Summary** Herbicide resistance in weeds, particularly in annual ryegrass (*Lolium rigidum*), has been a major challenge in Australian cereal cropping systems over the last few decades. As a result, cultural weed control methods have increased in importance to complement herbicide strategies. One such method has been to grow more competitive crop species and varieties. Triticale (× Triticosecale) has high tolerance to environmental stress but has received limited research attention for improving competitive ability with weeds. One factor that has shown potential for enhancing crop competitive ability is that of seed size or seed weight. A glasshouse experiment was conducted to study the effect of crop seed size of three triticale cultivars (Tuckerbox, KM10 and Bogong) and one comparative wheat cultivar (Spitfire) on their competitive ability with annual ryegrass. Seeds of each cultivar were divided into three seed size categories (small, standard and large), and grown in competition with ryegrass (weed-free (4:0), weedy (4:3 and 4:6) plants per pot) up to crop anthesis. The larger crop seed size increased crop height, leaf area and biomass, and significantly reduced ryegrass dry weight by 20%. In terms of crop or weed biomass, seed size did not interact with cultivar or competition level, indicating that the seed size effect was consistent. These findings suggest that selecting larger triticale seeds through more comprehensive grading is likely to confer the crop with greater competitive advantage against annual ryegrass (and presumably other weeds) during their early growth. Triticale was also less sensitive to ryegrass competition than the comparative wheat variety. More detailed studies are needed with a larger variety of wheat and triticale cultivars under both glasshouse and field conditions to better understand how, why and when large seeded crop plants are more competitive with weeds, and how they can be better utilised under Australian farming conditions.

**Keywords** Triticale, competition, annual ryegrass, seed weight

# Management of Globe Chamomile (*Oncosiphon piluliferum*)

Alexandra Douglas[1], David Nicholson[1]
[1]Department of Primary Industries and Regional Development, Australia
(alex.douglas@dpird.wa.gov.au)

**Summary** Matricaria (*Oncosiphon piluliferum*) is a serious weed in the eastern grain-belt of Western Australia. Effective management of matricaria in pastures is often poor as stock have a tendency to avoid grazing it and there are limited herbicide options available. This, plus the reduction in cultivation during crop establishment is thought to have favoured weed build-up. There are some herbicide options available for cereal crops, where herbicide efficacy is helped by crop competition. Field trials were conducted from 2016 to 2020 to investigate a range of herbicide options and application timings to control *O. piluliferum* in both medic and sub-clover based pastures. A number of highly effective options are available for herbicide management depending on the farming system. These include the application of selective herbicides. There are currently four selective herbicides available for matricaria control in pasture. Selective herbicides should be applied to small plants (six to eight leaves, 8cm rosette). Knockdown herbicides may work best when mixed with other herbicides to act as a 'spike' in fallow situations. The best time for applications targeting seed set are before plants are fully flowering, seed viability can be reduced (by up to 99%) when non-selective herbicides are applied during the flowering stage. Both glyphosate and paraquat can be used for seed set control.

**Keywords** Oncosiphon piluliferum, management, pastures, herbicides, seed viability

# Interactive effect of high temperature and reduced soil moisture availability on the morpho-physiological traits and glyphosate susceptibility of windmill grass (*Chloris truncata* R.Br.)

Arslan Masood Peerzada[1], Alwyn Williams[1], Chris O'Donnell[1], Steve Adkins[1]
[1]The University Of Queensland, Brisbane, Australia
(a.peerzada@uq.edu.au)

**Summary** Susceptibility to herbicides is not a constant trait of a species; rather, it varies with variation in both environmental and intrinsic factors. Therefore, two windmill grass (*Chloris truncata* R.Br.) biotypes with differential glyphosate tolerance were exposed to continuous ambient [aT – 28/20 ± 2°C] and elevated temperature [eT – 34/24 ± 2°C] in combination with either high [HSM – 90-100% of soil water holding capacity (WHC)], moderate (MSM – 50-60% of soil WHC) or reduced (RSM – 20-30% of soil WHC) soil moisture availability to quantify changes in their morpho-physiology and susceptibility to glyphosate. The primary objective was to illustrate the importance of combining multiple stressors to partially explain weed's herbicide tolerance in relation to climate change. In the current study, we demonstrated that the glyphosate susceptible biotype showed 4.9 times more tolerance to glyphosate at RSM under eT (LD50 = 600.5 g a.i. ha-1) as compared to HSM under aT (LD50 = 123.1 g a.i. ha-1). On the other hand, glyphosate within the recommended rate (740 g a.i. ha-1) was insufficient to suppress (>80%) tolerant biotype at both MSM and RSM under eT. Plants of both biotypes grown under hot, dry conditions produced fewer, smaller, and thicker leaves with increased leaf chlorophyll content and reduced stomatal conductance. The morpho-physiological changes, particularly in leaf surface characteristics, in response to RSM under eT, could have possibly impacted collectively on non-target-site mechanisms of herbicide tolerance (e.g., herbicide interception, distribution, absorption, translocation, and metabolism) and thus reduced glyphosate efficacy on *C. truncata*.

**Keywords** Chemical weed control; fallow weed management; climate change; herbicide efficacy

# Recent advances in field releases of environmental weed biocontrol agents

Ben Gooden[1], Gavin Hunter[1], Raelene Kwong[2], Andrew McConnachie[3], Hillary Cherry[4], Louise Morin[1]
[1]CSIRO Health & Biosecurity, Canberra, Australia,
[2]Department of Jobs, Precincts and Regions, Bundoora, Australia,
[3]Department of Primary Industries, NSW within Department of Regional New South Wales, Orange, Australia,
[4]Environment, Energy and Science, Department of Planning and Environment, Parramatta, Australia
(Ben.Gooden@csiro.au)

**Summary** The typical goals of biological control (biocontrol) programs for environmental weeds are to improve biodiversity and environmental benefits at site and landscape scales and reduce or eliminate the need and/or frequency of non-biocontrol weed management tactics. Since 2016, the NSW Environmental Trust ('The Trust') has supported an on-going research project for the biocontrol of environmental weeds that impact NSW. The project is overseen by a consortium of CSIRO (project lead), NSW Department of Primary Industries and NSW Department of Planning and Environment, and involves other research providers from Australia or overseas as required. The project has focused on previously identified, promising biocontrol agents for priority weeds, which have potential for use in Australia, but require additional research to demonstrate their safety before an application for their release in Australia can be submitted to the relevant authorities. In this poster, we present an update on the current subprojects that target environmental weeds for which biocontrol agents have recently been approved for release into the Australian environment: the smothering, shade-tolerant herb wandering trad (*Tradescantia fluminensis*) with a leaf-smut fungus (*Kordyana brasiliensis*); the coastal dune herb sea spurge (*Euphorbia paralias*; Euphorbiaceae) with a foliar blight fungus (*Venturia paralias*); the large, thorny shrub African boxthorn (*Lycium ferocissimum*; Solanaceae) with a rust fungus *Puccinia rapipes*; the invasive cactus Hudson pear (*Cylindropuntia pallida*; Cactaceae) with the cochineal insect agent *Dactylopius tomentosus* ('*californica* var. *parkeri*' lineage); and the emergent aquatic weed sagittaria (aka delta arrowhead, *Sagittaria platyphylla*, Alismataceae) with the fruit-feeding weevil *Listronotus appendiculatus*. The poster will summarise progress made to date with culturing and mass-releasing each agent into the Australian environment, and will be accompanied by leaflets with background information on biocontrol agent research and advice on how interested stakeholders can participate in release programs. Representatives from some of the subprojects will be in attendance to meet and greet with interested participants.

**Keynotes** Biological control, environmental weeds, prioritization, New South Wales

# Non-Chemical *Cyperus iria* weed management through rice densities and weed emergence times in dry-seeded rice eco-system

Tahir Awan[1], Bhagirath Chauhan
[1]Rice Research Institute Kala Shah Kaku, Lahore, Kala Shah Kaku, Pakistan
(tahirawanrri@gmail.com)

**Summary** *Cyperus iria* is one of the most threatening weeds of rice in Sri Lanka, India and the Philippines. A broad knowledge about ecology and fecundity of *C. iria* is important for its effective management. Field studies were conducted over two seasons {wet season (WS) and dry season (DS)} to understand the influence of *C. iria* densities (40 and 80 plants m-2) on its growth, survival, and fecundity, with varying emergence times of 3, 15, 30, and 45 d after rice emergence (DARE) during 2013. We hypothesized that (a) high plant density of weed produces more biomass and fertile seeds per unit area, (b) interference of rice decreases the biomass and fecundity of the weed, and (c) a delay in weed emergence reduces the % survival and soil seed bank. The results indicated that rice interference decreased *C. iria* growth and seed production as compared to the plants grown without rice interference. A linear decrease in the percent survival of *C. iria* without rice and sigmoid decrease with rice was observed during both the seasons. Plant height of *C. iria* was moderately affected up to 30 DARE and a significant reduction was observed at 45 DARE. Likewise, with a delay in emergence of each *C. iria* cohort relative to rice, tiller number and shoot biomass per plant declined in a linear manner in the DS and exponential manner in the WS. There was a linear relationship between *C. iria* shoot biomass and the number of seeds plant -1, across rice seeding rate, weed density, and emergence time. *C. iria* seed production, 1000-seed weight, and seed yield were greater when seedlings emerged with the crop (3 DARE), relative to the late-emerging weed cohorts. Under rice weed interference growth, production of viable seeds was completely stopped at 45 DARE. Seed germination of the first two *C. iria* cohorts (3, and 15 DARE) was 89% in DS and 49% in WS. Third cohort (30 DARE) produced viable and germinable seed in the DS and was unable to produce seed in the WS. The delay in emergence of *C. iria* up to 45 DARE was unable to produce seed in both seasons. The results of the current studies advocate that the emergence, weed biomass and seed production of *C. iria* can be checked by adopting suitable cultural weed management practices, which can delay the emergence of weed relative to rice These approaches that make the associated crop more competitive will be useful in integrated weed management programs, and are thus valuable for hindering seed rain to the seed bank by noxious weeds in the field.

# Before the bulldozers go in: reducing the risk of new and emerging weeds in Leigh Creek

Matthew Westover[1]
[1]SA Arid Lands Landscape Board, Port Augusta, Australia
(matthew.westover@sa.gov.au)

**Summary** At its peak, the mining town of Leigh Creek had a population of over 2,500. Following the closure of the Coal Mine in 2016 a plan was released in December 2021 to significantly reduce the footprint of the town to accommodate approximately 150 residents who remain. This will involve the removal of a large number of houses and fences. The majority of houses in Leigh Creek were vacated in 2016 and with much of the town earmarked for demolition, there is a risk that remaining garden plants could persist and escape into the landscape becoming weeds. Water was provided free of charge to residents which led to some extensive and elaborate gardens, home to a number of exotic plant species. Surveillance of surviving garden plants at Leigh Creek commenced in September 2019, with further surveys in October 2020 and June 2021. These were undertaken with Shannon Robertson (Facilitator newly established weeds, PIRSA) and Chris Brodie (Weeds Botanist, State Herbarium of South Australia). Surveys were conducted via inspections of abandoned yards in the township, roadsides, drainage lines and tourism sites around the town. Community engagement achieved through workshops, property visits and an open invite for community members to join the surveys created a local surveillance network. This resulted in community members bringing unknown plant species to our attention resulting in the detection of previously unknown weed populations or potential weeds. The surveys and community engagement resulted in a number of significant weed and potential weed species discoveries, including the first naturalised record for Australia of *Tephrocactus articulatus* (Pine Cone Cactus). Detection of plants provided the opportunity for their control early on in the invasion curve. Pine Cone Cactus and other discoveries have since been controlled and a long-term monitoring plan has been developed to re-visit high risk sites into the future.

**Keywords** New and emerging weeds, community engagement, Leigh Creek, surveillance, garden plant

# Weeding it out: The Australian online trade of invasive aquatic plants

Lisa Wood[1], Jacob Maher[1], Phill Cassey[1], Stephanie Moncayo[1], Oliver Stringham[1]
[1]University of Adelaide, Adelaide, Australia
(woodclisa@gmail.com)

**Summary** In Australia, the trade of plants via e-commerce websites is commonplace and widespread, yet the overall scale, diversity and underlying dynamics of the trade are poorly understood. Accordingly, the trade has rapidly emerged as a major pathway for the spread and establishment of weed species. Aquatic plants, in particular, have been identified as popular across several publicly accessible e-commerce websites, where they are sold primarily for use in ponds and aquariums. They are also a group with considerable invasion concern, with many species historically traded now recognised as invasive and declared in parts or all of Australia. We focused on the trade of five invasive aquatic species and one genus (*Eichhornia crassipes*, *Limnobium laevigatum*, *Pistia stratiotes*, *Salvinia molesta*, *Salvinia minima* and other *Salvinia* spp.). These species are declared in at least one State or Territory, and preliminary analysis indicated presence in the online trade. Trade data for these taxa was collected using webscraping technology that records advertisements from a popular Australian classifieds website. Search terms based on scientific, common, and trade names were matched with listing text to detect target taxa, with listing images used to visually confirm the detection. We found evidence of extensive online trade of the invasive aquatic plants, including 194 illegal advertisements. From all advertisements, most sellers only listed one or two advertisements, suggesting the majority of trade is done casually. However, the presence of a small number of sellers with many advertisements indicates that some people may be trading for profit. Our results indicate that the online trade of aquatic plants is an ongoing biosecurity risk for Australia. State legislation appears to decrease illegal trade, however given that illegal trade is still occurring, additional measures such as targeted education and enforcement of repeat offenders is recommended.

**Keywords** Online trade, e-commerce, web-scraping, aquatic weeds

# Maximising Control Effectiveness Using Prescribed Burning for Control of Sweet Pittosporum (*Pittosporum undulatum*)

Dale Fuller[1], Matt Russell, Ruby Wicks
[1]Parks Victoria, Lorne, Great Otway NP, Australia
(dale.fuller@parks.vic.gov.au)

**Summary** Based on the local fuel reduction burn schedule and working in collaboration with the Lorne Fire and Emergency Team, Project Officers were able to identify a large well established Sweet Pittosporum (*Pittosporum Undulatum*) infestation within a burn area. This infestation was mostly ranked as a 'heavy infestation' based on Weeds of Early Stages of Invasion (WESI) guidelines.

Due to the knowledge gained through consultation with industry experts around susceptibility of juvenile Sweet Pittosporum (less than 1.5m high) to fire events, contractors and Staff were able to focus works towards mature and dense stands, treating a larger total area. 53 hectares of Sweet Pittosporum was treated at this heavily infested site. The fuel reduction burn was an early season burn held in February with moderate intensity and approx 90% coverage. One year post burn the infestation has now been reclassified as a 'trace infestation'. This has enabled future control effort to come at a minimal cost and significantly less time to maintain low population levels with the aim of longer term eradication from the site. For the poster, the intent is to display photopoints of before control works and after control works, immediately post burn and 1 year post burn and a map of the site.

**Keywords** Sweet Pittosporum, *Pittosporum Undulatum*, fire

# Optimal barley grass management

Catherine Borger[1], Sam Stubna[2], Ben Whisson[3], Tiarna Kanny[4], Brad Joyce[5], Amy Bowden[6] and David Minkey[7]
Department of Primary Industries and Regional Development, PO Box 483, Northam WA 6401, Australia
South East Premium Wheat Growers Association, PO Box 365, Esperance WA 6450, Australia
Lakes Information and Farming Technology, PO Box 142, Kulin WA 6365, Australia
Mingenew Irwin Group, 54 Midlands Rd, Mingenew WA 6522, Australia
Kellerberrin Demonstration Group, PO Box 321, Northam WA 6401, Australia
FACEY Group, 40 Wogolin Rd, Wickepin WA 6370, Australia
Western Australian No-Till Farmers Association, Private Bag 5, Wembley WA 6913, Australia

**Summary** Grower group driven integrated barley grass management trials were conducted at five sites in Western Australia in 2021. Results indicate that it is possible to prevent seed production of barley grass, if management strategies include both pre-emergent and post-emergent tactics. Break crops or pastures offer an excellent opportunity to include a range of different herbicides and control tactics in a management plan.

**Keywords** *Hordeum leporinum*, *Hordeum glaucum*, panicles, seed production.

## INTRODUCTION

Barley grasses (*Hordeum leporinum* Link or *H. glaucum* Steud.) are one of the most prominent weeds of cropping and pasture systems in Western Australia (Borger et al. 2012). These species are increasingly difficult to control because a) herbicide resistance is increasing, b) many herbicides offer suppression rather than control and c) there is evidence that ecotypes are developing delayed or staggered emergence to avoid pre-sowing and pre-emergent herbicides (Borger et al. 2012, Gill et al. 2021). The staggered emergence indicates that management plans with late season control will be necessary to prevent seed production.

A national GRDC grower group driven project investigated optimal management strategies for local growers in WA, SA and NSW, from 2019 to 2021. In WA, the grower groups LIFT, SEPWA, KDG, MIG and FACEY/WANTFA investigated optimal pre-emergent herbicides, crop competitive ability, post-emergent herbicides, and break crops for optimal barley grass control.

## MATERIALS AND METHODS

Each grower group identified a trial site on farm in 2019 to determine the best strategy for optimal control of barley grass over three years (Table 1). Choice of crop rotation at each site was generally dictated by what each grower planned for the field in question. Each trial considered barley grass management plans ranging from inexpensive to more expensive options. All trials were replicated, in a randomised block design. Note that the FACEY/WANTFA trial failed due to poor barley grass establishment in 2019 and was run in 2020 and 2021. Unlike the other sites, the FACEY/WANTFA trial was in a new site each year. The site also used a split plot design, with sowing rate as the main plot factor and herbicides as the sub-plot factor.

For each trial, the weed control treatments applied in 2021 (including herbicide and dates) are specified in the tables in the results section (Table 2-4). Measurements included initial barley grass and crop density, and barley grass panicle number, from 4-10 quadrats per plot. Quadrat size varied from 10 by 25 cm to 50 by 50 cm, depending on weed density. Twenty mature panicles were collected from each plot, and seeds were manually removed and counted to determine seeds per panicle. This value and panicles $m^{-2}$ was used to determine seeds $m^{-2}$. The harvested seeds were after-ripened in an open glasshouse over summer, and then germinated in moist petri dishes in a germination cabinet with a 12 h temperature cycle of 12/20°C. After two weeks, ungerminated seeds were exposed to a tetrazolium chloride test, to determine total seed viability. Crop yield was determined by harvesting the entire plot.

This paper provides a summary of the third year of results, but full results of all trial sites including trial design, plot size and spray application methods can be found at GRDC Online Farm Trials (https://www.farmtrials.com.au/).

## RESULTS

**KDG group** In the 2021 pasture, the barley grass density and panicle number was not significantly reduced by the imazamox herbicide at the 2-4 leaf stage or Z31 (Table 2). However, barley grass seed production was reduced by the herbicide compared to the control, with less seed production following herbicide at Z31 compared to 2-4 leaf stage. Slashing removed all panicles before maturity and prevented seed set (Table 2). Pasture biomass was high in all treatments and was not affected by barley grass control.

**Table 1.** Trial site details including grower group, grower, location, and 2021 rainfall, crop sowing details, treatments and harvest date.

| Group | KDG | MIG | SEPWA | FACEY/ WANTFA* | LIFT |
|---|---|---|---|---|---|
| Grower | Gavin Morgan | Soullier family | Harris family | Gary Lang | Ashton Gray |
| Location | North Kellerberrin | Yandanooka | Esperance | Wickepin | Tarin Rock |
| 2021 annual (and growing season) rainfall | 326mm (223mm) | 433mm (361mm) | 385mm (269mm) | 464mm (370mm) | 426mm (356mm) |
| 2021 crop sowing details | Volunteer pasture (wheat and clover) | Canola cv. InVigor®, 2kg $ha^{-1}$, 16 Apr 2021 | Vetch cv. Volga, 40kg $ha^{-1}$, 22 Apr 2021 | Barley cv. Maximus, 40, 80 or 120kg $ha^{-1}$, 3 Jun 2021 | Oats cv. Wandering, 45kg $ha^{-1}$, 29 Apr 2021, 17 Jun 2021 |
| Treatments 2021 | 1. Untreated<br>2. Imazamox at 2-4 leaf<br>3. Imazamox at 2-4 leaf stage + slashed<br>4. Imazamox at Z31<br>5. Imazamox at Z31 + slashed<br>6. Slashed | 1. Glyphosate<br>2. Trifluralin, glyphosate<br>3. Glyphosate, quizalofop-p-ethyl<br>4. Trifluralin, glyphosate, quizalofop-p-ethyl | 1. Trifluralin, quizalofop-p-ethyl<br>2. Trifluralin+diuron, quizalofop-p-ethyl<br>3. Trifluralin+diuron, quizalofop-p-ethyl+clethodim<br>4. Carbetamide, quizalofop-p-ethyl+clethodim | Main plot: sowing rate of 40, 80 or 120 kg $ha^{-1}$.<br>Sub-plot:<br>1. Trifluralin 960 g a.i. $ha^{-1}$<br>2. Trifluralin 1440 g a.i. $ha^{-1}$ | 1. Early sowing<br>2. Late sowing |
| 2021 harvest | NA | 19 Oct 2021 | NA | NA (header fire) | 15 Dec 2021 |

*Note: in the original trial plan, the sub-plot factor was each rate of trifluralin with or without imazamox/imazapyr 12.3/5.6 g a.i. $ha^{-1}$ (Intervix®) post-emergence. However, the 2021 season was too wet to allow in-crop spraying at this site.

**MIG group** Barley grass density and subsequent seed production in the 2021 canola crop was lower in all treatments that included multiple herbicides, compared to two applications of glyphosate alone (Table 3). Barley grass panicle number or canola density and yield were not affected by herbicide treatments.

**SEPWA group** Barley grass density was similar in all treatments, but panicle and seed production were greatest following trifluralin pre-emergent and quizalofop-p-ethyl post-emergent, and reduced in subsequent treatments (Table 4). There was a slight reduction in vetch density following carbetamide pre-emergent and quizalofop-p-ethyl + clethodim post-emergent, but there was no difference in vetch biomass.

**FACEY/WANTFA group** The crop was sown into moist soil, and both rates of trifluralin provided a high initial rate of barley grass control. Crop density increased with seeding rate (94, 159 and 200 plants/$m^2$ at seeding rates of 40, 80 and 120kg/ha, P: 0.003, LSD: 34.4). Increasing crop density also reduced barley grass density (6.8, 2.3 and 2.5 barley grass/$m^2$, P: 0.015, LSD: 2.01). However, further barley grass cohorts emerged late in the season. As stated (see note in Table 1), seasonal conditions made it impossible to apply herbicide post-emergent, and so by the end of the season there was no difference in barley grass panicle and seed number between treatments.

**Table 2.** KDG barley grass density, panicles, and viable seeds, as well as pasture biomass for each treatment in the 2021 pasture. P and LSD values are included for separation of means. Note that barley grass density, panicle and seed production means are back-transformed from a $log_{10}$+1 transformation.

| Treatment* | Barley grass density $m^{-2}$ | Barley grass panicles $m^{-2}$ | Barley grass seeds $m^{-2}$ | Pasture biomass (t $ha^{-1}$) |
|---|---|---|---|---|
| Untreated | 102 | 597 | 2343 | 3.2 |
| Imazamox at 2-4 leaf stage | 32 | 486 | 691 | 2.8 |
| Imazamox at 2-4 leaf stage + slashed | 45 | 1 | 0 | 2.6 |
| Imazamox at Z31 | 109 | 357 | 218 | 1.6 |
| Imazamox at Z31 + slashed | 58 | 0 | 0 | 2.5 |
| Slashed | 101 | 0 | 0 | 2.3 |
| P | 0.201 | 0.004 | <0.001 | 0.246 |
| LSD | 3.2 | 369.1 | 229.1 | 2.51 |

*Herbicide formulations and application dates: imazamox 31.5 g a.i. $ha^{-1}$ (Raptor®) at 2-4 leaf stage of barley grass (11 July 2021) or at Z31 stage of barley grass (2 July 2021); slashing (20 September 2021).

**Table 3.** MIG canola density, barley grass density, panicles, and viable seeds, as well as yield in the 2021 canola crop. P and LSD values are included for separation of means. Note that barley grass panicle data is back-transformed from a square root transformation, and 'NS' indicates 'not significant'.

| Herbicide* | Canola density $m^{-2}$ | Barley grass density $m^{-2}$ | Barley grass panicles $m^{-2}$ | Barley grass seeds $m^{-2}$ | Canola yield (t $ha^{-1}$) |
|---|---|---|---|---|---|
| Glyphosate | 26 | 14.2 | 6.1 | 281 | 2.0 |
| Trifluralin, glyphosate | 24 | 6.9 | 1.3 | 49 | 2.0 |
| Glyphosate, quizalofop-p-ethyl | 27 | 4.2 | 0.5 | 12 | 1.9 |
| Trifluralin, glyphosate, quizalofop-p-ethyl | 26 | 1.7 | 0 | 0 | 1.9 |
| P | 0.860 | 0.004 | 0.114 | 0.045 | 0.522 |
| LSD | NS | 5.56 | NS | 195.8 | NS |

*Herbicide formulations and application dates: trifluralin 720 g a.i. $ha^{-1}$ (TriflurX®) pre-emergent (15 April 2021); glyphosate 621 g a.i. $ha^{-1}$ (Roundup Plantshield®) applied twice, at 2 leaf (14 May 2021) and tillering (8 June 2021); quizalofop-p-ethyl 50 g a.i. $ha^{-1}$ at 3-5 leaf (14 June 2021).

**Table 4.** SEPWA vetch density, barley grass density, panicles, and seeds, as well as biomass in the 2021 vetch pasture. P and LSD values are included for separation of means. Note that barley grass density, panicle and seed production data is back-transformed from a log+1 transformation, and 'NS' indicates 'not significant'.

| Herbicide* | Vetch density $m^{-2}$ | Barley grass density $m^{-2}$ | Barley grass panicles $m^{-2}$ | Barley grass seeds $m^{-2}$ | Vetch biomass (t $ha^{-1}$) |
|---|---|---|---|---|---|
| Trifluralin, quizalofop-p-ethyl | 47 | 1.5 | 19.3 | 588 | 4.6 |
| Trifluralin + diuron, quizalofop-p-ethyl | 45 | 0.9 | 7.1 | 185 | 3.6 |
| Trifluralin + diuron, quizalofop-p-ethyl + clethodim | 45 | 0 | 0 | 0 | 4.3 |
| Carbetamide, quizalofop-p-ethyl + clethodim | 41 | 0 | 0.5 | 4 | 5.0 |
| P | 0.031 | 0.261 | 0.001 | <0.001 | 0.196 |
| LSD | 3.88 | NS | 2.46 | 7.9 | NS |

*Herbicide formulations and application dates: trifluralin 576 g a.i. $ha^{-1}$ (Treflan®) pre-emergent (22 April 2021); diuron 450 g a.i. $ha^{-1}$ pre-emergent (22 April 2021), carbetamide 990 g a.i. $ha^{-1}$ (Ultro®) pre-emergent (22 April 2021); quizalofop-p-ethyl 25 g a.i. $ha^{-1}$ post-emergent (12 June 2021); clethodim 120 g a.i. $ha^{-1}$ post-emergent (12 June 2021).

**LIFT group** The time between early and late sowing in the 2021 oat crop was 7 weeks (Table 1). Initial oat density was much lower in the late sown plots, due to cold, water-logged conditions in June (average density of 115 and 51 oat plants m$^{-2}$, P<0.001, LSD: 11.5). As a result of this poor emergence, oat yield was lower for late sown plots (3.2 and 1.8 t ha$^{-1}$, P<0.001, LSD: 0.39).

Late sowing controlled all barley grass. Initial barley grass density was much lower in the late sown plots (275 and 0.3 barley grass m$^{-2}$, P<0.001, LSD: 98.8). As a result, late sowing reduced barley grass panicles and seed set to zero (178 or 0 panicles m$^{-2}$ in the early and late sown plots, P<0.001, LSD: 55.4, and 3466 or 0 seeds m$^{-2}$ in the early and late sown plots, P<0.001, LSD: 1328.7).

## DISCUSSION

This research concludes that multiple control tactics to target early and late season barley grass emergence is necessary to prevent seed production. The MIG and SEPWA sites both demonstrated that a combination of pre-emergent and post-emergent herbicides are required for zero seed set. The FACEY/WANTFA site failed to achieve seed set control because it was not possible to use post-emergent weed control. The LIFT group site achieved zero seed production using only delayed seeding, but seven weeks is an unusually large delay to seeding and was highly detrimental to crop yield. Earlier trials indicated that a four week delay was not sufficient to control barley grass (Borger and Whisson 2021).

Staggered cohorts and development of delayed emergence to avoid pre-seeding herbicides has previously been noted in barley grass in South Australia (Fleet and Gill 2012; Gill et al. 2021). Delayed emergence due to enhanced dormancy is less common in Western Australian populations of barley grass (Gill et al. 2021). However, dry sowing and warmer autumn conditions (reduced opportunity for early cold stratification of seeds) ensures that delayed and staggered emergence will become more common (Gill et al. 2021). If barley grass emergence is delayed, the competitive ability of this species will be reduced compared to other weed species. However, delayed emergence means that late season control is required for effective management of all barley grass cohorts (Gill et al. 2021).

The KDG found that late season control alone (slashing) in pasture was sufficient to prevent seed set. However, slashing or spray topping in pasture varies due to seasonal conditions and potential regrowth. The barley grass RIM model suggests these tactics only control 70% of the population (Monjardino et al. 2022).

In cereal crops, late season control is difficult. With the spread of resistance there are fewer options for selective control in cereal crops (Borger et al. 2012; Gill et al. 2021). Crop topping and harvest weed seed control has limited impact on barley grass due to variable maturity and early shedding (Gill et al. 2021). Break crops like canola or pasture/vetch rotations offer a wide range of herbicides, from different modes of action, including in-crop selective herbicides to control late emerging cohorts.

## ACKNOWLEDGMENTS

This research was made possible through the significant contributions of growers through trial cooperation and GRDC investment in the project UOA1903-004SAX Demonstrating and validating the implementation of integrated weed management strategies to control barley grass in low rainfall zone farming systems. Thanks are due to Nerys Wilkins and Pete Gray (DPIRD) for technical assistance, and staff at the Grower Groups running the trials including Michelle Handley, Niki Curtis and Andrea Carmody (SEPWA), Greg Warren and Richard Scott (Farm & General) Madi George (MIG), and Justine Tyson (ConsultAg).

## REFERENCES

Borger C, Whisson B (2021) Control of barley grass in a pasture-barley-oats rotation. https://www.farmtrials.com.au/trial/31015. Accessed 2 May, 2022

Borger CPD, Michael PJ, Mandel R, Hashem A, Renton M (2012) Linking field and farmer surveys to determine the most important changes to weed incidence. Weed Res. 52:564-574

Fleet B, Gill G (2012) Seed dormancy and seedling recruitment in smooth barley (Hordeum murinum ssp. glaucum) populations in southern Australia. Weed Sci. 60:394-400

Gill G, Borger C, Chauhan B (2021) Ecology of major emerging weeds. Kingston ACT, Australia: Grains Research and Development Corporation.

Monjardino M, Llewellyn R, Gill G, Kleemann S (2022) Barley grass RIM: barley grass integrated management. https://www.ahri.uwa.edu.au/our-research/wim-weed-management-models/barley-grass-rim/. Accessed 6 April, 2022

# Australian weed morphology and its potential impact on electric weed control application efficacy

Miranda Slaven[1] and Catherine Borger[1]
[1]Department of Primary Industries and Regional Development, PO Box 483, Northam WA 6401, Australia

**Summary** Weed management is essential throughout Australia, however, the implementation of current mechanical and chemical weed control methods has been impeded by numerous factors. Therefore, Australia needs to consider alternatives, including electric weed control. However, this technology is untested in Australian conditions and a multitude of variables will affect the technology's efficacy which are yet to be analysed. One such variable is the weed's morphology which may impact application efficacy. Results of two pot trials conducted by DPIRD indicate that volunteer crops and winter weed species may be harder to control based on their morphology. Therefore, while electric weed control offers a new alternative weed control method for Australian systems, the morphology of the weeds treated will need to be considered to obtain optimum efficiency.

**Keywords** *Electroweeding, electrophysical, weed management, non-chemical, weed morphology.*

## INTRODUCTION

Electric weed control or 'electroweeding' is where an electrical current is transferred through the target plant following plant-electrode contact (Vigneault and Benoit, 2001). This causes a pressure build up as the liquids inside the cells vaporise, causing them to rupture, killing the plant (Diprose et al., 1980). Electroweeding has proven a popular weed control method globally with numerous companies producing machines for various settings and interest is growing for their expansion into the Australian market. However, a multitude of variables will affect this technology's applicability and efficacy for Australia, and many are yet to be thoroughly researched.

One such variable is that of the weed's morphology, which alters the vegetative resistance and therefore, the energy threshold required to ensure the plant's complete death (Diprose and Benson, 1984, Diprose et al., 1980, Vigneault and Benoit, 2001). Morphological factors theoretically affecting efficiency of electric weed control include plant growth stage, shoot and root biomass as well as surface area (Diprose and Benson, 1984, Drolet and Rioux, 1983). More research is required on the characteristics of key Australian weed species to optimise electric weed control application.

Two pot trials were conducted between 2021-2022 to characterise morphological factors of common summer and winter weeds in Australia. This will allow us to predict the efficacy of electric weed control as a weed management option within Australia.

## MATERIALS AND METHODS

In the first 'winter' pot trial (2021), seven winter weed species were grown in controlled glasshouse conditions (12-hour temperature cycle of 10/20°C) in a fully randomised design. In the second 'summer' pot trial (2022), eleven summer weeds were grown in a screenhouse at standard summer temperatures (20-40°C), also in a fully randomised design. The winter species included wheat cv. Mace (*Triticum aestivum* L.), double gee (*Emex australis* Steinh.), blue lupin (*Lupinus cosentinii* Guss.), and brome grass (*Bromus diandrus* Roth). The summer species were windmill grass (*Chloris truncata* R. Br.), button grass (*Dactyloctenium radulans* R. Br.), Feathertop Rhodes grass (*Chloris virgata* Sw), caltrop (*Tribulus terrestris* L.), wild radish (*Raphanus raphanistrum* L.), Afghan melon (*Citrullus lanatus* Thunb.) and heliotrope (*Heliotropium europaeum* L.). Several species were also grown in both trials, including annual ryegrass (*Lolium rigidum* Gaudin.), sow thistle (*Sonchus oleraceus* L.), and kikuyu (*Pennisetum clandestinum* Hochst. ex Chiov.).

For both trials, four plants of each species were grown per pot (16 cm diameter, 16.5 cm height) with three replicates. Each pot was lined with plastic bags with six drainage holes and filled with sand to within 2 cm of the top. Small seeds were tickled into the surface, and large seeds were sown at a depth of approximately 1 cm. Irrigation was applied as required to ensure healthy growth.

Harvest was 3-4 weeks after seeding, and plant growth stage (number of leaves or tillers per plant) and the fresh root and shoot biomass (per pot) were recorded. To obtain root biomass, the roots were thoroughly washed clean of all soil material.

Scans of both the roots and shoots were then performed on the Epson Perfection V800 Photo Scanner. The analysis of these scans was completed using WinRHIZO PRO (2005, https://regentinstruments.com/assets/winrhizo_software.html) for the roots, and ImageJ (2021, https://imagej.nih.gov/ij/) for the shoots. The roots and shoots were then dried for a week at 60°C before their dry biomass was determined.

A one-way ANOVA using plant species as the factor was performed on each data set of each trial in Genstat (21st Edition), and graphic outputs of this data were graphed using R (version 4.1.3).

RESULTS

**Biomass** In the winter pot trial, blue lupins had the greatest shoot biomass (Figure 1B), followed by wheat ($P<0.001$, LSD=0.454). Wheat, blue lupins, and brome grass had the greatest root biomass ($P<0.001$, LSD=0.280), although the difference between the blue lupin and brome grassroot biomass was not significantly different to kikuyu (Figure 1D). Alternatively in the summer pot trial, no significant differences were found between the species' shoot ($P=0.084$, LSD=0.208) or root ($P=0.241$, LSD=0.181) biomass (Figures 1A and C).

**Shoot surface** Overall, the winter species (Figure 2B) had a greater root surface area than the summer species (Figure 2A). However, there was no consistent difference between the broadleaf or grass species.

In the winter trial, wheat and blue lupins had a greater shoot surface area than the other species ($P<0.001$, LSD:63.510). For the summer species Feathertop Rhodes grass and heliotrope had the greatest surface areas ($P<0.001$, LSD=29.330).

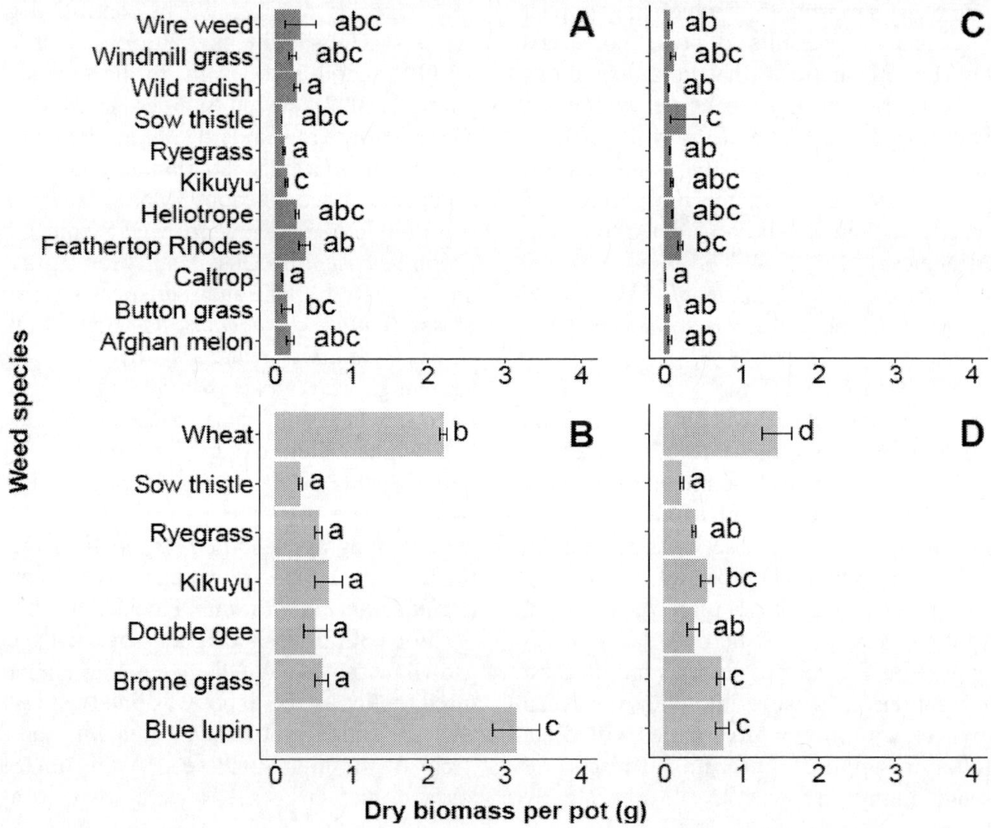


**Figure 1.** The mean dry biomass (g) per pot of the 'summer' (A) and 'winter' (B) shoots as well as of the 'summer' (C) and 'winter' root biomass (D) of each species in the pot trials. Letters on the columns indicate least significant differences between the means and the error bars indicate the standard error of 3 replications (3 pots of 4 plants).

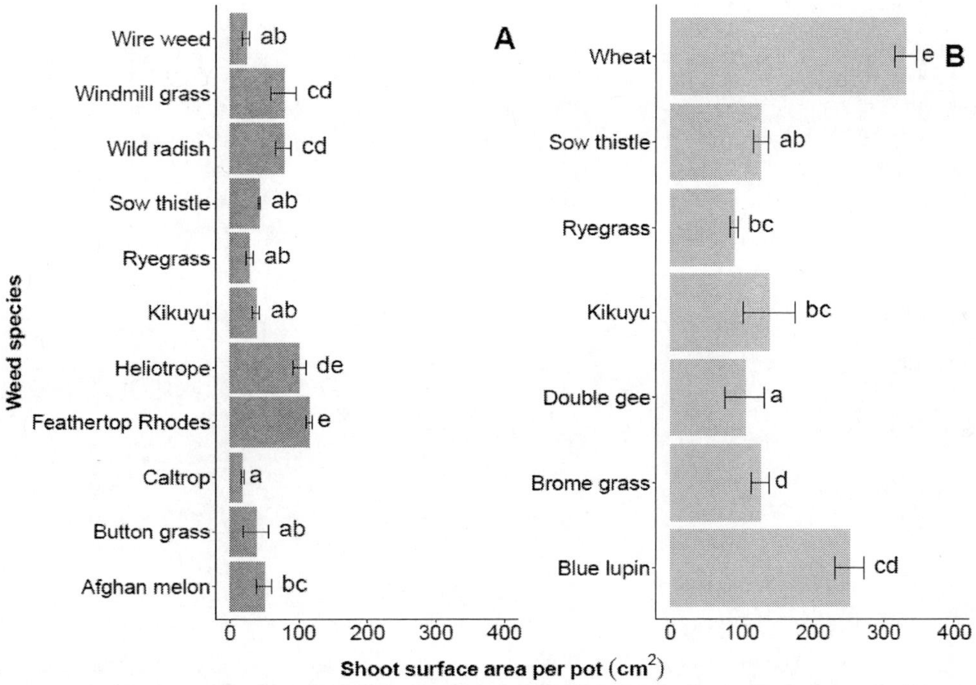


**Figure 2.** The mean shoot surface area (cm$^2$) per pot for each species in both the 'summer' (A) and 'winter' (B) pot trials. Letters on the columns indicate least significant differences between the means and the error bars indicate the standard error of 3 replications (3 pots of 4 plants).

**Root surface and diameter** The surface area of the roots (Figure 3) was greater than that of the shoots for most of the species, except caltrop and wild radish (Figure 2). However, the root surface area varied depending on the diameter of the root. Similar proportions of root surface area were found to occur in both the summer and winter pot trials between the diameter classes of 0-0.2 cm (Figures 3A and B) and 0.2-0.5 cm (Figures 3C and D). But these values were comparably lower for all species than the surface area of the thicker roots (>0.5 cm) (Figures 3E and F).

Across all root diameters, the winter species had a greater surface area when compared to those grown in summer, except for blue lupin roots in the 0-0.2 cm diameter range and double gee in the >0.5 cm range. There was no difference between grass and broadleaf species.

Out of all the species, wheat consistently had the greatest root surface area across all diameter classes. This was followed by brome grass. Yet, while the blue lupins returned comparably greater amounts of biomass and shoot surface area, the species was determined to have a very low proportion of finer roots (0-0.5cm) but the second highest proportion of thicker roots (>0.5cm) after wheat.

In the winter trial, a significant difference was found within the 0-0.2 cm ($P<0.001$, LSD=58.820), 0.2-0.5 cm ($P<0.001$, LSD=66.870) and >0.5 cm ($P<0.001$, LSD=160.100) diameter ranges.

In the summer trial, heliotrope had the greatest proportion of finer roots (0-0.5cm), while caltrop had the lowest. However, in the thicker roots (>0.5 cm), the greatest proportion of the surface area was found in Feathertop Rhodes grass. In this diameter class, caltrop still had the lowest proportion.

A significant difference between the species was found between both the 0-0.2 cm ($P<0.001$, LSD=14.180) and 0.2-0.5 cm ($P<0.001$, LSD=18.490). In the >0.5 cm diameter range, a significant difference was also found in the summer species, but to a lesser extent than seen within the other diameter ranges ($P=0.019$, LSD=54.720).

## DISCUSSION

Volunteer crops and winter weed species may be comparably harder to control with electric weed control, due to their faster growth habits and comparably greater shoot and root biomass, as well as surface area as reviewed in Vigneault and Benoit (2001). Yet, it is noted that field trials should be undertaken to verify these findings. Like most weed control methods, it is likely that electroweeding efficacy in Australia will be dependent on the specific weed species treated.

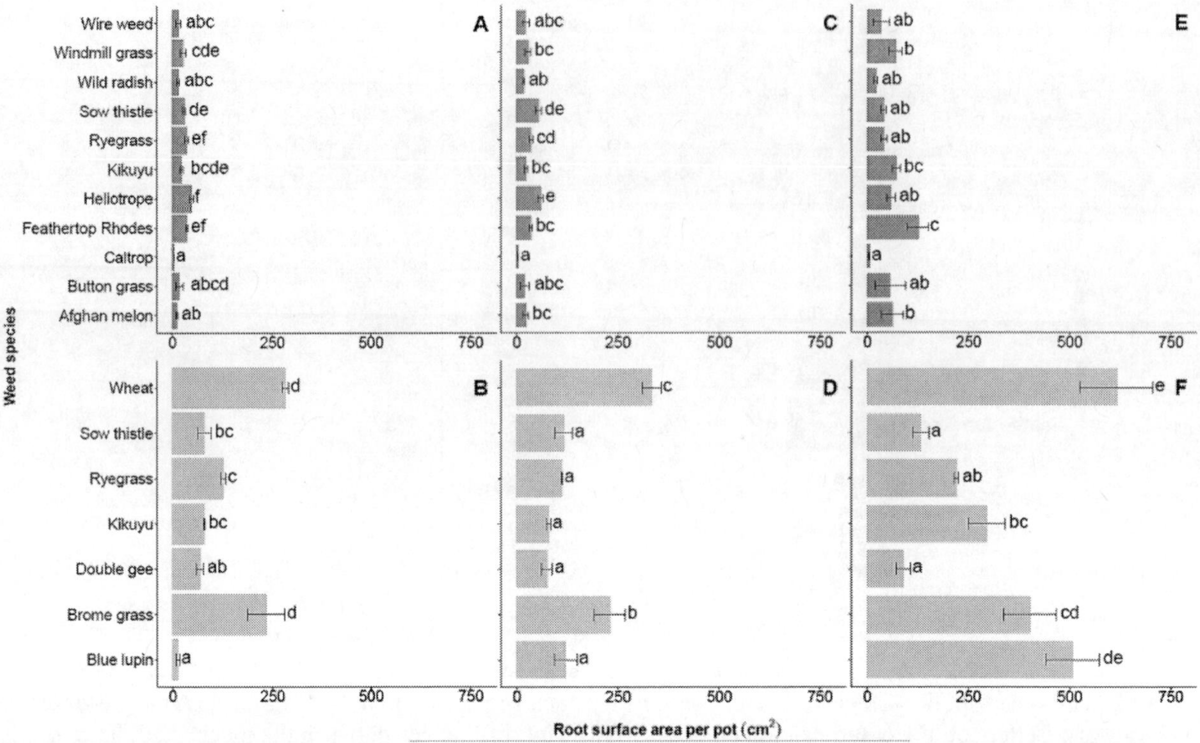


**Figure 3.** The mean surface area (cm$^2$) per pot of the roots between 0 – 0.2 cm diameter (A and B), 0.2 – 0.5 cm diameter (C and D) and >0.5 cm diameter (E and F) of each species in both the 'summer' (top graphs) and 'winter' (bottom graphs) pot trials. Letters on the columns indicate least significant differences between the means and the error bars indicate the standard error of 3 replications (3 pots of 4 plants).

Greater levels of biomass, as seen in the winter species of wheat and blue lupin as well as brome grass, have been indicated in the literature to reduce electroweeding efficacy. Often, only a portion of the plant is contacted by the electrode and the plant able to keep growing from the undamaged section (Diprose and Benson, 1984, Drolet and Rioux, 1983).

Studies have also indicated that extensive spreading or specialised root systems can allow for the treated plant to re-grow from undamaged root sections (Diprose and Benson, 1984, Drolet and Rioux, 1983). From these trials, it is indicated that this may be an issue with grasses with greater root surface area such as wheat and brome grass. However, limited research has occurred into the re-growth potential of these species following electroweeding and other weed control methods.

ACKNOWLEDGMENTS

This research was made possible through collaboration with CNH Industrial and their partnership with Zasso™. It is also supported by funding from the Grains Research and Development Corporation, Wine Australia and the Cotton Research and Development Corporation. Thanks are due to Andrew Vanburgel (DPIRD) for his statistical assistance as well as Nerys Wilkins and David Nicholson (DPIRD) for their technical assistance.

REFERENCES

Diprose, M. F. & Benson, F. A. 1984. Electrical methods of killing plants. *Journal of Agricultural Engineering Research,* 30, 197-209.

Diprose, M. F., Benson, F. A. & Hackam, R. 1980. Electrothermal control of weed beet and bolting sugar beet. *Weed research,* 20, 311-322.

Drolet, C. & Rioux, R. 1983. Evaluation d'une rampe utilisant un courant electrique pour le controle des mauvaises herbes.

Vigneault, C. & Benoit, D. 2001. Electrical Weed Control: Theory and Applications. *In:* Vincent, C., Panneton, B. & Fleurat-Lessard, F. (eds.) *Physical Control Methods in Plant Protection.* New York, USA: Springer-Verlag Berlin, Heidelberg.

Vigneault, C., Benoit, D. L. & McLaughlin, N. B. 1990. Energy aspects of weed electrocution. *Reviews of Weed Science,* 5, 15-26.

# Urban Waterway and Aquatic Weed Management - case studies from over 20yrs

Geraldene Dalby-Ball[1]
[1]Dragonfly Environmental, Kings Park, Australia
(gm@dragonflyenv.com.au)

**Summary** Waterways have again become central places for people to gather, to move along and in urban areas these are increasing the majority of the retained or replanted ' bushland'. Waterway development setbacks and creek lined corridors are mandatory in many new and renewed urban developments. As well as being recreation places with multi use pathways riparian zones often double as stormwater management locations with the inclusion of constructed wetlands. As with any built asset ongoing maintenance is required particularly in the area of weed management. Two key considerations are inhibiting weeds establishing, particularly in newly planted areas and controlling the weeds that are present. Of interest too is the prediction of cost and the required setting aside of funds from both the public and private sector to effectively manage weeds in urban waterways, wetlands and riparian zones. Aquatic weed management resourcing is generally under estimated and aquatic weed ID is generally low across the sector. This presentation provides case studies from over 25 years of working in urban waterways, riparian zones and wetlands. Wetlands and weeds can be synonymous if not managed – with 100 of $1000s being spent annually on waterway weed suppression. The aim is to provide information that assists with weed management including in designing these assets, preparing and costing the long term management of these areas, particularly when they go into community title management. A focus will be on successful actions to take in urban waterway and wetland management as well as predictive costing for budget allocations. Case studies are given with tools on wetland weed ID and how to minimise spread as well as the use of new technologies for larger areas such as Drones for the treatment of Salvinia in Penrith Lakes Sydney. Multiple considerations of managing weeds in wet environments are covered including high frequency repeat weed incursion, WHS considerations and ecological requirements including low/no chem in waterways.

**Keywords** Aquatic Weeds, Riparain zones, predicting budgets, techniques, on-ground successes

# Efficacy of Pre-emergent Herbicides on Ameliorated Soil

Bowen Zhang[1], Sultan Mia[1], Tom Edwards[1], Gaus Azam[1] and Catherine Borger[1]
[1]Department of Primary Industries and Regional Development, 75 York Rd, Northam, WA 6401
(bowen.zhang@dpird.wa.gov.au)

**Summary** A field trial investigating amelioration of water repellent soil was established in 2020, with treatments comprised of spading, mouldboarding and an untreated control. Intact soil cores were taken from this site in 2020 and 2021 to establish pot experiments that investigated the interaction of soil amelioration and pre-emergent herbicides on annual ryegrass control. Spading and mouldboarding allowed earlier emergence and better shoot and root growth of annual ryegrass seeds in the absence of herbicide. The efficacy of pre-emergent herbicides varied with herbicide type and rate but were not affected by soil amelioration.

## INTRODUCTION

Water repellent soils cover over two million hectares in Western Australia (Blake and Peltzer 2002). These soil types may delay seeding and reduce the yield potential of the following crop. Water repellence also results in staggered emergence, reducing the crop yield and making it difficult to control weeds at the appropriate growth stage (Blake and Peltzer 2002). The late emerging cohorts avoid control by pre-seeding, pre-emergent or early in-crop herbicides (Roper et al., 2015). Soil amelioration approaches such as deep ripping, deep spading and soil inversion alter physical and chemical soil properties and may bury crop residue, potentially changing the way pre-emergent herbicides affect weed or crop growth (Edwards et al. 2018). The behaviour of pre-emergent herbicides is dependent on multiple soil properties. The physical disturbance of the soil surface, burial of the weed seed or removal of the soil constraint could also affect weed seed emergence time and rate, as well as the weed's growth pattern (Chauhan *et al.* 2006). However, there is little research on the interaction of soil amelioration and the efficacy of pre-emergent herbicides. This study aims to investigate how the amelioration of water repellent soil influences weed emergence and pre-emergent herbicide behaviour.

## MATERIALS AND METHODS

A field trial was established on a site with water repellent soil in Esperance, WA, where plots had been spaded, mouldboarded or left intact (untreated control). Intact soil cores were removed from this site to conduct two pot experiments at the Department of Primary Industries and Regional Development (DPIRD) Northam screen house in 2020 and 2021. The soil cores were placed in pots of 12 cm diameter by 10 cm height. All pots were maintained in the screen house and watered via rainfall to ensure weed growth occurred in standard seasonal conditions.

**2020 experiment** The trial was arranged in a randomised block design with four replications of each herbicide-amelioration treatment. Pre-emergent herbicides included pyroxasulfone (480 g a.i. $L^{-1}$), trifluralin (480 g a.i. $L^{-1}$), prosulfocarb (800 g a.i. $L^{-1}$), triallate (500 g a.i. $L^{-1}$), or water (non-chemical control) applied using a spray cabinet calibrated to deliver 100 L/ha spray volume, at 2 bar pressure, from Hardi-Iso F-01-110 nozzles at 50 cm spacing. Herbicide treatments were sprayed at full label rate, half the label rate or a quarter of the label rate. Soil collected from the field was spaded, mouldboarded or undisturbed. Three seeds of annual ryegrass (*Lolium rigidum* Gaud.) (cv 'Safeguard' from Nutrien Ag Solutions, Midvale, WA) were sown into each pot at a depth of 0.5 cm. In total, 168 pots were included in the 2020 experiment.

**2021 experiment** The trial was arranged as a randomised block design with three replications of each herbicide-amelioration treatment. The pots were treated with the same herbicides (pyroxasulfone, trifluralin, prosulfocarb, triallate or water as the control) and three tillage types (mouldboarded, spaded and intact) used in 2021. However, in contrast to 2020, the herbicide treatments were only sprayed at full label rate and half the label rate. Each treatment was applied to a pair of pots to allow two separate harvest times (i.e. 81 pots for harvest one on July 23 and 81 pots for harvest two on 30 July). Ten seeds of annual ryegrass were sown in each pot at a depth of 0.5 cm. In total, 162 pots were included in the 2021 experiment.

**Data collection** For both experiments, annual ryegrass emergence data was recorded twice a week for three weeks and plant growth stage was estimated by counting the leaf number. At three weeks old plants were harvested. Fresh root and shoot weight of each plant and shoot length was recorded. Root scanning was then conducted to measure root length and surface area (using WinRHIZO™ 2019, Regent Instruments, www.regentinstruments.com). Samples were dried at 105°C for a week before assessing root and shoot dry weight.

**Statistical analysis** A two-way ANOVA was used to assess the effect of soil amelioration on herbicide efficacy and weed germination in both 2020 and 2021 using Genstat (VSN International 2021). A square root transformation was used to control for heteroskedasticity.

## RESULTS AND DISCUSSION

The emergence data for annual ryegrass in 2020 suggests that soil amelioration practices stimulated weed germination (Figure 1). Without any chemical control applied, annual ryegrass seedlings started to germinate 9 days after sowing in the pots with intact soil (1a), while the first germinations were seen 5 days and 6 days after sowing in mouldboarded soil (1b) and spaded soil (1c). In field conditions, spading or mouldboarding may bury 50-60% or 50-90% of the topsoil below 10 cm depth and place a layer of subsoil on the surface (Scanlan and Davies 2019). Therefore, both techniques may provide weed control by burying a portion of the weed seed (Roper et al., 2015). However, a proportion of seeds are likely to be left in the topsoil and may still be able to emerge (Scanlan and Davies 2019). In the current pot trials, all seeds were sown at a uniform depth, and emergence was more rapid in the ameliorated soil where water repellence was likely to be reduced. Rapid, uniform emergence will make the weeds easier to control with pre-emergent herbicides (Chauhan *et al.* 2006).

Pyroxasulfone and prosulfocarb achieved similar control in both years (Figure 2, Figure 3). In 2020 trifluralin and triallate were less effective while in 2021 only the half rate of triallate provided reduced control. Trifluralin applied in 2021 achieved 99% control but was less effective in 2020. Other growth parameters such as dry biomass and root length had a comparably similar pattern as seen in the emergence data (Table 1). The dry biomass of ryegrass treated with pyroxasulfone was close to zero in either intact or ameliorated soil. Except for prosulfocarb, other herbicide treatments were unable to show a significant reduction in terms of root length, root dry weight and shoot dry weight in the mouldboarded and spaded pots. In fact, the root length of trifluralin and triallate treated annual ryegrass increased 3-4-fold in ameliorated soil. On the other hand, prosulfocarb treatments had much shorter roots and suppressed biomass. It is clear there was an inconsistent effect of soil amelioration on herbicide efficacy and field trials on ameliorated sites are needed to clarify the impact. Several other studies showed varied results of herbicide efficacy in ameliorated soil. Buhler and Daniel (1988) concluded that control of giant foxtail (*Setaria faberi* Herrm.) was lower in a no-till system than a mouldboarded area.

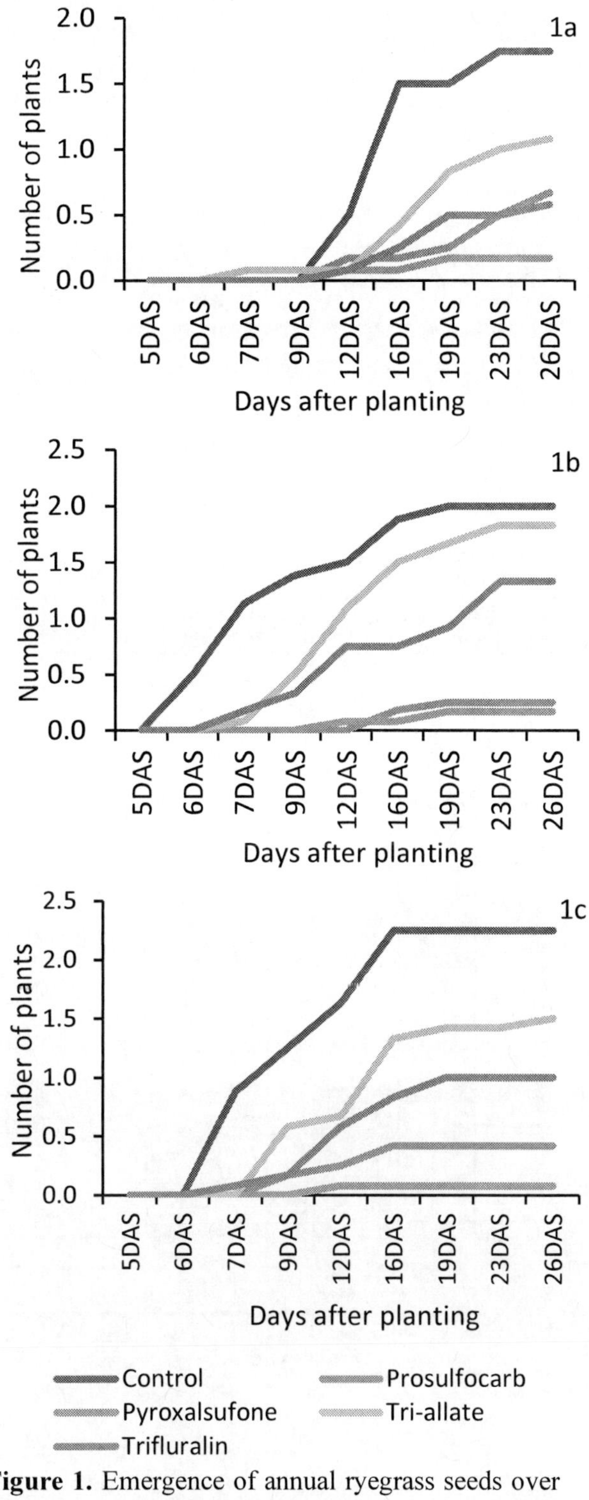


**Figure 1.** Emergence of annual ryegrass seeds over 26 days in pots with undisturbed soil with no soil amelioration (1a), mouldboarded soil (1b) or spaded soil (1c) in 2020. Annual ryegrass emergence was recorded and averaged over all herbicide treatments.

Comparably longer roots and more biomass can be seen in both ameliorated practices in the absence of herbicides due to improved soil structure and removal of water-repellent soil (Table 1).

**Table 1.** Annual ryegrass root length, root dry weight and shoot dry weight in 2021 collected from plants grown in intact, mouldboarded and spaded soil for three weeks. Data was averaged over herbicide rates. Letters indicate significantly different means when compared using ANOVA.

| Measurements | Root Length (cm) | Root Dry Weight (mg) | Shoot Dry Weight (mg) |
|---|---|---|---|
| **Intact** | **50.82** | **23.68** | **17.47** |
| Control | 389.93c | 75.77c | 41.9c |
| Prosulfocarb | 20.63b | 17.44b | 17.06bc |
| Pyroxasulfone | 0.38a | 1.41a | 4.82a |
| Tri-allate | 24.53b | 13.82b | 13.47b |
| Trifluralin | 16.44b | 9.95b | 10.08b |
| **Mouldboard** | **83.00** | **28.13** | **17.62** |
| Control | 378.30c | 100.67c | 51.9c |
| Prosulfocarb | 3.32ab | 3.63ab | 4.77a |
| Pyroxasulfone | 8.22ab | 0.95a | 3.35a |
| Tri-allate | 88.91bc | 25.39bc | 20.51bc |
| Trifluralin | 59.28bc | 10.03ab | 7.58ab |
| **Spaded** | **96.02** | **20.47** | **14.66** |
| Control | 421.48c | 93.72c | 43.47c |
| Prosulfocarb | 10.23ab | 5.21ab | 7.6ab |
| Pyroxasulfone | 0.12a | 1.36a | 4.22a |
| Tri-allate | 94.03bc | 16.88b | 13.52b |
| Trifluralin | 62.84bc | 5.63ab | 4.48a |

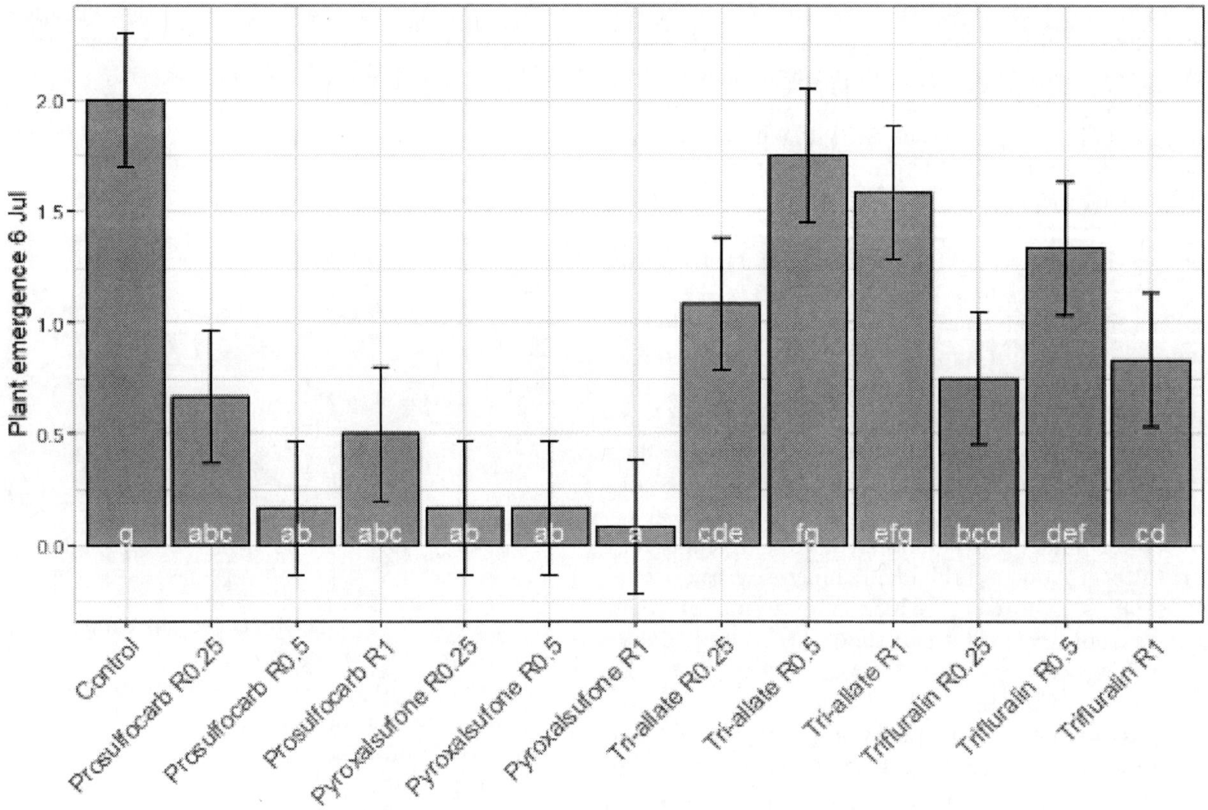


**Figure 2.** Annual ryegrass emergence in 2020, following pre-emergent herbicide treatments (control, pyroxasulfone, trifluralin, prosulfocarb and triallate) at full label rate, half label rate and quarter label rate, averaged over soil amelioration treatments (with 3 annual ryegrass seeds per pot).

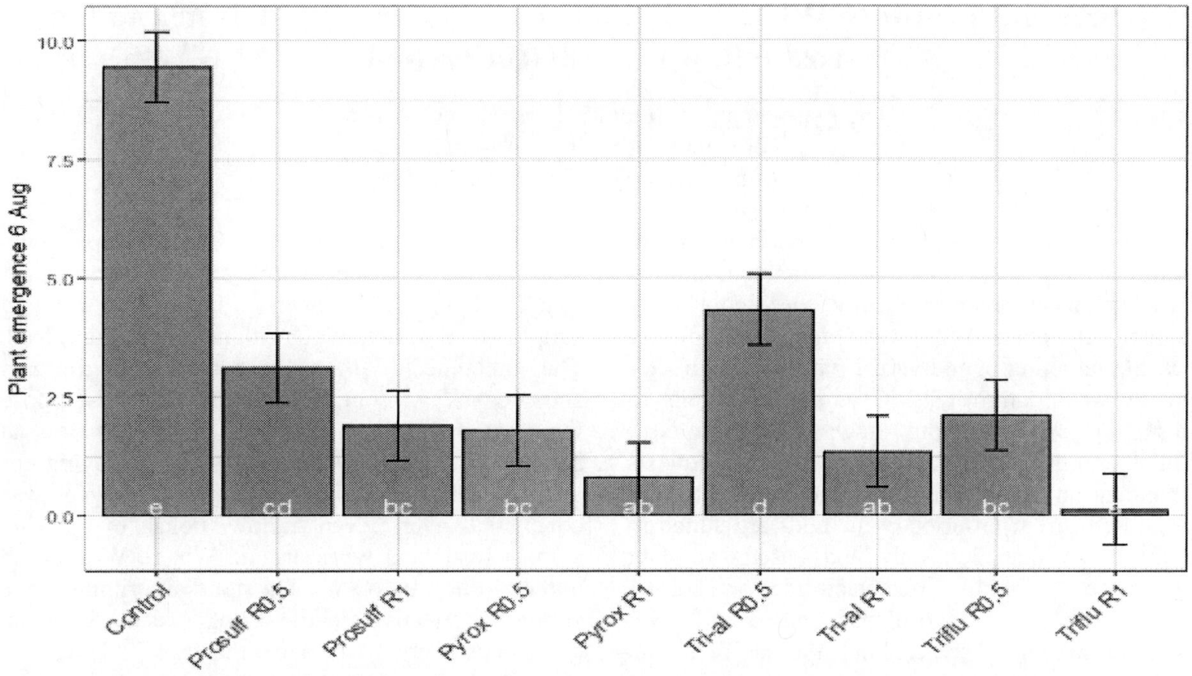


**Figure 3.** Annual ryegrass emergence in 2021, following pre-emergent herbicide treatments (control, pyroxasulfone, trifluralin, prosulfocarb and triallate) at full label rate and half label rate, averaged over soil amelioration treatments (with 10 annual ryegrass seeds per pot).


## ACKNOWLEDGEMENTS

The researchers acknowledge funding from the Boosting Grains Science Partnership projects. Field trials were supported by DPIRD and GRDC as part of the project DAW1901-006RTX: Increasing farming systems profitability and longevity of benefits on ameliorated soils in Western Australia. Thanks are due to Nerys Wilkins and Dave Nicholson (DPIRD) for technical assistance, staff at the Esperance Research Station for their assistance.


## REFERENCES


Blake, J. and Peltzer, S. (2002)13th Australian Weeds Conference. Weeds: threats now and forever? Perth, Western Australia 8-13 September 2002

Buhler, D.D. and Daniel, T.C. (1988). Influence of Tillage Systems on Giant Foxtail,*Setaria faberi*, and Velvetleaf,*Abutilon theophrasti*, Density and Control in Corn,*Zea mays*. *Weed Science*, 36(5), pp.642–647.

Chauhan, B.S., Gill, G.S. and Preston, C. (2006). Tillage system effects on weed ecology, herbicide activity and persistence: a review. *Australian Journal of Experimental Agriculture*, 46(12), p.1557.

Edwards T, Davies S, McDonald G, Hall D, Moore J (2018) Understanding interactions between pre-emergent herbicides and inversion tillage. 2018 GRDC Grains Research Updates, 26-27 February, Perth, Western Australia.

Liu, H., Crawford, M., Carvalhais, L.C., Dang, Y.P., Dennis, P.G. and Schenk, P.M. (2016). Strategic tillage on a Grey Vertosol after fifteen years of no-till management had no short-term impact on soil properties and agronomic productivity. *Geoderma*, 267, pp.146–155.

Matic R, Black I.D. (1990) Control of rigid brome (*Bromus rigidus*) in barley using tillage, trifluralin and metribuzin. In 'Proceedings of the 9th Australian weeds conference'. pp. 49–52.

Roper, M.M., Gupta, V.V.S.R. and Murphy, D.V. (2010). Tillage practices altered labile soil organic carbon and microbial function without affecting crop yields. *Soil Research*, 48(3), p.274.

Roper, M.M., Davies, S.L., Blackwell, P.S., Hall, D.J.M., Bakker, D.M., Jongepier, R. and Ward, P.R. (2015). Management options for water-repellent soils in Australian dryland agriculture. *Soil Research*, 53(7), p.786.

Scanlan, C.A. and Davies, S.L. (2019). Soil mixing and redistribution by strategic deep tillage in a sandy soil. *Soil and Tillage Research*, 185, pp.139–145.

VSN International (2021) GenStat for Windows. 19th ed. Hemel, Hempstead, UK: VSN International.

# OpenWeedLocator (OWL): An open-source, community-driven and low-cost fallow weed detection tool

Guy Coleman[1], William Salter, Michael Walsh
[1]University Of Sydney, Australia
(guy.coleman@sydney.edu.au)

**Summary** Fallow periods are an important tool to maximise crop yield potential in moisture limited environments and require effective weed control for optimal efficiency. With weed densities typically low, the targeting of individual plants through site-specific weed control (SSWC) is an opportunity to substantially reduce input usage and hence cost. Whilst promising, the efficacy of the technique is dependent on reliable and effective weed detection methods. Current proprietary methods are suited to fallow weed detection only and based on plant reflectance with optoelectronic sensors. Advancements in algorithms and small, yet powerful computers are enabling the use of digital images for weed detection and recognition in both fallow and more complex in-crop environments. The development of an image-based fallow weed detection tool, leaves the door open for future in-crop weed recognition use. The OpenWeedLocator (OWL) is an open-source, community-driven and low-cost device for image-based weed detection in fallow systems that acts as a practical educational tool and improves accessibility of the technology. The OWL uses a Raspberry Pi computer running simple green detection algorithms on a camera feed. The outcome of these colour-based algorithms activates relays connected to the general-purpose input/output (GPIO) pins on the board for an actionable response such as spot spraying or targeted tillage. Validation of the device was conducted over seven fallow fields of varying stubble loads and types around Wagga Wagga and outer Sydney in NSW. The four algorithms were similarly effective in detecting weeds with an average precision of 79% and recall of 52%, with up to 92% and 74% for precision and recall respectively at individual test sites. By taking a community-driven approach to image-based weed recognition technology, OWL is redefining the approach to site-specific weed control.

**Keywords** Site-specific weed control, weed detection, computer vision

# Assessing the fungal pathogen *Stagonospora tauntonense* as a biocontrol agent against rat's tail grasses

Eloise Martin[1]
[1]CSIRO,University of Queensland, Brisbane, Australia
(eloise.martin@csiro.au)

**Summary** *Stagonospora tauntonense* is a recently discovered fungal species, pathogenic on *Sporobolus natalensis*, commonly known as Giant Rat's Tail (GRT) grass. *S. natalensis* is one of five weedy Sporobolus grass (WSGs) species, which also include *S. pyramidalis*, *S.fertilis*, *S. jacquemontii* and *S. africanus*, all being weeds of significance in natural and agricultural systems. Previous surveys of fungal pathogens of these grasses show that Stagonospora is a favourable biological control agent (BCA) candidate; it has a large geographic distribution, is observed to cause a range of diseases in GRTs, has been successfully used as a BCA previously (*S. convolvuli* against large Bindweed) and has shown high levels of pathogenicity in initial bioassays. This investigation thoroughly explores the feasibility of this pathogen as a BCA, through systematic assessment of the plant-pathogen interaction between the 29 *Stagonospora tauntonense* strains and the 5 WSGs. The assessment begins with observation and ranking of the pathogenicity of each strain in bioassays (charcoal agar), to generate a complete and quantitative dataset regarding the disease interaction. DNA and RNA testing will be considered, to understand the biochemical underpinning of plant defence and pathogenesis pathways. Host-specificity, an integral feature of BCAs, will be assessed, specifically for native, agricultural and amenity species from the plant sub-family Chloridoideae. Life cycle testing will be carried out to identify optimal application timing, method and conditions, ideally using a shortlist of favourable strains. A key, overarching focus is enhancing sporulation, to improve production, commercialisation and application success. Thus far, it has not been successfully achieved in vitro, however factors of abiotic stressors, photoperiodism, nutrient/metabolite levels, substrate and pH are yet to be methodically investigated. This remains a challenging task, given the broad range of potential, and rather specific, conditions required by fungi for sporulation. In this investigation, we aim to better understand of the host-pathogen-environment interaction and pathways, and potentially for the development of an efficacious and sustainable replacement for traditional chemical herbicides, to control an invasive and damaging weed.

**Keywords** Rat's tail grass, biocontrol, biological control, *Stagonospora tauntonense*, sporobolous

# Trialling foot cleaning stations in Kosciuszko NP for biosecurity hygiene: How much do bushwalkers use them, what influences their use, and how do they respond to the cleaning stations?

Nicholas Gill[1], Leonie Miller[1]
[1]University Of Wollongong, Australia
(ngill@uow.edu.au)

**Summary** Recreational users of public lands can carry weed seed and other material on footwear, clothing and equipment. Cleaning stations are often used to reduce such biosecurity risks. Such biosecurity hygiene footwear cleaning stations were installed at Thredbo and Charlotte Pass in Kosciuszko NP, together with signs exhorting walkers to use the station. Between April 2019 and April 2021, we evaluated the use of these stations through intercept surveys with bushwalkers and covert observation of bushwalkers. We surveyed 1148 bushwalkers and made 17736 covert observations. Fieldwork was affected by bushfires and COVID, resulting in delays from 2019. This research builds on previous surveys prior to the installation of the cleaning stations. Our goals in this research were to gain insights into the extent to which KNP users (bushwalkers) will use installed cleaning stations and into their willingness or lack thereof to use cleaning stations; to determine barriers and facilitators of hygiene practice; to understand other factors that may shape users' decision to clean or not clean; to identify the extent of variability in hygiene practice across users; to test messaging (injunctive versus descriptive) to users regarding use of the stations; and to obtain feedback from KNP walkers regarding the implementation of footwear cleaning equipment in KNP. In this paper we present results from the survey and analysis to date and discuss insights from this research for future use of footwear cleaning stations.

**Keywords** Biosecurity hygiene; behaviour; public lands; prevention; recreation

# *Passiflora foetida*: prospects for biological control

Nagalingam Kumaran[1], Gavin Hunter[1], Giovanni Fichera[1], Mariano Maestro[2], Guillermo Cabrera Walsh[2], Alejandra Milena Clavijo-Giraldo[3], Paola Betancur García[3], Sandra Uribe-Soto[3], Juan Gonzalo Morales Osorio[3], Mauricio Salazar Yepes[3], Eduardo Adenesky-Filho[4], Marcelo Diniz-Vitorino[4], Bruce Webber[1], Louise Morin[1], S Raghu[1]

[1]CSIRO Health and Biosecurity, Australia,

[2]Fundación para el Estudio de Especies Invasivas, Argentina,

[3]Universidad Nacional de Colombia sede Medellín, Colombia,

[4]Universidade Regional de Blumenau, Brazil

(kumaran.nagalingam@csiro.au)

**Summary** *Passiflora foetida* L. (stinking passionflower), native to Central and South America, is an invasive weed across the Asia-Pacific region, including Australia. Its impacts span environmental and agricultural contexts in Australia. As an invasive herbaceous vine, it is commonly found in riparian ecosystems, forest margins, coastal habitats and ruderal areas (e.g. road verges and disturbed habitats) across tropical and subtropical parts of northern Australia. In the Pilbara and Kimberley regions of Western Australia, it has significant impacts on native vegetation through the formation of dense mats that smother native vegetation, and by carrying fires into tree canopies. Its climbing/smothering habit similarly results in negative impacts on tree crops (e.g. sandalwood plantations) and post-mining restoration efforts. Currently, the management of this species is largely dependent on physical (e.g. hand-pulling) and chemical (e.g. herbicides) control tactics, but these methods are not cost-effective and sustainable at the spatiotemporal scale of the weed's infestation. As a result, efforts are underway to investigate biological control options for this weed. To date surveys in the native range (Argentina, Brazil and Colombia), guided by ecophysiological and population genetics studies of *P. foetida* sensu lato, have identified a range of pathogens and insects that are being studied for their prospects as candidate biological control agents. These prospective agents are being screened in the native range, through a combination of field observations and laboratory host-specificity tests, for their ability to develop on commercial passionfruit cultivars. Those species that are unable to use commercial cultivars will subsequently be imported for further risk analyses that test their ability to use up to 50 other non-target species of increasing phylogenetic distance from *P. foetida*, in a quarantine laboratory in Australia. For candidate agents that pose negligible risk to non-target species, a release application will be prepared for review by Australian regulators.

# Making a difference to invasive grasses management on the ground

Stephen B. Johnson[1,2], Sarah. J. Baker[3,4], Bill Davidson[1,4], Carol A. Harris[3,5], Katrina Sinclair[3,6], Hanwen Wu[1,7] and Ali A. Bajwa[1,7]
[1]Weed Research Unit, Invasive Species, Biosecurity
[2]Orange Agricultural Institute, 1447 Forest Road, Orange 2800
[3]Pasture Systems, Livestock Systems, Agriculture
[4]Tamworth Agricultural Institute, 4 Marsden Park Road, Calala 2340
[5]444 Strathbogie Road, Glen Innes 2370
[6]Wollongbar Primary Industries Institute, 1243 Highway, Wollongbar 2477
[7]Wagga Wagga Agricultural Institute, PMB, Wagga Wagga 2650
Department of Primary Industries, New South Wales, Australia
(ali.bajwa@dpi.nsw.gov.au)

**Summary** Exotic, invasive, perennial grasses including African lovegrass, Chilean needle grass and serrated tussock have significant impacts on primary production (livestock) systems, rural communities, and the environment.

With prolific seed production and effective seed dispersal mechanisms, their spread to new areas continues mostly unabated. The dense growth habit of these grasses competes strongly with more desirable exotic and native pasture species, and negatively influences fire regimes, threatening ecosystem structure and function.

Management of these widespread invasive grass weeds is difficult to achieve at a regional scale. Impacts include reduced carrying capacity associated with decreased pasture productivity and quality, and contamination of hides, and wool. Affected farming-dependent communities are impacted by the economic cost of these weeds, potentially reducing community resilience (associated with a loss of services and facilities; human and social capital).

As part of a National Invasive Grasses Research, Development and Engagement Business Plan, New South Wales Department of Primary Industries (NSW DPI) have established three sites in NSW for the demonstration of best management practice (BMP) of African lovegrass, Chilean needle grass, and serrated tussock in stage 1 of this plan. The plan's second stage involves increasing the reach of the initial demonstration sites by working directly with land managers to establish 21 adaptation sites featuring co-designed BMP options for managing these invasive grasses. Formal and informal extension activities around these sites will raise land managers awareness of BMP options available to manage these weeds. These adaptation site networks will help build their knowledge, skills, and confidence to implement 'best management practices' to address invasive grasses on their properties. This staged demonstration and co-designed adaptation site approach is anticipated to accelerate uptake by land managers of effective BMP of difficult-to-control species, particularly these invasive grasses.

**Keywords** Adaptation, Adoption, Awareness, Best Management Practice, Coordination, Demonstration, Information Sharing, Weeds of National Significance.

## EXOTIC INVASIVE GRASS SPECIES

Exotic grass species have a long history of introduction into Australia with at least 2,250 species (and 53,278 accessions) introduced in the 1900s alone (Cook and Dias 2006). While this impressive total represents less than a quarter of global Poaceae (Jacobs 2022), the ongoing legacy of these efforts has been considerable. A number of these grass species form a critical basis for improved pasture and animal production systems, and are used as grain crops. Grasses have environmental benefits for soil conservation and revegetation and have the potential for carbon sequestration and use as biofuels (Jacobs 2022). Other important uses of these grasses include amenity purposes (including lawns) and decorative features in gardens.

In contrast, some of these grass species have become invasive, with more than 180 now recognized as "weeds" in Australia (Virtue *et al.* 2004). Two of the most invasive grasses in southern Australia are serrated tussock (*Nassella trichotoma* (Nees) Hack. ex. Arechav.) and Chilean needle grass (*N. neesiana* (Trin. & Rupr.) Barkworth). In recognition of the impacts these weeds cause, both have been listed and managed as Weeds of National Significance (WoNS, Weeds Australia 2022). Infestations of serrated tussock in southern and central New South Wales (NSW) were estimated to reduce gross margins by more than $26.5 million over two decades ago (Jones and Vere 1998), with current costs of control far exceeding $100 ha$^{-1}$ (Millar *et al.* 2016, AG DAWE n.d.). The potential distribution of serrated tussock in Australia is also considerable and estimated at 0.32-1.30 million $km^2$ (McLaren *et al.* 1998, Watt *et al.* 2011, Gallagher *et al.* 2013). Chilean needle grass is thought to incur similar costs (\$64-119 $ha^{-1}$, McLaren *et al.* 2002) but may have a far greater potential distribution at 0.40-2.19 million $km^2$ (McLaren *et al.* 1998, Bourdôt *et al.* 2012, Gallagher *et al.* 2013).

In contrast, little is known about the costs imposed by African lovegrass (*Eragrostis curvula* (Schrad.) Nees) with current research investigating this (Officer 2022). Despite the cultivar 'Consol' having been planted for pasture and soil conservation purposes in the past (Johnston 1989, Johnston *et al.* 2006), the distribution of African lovegrass is not yet well quantified. Conservatively, the species is known to occur across 10% of Queensland (Qld, Csurhes *et al.* 2016), and is recorded as particularly common across the eastern third of NSW (PlantNET 2022). It has considerable potential to spread in Qld, NSW, and from where it is currently growing in restricted areas of South and Western Australia, Tasmania, and Victoria (Csurhes *et al.* 2016, ALA 2022, VRO 2022). One estimate suggests that 11% of Australia or 2.99 million $km^2$ could be suitable (Gallagher *et al.* 2013). Despite some data being available, the impact of these species on environmental assets (Coutts-Smith and Downey 2006) and the community is not well known.

This paper will report on activities undertaken by NSW DPI for the National Invasive Grasses Research, Development and Engagement Business Plan, and the second stage of this plan.

## PROJECT STRUCTURE

It is difficult to manage the impacts caused by, and the spread of, invasive grass species like African lovegrass, Chilean needle grass and serrated tussock. In response, the Centre for Invasive Species Solutions (CISS), Meat and Livestock Australia (MLA), NSW DPI and Wild Matters engaged with all Australian jurisdictions and Commonwealth Scientific and Industrial Research Organisation (CSIRO) to develop a National Invasive Grasses Research, Development & Engagement Business Plan. NSW DPI have since partnered with CISS and commenced stage 1 implementation of this plan by establishing "BMP Proof/Demonstration sites", one for each of the three species. In consultation with land managers, demonstration sites are trialing a number of BMP treatments which can inform land managers of the options most appropriate to their situation.

The project briefly detailed in this paper forms stage 2 of the plan aiming to increase the reach of the initial demonstration sites. It will improve and update the knowledge base of land managers, industry and community in current BMP methods and tools. This will be achieved through raising awareness and providing sessions linked to the existing demonstration sites, expanding to seven additional adaptation sites for each species (21 adaptation sites in total, Figure 1). The adaptation sites will allow the application of BMP options across a broader set of geographies, properties and manager preferences. Again, in consultation with the land managers, BMP will be tailored to fit the preferences, resources, and limitations that are encountered by each land manager. Infestations will be quantified before and after treatment and monitored to assess treatment efficacy in both demonstration and adaption stages.

Provision of further funding would see this approach expanded to the national level.

## WHY THE PROJECT IS NEEDED

Land managers often lack the support and time to review management program outcomes and to refine future adaptive property management. A lack of capacity (including financial/resourcing) and activities/learning opportunities to build this capacity, including information-sharing and skill development, can also constrain the adoption of BMP options to manage these invasive weeds. Adequate resourcing, training and skills development can empower land managers to address these invasive grass weeds.

In stage 2 of this project, the approach is for NSW DPI weed and pasture experts, combined with land manager to co-design and implement best management across properties, landscapes and regions. This approach will build the capacity and skills of land managers to achieve practice change.

This project will seek practical answers to the following questions (but is not limited to just these questions):

i. Are parts of the infestation site arable or non-arable, and what BMP/herbicides could be used in such situations?

ii. What BMP is needed to prevent the evolution or progression of herbicide resistance in these weed species?

iii. Could herbicide applications be combined with grazing pressure and what time/s of grazing are best for suppressing/managing these weed species?

iv. What combination of BMP approaches can be used within the soil and financial constraints of certain land managers?

v. What is the role of competitive pastures, whether existing native or competitive, exotic species in suppressing these invasive grass weeds? and

vi. How can BMP practices optimize these pasture species and competition?

## SUMMARY

Using demonstration sites to inform BMP options that are then expanded into the adaptation sites, working with pasture and weed experts and codesign with land managers will be an effective way of enabling land managers to best manage their weeds.

Since our work is current, we hope to detail some of the questions explored and the answers derived from our work with various land managers at the presentation of this paper. Results arising from this project will appear in future publications.

**Figure 1.** Location of the 21 adaptation sites in south eastern NSW. The blue, red, and purple pointers represent African lovegrass (mainly south of Canberra), Chilean needlegrass (mainly north of Armidale) and serrated tussock (Orange-Yass), respectively (Credit A. Bajwa). The demonstration sites are not represented but are near Cooma, Tamworth and Yass, for the above species, respectively.

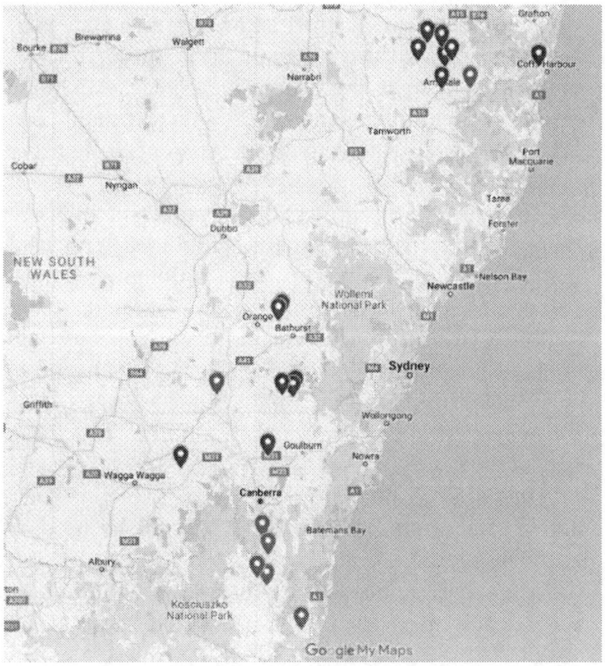


## ACKNOWLEDGMENTS

This research has been jointly funded by the Centre for Invasive Species Solutions, Department of Agriculture, Fisheries and Forestry of the Commonwealth Government of Australia and the NSW Government.

Drs David Officer, Sarah McDonald, and Graham Charles of NSW DPI, provided important corrections, information and comments during the paper writing process.

## REFERENCES

AG DAWE (Australian Government, Department of Agriculture, Water and the Environment, n.d.). Costs and benefits of serrated tussock control. Case Study 2: Small commercial landholder. 4 pp.

ALA, Atlas of Living Australia (2022). https://www.ala.org.au (Accessed 6 May 2022).

Bourdôt, G.W., Lamoureaux, S.L., Watt, M.S., Manning, L.K. and Kriticos, D.J. (2012). The potential global distribution of the invasive weed *Nassella neesiana* under current and future climates. *Biological Invasions* 14, 1545–56.

Cook, G.D. and Dias, L. (2006). Turner Review No. 12. It was no accident: deliberate plant introductions by Australia government agencies during the 20th century. *Australian Journal of Botany*, 54, 601–25.

Coutts-Smith, A.J. and Downey, P.O. (2006). 'Impact of weeds on threatened biodiversity in New South Wales'. Technical Series no. 11, CRC for Australian Weed Management, Adelaide. 97 pp.

Csurhes, S., Leigh, C. and Walton, C. (2016). African lovegrass: *Eragrostis curvula*. Invasive plant risk assessment. Queensland Government, Brisbane. 20 pp.

Gallagher, R.V., Englert Duursma, D., O-Donnell, J., Wilson, P.D., Downey, P.O., Hughes, L. and Leishman, M.R. (2013). The grass many not always be greener: projected reductions in climatic suitability for exotic grasses under future climates in Australia. *Biological Invasions* 15, 961–75.

Jacobs, S.W.L. (2022). Family Poaceae. The NSW Plant Information Network System. Royal Botanic Gardens and Domain Trust, Sydney. https://plantnet.rbgsyd.nsw.gov.au (Accessed 9 May 2022).

Johnston, B., McDonald, W., Freebairn, B. and Gardner, P. (2006). Consol lovegrass. Primefact 121. New South Wales Department of Primary Industries, Orange. 7 pp.

Johnston, W.H. (1989). Consol lovegrass (*Eragrostis curvula* complex) controls spiny burrgrass (*Cenchrus* spp.) in south-western New South Wales. *Australian Journal of Experimental Agriculture* 29, 37–42.

Jones, R.E. and Vere, D.T. (1998). The economics of serrated tussock in New South Wales. *Plant Protection Quarterly* 13, 70–6.

McLaren, D.A., Stajsic, V. and Gardener, M.R. (1998). The distribution and impact of South/North American stipoid grasses (Poaceae: Stipeae) in Australia. *Plant Protection Quarterly* 13, 62–70.

McLaren, D.A., Morfe, T.A. and Weiss, J. (2002). Distribution, economic impact and attitudes towards Chilean needlegrass (*Nassella neesiana* (Trin. & Rupr.) Barkworth) in Australia. Proceedings of the 13th Australian Weeds Conference, eds. H. Spafford Jacob, J. Dodd and J.H. Moore, pp. 749–52. (Plant Protection Society of Western Australia, Perth).

Millar, G., Simmons, A., Behrendt, K., Simpson, M. and Badgery, W. (2016). Determining the value of pasture to limit serrated tussock invasion. Unpublished report. New South Wales Department of Primary Industries, Orange. 21 pp.

Officer, D. (2022). Nomination of a candidate for biological control: African lovegrass. Unpublished report. New South Wales Department of Primary Industries, Grafton. 21 pp.

PlantNET (2022). *Eragrostis curvula* (Schrad.) Nees The NSW Plant Information Network System. Royal Botanic Gardens and Domain Trust, Sydney. https://plantnet.rbgsyd.nsw.gov.au (Accessed 6 May 2022).

Virtue, J.G., Bennett, S.J. and Randall, R.P. (2004). Plant introductions into Australia: how can we resolve 'weedy' conflicts of interest? Proceedings of the 14th Australian Weeds Conference, eds. B.M. Sindel and S.B. Johnson, pp. 42–8. (Weed Society of New South Wales, Wagga Wagga).

VRO, Victorian Resources Online (2022). African lovegrass (*Eragrostis curvula*). http://vro.agriculture.vic.gov.au/dpi/vro/vrosite.nsf/pages/weeds_perennial_african_lovegrass (Accessed 6 May 2022).

Watt, M.S., Kriticos, D.J., Lamoureaux, S.L and Bourdôt, G.W. (2011). Climate change and the potential global distribution of serrated tussock (*Nassella trichotoma*). *Weed Science* 59, 538–45.

Weeds Australia (2022). Weeds of National Significance. https://weeds.org.au/weeds-profiles/ (Accessed 5 May 2022).

# What is your state of vulnerability to climate change? Biosecurity and weed threats to New South Wales primary industries

Stephen B. Johnson[1,2], Jane M. Kelley[3,4], Bethany Ellis[3,4] and James Lawson[3,5]
[1]Weed Research Unit, Invasive Species, Biosecurity,
[2]1447 Forest Road, Orange 2800
[3]Vulnerability Assessment, Climate Change Unit, Agriculture
[4]11 Farrer Place, Queanbeyan 2620
[5]Locked Bag 26, Gosford 2250
Department of Primary Industries, New South Wales, Australia
(stephen.johnson@dpi.nsw.gov.au)

**Summary** Climate change in Australia is likely to result in increased temperatures, changes in rainfall patterns and a higher frequency of extreme climate events. An extensive review of 3280+ papers for the key terms 'climate change modelling' found that only 24 specifically examined weeds, almost all relevant to New South Wales (NSW) (Darbyshire *et al.* 2022). Based on these findings, the NSW Department of Primary Industries Vulnerability Assessment project is developing a series of Multi-Criteria Analysis (MCA) models to investigate the impact of climate change on two significant weeds, serrated tussock and parthenium weed.

Published literature and expert opinion were used to inform the development of these models. The process identified the key climatic variables (e.g. temperature, rain) for each life stage in the growth cycle. This data was then used to create the MCA hierarchical structure that underpins the climate suitability model. The model captures the extent to which the climate conditions satisfy seedling and reproductive growth requirements without considering other factors like management decisions, soil, and irrigation. The climate suitability was then mapped for NSW using historical climate data from 1970 to 2019 (50-years). The maps indicate areas of high and low climate suitability for the weeds. Data was extracted for incursion sites and established areas allowing the user to view the annual changes in climate suitability in NSW to validate the model.

Current and future state-wide climate suitability maps will be used to identify research, management and adaptation priorities that can be used strategically by industry, government and community to enhance weed risk assessment and management in NSW.

**Keywords** Impact Assessment, serrated tussock, *Nassella trichotoma*.

## INTRODUCTION

Australian primary producers operate in some of the worlds most naturally variable climates. Climate change has increased this variability, particularly with elevated temperatures and changed rainfall patterns (BoM 2022). Assessing the vulnerability of primary industries to future biosecurity/weed threats will be critical for adaptation and management.

We assessed the impacts of these threats on primary industries through a peer-reviewed (journal-only) literature search of 3280+ papers (Darbyshire *et al.* 2022). This search identified 188 relevant papers. Of the 55 papers that examined biosecurity, 24 specifically examined weeds, and almost all were relevant to New South Wales (NSW).

A second, broader review of weed species impacts on NSW was conducted (n=230+ papers). We know that NSW has 1750+ naturalised plant species, and many other non-naturalised plants (Johnson 2013). While the current/future habitat suitability of plant species is increasingly well known (e.g. the 700+ species listed at Weed Futures 2022), the degree to which their habitat suitability coincides with primary industry areas is largely unknown. We sought to partly address this as part of a broader agency study to examine the vulnerability of NSW primary industries to changing climate conditions. Two weeds: serrated tussock (*Nassella trichotoma* (Nees) Hack. ex Arechav.). and parthenium weed (*Parthenium hysterophorus* L. among 12 other biosecurity threats), were selected for their relevance to livestock and cropping commodities in the broader study. This paper reports only on the MCA model and outcomes for serrated tussock.

## PROJECT BACKGROUND

This project forms part of the NSW Primary Industries Climate Change Research Strategy (CCRS) (NSW DPI 2022), an investment of $29.2 M in projects to support the State's primary industries sector to find mitigation and adaptation strategies for primary producers to climate change. Part of this strategy is the Vulnerability Assessment (VA) project which aims to deliver a consistent and state-wide understanding of climate change risks and adaptation options for a broad range of industries and some of the biosecurity risks threatening them. The

VA project will provide strategic information for policy-makers, the government and the community.

## MATERIALS AND METHODS

**Climate suitability** This was determined using a Multi-Criteria Analysis (MCA) modelling approach. The methodology was used across 28 primary industry commodities and 14 biosecurity threats. While further details of MCA are outlined elsewhere (Saaty 1978, Romeijn *et al.* 2016); a key component is the use of an expert panel/focus group. This group critiqued the criteria within the MCA hierarchy and set the variable weightings, reflecting the importance of the variable to the weed species. Weightings are applied to each element of the model, such that each level of the model hierarchy is calculated as the weighted sum of elements in the level below. The model weightings were determined through a standard analytic hierarchy process (Saaty 1978).

**Serrated tussock** Serrated tussock is a Weed of National Significance. It is an invasive perennial grass weed that invades cool-season and temperate grasslands. Often forming near monocultures it reduces pasture productivity and palatability, and hence animal carrying capacity (Weeds Australia 2022). It contaminates meat, hides and wool. It readily invades both disturbed pastoral and threatened grassland ecosystems threatening non and endangered plant and animal species (Coutts-Smith and Downey 2006), ecosystem structure and function. Further, serrated tussock and similar grasses reduce community resilience. The costs associated with managing these species and their spread result in a loss of services and facilities, and human and social capital.

**Serrated tussock MCA** An MCA was designed to reflect the variables that influence the climate suitability of serrated tussock in NSW. This MCA used a range of national and international literature (e.g. Kriticos *et al.* 2004, Millar *et al.* 2016, Humphries *et al.* 2018, Ruttledge *et al.* 2020) and focus group observations where data was deficient. Due to Covid-19, an online focus group (see acknowledgements for membership) was convened after the initial MCA structure was designed.

The focus group recommended structural changes to the serrated tussock MCA, climate range threshold values, and participated in the analytic hierarchy process to weight the MCA variables. The final MCA structure is illustrated and was run seasonally (i.e. 3-monthly, Figure 1). It contains only the critical seedling and reproductive growth lifecycle stage arms, with the seedling arm further broken down into sub-component Establishment and Survival stages. The impact of mean temperature and cumulative rainfall on the two seedling stages is expressed in a four (rainfall) x five (temperature) matrix for each of these stages (not illustrated). Similar temperature and rainfall categories were used in the reproductive growth stage. Hot 'bombs' were employed as biological species limits.

The MCA model outputs were iteratively checked and reviewed by the VA team and focus group participants to explore and examine the implications of the MCA structure and consider whether the MCA variables and resulting output were consistent with expert knowledge.

**Interpreting model outputs** The serrated tussock MCA is a *climate suitability* model. *Climate suitability* is derived from climate data as a unitless index scaled between zero and one. *Climate suitability* is defined as the extent to which climate conditions satisfy the requirements of plant or animal growth without considering other limiting factors. The MCA does not account for a range of factors, including management decisions, soil, irrigation, topography and climatic extremes (e.g. drought, bushfires, flooding, hail, and consecutive hot or cold days unless specified). *Climate suitability* allows us to determine the weather conditions that are highly unfavourable (0) through to highly favourable (1) for serrated tussock growth (Figure 2). Sub-optimal conditions do not mean that the weed cannot grow, only that these conditions are less suitable.

Climate suitability results were mapped for each season (i.e. summer, autumn, winter and spring). Climate suitability was considered spatially across NSW for the years 1970 to 2019. The invaded sites were predominantly higher rainfall (600+ mm mean annual), cool temperate pasture/grazing sites, including Armidale (Northern Tablelands); Orange (Central Tablelands); Yass (Southern Tablelands/South-Western Slopes), and Young (Central Western Slopes). Three outlier sites: Tibooburra, Alstonville, and Tumbarumba were used to check model calibration. These sites were selected because Tibooburra best represents a "hot/dry extreme" area, Alstonville a "warm/wet extreme area" and Tumbarumba a "cold extreme" area.

## RESULTS AND DISCUSSION

**Weightings in the serrated tussock MCA model** Conditions at the seedling lifecycle stage were weighted heavily at 0.88 (out of 1) as seedling growth contributed a large amount to successful serrated tussock invasion (Figure 1). Reproductive growth was weighted far lower at 0.12. The overall seedling stage was composed of two sub-components, establishment (0.13) and survival

(0.87). The substantial differences between the seedling survival and establishment ratings reflects the relative importance of the survival lifecycle stage (and sub-component) to the serrated tussock MCA model and the invasion success (Figure 2).

**Validation of the historic serrated tussock MCA outputs, and future projections** The historical analysis was carefully validated by analysing climate data at previous invaded and non-invaded areas from 1970 to 2019. The accuracy of these historical analyses provide confidence in the models performance to which we are now applying climate projection data (from 2036 to 2065, centered on 2050). These projection layers will be overlaid with areas of potentially impacted primary industries, particularly high rainfall pastures, mixed cropping/grazing pasture systems, sheep and cattle production based on these systems, and select horticultural industries, to assess future interactions between the weed and production systems.

**Serrated tussock MCA model outputs** Historic climate suitability for serrated tussock in NSW extends throughout the Northern, Central, and Southern Tablelands (Figure 2). These areas include high-value pasture and wool production systems. Suitability extends to the South Coast (southern polygon) and to higher altitude parts of the North Western, Central Western, and South Western Slopes (west of the polygons). These areas are part of the highly productive mixed wheat-sheep zone. The area between the two polygons (the Hunter Valley to the east and west of Tamworth) are active invasion fronts (Auld and Johnson 2014, ALA 2022). In contrast, much of western NSW is of lower suitability.

There is an area of relatively high suitability to the west of Kosciuszko National Park (KNP) and Tumbarumba. The Tumbarumba area is a high-value mosaic-area of horticultural crops interspersed with forestry (pine) and high rainfall cattle grazing. There are almost no records of serrated tussock from KNP, east of Tumbarumba but west of the boundary of the southern polygon. This cannot easily be explained because of the suitability of similar invaded areas around Yass, and into the Australian Capital Territory) and that the weed has had 80+ years to spread from Yass where it was first identified in 1937 from over 20,000 ha (Blackmore 2019).

IMPLICATIONS – HISTORICAL ANALYSES

Serrated tussock has a long history in NSW, with invasions in the Southern and Central Tablelands before 1937, and the Northern Tablelands before 1955 (Blackmore 2019). Despite this, the weed's continued and inevitable spread should not be

**Figure 1.** High-level serrated tussock MCA structure. Values denoted by 'W=' represent model weightings.

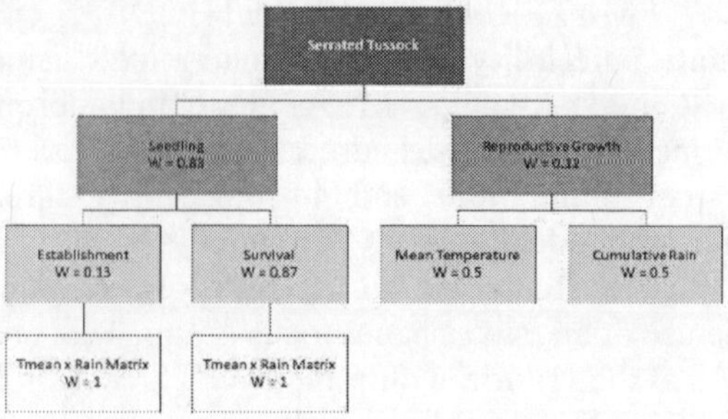


**Figure 2.** Average historical climate suitability for serrated tussock in NSW (1970 to 2019). The polygons enclose 95+ % of herbarium incidence data (from ALA 2022,not illustrated here). North is at the top of the map.

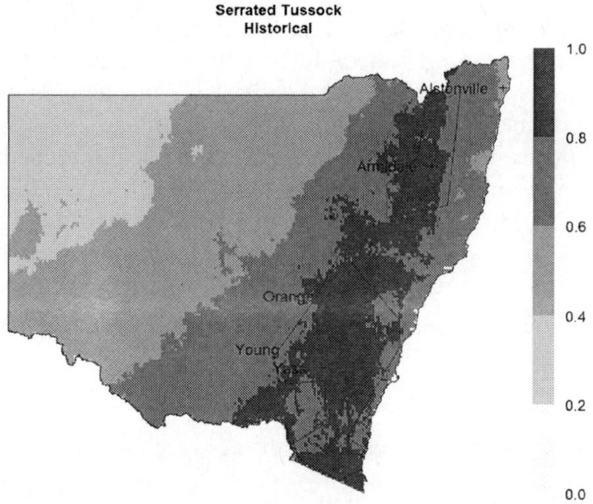


assumed, particularly under the combined conditions of drought, high grazing pressure, and disturbance. Various studies have indicated that the future distribution of serrated tussock will decrease within NSW and Australia (e.g. Kriticos *et al.* 2004, Watt *et al.* 2011, Gallagher *et al.* 2013, Millar *et al.* 2016, Weed Futures 2022). As such management would be best focused on eradicating outlying infestations in the Hunter, Northern Tablelands and on the North Western Slopes of NSW, and on containment further south. Opportunities also exist to revegetate both areas where serrated tussock has been managed and areas prone to invasion, particularly with more desirable and competitive native and/or exotic pasture species (while avoiding new weedy species).

More broadly, there is an opportunity to examine further weeds, and identify and examine a broader range of biosecurity threats, including possible new threats to Australian and NSW primary industries using this method. A systematic, nation- and industry-wide analysis of impacts and adaptation activities would help prepare Australian primary industries for future biosecurity risks and management changes. Coordinated risk assessment/management and nation-building-size investment by industry, government and community will be needed.


ACKNOWLEDGMENTS

We are indebted to our NSW DPI focus group members: Dr Warwick Badgery (Research Leader Rangelands and Tropical Pastures), Dr Aaron Simmons (Technical Specialist Climate Research) both former scholars on the species; and Mr Mich Michelmore (State Priority Weed Coordinator and National Serrated Tussock Management Group representative) for their considerable expertise and guidance in this work. Our considerable thanks also go to Dr Joanna Pardoe, Mr Gary Allan, Ms Mary Timms and Mr Lachlan Phillips (current VA members) and Dr Rebecca Darbyshire (a former member, now with CSIRO) and the peer reviewers. Project funding was part of the NSW DPI CCRS.


REFERENCES


ALA, Atlas of Living Australia (2022). https://www.ala.org.au (Accessed 17 May 2022).

Auld, B.A. and Johnson, S.B. (2014). Invasive alien plant management. *CAB Reviews: Perspectives in Agriculture, Veterinary Science, Nutrition and Natural Resources* 9, No. 37. 12 pp. doi: 10.1079/PAVSNNR20149037

BoM, Bureau of Meteorology (2022). State of the Climate 2020. http://www.bom.gov.au/state-of-the-climate/australias-changing-climate.shtml (Accessed 20 May 2022).

Blackmore, P. (2019). Successes and pitfalls – a review of four historic weed incursions in NSW. Proceedings of the 20th NSW Biennial Weeds Conference, eds. K. Hignell, H. Wu and M. Walsh, pp. 21–7. (Weed Society of New South Wales, Newcastle).

Coutts-Smith, A.J. and Downey, P.O. (2006). 'Impact of weeds on threatened biodiversity in New South Wales'. Technical Series no. 11, CRC for Australian Weed Management, Adelaide. 97 pp.

Darbyshire, R.O., Johnson, S.B., Anwar, M.R., Ataollahi, F., Burch, D., Champion, C. *et al.* (2022). Climate change and Australia's primary industries: factors hampering an effective and coordinated response. *International Journal of Biometeorology*, https://doi.org/10.1007/s00484-022-02265-7

Gallagher, R.V., Englert Duursma, D., O-Donnell, J., Wilson, P.D., Downey, P.O., Hughes, L. and Leishman, M.R. (2013). The grass may not always be greener: projected reductions in climatic suitability for exotic grasses under future climates in Australia. *Biological Invasions* 15, 961–75.

Johnson, S.B. (2013). Some weeds have no boundaries. What are the next steps we need to take with species that jump the fence? Proceedings of the 17th Biennial NSW weeds conference, ed. H. Wu, pp. 61–9. (Weed Society of New South Wales, Corowa).

Kriticos, D.J., Lamoureaux, S., Bourdôt, G.W. and Pettit, W. (2004). *Nassella* tussock: Current and potential distributions in New Zealand. *New Zealand Plant Protection* 57, 81–88.

McLaren, D.A., Stajsic, V. and Gardener, M.R. (1998). The distribution and impact of South/North American stipoid grasses (Poaceae: Stipeae) in Australia. *Plant Protection Quarterly* 13, 62–70.

Millar, G., Simmons, A., Behrendt, K., Simpson, M. and Badgery, W. (2016). Determining the value of pasture to limit serrated tussock invasion. Unpublished report. New South Wales Department of Primary Industries, Orange.

NSW DPI (2022). Climate Change Research Strategy. https://www.dpi.nsw.gov.au/dpi/climate/about-dpi-climate/climate-change-research-strategy (Accessed 9 May 2022).

Romeijn, H., Faggian, R., Diogo, V. and Sposito, V. (2016). Evaluation of deterministic and domplex analytical hierarchy process methods for agricultural land suitability analysis in a changing climate. *ISPRS International Journal of Geo-Information* 5, 99.

Ruttledge, A., Whalley, R.D.B., Falzon, G., Backhouse, D. and Sindel, B.M. (2020). The role of soil temperature and seed dormancy in the creation and maintenance of persistent seed banks of *Nassella trichotoma* (serrated tussock) on the Northern Tablelands of New South Wales. *The Rangeland Journal* 42, 85–95.

Saaty, R.W. (1987). The analytic hierarchy process—what it is and how it is used. *Mathematical Modelling* 9, 161–76.

Watt, M.S., Kriticos, D.J., Lamoureaux, S.L and Bourdôt, G.W. (2011). Climate change and the potential global distribution of serrated tussock (*Nassella trichotoma*). *Weed Science* 59, 538–45.

Weed Futures (2022). Determining the current and future weed threats in Australia. http://weedfutures.net/ (Accessed 9 May 2022).

Weeds Australia (2022). Weeds of National Significance. https://weeds.org.au/weeds-profiles/ (Accessed 5 May 2022).

# Getting the most out of herbicides and combinations for lentil (*Lens culinaris*) crop safety and broadleaf weed control on light to medium textured soils

Navneet Aggarwal[1,2], Sam Trengove[3], Jordan Bruce[3], Stuart Sherriff[3], and Penny Roberts[1,2]
[1]South Australian Research and Development Institute, 70 Farrell Flat Road, Clare-5453, South Australia.
[2]The University of Adelaide, Waite Campus, Urrbrae-5064, South Australia.
[3]Trengove Consulting, PO Box 6, Bute-5560, South Australia.
(navneet.aggarwal@sa.gov.au)

**Summary:** Crop safety and effective weed control with Reflex® (fomesafen 240 g $L^{-1}$), Group 14 herbicide and its combination with other registered herbicides was assessed in lentil on different soil types at four sites in South Australia. Lentil crop safety varied between acidic and alkaline sands with the use of Reflex®, diuron, metribuzin and terbuthylazine herbicides, with alkaline sand sites incurring more herbicide damage than the acidic sand site. Crop damage with Reflex® on alkaline sands was rate responsive, with yield loss increasing from 17% when applied at 0.5L $ha^{-1}$ to 54% when applied at 1L $ha^{-1}$. Crop damage on alkaline sands was cumulative when Reflex® was applied in combination with a Group 5 herbicide, such as diuron. Effective control of Bifora (*Bifora testiculata*), common sowthistle (*Sonchus oleraceus*), Indian hedge mustard (*Sisymbrium orientale*) and wild turnip (*Brassica tournefortii*), including populations resistant to imidazolinone herbicides, was achieved with Reflex®. Wide-spectrum broadleaf weed control was achieved using Reflex® in combination with other registered Group 2, 5 and 12 herbicides.

**Keywords:** herbicide efficacy, herbicide tolerance,
lentil, sandy soils.

## INTRODUCTION

Effective broadleaf weed control is a major constraint to achieving full yield potential in pulse crops. The adoption of herbicide tolerant pulse crops has improved broadleaf weed control options. However, it has resulted in over-reliance on a few modes of action, particularly Group 2 imidazolinone (IMI) herbicides. Reflex® (fomesafen) herbicide has been recently registered for chickpea, narrow leaf lupin, lentil, field pea, faba bean and vetch to control range of broadleaf weeds. Of all the pulses with a Reflex® registration, lentil is the most sensitive, with a maximum rate of 1L $ha^{-1}$ incorporated by sowing (IBS) only, whilst other legume species have a maximum rate of 1.25 L $ha^{-1}$ post-sowing and pre-emergence (PSPE) (except vetch, maximum 0.9L $ha^{-1}$ PSPE) or 1.5L $ha^{-1}$ IBS. A new mode of action registered in lentils will provide herbicide rotation options for both conventional and herbicide tolerant cultivars, and will be particularly useful where herbicide resistance is developing or already present for Group 2 herbicides.

Previous research studies have investigated lentil crop safety and weed control on sandy soils of the Northern Yorke Peninsula for Group 2, Group 5 and Group 12 herbicides. This work highlighted the heightened risk of crop damage from soil residual herbicides on these soil types, in particular the Group 2 and 5 herbicides (Trengove et al. 2021). The current research studies have extended this work, including Reflex®, investigating herbicide crop safety on a range of soil types, including differences in soil texture and pH.

## MATERIALS AND METHODS

A total of four experiments were established in 2021 to assess herbicide tolerance and broadleaf weed control on imidazolinone (IMI) tolerant lentils (Table 1). Two of these four experiments were established at Alford and Bute 1 (Northern Yorke Peninsula) on sandy soils with either high or low soil pH to assess crop safety when using Group 2, 5, 12 and 14 pre-emergent and/or post-emergent herbicides. Other two experiments were established at Bute (2 & 3) to develop strategies for controlling broadleaf weeds on loamy soils, and sandy alkaline soils. The treatments included combinations from Group 2 (Intercept®), 5 (metribuzin, diuron and Terbyne®) and 14 (Reflex®) in a randomised complete block design with three replicates (Tables 2 and 3).

Experiments were sown to PBA Hurricane XT^A using knife points and press wheels between 26 May and 4 June, 2021. Two major rainfall events occurred after seeding, with 27.6 mm and 24.0 mm of rainfall received within the first and second week, respectively. A total of 278 mm was received between seeding and harvest at Bute. Post-emergent herbicide treatments were applied at 5-6 crop node stage. Herbicides were applied using hand boom equipment delivering 100 L ha$^{-1}$ water at a pressure of 200 kPa. Plots at the herbicide tolerance sites were rolled post-emergent compared to the weed control experiments which were rolled with seeding. Broadleaf weed counts from herbicide tolerance experiments (Alford and Bute 1) were taken four weeks after post-emergent herbicide treatments and were removed by hand after counting to determine herbicide effects in the absence of weeds. Broadleaf weed pod/seed set counts (Bute 2 and Bute 3) were taken 130 days after herbicide treatments using a 0.25 m$^2$ quadrat placed at three random locations in each plot.

**Table 1:** Descriptions for the four trial sites established in 2021.

| Location | Site | 0-10 pH ($CaCl_2$) | 0-10 pH ($H_2O$) | ECEC Cmol kg$^{-1}$ | OC (%) | Texture |
|---|---|---|---|---|---|---|
| Alford | Alkaline herbicide tolerance | 7.7 | 8.4 | 11.7 | 0.94 | Sand |
| Bute 1 | Acidic herbicide tolerance | 4.7 | 5.8 | 3.09 | 0.76 | Sand |
| Bute 2 | Loam weed control | 7.5 | 8.1 | - | 1.33 | Loam |
| Bute 3 | Sand weed control | 6.8 | 8.1 | - | 0.82 | Loamy sand |

## RESULTS AND DISCUSSION

*Crop safety*

At the alkaline site (Alford), the Group 5 herbicides diuron and Terbyne® caused significant herbicide damage with scores for necrosis reaching 6.2 out of 9 from the application of Terbyne® (Figure 1). Reflex® caused significant damage at this site but in the form of leaf chlorosis. The combination of Group 5 and 14 herbicides did not lead to increased leaf necrosis or chlorosis damage. In contrast, at the acidic site (Bute 1), there were only minor leaf necrosis symptoms evident in association with the application of diuron, and no other herbicide was significantly different from the control treatment. Reflex® caused stunting in lentil as the rate increased from 500 to 1000 mL ha$^{-1}$ in weed control experiments (Bute 2 & 3) (data not shown) and the effect was more pronounced in alkaline sands than in loamy soils.

**Figure 1.** Leaf necrosis (left) and chlorosis (right), scored 13 July at Alford (alkaline sand) and 20 July at Bute (acidic sand) (0 = no chlorosis, 9 = death) of PBA Hurricane XT$^A$ . Lower case letters and upper-case letters denote significant differences for each site, P values = <0.001.

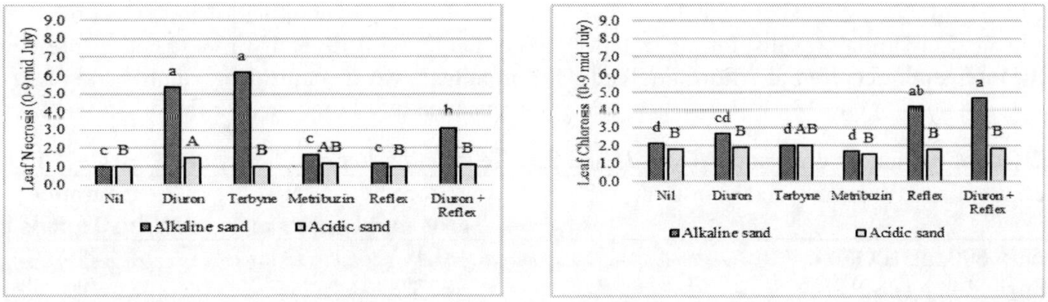


Herbicide damage on this sandy soil resulted in growth and biomass reduction (data not shown) and led to decreased yields. Grain yield was significantly reduced in response to the application of some herbicide treatments at the alkaline sand trial site, consistent with earlier herbicide damage scores (Table 3). Diuron and Reflex® treatments both reduced grain yields by 20% when applied alone, and Terbyne® reduced yield by 51%. This contrasts with the acidic sand site where no significant yield differences occurred in response to the application of any individual herbicide.

Diuron and Reflex® were applied in combination at the Alford alkaline site increased yield loss to a 52% compared to the untreated control. Post-emergent herbicides Intercept® and diflufenican (DFF) did not cause yield loss at either site, which is consistent with results of Trengove et al. (2021) for similar soil types. Generally, DFF and Intercept® were also safe to apply following application of either diuron or Reflex® IBS. Where these had caused damage at the alkaline sand site, the post-emergent applied herbicides did not exacerbate the damage. However, the most damaging combination of herbicide at the alkaline sand site was diuron plus Reflex® applied IBS followed by DFF post-emergent and reduced grain yield by 79%. Grain yield loss at the alkaline sand trial (Bute 3) varied depending on the Reflex® rate applied with the 500 mL, 750 mL and 1000 mL ha$^{-1}$ rates yielding

83%, 76% and 46% of the untreated plot, respectively. This indicates that if rates can be reduced and weed control is still maintained, the crop safety margin can be improved.

Reflex® was also included in 2020 experiments and, whilst similar herbicide damage symptoms were present on an alkaline sand, this did not translate into any yield loss in 2020. There were no herbicide damage symptoms or yield loss at the acidic sand site in 2020. A reason for the increased herbicide damage in the 2021 season may be due to more rainfall in the weeks following sowing, which may have moved the herbicide further into the soil profile and laterally into the crop row, with June 2021 rainfall recorded at 56 mm compared to 19 mm in June 2020. Greater spring rainfall in 2020 is also likely to have contributed to better crop recovery.

*Broadleaf weed control*

Reflex® was effective in controlling 94-98% of bifora at rates of between 500 and 1000 mL $ha^{-1}$ (Table 2). Intercept®, on its own or in combination with Reflex®, provided excellent control of bifora, reducing seed set to <1 bifora seed $m^{-2}$ compared to existing pre-emergent herbicide options metribuzin and Terbyne® with 323 and 1672 bifora seeds $m^{-2}$, respectively. Similarly, the combination of Reflex® + Intercept® provided high levels of common sowthistle control at all sites (Tables 2 and 3). Intercept® did not provide adequate control of Indian hedge mustard (IHM) and was not different to the untreated control at the loam site (Bute 2) (Table 2). Similar results for poor IHM control with Intercept® occurred at Alford and Bute 1 (Table 3), and Bute 3 (data not shown) sites. However, wild turnip was effectively controlled with Intercept®. This poor control of IHM may be explained by the increase of IHM populations resistant to imidazolinone herbicides in this area.

Reflex® was effective at controlling IMI resistant IHM populations. The level of weed control improved with increasing Reflex® rates from 500 ml $ha^{-1}$ (217 IHM pods $m^{-2}$) to 1000 ml $ha^{-1}$ (24 IHM pods $m^{-2}$) (Table 2). Most of the surviving IHM plants in Reflex® plots were found in the in-row spaces, from where the applied herbicide was likely moved out by the seeding operation. When Reflex® IBS was followed by a Group 5 herbicide, metribuzin/Terbyne® applied as PSPE, the surviving weeds in the in-row area were mostly controlled. Herbicide combinations by including Intercept® proved effective for medic control (Table 3).

The availability of the new Group 14 herbicide Reflex® has increased the options for achieving improved broadleaf weed control in lentil, including weeds resistant to IMI herbicides. Careful decisions regarding safe dosage rates of Reflex®, governed by the soil type, and a follow-up application of Group 2, 5 and Group 12 herbicides provide broad-spectrum broadleaf weed control in lentil. IMI herbicides will continue to be a valuable tool for broadleaf weed control in lentil, especially for weeds that have not evolved resistance to this mode of action, and the weeds such as medics that are not effectively controlled with other herbicides. Using Reflex® in conjunction with Group 2, 5 and 12 herbicides will diversify the selection pressure for broadleaf weed control in lentil and delay the resistance build up to a specific mode of action.

**Table 2.** Effect of herbicides on broadleaf weeds and their seed set on loam soils at Bute 2, 2021.

| Herbicide treatment (commercial product rate $ha^{-1}$) | Bifora seeds $m^{-2}$ | IHM pods $m^{-2}$ | Common sowthistle pods $m^{-2}$ |
|---|---|---|---|
| 1. Intercept® 600 mL (POST) | $0.4^{c}$ | $731^{a}$ | $4^{de}$ |
| 2. Metribuzin 200 g (PSPE) | $323^{b}$ | $1^{de}$ | $0^{f}$ |
| 3. Reflex® 500 mL (IBS) | $35^{c}$ | $217^{bc}$ | $12^{bcd}$ |
| 4. Reflex® 500 mL (IBS) + Intercept® 600 mL (POST) | $0^{c}$ | $409^{ab}$ | $1^{ef}$ |
| 5. Reflex® 500 mL (IBS) + Metribuzin 200 g (PSPE) + Intercept® 600 mL (POST) | $0^{c}$ | $24^{de}$ | $0^{f}$ |
| 6. Reflex® 500 mL (IBS) + Terbyne® 1000 g (IBS) + Intercept® 600 mL (POST) | $0.4^{c}$ | $0^{e}$ | $0^{f}$ |
| 7. Reflex® 750 mL (IBS) | $7^{c}$ | $64^{cde}$ | $15^{abc}$ |
| 8. Reflex® 750 mL (IBS) + Intercept® 600 mL (POST) | $0^{c}$ | $81^{cde}$ | $1^{ef}$ |
| 9. Reflex® 750 mL (IBS) + Metribuzin 200 g (PSPE) + Intercept® 600 mL (POST) | $0^{c}$ | $0^{e}$ | $0^{f}$ |
| 10. Reflex® 750 mL (IBS) + Terbyne® 1000 g (IBS) + Intercept® 600 mL (POST) | $0^{c}$ | $10^{de}$ | $0^{f}$ |
| 11. Reflex® 1000 mL (IBS) | $21^{c}$ | $24^{de}$ | $21^{ab}$ |
| 12. Terbyne® 1000 g (IBS) | $1672^{a}$ | $105^{cd}$ | $5^{cde}$ |
| 13. Unweeded control | $1987^{a}$ | $836^{a}$ | $29^{a}$ |

**Table 3**. Broadleaf weed control on an alkaline and acidic sandy soil at Alford and Bute 1, respectively, in 2021.

| Herbicide (commercial product rate $ha^{-1}$) | Alkaline sand at Alford % weed control | | | | Grain | Acidic sand at Bute 1 % weed control | | | | Grain |
|---|---|---|---|---|---|---|---|---|---|---|
| | Medic | IHM | Wild turnip | Common sowthistle | yield ($t\ ha^{-1}$) | Medic | IHM | Wild turnip | Common sowthistle | yield ($t\ ha^{-1}$) |
| 1. Nil | $0^{a}$ | $0^{a}$ | $0^{a}$ | $0^{a}$ | $1.61^{abc}$ | $0^{a}$ | $0^{a}$ | $0^{a}$ | $0^{a}$ | $1.0^{abc}$ |
| 2. Diuron 830 g (IBS) | $63^{b}$ | $93^{cd}$ | $100^{c}$ | $96^{b}$ | $1.29^{cd}$ | $38^{abc}$ | $80^{b}$ | $82^{b}$ | $84^{cd}$ | $1.04^{abc}$ |
| 3. Terbyne® 750 g (IBS) | $78^{bc}$ | $96^{cde}$ | $100^{c}$ | $96^{b}$ | $0.79^{ef}$ | $30^{abc}$ | $87^{bc}$ | $78^{b}$ | $90^{cd}$ | $0.86^{bc}$ |
| 4. Metribuzin 180 g (IBS) | $53^{ab}$ | $74^{ab}$ | $98^{b}$ | $48^{a}$ | $1.74^{ab}$ | $4^{ab}$ | $73^{b}$ | $76^{b}$ | $57^{b}$ | $0.91^{abc}$ |
| 5. Intercept® 500 mL (POST) | $93^{d}$ | $70^{ab}$ | $98^{b}$ | $85^{b}$ | $1.60^{abc}$ | $80^{ef}$ | $0^{a}$ | $100^{d}$ | $84^{c}$ | $0.91^{abc}$ |
| 6. Diuron 830 g (IBS) + Intercept® 500 mL (POST) | $89^{cd}$ | $96^{de}$ | $100^{c}$ | $96^{b}$ | $1.14^{de}$ | $70^{bcde}$ | $70^{b}$ | 99d | $97^{ef}$ | $1.09^{abc}$ |
| 7. Diflufenican 150 mL (POST) | $53^{ab}$ | $100^{g}$ | $100^{c}$ | $93^{b}$ | $1.78^{a}$ | $42^{abcd}$ | $100^{e}$ | $100^{d}$ | $90^{cd}$ | $0.91^{abc}$ |
| 8. Diuron 830 g (IBS) + Diflufenican 150 mL (POST) | $58^{b}$ | $100^{g}$ | $99^{c}$ | $100^{b}$ | $1.26^{cd}$ | $90^{efg}$ | $100^{e}$ | $100^{d}$ | $100^{f}$ | $0.98^{abc}$ |
| 9. Diuron 830 g (IBS) + Diflufenican 150 mL (POST) + Intercept® 500 mL (POST) | $91^{d}$ | $100^{g}$ | $100^{c}$ | $96^{b}$ | $0.76^{f}$ | $98^{fg}$ | $100^{e}$ | $100^{d}$ | $100^{f}$ | $0.85^{c}$ |
| 10. Reflex® 1000 mL (IBS) | $54^{ab}$ | $96^{de}$ | $99^{c}$ | $93^{b}$ | $1.29^{cd}$ | $0^{a}$ | $93^{cd}$ | 94c | $94^{de}$ | $1.14^{ab}$ |
| 11. Reflex® 1000 mL (IBS) + Intercept® 500 mL (POST) | $88^{cd}$ | $96e^{fg}$ | $100^{c}$ | $100^{b}$ | $1.41^{bcd}$ | $80^{cde}$ | $97^{d}$ | $100^{d}$ | $100^{f}$ | $0.92^{abc}$ |
| 12. Reflex® 1000 mL (IBS) + Diflufenican 150 mL (POST) | $57^{b}$ | $100^{g}$ | $100^{c}$ | $100^{b}$ | $1.54^{abc}$ | $78^{def}$ | $100^{e}$ | $100^{d}$ | $100^{f}$ | $0.88^{bc}$ |
| 13. Reflex® 1000 mL (IBS) + Diflufenican 150 mL (POST) + Intercept® 500 mL (POST) | $93^{d}$ | $100^{g}$ | $100^{c}$ | $100^{b}$ | $1.44^{abcd}$ | $100^{g}$ | $100^{e}$ | $100^{d}$ | $100^{f}$ | $1.06^{abc}$ |
| 14. Diuron 830 g (IBS) + Reflex® 1000 mL (IBS) | $72^{bc}$ | $97^{def}$ | $99^{c}$ | $93^{b}$ | $0.78^{f}$ | $18^{ab}$ | $96^{d}$ | 97c | $91^{cd}$ | $1.19^{a}$ |
| 15. Diuron 830 g (IBS) + Reflex® 1000 mL (IBS) + Intercept® 500 mL (POST) | $87^{cd}$ | $96^{de}$ | $100^{c}$ | $96^{b}$ | $0.88^{ef}$ | $88^{efg}$ | $97^{d}$ | $100^{d}$ | $100^{f}$ | $1.0^{abc}$ |
| 16. Diuron 830 g (IBS) + Reflex® 1000 mL (IBS) + Diflufenican 150 mL (POST) | $75^{bc}$ | 100g | $100^{c}$ | $100^{b}$ | $0.33^{g}$ | $66^{cde}$ | $100^{e}$ | $100^{d}$ | $100^{f}$ | $0.97^{abc}$ |
| 17. Diuron 830 g (IBS) + Reflex® 1000 mL (IBS) + Diflufenican 150 mL (POST) + Intercept® 500 mL (POST) | $93^{d}$ | $100^{fg}$ | $100^{c}$ | $100^{b}$ | $0.52f^{g}$ | $86^{efg}$ | $100^{e}$ | $100^{d}$ | $100^{f}$ | $0.85^{c}$ |
| Weed density in nil (weeds $plot^{-1}$) | 159 | 91 | 56 | 9 | - | 16 | 119 | 87 | 22 | - |
| Weed density in nil (weeds $m^{-2}$) | 10.6 | 6.1 | 3.7 | 0.6 | - | 1.1 | 7.9 | 5.8 | 1.5 | - |

ACKNOWLEDGEMENTS

The research undertaken as part of this project is made possible by the significant contributions of growers through both trial cooperation and the support of the GRDC and SAGIT, the authors would like to thank them for their continued support.

REFERENCES

Trengove, S., Sherriff, S. and Bruce, J. (2021). Increasing reliability of lentil production on sandy soils. *Proceedings GRDC Grains Research Update*, Adelaide, February 2021, pp. 99-106.

# HARDI GeoSelect

Heath Thompson[1], Andrew Snowball[1]
[1]Hardi Australia Pty Ltd, Cavan, Australia
(HAU.innovation@hardi.com)

**Summary** In most agricultural, horticultural and viticulture practices, spraying is undertaken on a 'per area' process. For example, a broadacre paddock is usually sprayed with the same rate of chemical across the entire area, regardless of the variability of the intended target/s across that paddock. As weeds and other targets are rarely evenly distributed across areas and only makeup a fraction of the total area, there are large amounts of inefficiencies and wastage applying a blanket treatment. To combat this, various processes of selective spraying targets have been analyzed and refined. We have compared the ability of existing selective sprayers, and then refined the most viable into a working prototype. The various selective spraying technologies will be compared in the following parameters.

- Target detection rate (percentage of actual targets)
- Treatment efficiency of target (percentage of actual targets)
- Speed of treatment (area per hour)
- Other strategic advantages
- Disadvantages
- Cost of technology

Our results show there are significant disadvantages in the existing selective spraying technology, most of which comprise of some type of camera or image sensor directly mounted to the sprayer. As the sprayer passes over the intended target, the system records a detection and then actuates the spray for that area. Our newly refined system to remotely fly the image sensor over the area prior to the sprayer negates the need for multiple image sensors, speed of treatment limitations and provides an overview of the area so that strategic decisions can be made with up-to-date information. This study emphasizes the need for a broader look into how weeds and other targets are treated from a workflow practice perspective in agricultural, horticultural and viticulture settings. Furthermore, what limitations there are with existing selective spraying systems and how they can be improved.

**Keywords** Selective spraying, spatial data, geoselect, Hardi, sprayer

# Siam weed and the Dust Devils. Managing *Chromolaena odorata* in the Northern Territory

Thomas Price; Northern Territory Government - Department of Environment, Parks and Water Security, PO Box 496 Palmerston, Northern Territory, Australia.

**Summary** Chromolaena (*Chromolaena odorata*) was discovered for the first time in the Northern Territory in 2019. Apart from two small roadside outliers, the following two years of survey found all plants within ~10km from 'ground zero'. If this remained all that was present in the NT the case for eradication was considered ambitious, but feasible.

Delimitation of a weed in the landscape is key to eradication. Debate ensued on the best allocation of limited survey resource to do so.

Four weeks of helicopter survey took place in July 2021, crossing five major river catchments across approximately 9,500 km$^2$. The results of the survey changed the direction of management in the NT. Eradication was now deemed unachievable.

Locations of new infestations, many in areas unrelated to infrastructure or animal tracks, can only be attributed to long-distance spread by wind.

The survey results underpin the new reality of managing chromolaena in the Northern Territory. Focus has shifted from an in-house eradication program to land holder extension; with a view to containment, early intervention of new infestations and mitigation of spread. Effort is being channeled into control research and remote survey techniques, ultimately providing best practice tools for management applicable to the Northern Australian savanna.

**Keywords** Eradication, Chromolaena odorata, Siam weed, Remote sensing, Dust Devils

## THE PREMIER WEED THREAT TO NORTHERN AUSTRALIA

'***You might have gamba and mimosa, but at least you don't have Siam weed'***- was a phrase used to warn Northern Territory (NT) weed managers by 'those in the know'; Land managers, scientists and researchers familiar with chromolaena in Asia and Northern Queensland.

Commonly known as Siam weed, chromolaena (*Chromolaena odorata* (L.) R.M. King & H. Rob) is widely regarded as one of the world's worst tropical weed species. It grows as a 3-4 metre tall sprawling shrub that forms dense, entwined thickets. If supported, chromolaena's scrambling branches can climb to 20 metres. Exuding allelopathic compounds the plants smother and suppress all vegetation below.

During seeding, the plants are adorned with fluffy balls of barbed parachute-like seeds that float from the plant with the slightest knock or puff of wind. Across Asia, Africa and the Pacific islands, the weed is renowned for out-competing horticulture, native vegetation and reducing available pasture. It is an allergen to many humans and toxic to livestock-leading to aborted calves and, in extreme cases, death (Parsons & Cuthbertson, 2001).

During a trip to Timor and witnessing the vast understory of chromolaena throughout the landscape, Wilson (1994) could not help but note the similarities in the climate and eucalypt savannah vegetation with the Top End. His report warns of chromolaena as "*the premier weed threat to Northern Australia*" and provides a depressing vision that unless action is taken to prevent chromolaena establishing he could see a Top End covered in mimosa, chromolaena and rubber vine.

The close proximity of major chromolaena infestations across Asia, and now sadly northern Queensland (QLD), meant the likelihood of it finding its way to the NT was feasible, and somewhat expected. Not so much a matter of if, but how, when and where?

Historically, major efforts have been undertaken to prevent chromolaena spreading to our shores. It was considered such a threat that following the INTERFET campaign in Timor Leste, Australian Quarantine staff were posted overseas to inspect the wash-down of returning military equipment. The effort was of a massive scale. Reportedly up to 300 Defence personnel operated 20 wash stations, 18 hours a day for three months (Waterhouse & Zeimer, 2000).

Though the NT is remote place, the pathways for weed spread generally remain the same as elsewhere. New incursions tend to be associated with deliberate plantings or follow movement of livestock, machinery, vehicles or soils. Following this logic, generally, they appear in gardens or places associated with disturbance. Roadsides, culverts, industrial areas, cattle yards, creek crossings, clearings or cropping paddocks.

So, in July 2019 it was not so much a surprise that chromolaena was detected in the NT, but where it was detected. On a remote cattle station to the west of Darwin, hidden amongst the unproductive and densely timbered uplands of the Reynolds River floodplains.

Ground zero, as it is referred to, finds the chromolaena positioned amongst rocky outcrops, supporting stands of large old growth banyan figs, bombax and milkwoods. Chromolaena was also found along seasonal springs and drainage lines that feed melaleuca swamps and the vast seasonally inundated floodplains.

Whilst the origin of the infestations discovered in QLD in 1994 are recognised as contaminated pasture seed imported from Brazil in the 1960's (Scott, 1996), how it found its way to the NT remains a mystery. What is known, similar to the timeline in QLD, is that chromolaena is believed to have been present in the NT for ~10-20 years prior to its discovery (McKenna, 2019).

## OUT OF SIGHT, OUT OF MIND

*Mimosa pigra* (mimosa) has long been a scourge on the floodplains of the NT. Infestations severely reduce carrying capacity, hinder stock movement and reduce available pasture (NTG, DoR, 2013). The floodplains of the Top End are highly prized for finishing cattle ready for market. As such, the properties surrounding ground zero spend many hundreds of thousands of dollars annually on mimosa control. Doing so sees a direct financial return for effort, as well managed floodplain grazing is very profitable enterprise, with the resultant live-weight gains of 0.5kg to 1kg per day during the dry season to early wet (NTG, DoR, 2013).

Conversely, the surrounding uplands areas do not support nutritious fodder. They cannot maintain large numbers of cattle held over the wet season and are easily subject to overgrazing. Uplands are considered of lower economic importance, producing little, if no, financial return (NTG, DoR, 2013). As such, the bushland areas are relatively disregarded and seldom visited. Often, the only real management being asset protection burns to firebreak the floodplains, or back burning to counteract coming wildfire. It is easy country for a weed to remain out of sight and out of mind.

## DELINEATION IS KEY TO ERADICATION

The discovery in 2019 elicited an emergency response led by the Weed Management Branch of the Northern Territory Government Department of Environment, Parks and Water Security (DEPWS).

In an emergency response the initial goal is almost always eradication. After all, to find a weed you had been cautioned about all your career and not try to kill at all costs is a very hard urge to quash. The feeling is to go hard and go fast, with a kill 'em all attitude - prevent seeding at all costs and begin the countdown to eradication.

An Incident Management Team (IMT) was set up and liaised widely with the pastoral industry, primary industries, Aboriginal groups, Parks and other land managers. Emergency funding for immediate survey, treatment, communications and extension was secured. Importantly the IMT sourced knowledge about the habits of the weed and the issues learnt in running a chromolaena eradication program. Much information, good and bad, was garnered from researchers and land management officers involved in the ultimately unsuccessful QLD Siam weed eradication program (QSWEP). The key messages being to look far and wide and don't concentrate all your resources in one area as the weed may be hiding on the other side of the hills.

Barring two small roadside outliers, over the first two years of survey all chromolaena found sat within a diameter of 10km from 'ground zero.' All known plants remained on the original pastoral property and the neighboring Aboriginal Land Trust. Eradication, though considered ambitious, remained feasible in the minds of officers involved.

One of the main lessons learned through QSWEP is that constant delineation is the key to eradication (Jeffery, 2012). Though thought to be contained in a relatively small area, like the QSWEP experience, the population of chromolaena in the NT had a long head start.

There was a consensus for the need to conduct significant surveillance from the air. However, the rationale, cost and scale was open to conjecture and debate. Whilst one approach is to focus intensively around the known infestations and radiate out as new finds are revealed, the other theory is to look as far and wide as possible. A problem then arises how far, how wide and where?

The COVID pandemic meant that across DEPWS little remote fieldwork was able to take place. This left a surplus across operational budgets and as a result funding became available for a once-off aerial survey.

The IMT had maintained the idea of eradication. As such, the announcement to industry partners of large-scale aerial survey over country with no connection to the known chromolaena infestations caught the ire of several, with one prominent industry member stating '*taxpayer money should be used for weed control - not scenic flights'*.

Though survey is critical to any eradication program it can be seen as an easy option – wasting money whilst avoiding laborious control work. Funding for large scale aerial surveillance is not cheap and is often hard to come by. At ~$1000-$1500 per hour for helicopter hire alone, the cost of aerial survey often far outweighs expenditure on seasonal control.

Concerns aside, the final consensus was to look as far and wide as possible. The approach taking a big risk that nothing would be found. The method of survey utilised a grid approach around the known infestations, and then followed a more fluid route, targeting likely habitats and paths of spread across the remainder of the survey. The experience with chromolaena to that point had indicated a preference for rocky outcrops, riparian edges and other habitats where it could establish protected from fire in its initial year.

Coinciding with peak flower in July 2021, survey took place over four weeks covering ~9,500km$^2$ of the western Top End. It crossed five major river catchments and covered numerous pastoral properties, Aboriginal Land Trusts, National Parks, Defence land and many private properties.

The results changed the narrative of chromolaena management in the NT. Chromolaena is now known to be present on 10 properties with the farthest 120km from ground zero. Large range extensions were found on the two known infected premises and another 'core' infestation was discovered on a separate cattle station approximately 40km from ground zero.

Chromolaena proved itself to be less selective than initially thought. It was found across many differing land types; Open savannah woodland, pandanus grasslands, coastal scrublands, paperbark swamps and floodplain margins. Persisting in areas of thick clean pasture and holding its own amongst other formidable weeds, notably mimosa and bellyache bush (*Jatropha gossypiifolia)*. In fact the only habitat it was not found on was the seasonally inundated floodplains.

The locations of new infestations surprised all experts involved in the program, many occurring in areas unrelated to other known populations. Having no connection to infrastructure, downstream movement of water or animal tracks. The only logical method of long-distance dispersal over the landscape being attributed to wind.

DISPERSAL BY DUST DEVIL?

Chromolaena produces massive quantities of seed with records of up to 86,000 seeds per plant (Gautier, 1993). Being apomictic - having no need for seed fertilization, one plant is all that is required to begin an infestation.

Each seed is adapted for spread by various means. A fine pappus, which enables the seeds to drift in the wind; rows of fine barbed hairs along the stem which stick to clothing, fur and machinery; and the ability to float, or be immersed in fresh, or saline water, for extended periods of time without degrading viability (Brooks, 2017), ensures once a plant seeds it has many avenues to spread and proliferate.

Wind is accepted as the main vector for short distance spread, but it had been considered unlikely to disperse seed more than 100m from the parent plant without significant updraft. Rather, long distance dispersal being generally attributed to water, or hitchhiking on animals, humans and vehicles (Brooks, 2017).

Seeding occurs August - October, coinciding with the climax of the Top End dry season and the more unstable weather of the build-up. The time of year is synonymous with hot blustery winds and large dust devils or willy-willys. These are vortexes of hot wind which roll across the landscape picking up soil, ash, leaves and debris propelling them into the air.

Dust devils range from small fleeting swirls in the grass to huge, long-lasting columns of debris visible across the horizon. It was during ground surveys in 2021 a seeding Chromolaena bush was observed being run down and buffeted by a dust devil.

Seed rain across Greater St Lucia Wetland Park in South Africa has been previously hypothesized, and somewhat debunked by Blackmore (1998) However; the spread across the NT suggests there are few other methods that could enable such a random distribution across lands disconnected by movement of animals, vehicles or flow of water.

It is a hard to prove, yet a plausible theory, that seed could be picked up from the plants and dragged up high into the air where it is at the mercy of the prevailing winds.

Providing more weight to the theory, the Finniss-Reynold's floodplains are well known for predictable south easterly dry winds during much of the day, with a westerly sea breeze in the afternoon. If the dust devils can lift the seed high enough, there is no reason preventing seed dispersal many kilometres in multiple directions. If the theory proves correct, it has to be accepted that some methods of spread are unavoidable.

## REALITY BRINGS NEW FOCUS

The sporadic nature of chromolaena distribution over the landscape and the distances between known infestations suggest there is more chromolaena that remains unaccounted for. To paraphrase Donald Rumsfeld's analogy 'there are unknown knowns'. The reality being that eradication in the NT is no longer feasible.

Efforts of the NT Chromolaena program are now transitioning from an in-house eradication program to land holder extension and research into techniques and controls most applicable to the Northern Australian savanna with a view to early intervention, containment and mitigation of further spread.

Working with Biosecurity QLD, building on their research conducted during and following QSWEP, the chromolaena program in the NT has shifted focus to giving land managers a variety of best practice tools to manage the weed, if and when, the need should arise

Paramount to this is to provide cost effective options for treatment, be that ground, foliar, soil applied or aerial solutions for treatment through different times of the year.

Poor roads, rugged terrain, seasonal flooding and the presence of a healthy population of crocodiles, (seemingly in every available body of water), mean access to many of the chromolaena infestations on ground is limited for much of December through to July. This means conducting ground control using conventional foliar spraying, splatter guns or hand removal is often impractical and unable to be conducted by land holders over a large area. As such there is major emphasis placed on developing a cost-effective methodology for aerial control using conventional equipment.

## FIRE INTEGRATION

Fire's prevalence in the landscape necessitates any research into control methods integrate the timing of traditional burning practices. Traditionally fire is used to 'open up' the country, promote a new flush of growth, and protect assets and susceptible country from late season wildfires.

Chromolaena is a fire tolerant species once the root ball has established. However, well managed fire in the lead up to or during flowering has been shown to be a very handy tool for preventing plants from setting seed. Fire also allows better access and follow up control on the new flush of growth.

Conversely, burning the plants too soon after chemical treatment has shown to nullify the effects of the chemical control. As such, the program is working on techniques to allow longer intervals between control and fire to ensure the best results.

## REMOTE SENSING

Remote sensing is being explored in partnership with Charles Darwin University. Investigation thus far has looked into hyperspectral bandwidth analysis, spectral reflectance curve, map based, pixel based and single image object based machine learning.

The most effective method thus far is using single image object-based machine learning, utilising geolocated, UAV captured hi-res RGB imagery. The machine learning model demonstrated a very high rate of success in identifying flowering chromolaena amongst background vegetation.

NT-based field trials will take place in July 2022. If successful, it is hoped to upscale the carrier aircraft from UAV to helicopter and ultimately a light aircraft. The latter being much faster and able to economically cover large areas of terrain while the plant is in flower.

As the machine learning system is almost fully automated it does not require spotters… less time for scenic flights and more time for control!

## THE NT NEEDS MORE FLIES

Fortuitously, QLD Department of Agriculture and Fisheries Biosecurity completed Australian host specificity testing and risk analysis releasing the chromolaena biocontrol agent *Cecidochares connexa* just prior to the NT chromolaena incursion being discovered.

Through support from Biosecurity QLD, the NT has been rearing and establishing colonies on the 'core 'infestations since receiving the first flies in 2019.

The flies lay eggs in the growth tips of the plant. The galls form on the stems as the plant grows reducing vigor and seeding potential. A core component of the project moving forward is increasing the production and ramping up releases onto larger infestations to aid in seeding suppression. The flies are currently established at two separate locations, with new galls appearing on plants up to 300m from the release sites.

## ALL IS NOT LOST

Acceptance of moving from eradication to management has been hard to take. Though it also needs to be acknowledged that continuing down a fruitless path can be demoralising and demanding on those involved. Accepting the change and working to develop awareness, tools and the best practice required to tackle chromolaena will hopefully provide better outcomes for NT land managers and the wider northern Australia into the future.

## ACKNOWLEDGMENTS

The author would like to thank all those involved in the NT Siam weed project thus far. In particular Simon Brooks, Paige Richter, Dave Green, Dan Chapman, Chris Collins, Michelle Franklin, Chris Parker, Deepak Gautam and Lou Elliott.

## REFERENCES

Blackmore, A. (1998). Seed dispersal of Chromolaena odorata reconsidered. IN; Ferrer, P. Muniappan, R. and Jayanth, KP (eds) Proceedings of the *The 4th international Workshop on the Biological control and management of Chromolaena odorata,*, (pp. 16 -21). University of Guam, Mangilao, Bangalore, India.

Brooks, S. J. (2017). Siam weed disperal mechanisms. *14th Queensland Weed symposium*, (pp. 153-158).

Gautier, L. (1993). Reproduction of a pantropical Weed. Chromolaena odorata. *Candollea 48*, 179-193.

Jeffery, M. (2012). Eradication: lessons learnt from 17 years of the National Siam Weed. *Eighteenth Australasian Weeds Conference.* Melbourne.

McKenna, S. (2019, December). Personal Communication - Gathering Siam Weed Genetic Material. (N. W. Branch, Interviewer)

NTG, DoR. (2013). *Cattle and land management best practices in the Top End Region.* Darwin: Northern Territory Government.

Parsons, W. T. (2001). *Noxious weeds of Australia (Ed.2).* Collingwood Melbourne: CSIRO Publishing,.

Waterhouse, B., & Zeimer, O. (2000). 'On the brink': The status of chromolaena odorata. *Proceedings of the Fifth International Workshop on Biological Control and Managementof Chromolaena odorata, Durban, South Africa, Zachariades, C.R. Muniappan and L.W. Strathie (eds)*, (pp. 29-33).

Wilson, C. (1994). *I'm dreaming of a white Christmas bush; Biological control of Chromolaena odorata in Timor.* Darwin : Northern Territory Govenment .

# Efficacy of Steam Weed Management on Mother of Millions (*Bryophyllum* species), Singapore Daisy (*Sphagneticola trilobata*) and Butterfly Heaven (*Dyschoriste depressa*)

Kerrie Guppy[1], Tali Shelley[2]
[1]Aus Eco Solutions, Australia
(kerrie@ausecosolutions.com.au)

**Summary** Mother of Millions (*Bryophyllum* species) and Singapore Daisy (*Sphagneticola trilobata*) are both significant weeds in NSW and Queensland, with the potential to be a significant weed in Victoria due to their ability to rapidly spread. Dyschoriste or Butterfly Heaven (*Dyschoriste depressa*) is an emerging environmental weed in Queensland that has spread at an alarming rate in the last 20 years across Brisbane. Mother of Millions is also toxic to livestock, and potentially in large enough doses to humans and pets. All these weeds are challenging to control as the smallest cutting can regrow and/or herbicides have been found to have limited effect.

This paper explores the efficacy of saturated steam weed management on Mother of Millions, Singapore Daisy and Dyschoriste across a variety of environments. Saturated steam weed management has been found to have varying degrees of success depending on the soil structure i.e. mulched or sandy sites. In some cases, the weeds are effectively eradicated with one treatment and in other cases require multiple treatments with saturated steam. In the case of Dyschoriste, saturated steam has been demonstrated in the field to be the most effective treatment method. Some of the coastal sites in NSW were severely impacted by the 2019-2020 Black Summer Fires and these impacts on weeds, spread and saturated steam weed management will be covered. This paper also explores factors to improve success rates. Cited case studies include urban bush regeneration sites in Brisbane, coastal sites at Morton Bay in Queensland, and coastal sites at Wallabi Point in NSW.

# Author Index

## A



## B



## C

# I



# J



# K



# L



# M

## N



## O



## P



## Q



## R

## S



## T



## U



## V

## W



## X



## Y



## Z

**Council of Australasian Weed Societies Inc. (CAWS)**
Massey University, Private Bag
11-222 Palmerston North 4442
New Zealand

ISBN 978-1-7138-6247-5

CURRAN ASSOCIATES INC.
proceedings.com